The Interferons

Edited by
Anthony Meager

Related Titles

Frosch, M., Maiden, M. C. J. (eds.)

Handbook of Meningococcal Disease

Infection Biology, Vaccination, Clinical Management

apprcx. 600 pages with approx. 90 figures
Hardcover
ISBN 3-527-31260-9

Lutz, M. B., Romani, N., Steinkasserer, A. (eds.)

Handbook of Dendritic Cells

Biology, Diseases, and Therapies

approx. 1250 pages in 3 volumes
Hardcover
ISBN 3-527-31109-2

Pollard, K. M. (ed.)

Autoantibodies and Autoimmunity

Molecular Mechanisms in Health and Disease

approx. 700 pages
2005
Hardcover
ISBN 3-527-31141-6

Kaufmann, S. H. E. (ed.)

Novel Vaccination Strategies

670 pages with 82 figures and 50 tables
2004
Hardcover
ISBN 3-527-30523-8

Kalden, J. R., Herrmann, M. (eds.)

Apoptosis and Autoimmunity

From Mechanisms to Treatments

392 pages with 50 figures and 25 tables
2003
Hardcover
ISBN 3-527-30442-8

Stuhler, G., Walden, P. (eds.)

Cancer Immune Therapy

Current and Future Strategies

434 pages with 57 figures and 29 tables
2002
Hardcover
ISBN 3-527-30441-X

Kropshofer, H., Vogt, A. B. (eds.)

Antigen Presenting Cells

From Mechanisms to Drug Development

651 pages with 63 figures and 16 tables
2005
Hardcover
ISBN 3-527-31108-4

The Interferons

Characterization and Application

Edited by Anthony Meager

WILEY-VCH

WILEY-VCH Verlag GmbH & Co. KGaA

The Editor

Dr. Anthony Meager
National Institute for Biological
Standards and Control
Division of Immunology and
Endocrinology

■ All books published by Wiley-VCH are carefully produced. Nevertheless, authors, editors, and publisher do not warrant the information contained in these books, including this book, to be free of errors. Readers are advised to keep in mind that statements, data, illustrations, procedural details or other items may inadvertently be inaccurate.

Library of Congress Card No.: applied for
British Library Cataloguing-in-Publication Data:
A catalogue record for this book is available from the British Library.

Bibliographic information published by Die Deutsche Bibliothek
Die Deutsche Bibliothek lists this publication in the Deutsche Nationalbibliografie; detailed bibliographic data is available in the Internet at ⟨http://dnb.ddb.de⟩.

© 2006 WILEY-VCH Verlag GmbH & Co. KGaA, Weinheim

All rights reserved (including those of translation into other languages). No part of this book may be reproduced in any form – by photoprinting, microfilm, or any other means – nor transmitted or translated into a machine language without written permission from the publishers. Registered names, trademarks, etc. used in this book, even when not specifically marked as such, are not to be considered unprotected by law.

Printed in the Federal Republic of Germany.
Printed on acid-free paper.

Typesetting   Asco Typesetters, Hong Kong
Printing   betz-Drude GmbH, Darmstadt
Binding   Litges & Dopf Buchbinderei GmbH, Heppenheim

ISBN-13: 978-3-527-31180-4
ISBN-10: 3-527-31180-7

Contents

Preface *XIII*

List of Contributors *XV*

Color Plates *XIX*

Section A Molecular Aspects, Introduction and Purification *1*

1 Type I Interferons: Genetics and Structure *3*
Shamith A. Samarajiwa, William Wilson and Paul J. Hertzog
1.1 Introduction *3*
1.2 The Type I IFN Genetic Locus *4*
1.3 Type I IFN Genes *7*
1.3.1 IFN-α *7*
1.3.2 IFN-β *7*
1.3.3 IFN-ω *8*
1.3.4 IFN-κ *8*
1.3.5 IFN-ε *9*
1.3.6 IFN-ζ (Limitin) *9*
1.3.7 IFN-τ *9*
1.3.8 IFN-δ *10*
1.3.9 IFN-ν *10*
1.4 Type I IFN Gene-regulatory Regions *11*
1.5 Evolution of the Type I IFNs *15*
1.5.1 Vertebrate IFN Genes *15*
1.5.2 The Expansion and Divergence of the IFN Genes *20*
1.6 Natural and Induced Mutations in IFN Genes *21*
1.7 Secondary Structural Features of Type I IFNs *22*
1.7.1 Conserved Amino Acid Residues *22*
1.7.2 Post-translational Modifications of Type I IFNs *23*
1.7.3 Conserved Cysteine Residues and Disulfide Bond Formation *24*
1.8 The Structure of Type I IFNs *24*
References *27*

The Interferons: Characterization and Application. Edited by Anthony Meager
Copyright © 2006 WILEY-VCH Verlag GmbH & Co. KGaA, Weinheim
ISBN: 3-527-31180-7

2	Activation of Interferon Gene Expression Through Toll-like Receptor-dependent and -independent Pathways 35
	Peyman Nakhaei, Suzanne Paz and John Hiscott
2.1	Introduction 35
2.2	IFN-β Gene Transcription 39
2.3	IRF Family Members 40
2.4	Role of IRFs in Virus-mediated IFN Activation 41
2.4.1	IRF-3 41
2.4.2	IRF-5 47
2.4.3	IRF-7 48
2.5	IFN Signaling Pathways 49
2.5.1	TLR-dependent Signaling to IFN Activation 49
2.5.1.1	TLR Overview 49
2.5.1.2	TLR-3 Signaling 51
2.5.1.3	TLR-4 Signaling 53
2.5.1.4	TLR-7 Signaling 57
2.5.1.5	TLR-9 Signaling 57
2.5.2	TLR-independent Signaling 58
2.5.2.1	Retinoic-inducible Gene (RIG)-I Signaling 58
2.5.2.2	Melanoma Differentiation-associated Gene-5 (*mda-5*) 59
2.6	Conclusions 61
	References 61

3	Interferon Proteins: Structure, Production and Purification 73
	Dimitris Platis and Graham R. Foster
3.1	Introduction 73
3.2	The Structure of Type I IFNs 73
3.3	Production and Purification of Type I IFNs 75
3.3.1	Leukocyte-derived IFN – First Steps in Producing Commercial IFN 75
3.3.2	Lymphoblastoid IFN – Towards more Reliable Supplies of IFN 76
3.3.3	Cloned Type I IFNs – An Inexhaustible Supply of Therapeutic Material 76
3.4	Long-acting IFNs 78
3.5	Summary 79
	References 80

4	Interferon-γ: Gene and Protein Structure, Transcription Regulation, and Actions 85
	Ana M. Gamero, Deborah L. Hodge, David M. Reynolds, Maria Cecilia Rodriguez-Galan, Mansour Mohamadzadeh and Howard A. Young
4.1	Introduction 85
4.2	IFN-γ Gene Structure and Regulation 86
4.2.1	Transcriptional Regulation 86
4.2.2	Epigenetic Regulation 88
4.2.3	Post-transcriptional Regulation 90

4.3	IFN-γ Signal Transduction	90
4.3.1	The JAK–STAT Signaling Pathway	91
4.3.2	Activation of Alternate Signaling Pathways	92
4.3.3	Regulation of IFN-γ Signaling	93
4.4	IFN-γ in T_h Cell Development	95
4.4.1	Signaling Pathways Involved in T Cell Development	95
4.5	IFN-γ and DCs	97
4.5.1	IFN-γ and T Cell–DC Crosstalk	97
4.5.2	Signals through Toll-like Receptors (TLRs) Activate DCs and Influence IFN-γ Expression	99
4.6	IFN-γ – Role in Tumor Development and Growth	100
4.6.1	IFN-γ in Tumor Growth and Survival	100
4.6.2	Inhibition of Angiogenesis by IFN-γ	101
4.6.3	Role of IFN-γ in Promoting Immune Responses against Tumors	102
4.7	Summary	103
	References	104

5	**Interferon and Related Receptors**	**113**
	Sidney Pestka and Christopher D. Krause	
5.1	Introduction	113
5.2	IFNs and IFN-like Molecules in Brief	113
5.3	The Receptors	114
5.3.1	Receptor Nomenclature	115
5.4	The Type I IFN Receptor	116
5.4.1	Discovery of the Type I Receptor Complex	117
5.4.2	Diversity of the Interaction of Type I IFNs with the Receptor	119
5.5	The Type II IFN (IFN-γ) Receptor	121
5.5.1	Chromosomal Localization of the IFN-γ Receptor Ligand-binding Chain and Discovery of Two Chains Required for Activity	121
5.5.2	The IFN-γR1 Chain	123
5.5.3	The Second Receptor Chain (IFN-γR2)	123
5.5.4	The Functional IFN-γR Complex	124
5.5.4.1	Specificity of Ligand Binding	124
5.5.4.2	Specificity of the Interactions of the Two Chains	124
5.5.5	Specificity of Signal Transduction	124
5.5.6	Receptor Structure	125
5.5.7	Preassembly of the Receptor Complex	126
5.6	The IL-28R1 and -10R2 Receptor Complex	127
5.7	Overview of Multichain Receptors	128
5.8	Global Summary	128
	References	129

6	**Type III Interferons: The Interferon-λ Family**	**141**
	Sergei V. Kotenko and Raymond P. Donnelly	
6.1	Introduction	141

6.2	The Class II Cytokine Receptor Family (CRF2) and their Ligands	142
6.3	Genomic Structure	145
6.4	Receptor Complex and Signaling	148
6.5	Biological Activities	152
6.6	The Murine IFN-λ Antiviral System	154
6.7	Evolution of the IFN Family	156
6.8	Therapeutic Potential	157
6.9	Conclusions	158
	References	158

Section B Biological Properties 165

7 Biological Actions of Type I Interferons 167
Melissa M. Brierley, Jyothi Kumaran and Eleanor N. Fish

7.1	Introduction	167
7.2	Sources of Type I IFN Production and Secretion	167
7.3	Type I IFN Interactions with the Receptor Complex	168
7.3.1	Structure and Functional Regions of Type I IFNs	168
7.3.2	IFN Domains Mediating Interactions with IFNAR-2	168
7.3.3	IFNAR-2 Domains Mediating Interactions with IFNs	170
7.3.4	IFN-α Interaction with IFNAR-2	172
7.3.5	IFN-β Interaction with IFNAR-2	173
7.3.6	Type I IFN Interactions with IFNAR-1	174
7.3.7	IFNAR-1 Receptor Interaction with Glycosphingolipids	174
7.3.8	Residues in Type I IFNs that Mediate Biological Responses	175
7.4	Type I IFN-induced Signaling Cascades	175
7.4.1	The JAK–STAT (Signal Transducers and Activators of Transcription) Pathway	175
7.4.1.1	The Importance of STAT-1 and -2 in Type I IFN Signaling	178
7.4.2	Other IFN-inducible Signaling Cascades	179
7.4.2.1	The CrkL Pathway	179
7.4.2.2	The IRS Pathway	179
7.4.2.3	The p38 Mitogen-activated Protein Kinases (MAPK) Pathway	181
7.4.2.4	The Vav Protooncogene and IFN Signaling	182
7.4.2.5	The Protein Kinase C (PKC) Family and IFN Signaling	182
7.5	IFN-inducible Biological Responses	183
7.5.1	IFN-inducible Antiviral Responses	183
7.5.1.1	PKR	186
7.5.1.2	$2'$–$5'$-OAS/RNase L	186
7.5.1.3	Mx	187
7.5.1.4	Other IFN-inducible Antiviral Effectors	188
7.5.2	IFN-inducible Growth-inhibitory Responses	188
7.5.3	IFN-inducible Immunomodulation	190
7.6	Summary	192
	References	193

8	**Interferons and Apoptosis – Recent Developments** 207
	Michael J. Clemens and Ian W. Jeffrey
8.1	Introduction 207
8.2	The Role of IFN-regulated Genes in the Control of Apoptosis 209
8.2.1	Proapoptotic Genes Induced by IFNs 209
8.2.2	p53 and IFN-induced Apoptosis 209
8.2.3	The Ribonuclease (RNase) L System 210
8.3	The Protein Kinase PKR and the Phosphorylation of Polypeptide Chain Initiation Factor eIF2α 212
8.3.1	Regulation of Apoptosis by PKR 212
8.3.2	The Role of eIF2α Phosphorylation 212
8.4	IFNs and the Apoptotic Effects of TRAIL 215
8.4.1	TRAIL Induction 215
8.4.2	TRAIL Activity 215
8.5	Signal Transduction Pathways for IFN-mediated Effects on Apoptosis 218
8.6	The Antiapoptotic Effects of IFNs 219
8.7	Conclusions 220
	Acknowledgments 220
	References 221
9	**Viral Defense Mechanisms against Interferon** 227
	Santanu Bose and Amiya K. Banerjee
9.1	Introduction 227
9.2	Innate Immune Antiviral Defense Mechanisms of Host Cells 227
9.3	Evasion of IFN-mediated Antiviral Response 231
9.3.1	Nonsegmented Negative-strand RNA Viruses 232
9.3.1.1	Paramyxoviruses 232
9.3.1.2	Filovirus 242
9.3.1.3	Rhabdovirus 242
9.3.2	Segmented Negative-strand RNA Viruses 242
9.3.2.1	Orthomyxovirus 243
9.3.2.2	Bunyaviruses 243
9.3.3	Positive-sense ssRNA Viruses 244
9.3.3.1	Picornaviruses 244
9.3.3.2	Flaviviruses 244
9.3.3.3	Coronavirus 246
9.3.4	dsRNA Viruses 247
9.3.5	RNA and DNA Reverse-transcribing Viruses 247
9.3.5.1	Retrovirus 247
9.3.5.2	Hepadnavirus 248
9.3.6	dsDNA Viruses 248
9.3.6.1	Adenovirus 248
9.3.6.2	Poxvirus 248
9.3.6.3	Herpesvirus 250

	9.3.6.4	Papillomavirus 252
9.4		Concluding Remarks 253
		References 254

Section C Clinical Applications 275

10		**Overview of Clinical Applications of Type I Interferons** 277
		Frank Müller
10.1		Introduction 277
10.2		Biological Effects 277
10.3		Type I IFN Products Currently Available or Under Development 278
10.4		Pharmacokinetics 284
10.4.1		IFN-α 284
10.4.2		IFN-β 285
10.5		Clinical Applications of Type I Interferons 285
10.5.1		Chronic hepatitis B (CHB) 286
10.5.2		Chronic hepatitis C (CHC) 288
10.5.3		Chronic hepatitis D (CHD) 290
10.5.4		Hairy Cell Leukemia (HCL) 290
10.5.5		Renal Cell Carcinoma (RCC) 291
10.5.6		Basal Cell Carcinoma (BCC) 292
10.5.7		Malignant Melanoma 292
10.5.8		Kaposi Sarcoma (KS) 292
10.5.9		Multiple Myeloma 293
10.5.10		Chronic myelogenous leukemia (CML) 294
10.5.11		Non-Hodgkinis lymphoma (NHL) 295
10.5.12		Laryngeal Papillomatosis 295
10.5.13		Mycosis Fungoides (MF) 296
10.5.14		Condyloma Acuminata 296
10.5.15		Multiple Sclerosis (MS) 297
10.6		IFN Toxicity 298
10.7		Type I IFNs in the Future 299
		References 300

11		**Clinical Applications of Interferon-γ** 309
		Christine W. Czarniecki and Gerald Sonnenfeld
11.1		Introduction 309
11.2		IFN-γ – The Molecule 311
11.3		FDA-approved Indications: Established Benefit and Risks 311
11.3.1		Chronic Granulomatous Disease (CGD) 311
11.3.2		Osteopetrosis 313
11.3.3		Adverse Reactions 314
11.4		Infectious Diseases 314
11.4.1		Mycobacterial Infection 314
11.4.1.1		Leprosy 314

11.4.1.2   *Mycobacterium avium* Infection   *315*
11.4.1.3   Tuberculosis (TB)   *315*
11.4.2   Leishmaniasis   *317*
11.4.3   Opportunistic Infections in HIV Disease   *317*
11.5   Infection Following Serious Trauma   *318*
11.6   Atopic Dermatitis (AD)   *321*
11.7   Idiopathic Pulmonary Fibrosis (IPF)   *322*
11.8   Systemic Sclerosis (SSc)   *326*
11.9   Radiation-induced Fibrosis   *326*
11.10   Chronic Hepatitis   *327*
11.11   Oncology Indications: Ovarian Cancer   *329*
11.12   Conclusions   *330*
         References   *331*

Section D   Measurement of Interferons and Anti-Interferons   *337*

12   Measurement of Interferon Activities   *339*
         Tony Meager
12.1   Introduction   *339*
12.2   The IFNs: Mechanisms of Action, Protein Induction and Biological Activities   *340*
12.2.1   Mechanisms of Action   *340*
12.2.2   Protein Induction   *342*
12.2.3   Biological Activities   *344*
12.2.3.1   Antiviral Activity   *344*
12.2.3.2   Antiproliferative Activity   *344*
12.2.3.3   Immunomodulatory Activity   *345*
12.3   Measurement of Biological Activities of IFNs   *345*
12.3.1   General Considerations   *345*
12.3.2   Biological Standards for IFNs   *346*
12.3.3   Practical Considerations for IFN Preparations   *348*
12.3.4   Antiviral Assays   *349*
12.3.5   Antiproliferative Assays   *351*
12.3.6   IFN-inducible Protein Assays   *354*
12.3.6.1   Bioimmunoassays   *354*
12.3.6.2   Enzyme Expression and Reporter Gene Assays   *356*
12.3.7   Assays Based on Intracellular Signaling Intermediates   *361*
12.3.8   Assay Design and Data Analysis   *363*
12.4   Regulatory Landscape   *365*
         Acknowledgments   *366*
         References   *366*

| 13 | **The Development and Measurement of Antibodies to Interferon** | 375 |

Sidney E. Grossberg and Yoshimi Kawade

- 13.1 Introductory Perspective 375
- 13.1.1 Immunological Perspective 375
- 13.1.2 Antibodies to Self-antigens 376
- 13.2 NAbs 377
- 13.2.1 Neutralization Bioassay Design 378
- 13.2.1.1 The Constant IFN Method 379
- 13.2.1.2 The Constant Antibody Method 380
- 13.2.2 Theoretical Analyses 382
- 13.2.3 Experimental Analyses 383
- 13.2.4 Standardization and the Reporting of Neutralization Results 383
- 13.2.5 A Solution to the Problem of NAb Unitage 385
- 13.3 Immunoassays for Non-NAbs 387
- 13.4 Epitope Analysis 388
- 13.5 Development of Antibodies during IFN Therapy 390
- 13.5.1 Antibodies to IFN-α 391
- 13.5.2 Antibodies to IFN-β 392
- 13.6 Summary 394
- References 395

Index 401

Preface

The Interferon field has its origins nearly 50 years ago with the discovery of an antiviral factor produced by virally infected chick cells. This factor, designated "The Interferon", provided the first evidence that antiviral defence mechanisms could be triggered by secreted cellular factors. Since interferon (IFN) was found to protect against many viruses, scientific interest was high. A research dynasty was literally founded then that has actively pursued the characterization and potential clinical applications of IFN. Initially, the field was sustained by dedicated pioneering scientists trying to understand what IFN was and how it worked. At the moment when their research was beginning to pay off with the purification of IFNs to homogeneity and increasing knowledge of their biological activities, the field was catalysed by the advent of recombinant DNA technology. The Pharmaceutical Industry too was drawn in by the "promise" of IFN's broad therapeutic activity against a range of tumours and viruses. Although the clinical success of IFNs proved to be much more limited than hoped for, the research that was generated in those heady times gave a tremendous boost to our understanding of the molecular structure of IFN' genes and proteins, their cellular receptors, their mechanism of action and their biological activities, both *in vitro* and *in vivo*. Despite the sombre assessment of the clinical worth of IFNs two decades ago, they have come back as strong market leaders for the treatment of chronic hepatitis C virus infection (IFN-alpha) and multiple sclerosis (IFN-beta). Research studies within the IFN field from then on have proved invaluable to the elucidation of IFN' induction and connected intracellular signalling pathways, cellular defence mechanisms, and the evasion mechanisms of viruses and tumour cells to IFNs.

The Interferons: Characterization and Application covers many aspects of our current knowledge of IFNs. This includes the structure and functions of all known IFN types, evolution and structure of their genes, their receptors and signalling pathways, their induction and biological activities and mechanisms whereby viruses evade their antiviral actions. In addition, coverage of the clinical applications of type I and II IFNs, together with methodologies to measure biological activities of IFNs and the antibodies that may develop against them as a consequence of IFN therapies, is provided. I believe there is a serious need for this publication, even in view of the vast amount of information available in the scientific literature and of the World Wide Web. I feel that there is no substitute to an up-to-date monograph on the IFN field that embraces an integrated and well-selected approach to the subject. It is my hope this will provide a comprehensive foundation to the professional scientific and medicinal research community, especially newcomers to the field, and will promote further advances in the field.

I gratefully acknowledge the authors for their time, motivation and dedication in preparing their contributions, without which this book would not have been possible. I also thank Andreas Sendtko and his colleagues at Wiley-VCH for outstanding support throughout the planning and preparation of this book.

Potters Bar, October 2005 *Anthony Meager*

ISAACS A, LINDEMANN J. Virus interference. I. The interferon.
Proc Roy Soc B **1957**, *147*, 258–267.

List of Contributors

Melissa M. Brierley
Toronto General Research Institute
University Health Network
Toronto, Ontario M5G 2M1
Canada

Amiya K. Banerjee
Virology Section
Department of Molecular Genetics
Lerner Research Institute
The Cleveland Clinic Foundation
Cleveland, OH 44195
USA

Santanu Bose
Department of Microbiology and Immunology
University of Texas Health Science Center at San Antonio
San Antonio, TX 78229
USA

Michael J. Clemens
Translational Control Group
Department of Basic Medical Sciences
St George's Hospital Medical School
London SW17 0RE
UK

Christine W. Czarniecki
Division of Allergy, Immunology and Transplantation
National Institute of Allergy and Infectious Disease
National Institutes of Health
Bethesda, MD 20817
USA

Raymond P. Donnelly
Division of Therapeutic Proteins
Center for Biologics Evaluation and Research
Food and Drug Administration
Bethesda, MD 20892
USA

Eleanor N. Fish
Toronto General Research Institute
University Health Network
Toronto, Ontario M5G 2M1
Canada

Graham. R. Foster
Hepatobiliary Group
Institute of Cellular and Molecular Science
Queen Mary's School of Medicine and Dentistry
Barts and The London
London E1 2AT
UK

Ana M. Gamero
Laboratory of Experimental Immunology
Center for Cancer Research
National Cancer Institute – Frederick
PO Box B, Bldg. 560, Rm. 31-18
Frederick, MD 21702-1201
USA

Sidney E. Grossberg
Department of Microbiology and Molecular Genetics
Medical College of Wisconsin
Milwaukee, WI 53226
USA

John Hiscott
Terry Fox Molecular Oncology Group
Lady Davis Institute for Medical Research
Departments of Microbiology & Immunology and Medicine
McGill University
Montreal, Quebec H3T 1E2
Canada

List of Contributors

Paul J. Hertzog
Center for Functional Genomics and Human Disease
Monash Institute of Medical Research
Monash University
Clayton, Victoria 3168
Australia

Deborah L. Hodge
Laboratory of Experimental Immunology
Center for Cancer Research
National Cancer Institute – Frederick
PO Box B, Bldg. 560, Rm. 31-18
Frederick, MD 21702-1201
USA

Ian W. Jeffrey
Translational Control Group
Department of Basic Medical Sciences
St George's Hospital Medical School
London SW17 0RE
UK

Yoshimi Kawade
Kyoto University
Kyoto 43-6 Ozazaki-Minamigosho-cho,
Sakyo-ku Kyoto 606-8334
Japan

Jyothi Kumaran
Toronto General Research Institute
University Health Network
Toronto, Ontario M5G 2M1
Canada

Sergei V. Kotenko
Department of Biochemistry and Molecular Biology
University of Medicine and Dentistry
New Jersey Medical School
Newark, NJ 07103
USA

Christopher D. Krause
University of Medicine and Dentistry
Department of Molecular Genetics,
Microbiology and Immunology
Robert Wood Johnson Medical School
675 Hoes Lane, Room 801
Piscataway, NJ 08854-5635
USA

Tony Meager
Division of Immunology and Endocrinology
The National Institute for Biological Standards and Control
South Mimms EN6 3QG
UK

Mansour Mohamadzadeh
Senior Investigator
USAMARIID
1425 Porter Street
Frederick, MD 21702
USA

Frank Müller
Rhein Minaphasm Biogenetics
El Bardissi Street
2T Taksseem Asmaa Fahmy
Heliopolis, Cairo
10th of Ramadan City

Peyman Nakhaei
Terry Fox Molecular Oncology Group
Lady Davis Institute for Medical Research
Departments of Microbiology & Immunology
McGill University
Montreal, Quebec H3T 1E2
Canada

Suzanne Paz
Terry Fox Molecular Oncology Group
Lady Davis Institute for Medical Research
Departments of Microbiology & Immunology
McGill University
Montreal, Quebec H3T 1E2
Canada

Sidney Pestka, M.D.
University of Medicine and Dentistry
Department of Molecular Genetics,
Microbiology and Immunology
Robert Wood Johnson Medical School
Chairman and Professor
G75 Hoes Lane, Room 727
Piscataway, NJ 08854-5635
USA

Sidney Pestka, M.D.
PBL Biomedical Laboratories
Chief Scientific Officer
131 Ethel Road West, Suite 6
Piscataway, NJ 08854-5900
USA

Dimitris Platis
Agricultural University of Athens
Department of Biotechnology
Laboratory of Enzyme Technology
118 55 Athens
Greece

David M. Reynoldsl
Laboratory of Experimental Immunology
Center for Cancer Research
National Cancer Institute – Frederick
PO Box B, Bldg. 560, Rm. 31-18
Frederick, MD 21702-1201
USA

Maria Cecilia Rodriguez-Galan
National University of Cardoba
Faculty of Chemistry
Hayas de la Torre y Medina Allende,
Cuidad Universitaria
CP (5000)
Ciudad de Cardoba, Cardoba
Argentina

Shamith A. Samarajiwa
Center for Functional Genomics and Human Disease
Monash Institute of Medical Research
Monash University
Clayton, Victoria 3168
Australia

Gerald Sonnenfeld
Binghamton University
State University of New York
Binghamton, NY 13902
USA

Howard A. Young
Laboratory of Experimental Immunology
Center for Cancer Research – Frederick
National Cancer Institute
PO Box B, Bldg. 560, Rm. 31-23
Frederick, MD 21702-1201
USA

William Wilson
CSIRO Mathematical and Information Sciences
Bioinformatics for Human Health
Clayton, Victoria 3168
Australia

Color Plates

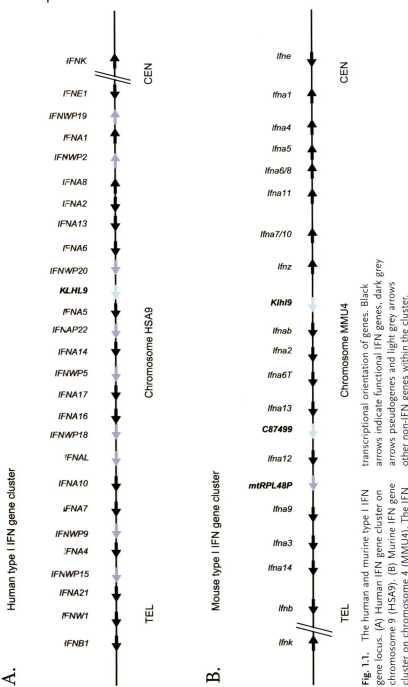

Fig. 1.1. The human and murine type I IFN gene locus. (A) Human IFN gene cluster on chromosome 9 (HSA9). (B) Murine IFN gene cluster on chromosome 4 (MMU4). The IFN gene locus was drawn using current genomic information available on NCBI human chromosome 9 (HSA9) contig (contig ID; NT_008413.16) and mouse chromosome 4 (MMU) contig (contig ID; NT_039260.4). Arrows indicate relative position and transcriptional orientation of genes. Black arrows indicate functional IFN genes, dark grey arrows pseudogenes and light grey arrows other non-IFN genes within the cluster. Orientation of the chromosome is indicated by TEL (telomeric end) and CEN (centromeric end). Nomenclature suggested by van Pesch et al. [32] is used in naming murine IFNs. (This figure also appears on page 5.)

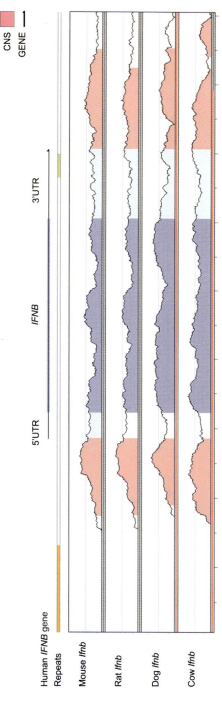

Fig. 1.2. Human type I IFN promoter sequence alignment. (A) VISTA alignment of human, mouse, rat, dog and cow IFN-β genes. The figure represents an AVID genome alignment [58] and VISTA visualization [59] of mouse, rat, dog and cow *Ifnb* genomic sequences compared to the human *IFNB* genomic sequence. At the top, the human *IFNB* gene is depicted by a dark blue line with an arrow indicating transcriptional orientation, together with 5′- and 3′-UTR regions depicted as light blue lines. The line immediately below indicates the location of repetitive elements within the genomic sequence. The graphed regions below demonstrate homology of mouse, rat, dog and cow *Ifnb* genomic regions to the human *IFNB* gene. Exon regions with more than 65% sequence identity over 100 bp are indicated by dark blue, whereas conserved noncoding sequences (CNS) and 5′- and 3′-UTRs with over 65% sequence identity are depicted in orange and light blue, respectively. (This figure also appears on page 12.)

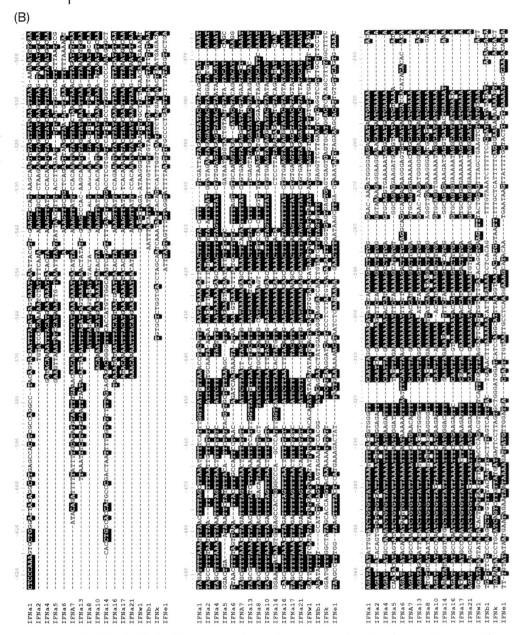

Fig. 1.2. (B) Nucleotide sequence alignment of human IFN promoter. Alignment was performed using ClustalW algorithm [79] of 500 bp 5′-flanking sequences containing the promoter elements of human IFN genes. The regulatory elements present in this region are indicated by boxed regions. (This figure also appears on pages 14–15.)

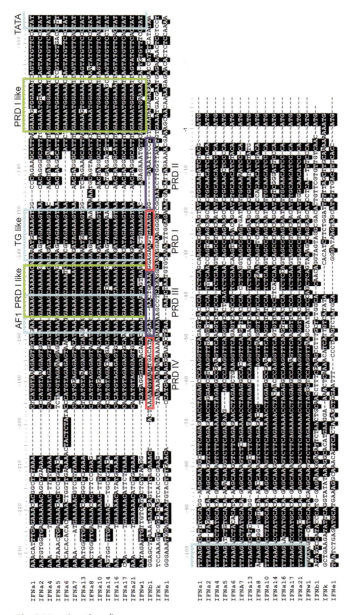

Fig. 1.2B. *(continued)*

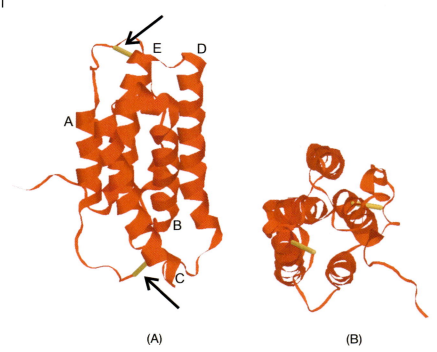

Fig. 1.4. Representative tertiary structure of human IFN-α. Tertiary structure of human IFN-α2 represented as a ribbon diagram, showing the five α-helices connected by loop regions. The disulfide bonds between cysteine residues are indicated by arrows. (A) Side view of the molecule. (B) View from above. (This figure also appears on page 25.)

Fig. 2.1. Schematic representation of the IFN-β promoter enhanceosome. This diagram shows the composition and organization of the human IFN-β enhanceosome. The nucleotide sequence (−110 to −36) is situated upstream from the starting transcription site. The positive regulatory sites as well as the negative regulatory domain are named PRDs and NRD, respectively. The transcriptional proteins binding to the PRDs are also shown (c-Jun, ATF-2, IRFs and NF-κB), as well as the architectural proteins HMGI(Y) and the transcriptional coactivators proteins p300 and CBP. The multiple protein–protein interactions required for transcriptional synergy of the IFN-β promoter derived from the cooperative assembly of the enhanceosome, cooperative assembly between the enhanceosome and the USA/BAD complex, and interaction with the RNA polymerase II complex [32]. (This figure also appears on page 38.)

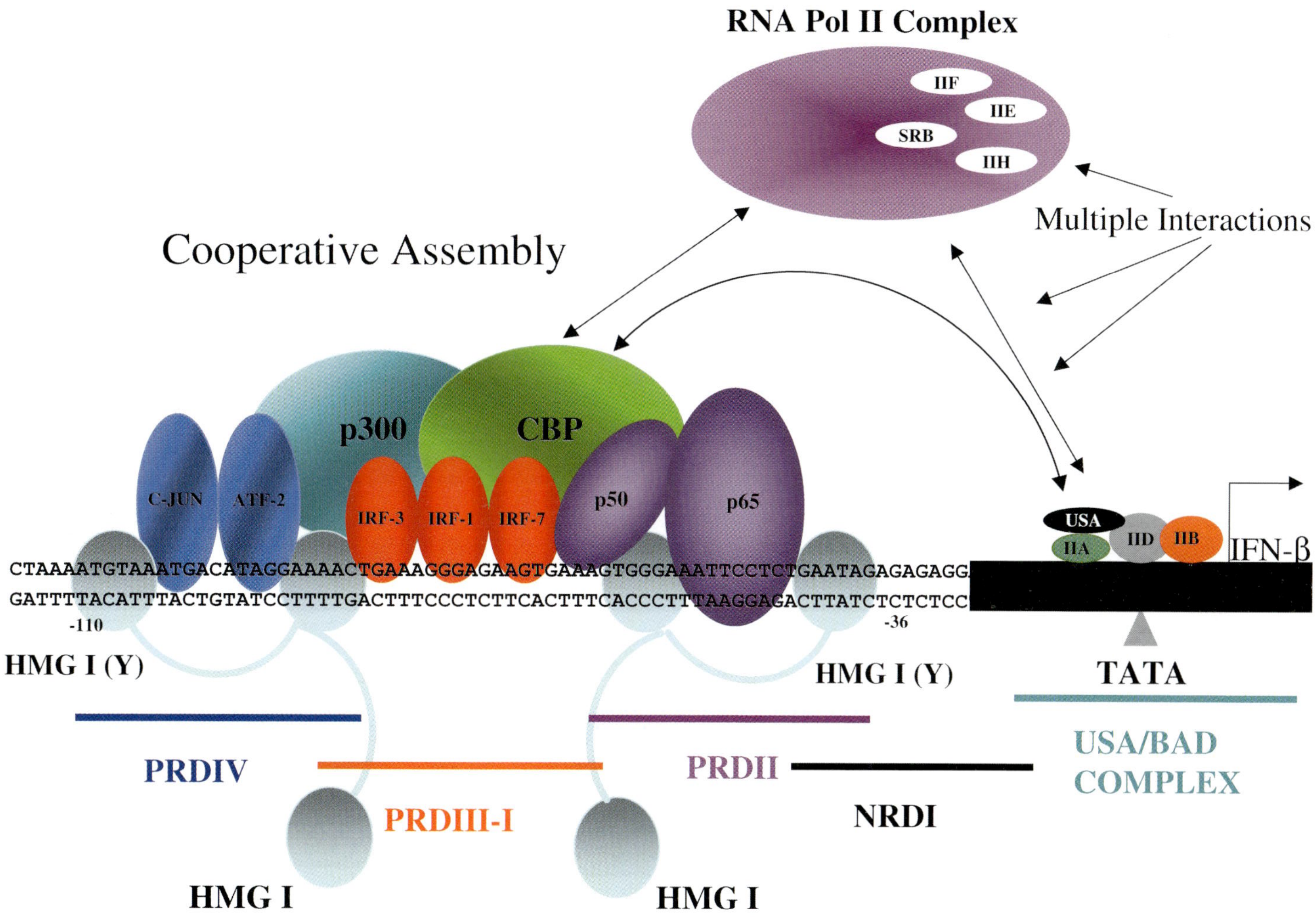

Color Plates | XXV

RNA Pol II Complex

IIF IIE
SRB IIH

Multiple Interactions

IIB
IID
USA IIA

IFN-β

TATA

USA/BAD COMPLEX

Cooperative Assembly

C-JUN ATF-2   p300   IRF-3 IRF-1 IRF-7   CBP   p50 p65

CTAAAATGTAAATGACATAGGAAAACTGAAAAATGGAAGAAGTGAAAGTGGGAAATTCCTCTGAATAGAGAGAGG
GATTTTACATTTACTGTATCCTTTTGACTTTCCCTCTTCACTTTCACCTTTAAGGAGACTTATCTCTCC
-110   -36

HMG I (Y)   HMG I (Y)

PRDIV   PRDIII-I   PRDII   NRDI

HMG I   HMG I   HMG I

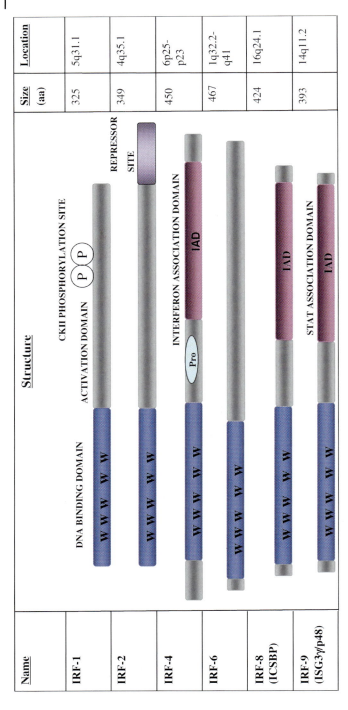

Fig. 2.2. Schematic representation of IRF-1, -2, -4, -6, -8 and -9 transcription factors. This diagram demonstrates the general structural homology amongst the IRF family. Different domains are shown: DBD, CKII phosphorylation site, repressor site and IAD. The size of each IRF as well as their chromosomal location is also described. (This figure also appears on page 44.)

Color Plates | XXVII

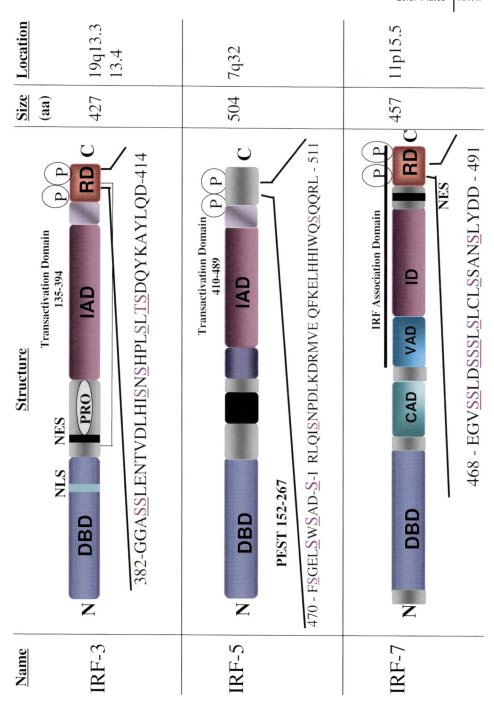

Fig. 2.3. Schematic representation of IRF-3, -5 and -7 transcription factors. Different domains are shown: NLS, NES, DBD, CAD, VAD, IAD and the signal response domain (RD). The sequence of amino acids 382–414 (IRF-3) and the sequence of amino acids 468–491 (IRF-7) are amplified below the schematic. Important amino acids are shown in larger letters with the respective position number. (This figure also appears on page 48.)

Fig. 2.4. Overview of the TLR dependent signaling of the IFN response. TLR3 stimulation with dsRNA on the cell membrane or the endosomal compartment leads to the recruitment of the TRIF adaptor molecule and the activation of TBK1 and IKKε kinases that phosphorylate IRF-3 and induce the transcription of IFN-β. TRAF6 association with the TRIF complex results in NF-κB induction via IKKα/β/γ. IFN-β is released from infected cells, recognized by specific IFN receptors on adjacent cells and triggers the JAK/STAT pathway. Among the genes activated by the JAK/STAT pathway, IRF7 is induced and activated by phosphorylation. IRF7 expression results in the production of multiple IFN-α subtypes and amplification of the IFN response. pDCs express on their endosomes hTLR7/mTLR8 and TLR9 that recognize ssRNA and CpG DNA, respectively. The recognition of specific PAMPs results in the recruitment and formation of a complex composed of MyD88, IRAK4, IRAK1 leading to IRF7 phosphorylation and induction of type I IFN in a MyD88-dependent fashion. TRAF6 association with the MyD88 complex leads to the activation of the signalosome complex (IKKα, IKKβ, IKKγ), which initiates the NF-κB signaling cascade and subsequent inflammatory response. (This figure also appears on page 52.)

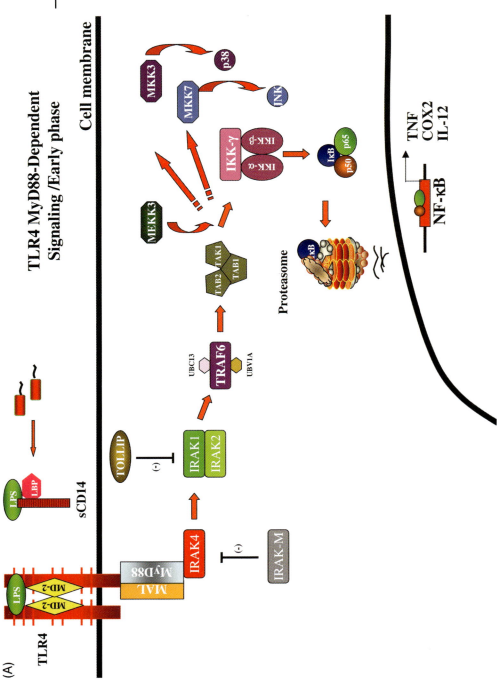

Fig. 2.5. (a) TLR-4 MyD88-dependent pathway – early phase. LPS from the outer membrane of Gram-negative bacteria is presented by the CD14–LBP complex to the TLR-4 receptor complex to initiate a MyD88-dependent signaling pathway. LPS signaling leads to the early activation of NF-κB. After activation and phosphorylation of IRAK, TRAF6 becomes activated and gives rise to the production of pro-inflammatory cytokines such as TNF, COX2 and IL-12. (This figure also appears on page 54.)

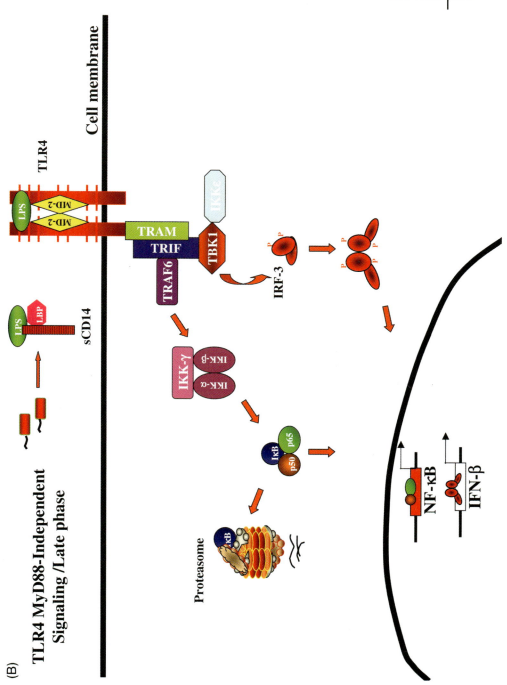

Fig. 2.5. (b) TLR-4 MyD88-independent pathway – late phase. LPS from the outer membrane of Gram-negative bacteria is presented by the CD14–LBP complex to the TLR-4 receptor complex to initiate a MyD88-independent signaling pathway. LPS signaling leads to the late activation of NF-κB as well as IFN-β induction. Following recruitment of TRIF to the receptor complex, TBK1 associates and initiates phosphorylation of IRF3, whereas TRAF6 associates with TRIF leading to the activation of NF-κB. (This figure also appears on page 55.)

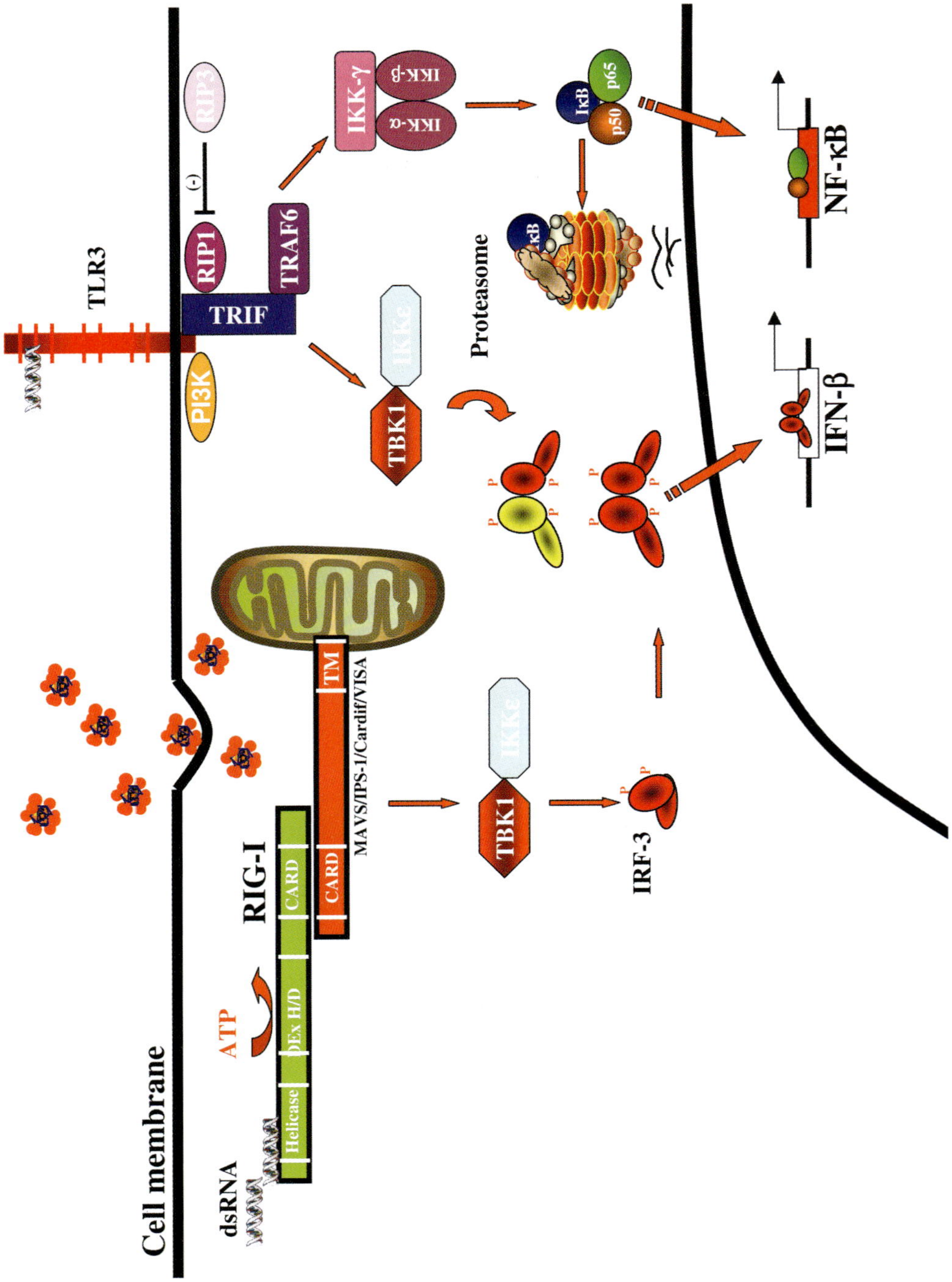

Fig. 2.6. Recognition of dsRNA and/or incoming cytoplasmic ribonucleoprotein complexes by RIG-I leads to the recruitment of the MAVS/IPS-1/VISA/Cardif. CARD domain interactions between RIG-I and MAVS lead to the activation of NF-κB and IRF dependent pathways through IKKα/β and TBK1/IKKε kinases respectively. The TLR3 dependent pathway also leads to the induction of NF-κB and IRF dependent pathways activating the same molecules involved in RIG-I pathway with the exception of recruitment of the TRIF, TRAF6, and RIP1 molecule to the TIR domain of TLR3 for NF-κB activation. PI3K is required for the full activation of IRF-3 via TLR3. (This figure also appears on page 60.)

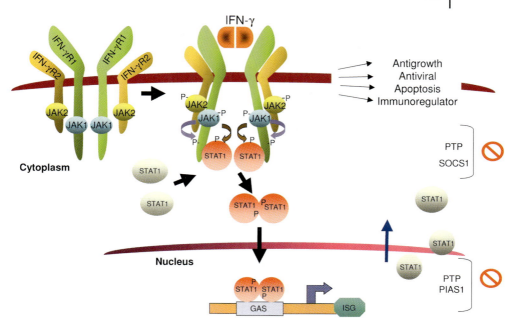

Fig. 4.2. Signal transduction of IFN-γ. IFN-γ binding to its receptor causes JAK-1 and -2 auto and *trans* tyrosine phosphorylation, respectively. Activated JAK-1 phosphorylates IFN-γR1 and STAT-1 becomes recruited to the receptor to be phosphorylated by JAK-1. Activated STAT-1 dissociates from the receptor, dimerizes with another STAT-1 and translocates to the nucleus to drive ISG expression. Inactivation of the JAK–STAT pathway is controlled by PTPs, SOCS-1 and PIAS-1. (This figure also appears on page 91.)

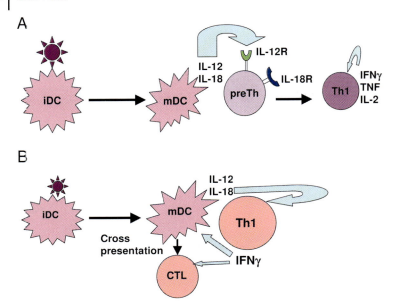

Fig. 4.3. (A) DCs capture, internalize and process pathogens in endosomal compartments. Their primary sequence is then presented as small peptides via MHC class II molecules to naive T cells. In parallel, DCs can be induced to secrete IL-12 and -18, which skew precursor T cells toward T_h1 polarization. (B) Immature DCs also internalize immunogenic antigens, but this cannot be cross-presented via MHC class I to $CD8^+$ T cells. IFN-γ secreted by T cells can significantly improve this process where mature DCs can then cross-present immunogenic peptides to rare $CD8^+$ T cells. In addition, IFN-γ upregulates the expression of IL-2 receptors that facilitate the expansion of CTL. (This figure also appears on page 98.)

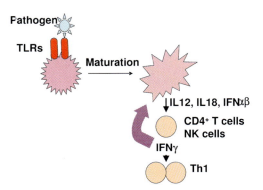

Fig. 4.4. Recognition of conserved subunits of infectious pathogens by TLRs induces DC maturation resulting in critical pro-inflammatory cytokine production (e.g. IL-12 or -18, or IFN-α/β). Such cytokines induce the expression of IFN-γ in T and NK cells. This IFN-γ can affect further DC and T activation. (This figure also appears on page 99.)

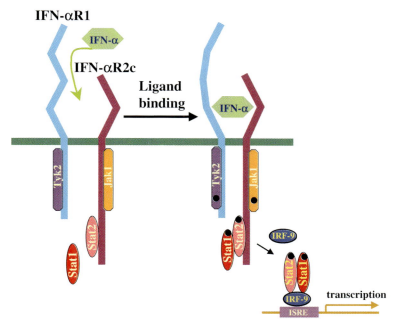

Fig. 5.1. Model of the Type I IFN receptor complex. The IFN-α receptor complex consists of two different chains, IFN-αR1 and -αR2c [67, 70, 88, 93, 94]. The ligand IFN-α is a monomer that binds to the two-chain complex [153]. Upon entry of IFN-α into the complex, STAT-2 binds to IFN-αR2c and recruits STAT-1. STAT-1 and -2 are phosphorylated, then released, associate with IRF-9, translocated to the nucleus and activate genes that contain the ISRE as described [26]. It is reported that limitin uses the Type I IFN receptor complex [154, 155]; however, although limitin requires TYK-2, it does not require STAT-1 for B cell growth inhibition where Daxx may replace the need for STAT-1 [155, 156]. Based on homology with the IFN-γ receptor complex [143], we believe that the Type I IFN receptor complex is preassembled and that a conformational change accompanies the activation of the receptor complex by Type I IFN. (This figure also appears on page 117.)

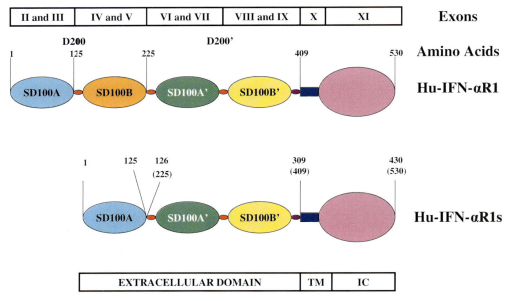

Fig. 5.2. Schematic Illustration of the human IFN-αR1 chain and the splice variant human IFN-αR1s. The domains of the human IFN-αR1 chain are shown. The splice variant chain, human IFN-αR1s, lacks exons IV and V. (This figure also appears on page 120.)

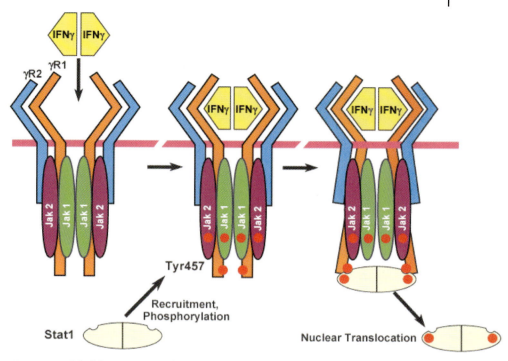

Fig. 5.3. Model of the IFN-γR complex. The IFN-γR complex consists of two different chains, IFN-γR1 and -γR2 [9, 33, 85, 99–101, 110, 120, 121, 152, 157–164]. The ligand IFN-γ is a dimer that binds to two IFN-γR1 chains, but does not directly bind to the IFN-γR2 chain in the absence of the IFN-γR1 chain. The two receptor chains are preassembled prior to binding of IFN-γ [143]. Upon ligand binding, the JAK kinases cross-phosphorylate each other (solid circles). The activated JAK kinases then phosphorylate Tyr457 of each IFN-γR1 chain that serves as the recruitment site for STAT-1α which, in turn, attaches to phosphotyrosine Tyr457 of IFN-γR1, moves the receptor chains apart [143] and is phosphorylated by the JAK kinases. Once phosphorylated, the phosphorylated STAT-1α proteins detach from each IFN-γR1 chain, forming the transcription factor that is translocated to the nucleus to activate IFN-γ regulated genes. (This figure also appears on page 122.)

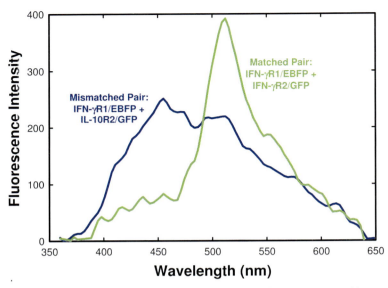

Fig. 5.4. Comparison of fluorescence emission spectra of cells expressing the matched and mismatched pair of receptor chains. The matched receptor chains are FL-IFN-γR2/GFP and IFN-γR1/EBFP (green curve); the mismatched receptor chains are IFN-γR1/EBFP and FL-IL-10R2/GFP (blue curve). The fluorescence emission spectra in response to two-photon excitation at 760 nm are shown. (Figure modified from data of Krause et al. [143].) (This figure also appears on page 125.)

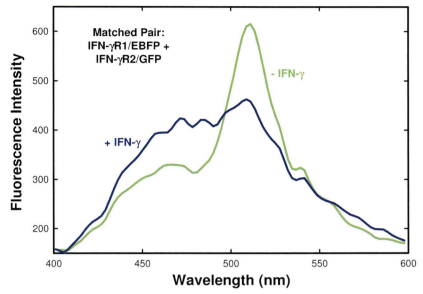

Fig. 5.5. Comparison of fluorescence spectra of cells expressing the matched pair of receptor chains in the presence and absence of IFN-γ. The matched receptor chains are FL-IFN-γR2/GFP and IFN-γR1/EBFP. The spectrum in green was taken in the absence of IFN-γ. IFN-γ (3500 U mL^{-1}) was then added to the medium and the spectrum taken (blue curve) of the same region in the same cell. The fluorescence emission spectra in response to two-photon excitation at 760 nm are shown. (Figure modified from data of Krause et al. [143].) (This figure also appears on page 126.)

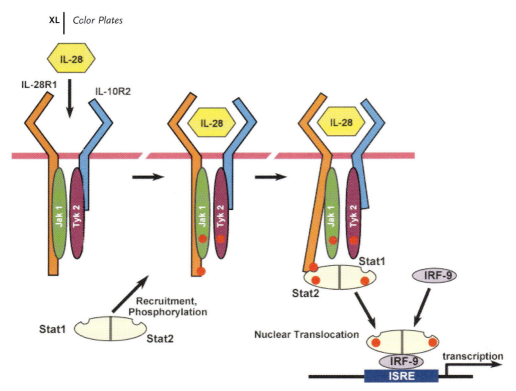

Fig. 5.6. Model of the IL-28R1/10R2 receptor complex. This receptor complex consists of two chains, IL-28R1 and -10R2, that are associated with JAK-1 and TYK-2, respectively, as shown. The STAT-1/2 heterodimer appears to function together with IRF-9 as the latent transcription factor. Based on analogy with the IFN-γR complex, we believe that this receptor complex is preassembled and undergoes a conformation change after its contact with ligand. (This figure also appears on page 127.)

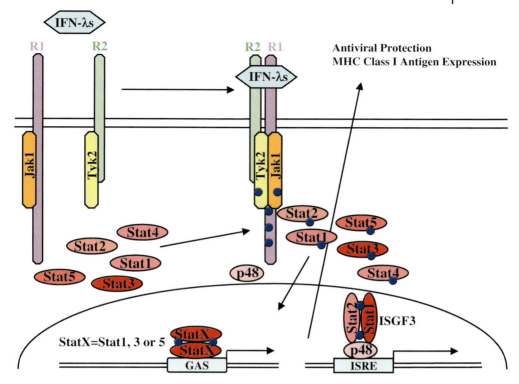

Fig. 6.3. Model of the IFN-λ receptor system. IFN-λs are likely to be monomers. The functional IFN-λ receptor complex consists of two receptor chains – the unique IFN-λR1 chain and the IL-10R2 chain. The IL-10R2 chain is a shared common chain in four receptor complexes, the IL-10, -22 and -26, and IFN-λ receptor complexes. Expression of both chains of the IFN-λ receptor complex is required for ligand binding and for assembling of the functional receptor complex. Ligand binding leads to the formation of the heterodimeric receptor complex, and to the initiation of a signal transduction cascade involving members of the JAK protein kinase family and the STAT family of transcription activators. The IL-10R2 chain is associated with TYK-2 [60] and the IFN-λR1 chain is likely to interact with JAK-1. Upon the ligand-induced heterodimerization of IFN-λ receptor chains receptor-associated JAKs crossactivate each other, phosphorylate the IFN-λR1 intracellular domain and, thus, initiate the cascade of signal transduction events. STAT-1, -2, -3, -4 and -5 are activated by IFN-λ leading to activation of biological activities, such as upregulation of MHC class I antigen expression and induction of antiviral protection. (This figure also appears on page 149.)

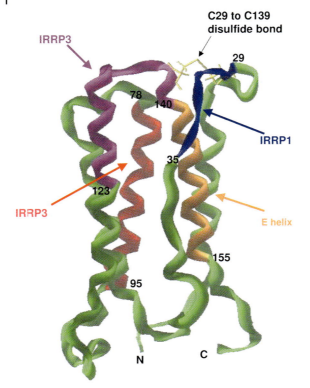

Fig. 7.1. Ribbon diagram representation of human IFN-α2 (protein data base ID 1ITF). Regions mediating binding interactions with the IFNAR subunits are highlighted. IRRP-1 (residues 29–35) is represented as a blue ribbon, IRRP-2 (residues 78–95) as a red ribbon, IRRP-3 (residues 123–140) as a purple ribbon and the E helix (residues 141–155) as an orange ribbon. The Cys29–Cys139 residues involved in the disulfide bond are represented as yellow sticks. The N- and C-termini are indicated. (This figure also appears on page 169.)

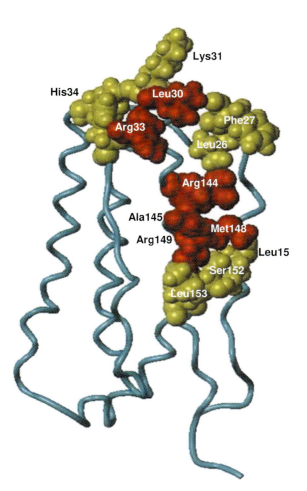

Fig. 7.2. Key residues in human IFN-α2 that contribute the most binding energy during interaction with IFNAR-2. Amino acids are represented in space-filling format. Residues that contribute more than 2 kcal mol^{-1} of energy are shaded red, and residues that contribute between 0.5 and 2 kcal mol^{-1} of energy are shaded yellow. (This figure also appears on page 170.) Adapted from [32].

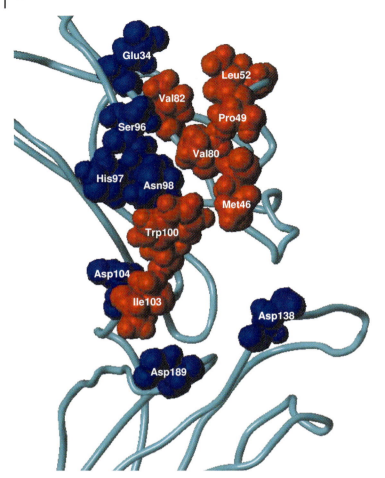

Fig. 7.3. Space-filling representation of the IFN binding surface on IFNAR-2. The elongated hydrophobic patch is comprised of residues colored in red surrounded by a ring of pclar and charged residues colored in blue. (This figure also appears on page 171.) Adapted from [21].

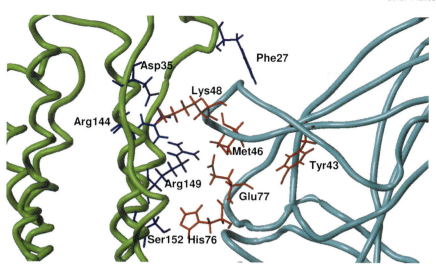

Fig. 7.4. Residues in the human IFN-α2–IFNAR-2 interface that mediate interactions between IFN and the receptor subunit. IFN-α2 is represented as a green ribbon with blue residue side-chains and IFNAR-2 is represented as a cyan ribbon with red residue side-chains. (This figure also appears on page 172.) Adapted from [21, 32].

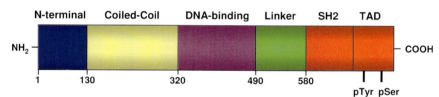

Fig. 7.5. Domain structure of STAT proteins. STAT proteins are a family of latent cytoplasmic transcription factors that serve as important mediators of cytokine, hormone and growth factor signal transduction. There are seven mammalian members of this family, STAT-1, -2, -3, -4, -5a, -5b and -6, all of which share a conserved domain-like structure. STAT proteins range in size from 748 to 851 amino acids (90–115 kDa) and consist of six different domains, each with its own defined function. The N-terminal domain (residues 1 to ~130) is involved in stabilizing STAT dimer–dimer interactions. The coiled-coil domain (~130 to ~320) is important for protein interactions. The central DNA-binding domain extends from amino acids ~320–490 and contains several residues conserved in all members of the STAT family. A linker domain exists between residues 490 and 580, and separates the DNA-binding domain from the SH2 domain. This area is comprised primarily of α-helices and appears to play a role in mediating transcription. The phosphotyrosine-binding SH2 domain is required for receptor binding and dimerization. Within this domain is a conserved tyrosine residue. Phosphorylation of this tyrosine activates the STAT molecule, allowing it to interact with the SH2 domain of another STAT. At the C-terminal end a transcriptional-activation domain (TAD) modulates the transcriptional functions of the various STAT proteins. The TAD mediates interactions of the STAT protein with a number of nuclear coactivators, facilitating chromatin modifications and transcriptional activation. (This figure also appears on page 177.)

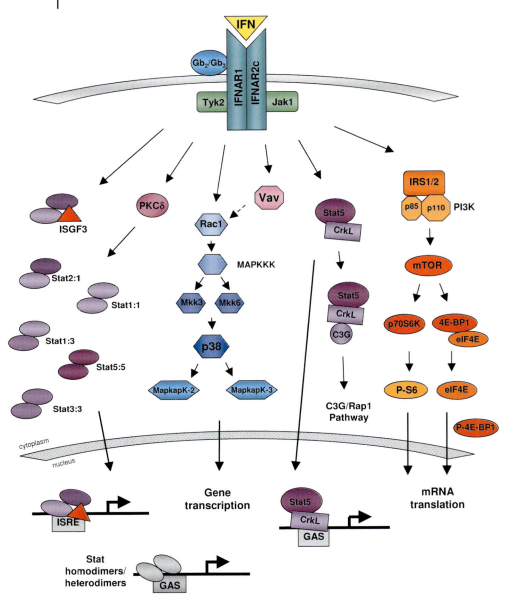

Fig. 7.6. Type I IFN-induced signaling. Engagement of the type I IFN receptor complex leads to the phosphorylation of JAK-1 and TYK-2, and activation of the JAK–STAT, p38MAPK, CrkL and IRS–PI3K pathways. (This figure also appears on page 180.)

Color Plates | XLVII

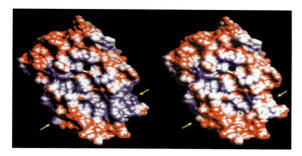

Fig. 10.2. A comparison of the surface area of the wild-type IFN-β and the modified IFN-β (Soluferon). www.igb.fraunhofer.de. (This figure also appears on page 284.)

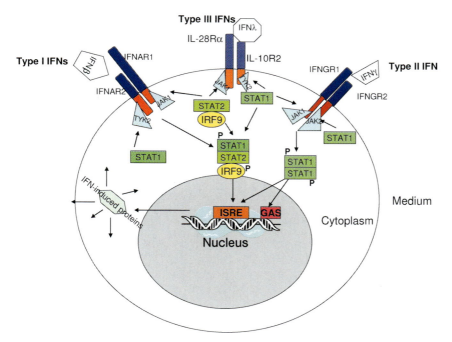

Fig. 12.1. Schematic diagram illustrating the IFN signal transduction pathways. IFNs bind to extracellular domains of their cognate class II cytokine receptors leading to activation of JAK-1 and -2, and TYK-2 tyrosine kinases associated with receptor endodomains. Subsequent phosphorylation of endodomains leads to recruitment of STAT-1 and/or -2–IRF-9, phosphorylation and dimerization of STATs followed by their translocation into the nucleus. These transcription factor complexes bind to the ISRE and/or GAS present in the promoter region of IFN-inducible genes to activate their transcription and the synthesis of IFN-inducible proteins. (This figure also appears on page 341.)

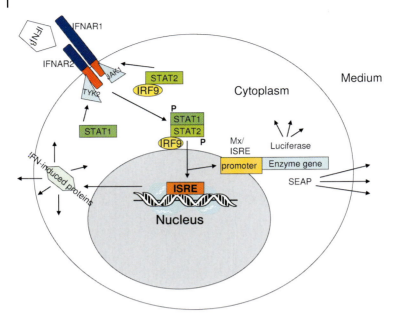

Fig. 12.4. Schematic diagram illustrating the type I IFN signal transduction pathway and basis for reporter gene assays. IFN-β binds to extracellular domains of its cognate class II cytokine receptor leading to activation of JAK-1 and TYK-2 tyrosine kinases associated with receptor endodomains. Subsequent phosphorylation of endodomains leads to recruitment of STAT-1 and -2, phosphorylation and dimerization of STATs, followed by their translocation into the nucleus. There, STATs combine with another DNA-binding protein, IRF-9 or p48, and this transcription factor complex binds to the ISRE present in the promoter region of IFN-β-inducible genes to activate their transcription. For reporter gene assays, the ISRE, or other IFN-inducible promoter, e.g. Mx, are present in DNA plasmids and are fused to enzyme genes, such as luciferase or SEAP. Binding of the transcription factor complex to these plasmid-encoded IFN-inducible promoters leads to synthesis of enzyme mRNA and enzyme protein, the latter being quantified by addition of appropriate substrates. (This figure also appears on page 358.)

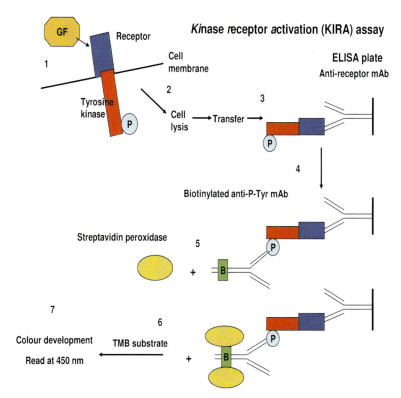

Fig. 12.5. Schematic flow diagram of the KIRA assay. Receptor activation is triggered by growth factor (GF) binding resulting in autophosphorylation of tyrosine residue(s) in the receptor endodomain (step 1). Following a brief incubation, cells are lysed (step 2) and lysates transferred to an ELISA plate coated with anti-receptor MAb (step 3). Phosphorylated receptor is quantified by addition of biotinylated anti-P-Tyr MAb (step 4), followed by addition of streptavidin peroxidase conjugate (step 5), addition of peroxidase substrate (TMB; step 6) with appropriate washings between steps. Following color development and termination of enzyme activity with 2 M sulfuric acid, absorbances are read at 450 nm (step 7). (This figure also appears on page 362.)

(A) Areas for IFNAR-1 and IFNAR-2 binding

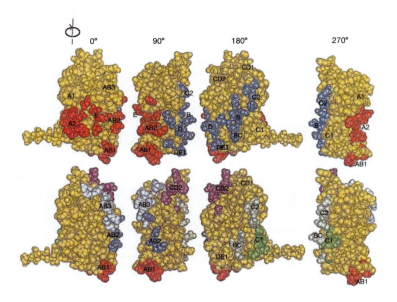

(B) Mutated residues reducing mAb binding

Fig. 13.4. Location of the binding sites of receptor and of anti-IFN-β MAbs on the IFN-β molecule. The three-dimensional crystal structure of IFN-β is shown in space-filling models. Four different views of the molecule are shown, as indicated at the top. (A) The positions of amino acid residues important for IFNAR-1 and -2 receptor chain binding are shown in blue and red, respectively. The regions occupied by the various alanine substitution mutants (A1–E) are labeled on the molecule. (B) The positions of residues, which when mutated to alanine abrogate or reduce the binding of anti-IFN-β MAbs, are highlighted in different colors on the IFN-β molecule. Shown in pale and dark blue are those that respectively abrogate or reduce the binding of MAbs Bio 1, Bio 2, Bio 5, A1 and A5. Those MAbs abrogating binding (Bio 4, A7 or Bio 6) are shown, respectively, in red, dark green, and purple. The sites abrogating (BC and C1) or reducing (C2) by MAb B-02 binding are shown in pale and dark green. (Reproduced with modifications from [78] with permission.) For identity of the MAbs and mutants the reader is referred to the original publication [78]. (This figure also appears on page 389.)

**Section A
Molecular Aspects, Introduction
and Purification**

1
Type I Interferons: Genetics and Structure

Shamith A. Samarajiwa, William Wilson and Paul J. Hertzog

1.1
Introduction

Since their discovery about 50 years ago [1], the interferons (IFNs) have been extensively studied, they are used clinically and their multiple functions are being elucidated at the molecular level. Recent sequencing of numerous vertebrate genomes, including the human and mouse genomes, has resulted in rapid expansion of known IFN gene and promoter sequences, and detailed characterization of the IFN gene cluster. Appreciation of the diversity of the IFN gene family and its origins, coupled with progress in structural biology of IFNs, has advanced our understanding of their functions.

The type I IFN genes were first located to specific chromosomes using cytogenetic methods based on aneuploid human cells or cross-species hybrid cells. While these initial studies led to the incorrect assignment of the fibroblast IFN gene to chromosome 2 and 5 [2, 3]; subsequently, analyses of a large number of hamster–human and mouse–chimpanzee somatic cell hybrid clones led to the localization of human and chimpanzee genes encoding IFN-β onto chromosome 9 [4]. During the same period, the unusual absence of introns within genes encoding IFN-α and -β was discovered by two separate groups [5, 6]. Leukocyte and fibroblast IFN cDNA clones were used to generate radioactive probes for hybridization experiments on human cells and human–mouse hybrid cells. These enabled the identification of multiple *IFNA* genes and a single *IFNB* gene on human chromosome 9 [7]. *IFNA* and *IFNB* genes were then mapped onto chromosome 9p and blot hybridization detected a cluster of at least 10 genes encoding IFN-α subtypes and a single gene encoding IFN-β on chromosome 9p ter-q12 [8]. Multiple functional genes and pseudogenes encoding IFN-α and -ω, and a single gene encoding IFN-β were identified in this region, and their relative position and transcriptional directions determined [9, 10]. During this time, the existence of multiple IFN-α subtypes was also discovered at the protein level [11]. Complete sequencing, extensive analysis and annotation of human chromosome 9 was completed recently, and the human type I IFN gene cluster was extensively characterized [12, 13].

The Interferons: Characterization and Application. Edited by Anthony Meager
Copyright © 2006 WILEY-VCH Verlag GmbH & Co. KGaA, Weinheim
ISBN: 3-527-31180-7

IFNs are grouped into three separate classes: types I (α/β IFNs, etc.), II (IFN γ) or III (λ IFNs), based on their sequence, receptor specificity, chromosomal location, physicochemical properties and structure. Type I IFNs are unique among the IFN family with unusual physicochemical properties of heat (65 °C) and acid stability (pH 2) [14]. Originally, type I IFNs were named leukocyte IFN (IFN-α) and fibroblast IFN (IFN-β) based on the perceived cellular origins of these cytokines. However, it is now accepted that most nucleated cells are able to produce IFN [15, 16], but specialized IFN-producing cells (IPC) such as plasmacytoid dendritic cells can produce 1000-fold higher levels of IFN when stimulated by an appropriate inducer [17]. Microorganisms such as viruses, bacteria, mycoplasma, protozoa, their products such as viral glycoprotein, bacterial lipopolysaccharide (LPS), CpG DNA, double-stranded RNA, host-derived molecules such as mitogens, other cytokines and a variety of other stimuli induce type I IFN production [18, 19].

In addition to IFN-α and -β, other members of the type I IFN gene family, such as IFN-ε and -κ are present in most mammalian genomes, while IFN-ω is present in primate and some mammalian genomes, and IFN-ζ and -ν are only present in rodent and feline genomes, respectively [20, 21]. Other type I IFNs such as IFN-τ and -δ that are involved in reproduction events have been identified, and are unique to ungulate ruminants and pigs [22–24]. This entire repertoire of type I IFNs is believed to bind to and signal through the cell surface IFNAR receptor complex, consisting of IFNAR-1 and -2 transmembrane receptor chains. However, binding of IFN-κ, -ε and -ν to the IFNAR receptors has not been determined. The genes encoding other types of IFN are located on different chromosomes and bind different receptor complexes.

1.2
The Type I IFN Genetic Locus

The human type I IFNs are encoded by a multigene family clustered over a 350-kb region on human chromosome 9p21 [10, 12, 13, 25, 26]. This gene family consists of 13 nonallelic IFN-α genes, at least five pseudogenes, and single functional IFN-β, -ε, -κ and -ω genes. Analysis of public sequence databases [27, 28] shows that almost all type I human IFNs are clustered together on chromosome 9 (9p21.1–9p21.2) between *IFNB1* and *IFNE* genes, which are positioned towards the telomere and centromere, respectively (Fig. 1.1A). The exception is *IFNK*, which is located approximately 6.4 Mb from *IFNE* towards the centromere and is transcribed towards the centromere. Within the cluster, genes are located in the following order: *IFNB1, IFNW1, IFNA21, IFNWP15* (pseudogene), *IFNA4, IFNWP9* (pseudogene), *IFNA7, IFNA10, IFNAL* (pseudogene), *IFNWP18* (pseudogene), *IFNA16, IFNA17, IFNWP5* (pseudogene), *IFNA14, IFNAP22* (pseudogene), *IFNA5, KLHL9* (kelch-like protein 9), *IFNWP20* (pseudogene), *IFNA6, IFNA13* and *IFNA2*; all of which are transcribed in the same direction (towards the telomere). *IFNA8*,

1.2 The Type I IFN Genetic Locus | 5

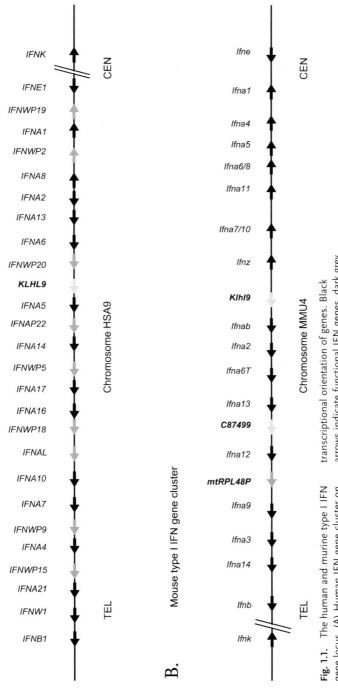

Fig. 1.1. The human and murine type I IFN gene locus. (A) Human IFN gene cluster on chromosome 9 (HSA9). (B) Murine IFN gene cluster on chromosome 4 (MMU4). The IFN gene locus was drawn using current genomic information available on NCBI human chromosome 9 (HSA9) contig (contig ID; NT_008413.16) and mouse chromosome 4 (MMU) contig (contig ID; NT_039260.4). Arrows indicate relative position and transcriptional orientation of genes. Black arrows indicate functional IFN genes, dark grey arrows pseudogenes and light grey arrows other non-IFN genes within the cluster. Orientation of the chromosome is indicated by TEL (telomeric end) and CEN (centromeric end). Nomenclature suggested by van Pesch et al. [32] is used in naming murine IFNs. (This figure also appears with the color plates.)

IFNWP2 (pseudogene), *IFNA1* and *IFNWP19* (pseudogene) are transcribed in the opposite orientation (towards the centromere), with *IFNE* transcribed towards the telomere. The relative position of IFN genes within the gene cluster and their transcription orientations is shown in Fig. 1.1(A). The only non-IFN gene within the human cluster, *KLHL9*, is syntenic on human and mouse chromosome HSA9 and MMU4, respectively, and is a homolog of the *Drosophila kelch* gene. These are implicated in embryogenesis, tissue development and carcinogenesis, and also mediate cytoskeleton organization [29, 30].

Murine type I *Ifn* genes were mapped to a centromere proximal region on (42.6 cM) *Mus musculus* (MMU) chromosome 4, and almost all of these *Ifn* genes are clustered on a 450-kb region [31] (Fig. 1B). The overall order of genes in the type I *ifn* locus is conserved between human and mouse genomes. This region contains at least three pseudogenes, 14 functional *Ifna* genes, and single *Ifnb*, *Ifnk*, *Ifne* and *Ifnz* (limitin) genes. Due to the low level of sequence resolution and the possibility of assembly artifacts in the *Ifnz* region, the exact number of *Ifnz* genes is not known. In a recent study, Hardy et al. [13] reported the presence of at least two genes and van Pesch et al. [32] detected a tandem array of 16 consecutive, almost identical *Ifnz* genes on the National Center for Biotechnology Information (NCBI) contig assembly. Within the main cluster of IFN genes, *Ifnb* is located more towards the telomere, whereas *Ifne* is located more towards the centromere. Murine *Ifnk* is located approximately 52 Mbp distal to *Ifnb* and the main IFN gene cluster, and is transcribed towards the centromere. Other IFN genes are located between these *Ifnb* and *Ifne* genes in the following order: *Ifnb*, *Ifna14*, *Ifna3*, *Ifna9*, *Ifna12*, *Ifna13*, *Ifna6T*, *Ifna7/10*, *Ifna2* and *Ifnab*; all transcribed in a single direction with their 3′-end towards the telomere (Fig. 1B). *Ifnz*, *Ifna7/10*, *Ifna11*, *Ifna6/8*, *Ifna5*, *Ifna4*, *Ifna1* and *Ifne1* form the remainder of the IFN gene cluster, and except for *Ifne* these genes are arranged in tandem in the opposite orientation and transcribed towards the centromere. Three non-IFN genes are located within the *Ifn* gene cluster and include the *klhl9* gene as in the human genome.

Interestingly, mammalian and avian type I IFNs lack introns – an unusual property shared with a number of other genes such as those that encode histones and G-protein-coupled receptors [33]. Recent genome analysis indicates that intronless genes are more common in human and other eukaryotic genomes than previously believed, with at least 12% of human genes being intronless [34–37]. The absence of introns in IFN genes may have been a property of the common ancestral gene or may have resulted from retro-transposed copying of an-intron encoded gene. The existence of introns in fish IFNs and in distantly related IFN-λ genes, and the presence of a single intron in the 3′-untranslated region (UTR) of IFN-κ suggests that IFN genes may have lost their introns due to retro-transposition. Their intronless gene structure and chromosomal colocalization, together with the multiplicity of IFN genes and extensive sequence conservation, indicates that this gene family arose by gene duplication. A similar expansion of genes encoding IFN-α is found in genomes of other mammals and in avian genomes, supporting the biological importance of this gene family.

1.3
Type I IFN Genes

1.3.1
IFN-α

The 13 human *IFNA* genes encode 12 different IFN-α proteins with *IFNA1* and *IFNA13* encoding an identical mature protein [26]. The IFN-α proteins share 76–99% amino acid identity and include a hydrophobic, 23-amino-acid signal peptide plus a 166-amino-acid mature peptide sequence. The exception is IFN-α2, which has a deletion at position 44 and encodes a 165-amino-acid protein. In addition, some of these subtypes exist in variant polymorphic forms such as IFN-α2a (K23, H34), -2b (R23, H34) and -2c (R23, R34) [38].

When murine *Ifna* gene sequences submitted to public databases were compared to the genomic sequence, discrepancies in nomenclature of some *Ifna* genes were identified. Certain *Ifna* genes such as *Ifna6* and *Ifna8* were identified as the same gene, as were *Ifna7* and *Ifna10* [32]. Therefore, these were named *Ifna6/8* and *Ifna7/10*. Most corresponding murine *Ifna* genes from different mouse strains such as 129/Sv, C57BL/6 and BALB/c showed 99% sequence identity; however, interestingly, genes encoding IFN-α6/8, -α7/10, -α11 and -α1 diverge substantially between mouse strains [32]. Except for *Ifna1*, the divergent genes all cluster together, suggesting that even though these may have originated from a cluster of common ancestral genes, they were subjected to independent evolution in separate mouse strains [32]. The IFN-α subtypes are produced in response to infection by diverse microorganisms and represent a first-line defense, particularly in viral infections. Individual subtypes are differentially expressed depending on the inducer and the producing cell [18, 39].

1.3.2
IFN-β

Unlike genes encoding IFN-α, duplication and expansion of the *IFNB* gene has not occurred in most mammalian genomes. A single gene encoding IFN-β is present in most mammalian genomes with the exceptions of ruminant (bovine, equine) and porcine genomes, where more than one copy of this gene is present. In most species extensively examined so far, only a single *IFNB* gene is present, although in a study of 25 Caucasian families, the *IFNB* gene segregated as a single copy, but duplication of the *IFNB* gene was seen in some members of two of the families and is believed to be a recent event [39].

Human IFN-β consists of 166 amino acids and shows 25–32% identity to human IFN-αs, whereas murine IFN-β consists of a 161-amino-acid mature protein and displays only 19–23% identity to murine IFN-αs [26]. IFN-β is reportedly expressed during myeloid cell differentiation and has a function there. Otherwise, it is produced like IFN-αs in response to infection, notably in response to Gram-negative bacteria or their constituent LPS, which interestingly induces no IFN-α [40].

1.3.3
IFN-ω

IFN-ω is a type I IFN that is distinct from α and β IFNs [41]. There are multiple *IFNW* genes in the human genome, but only one of them is a functional gene, while the rest are pseudogenes [42]. Orthologs of IFN-ω have been identified in the genomes of cattle, sheep and horses, but not in mice [43, 44].

The mature human IFN-ω consists of 172- and 174-amino-acid polypeptides, the latter being a minor species generated by alternative signal peptide cleavage. Similar to all the other IFNs, this subtype signals through the IFNAR receptor complex [42, 45]. These proteins are N-glycosylated at a position (Asn78) corresponding to human IFN-β and murine IFN-α, but are more related to α subtype IFNs with 55–60% identity [46]. IFN-ω is produced in response to viral infection, like IFN-α subtypes [47].

1.3.4
IFN-κ

The gene encoding IFN-κ is located at least 60 Mb telomeric from the rest of the human IFN gene cluster and its isolation from the rest of the cluster suggests separate evolution. This gene may have diverged early to play a specific role in mammals. Mouse, rat and chimpanzee counterparts of the *Ifnk* gene show a similar solitary location. While the mammalian type I IFN genes usually lack introns, the *IFNK* gene is the exception because it has a single intron in the 3'-UTR immediately following the stop codon. This may be important in the transcriptional regulation of this gene.

The single human IFNK does not show preferential identity with any other single IFN subtype, having only 27–32% identity to IFN-α subtypes, 34% identity to IFN-β, and 28% identity to IFN-ω and -ε. This type I IFN encodes a 180-amino-acid protein which is slightly larger than other IFNs due to an insertion in the CD loop.

IFN-κ is constitutively expressed by keratinocytes and other cells of the innate immune system, such as monocytes and dendritic cells, and shows some similarities to IFN-β [48] in binding to heparin with a high affinity. This binding may assist in maintaining higher local concentrations of IFN-κ [49].

1.3.5
IFN-ε

A single IFN-ε-encoding gene is present in a syntenic region of mouse, rat and human genomes. This *IFNE* gene is located in the extreme centromeric region of the human IFN gene cluster on 9p21 and the protein encoded by this gene is more structurally related to IFN-β than any other type I IFNs [13]. The mature protein is a 185-amino-acid polypeptide and contains a C-terminal extension relative to other type I IFNs. The human IFN-ε polypeptide is 15 amino acids longer than mouse IFN-ε, and shares 54% amino acid sequence identity and 69% similarity [13]. Murine and human IFN-ε possess reproductive hormone-regulatory elements on their promoter sequence, and are constitutively expressed in murine placental and ovarian tissue, indicating a possible reproductive role in placental mammals, perhaps similar to IFN-τ in ungulate ruminants [13].

1.3.6
IFN-ζ (Limitin)

Another IFN-like cytokine was recently identified in the mouse IFN gene cluster [50, 51]. Due to its growth-limiting ability it was named limitin; however, further analysis suggested that this protein belonged to the type I IFN family and led to this protein being named IFN-ζ [52]. The exact number of *Ifnz* genes in the mouse genome is undetermined, no homologs have been identified in humans and only an *Ifnz* pseudogene is present in the rat genome. The *Ifnz* gene consists of two potential open reading frames that can encode two different proteins, but there is preferential use the second ATG that encodes a full-length biologically active cytokine.

This IFN subtype encodes a 21-amino-acid signal peptide and a 182-amino-acid mature protein, with a single N-glycosylation site at Asn68. It has 25–28% amino acid identity to IFN-α, 21% identity to IFN-β and 30% identity to IFN-ω. Similar to IFN-β, this IFN is also a heparin-binding protein [51, 53]. This protein is constitutively produced, has potent B lymphopoiesis activity and shows a limited expression profile [53]. IFN-ζ has very high antiviral, immunomodulatory and antitumor activity compared to other IFN-αs, and it does not cause fever or myelosuppression, which are common clinical side-effects of other type I IFNs [54]. Its unique properties indicate that it may be useful as a novel therapeutic without toxic side-effects common to other IFNs.

1.3.7
IFN-τ

In addition to the IFNs commonly seen in murine and primate organisms, IFN-τ is a pregnancy-related IFN found only in ungulate ruminants (such as sheep, cattle, goats, musk ox, red deer, giraffe, etc.). Multiple genes encoding IFN-τ are seen in ruminant genomes. This unique physiological IFN is believed to have evolved by duplication from an *Ifnw* gene which acquired a promoter region that imparted trophoectoderm specific expression. Phylogenetic analysis indicates that this duplication event occurred about 36 million years ago (MYA) [55]. This protein is believed to be involved in the maternal recognition of pregnancy by prolonging the lifespan of the corpus luteum [56]. Unlike other IFNs, IFN-τ is not induced by viruses and is constitutively expressed by the embryo, with maximal expression detected prior to implantation [56]. Similar to IFN-ω, these IFNs contain a 6-amino-acid extension at the C-terminus. Even though IFN-τ genes have a high degree of conservation, sequence identity within a species is greater than between species, suggesting continued independent duplication of genes in different lineages of ruminants [57]. In bovine species these genes are clustered close to IFN-α and -ω genes on chromosome 8, which is syntenic to human chromosome 9 and mouse chromosome 4.

1.3.8
IFN-δ

IFN-δ is another pregnancy-related IFN subtype which was identified in pigs and encodes a 149-amino-acid protein. The porcine genome encodes two intronless nonallelic loci with strong homology to IFN-δ cDNA, one of these was a pseudogene with a premature stop codon and the other encoded the IFN-δ protein. The 5′ promoter region of this gene is devoid of regulatory and transcription factor binding elements needed for virus inducibility, consistent with the lack of viral inducibility of this gene. It has no preferential homology to other IFN subtypes with only 42% identity with human IFN-α2 and 27% identity to murine IFN-β [24]. IFN-δ is more divergent than any other IFNs and shows very little homology to the ruminant trophoblast IFNs. The gene encoding IFN-δ is believed to have diverged from the common ancestral sequence 180 MYA and before the mammalian radiation [24]. There are no human or murine orthologs of IFN-δ. Due to its shorter length than other IFNs it was originally called spI IFN (short porcine type I). The reduced length of the molecule is due to a 7-amino-acid deletion in the C-terminal end. This central deletion occurs in a region with the highest variability among the different IFNs and corresponds to a loop region.

1.3.9
IFN-ν

IFN-ν is a novel type I IFN subtype recently described by Krause and Pestka [21]. Homologs of this gene are present in most eutherian mammals such as cat, mouse, human, dog, pig, olive baboon and chimpanzee. However, only the feline genomes encode a functional *Ifnv* gene, while in all other sequenced mammalian genomes *Ifnv* is a pseudogene [21]. This gene is located about 25 kb downstream of IFN-β and is transcribed in the same orientation as IFN-β [21]. The function of this gene in cats is yet to be elucidated.

1.4
Type I IFN Gene-regulatory Regions

As mentioned above, the individual type I IFNs show differences in both "constitutive" and inducible expression, and therefore different functions associated with when and where they are produced. Since the type I IFNs are considered the first line of defense against infections, they require rapid and controlled regulation of gene expression. Comparative global genome analyses by programs such as AVID and VISTA [58, 59] have been effectively used to identify regulatory regions based on sequence homology across species. Such analyses of the 5' sequences in type I IFN genes indicates that the predicted regulatory promoter region of all these genes resides within 500 bp upstream of the ATG (exemplified by the IFN-β genes of human, mouse, rat dog and cow in Fig. 1.2A).

The 5' promoter region of type I IFN subtypes shows a high degree of sequence identity (Fig. 1.2B) and conserved transcription factor binding motifs in most mammalian IFN promoters are consistent with their common functions. In the 5' putative promoter region of eutherian mammals, multiple copies of GAAA tetranucleotide elements are present usually preceded by a spacer or AA dinucleotide elements and form part of the core IFN-regulatory factor (IRF) binding sites. A number of different transcription factors binding sites such as nuclear factor (NF)-κB, Jun, ATF and IRFs are present in type I IFN promoters, and the activation of these transcription factors during viral infection has been studied in detail.

The IFN-β promoter has been well characterized and serves as a paradigm for the study of transcriptional control of gene expression [60]. The regulatory sequences that form the IFN-β promoter are located within a 110-bp region immediately 5' of the transcription initiation start site [61, 62]. Nucleosome remodeling occurs in this promoter region prior to transcription [63]. Virus inducibility of IFNs is conferred by a virus specific enhancer region known as the virus response element (VRE), present in the promoters of both IFN-α and -β genes (Fig. 1.2B). Within the VRE of IFN-β there are four positive regulatory domains (PRDs) designated PRDI–IV to which transcription factors bind during virus-induced gene induction [64–66]. A negative regulatory element is also located in the region partially overlapping PRDII. Transcription factors such as IRF-1, -2 and -3 bind to PRDI, NF-κB binds to PRDII, IRF-3 binds to PRDIII, and activating transcription factor (ATF)-2/c-Jun binds to PRDIV [67, 68]. Within this promoter region, there are two binding sites within the minor groove of AT-rich sequences for the high-mobility-group protein HMGI(Y) – one near the ATF-2/c-Jun site and the other near the NF-κB site [69, 70]. These PRD regions interact synergistically to activate IFN-β gene transcription.

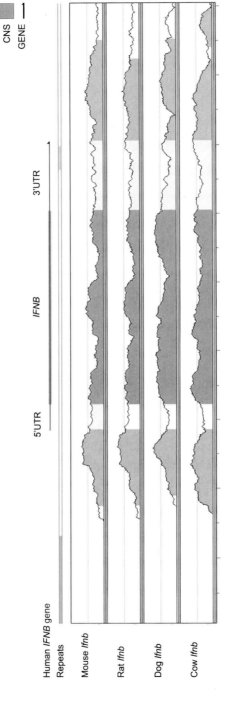

Fig. 1.2. Human type I IFN promoter sequence alignment. (A) VISTA alignment of human, mouse, rat, dog and cow IFN-β genes. The figure represents an AVID genome alignment [58] and VISTA visualization [59] of mouse, rat, dog and cow *Ifnb* genomic sequences compared to the human *IFNB* genomic sequence. At the top, the human *IFNB* gene is depicted by a dark blue line with an arrow indicating transcriptional orientation, together with 5′- and 3′-UTR regions depicted as light blue lines. The line immediately below indicates the location of repetitive elements within the genomic sequence. The graphed regions below demonstrate homology of mouse, rat, dog and cow *Ifnb* genomic regions to the human *IFNB* gene. Exon regions with more than 65% sequence identity over 100 bp are indicated by dark blue, whereas conserved noncoding sequences (CNS) and 5′- and 3′-UTRs with over 65% sequence identity are depicted in orange and light blue, respectively.

In addition to the similarities in the regulatory regions of IFN genes, there are also differences and these sequence differences are likely to explain the differences in expression of particular type I IFN genes in different circumstances. Transcription factors such as NF-κB and IRF-3 are involved in inducing IFN-β transcriptional activity in infected cells, and IRF-3 and -7 are important for IFN-α transcription. Accordingly, NF-κB sites are found in the promoters of IFN-β, -ε and -α14 genes only. Unlike the IFNB promoter, the IFNA VRE does not have NF-κB sites, but contains several PRDI-like elements (Fig. 1.2B). The VRE of IFN-α1 and -α4 also contains binding sites for IRF-1, TG protein and αF-1 binding proteins. Cooperation of these factors is necessary for efficient virus-mediated induction of IFN-α. Selective binding of transcription factors to different regulatory sequences within the IFN-α and -β promoters determines the specificity of induction of these cytokines. Even though there is a high degree of sequence conservation among the 5' promoter regions of human *IFNA* genes, due the presence of different types of transcription factor-binding elements, and the location and number of these elements result in differences between these IFN promoters. Similarly, the differences in promoter sequences among the type I IFNs will enable the binding of different transcriptional regulators according to the cell, stimulus or mechanism of activation.

In addition to regulatory elements seen in promoters, other elements within IFN genes are also important in regulation of IFN gene expression. The AU-rich elements present in the 3'-UTR of mammalian IFN genes and *cis*-acting CRID (coding region instability determinant) elements in transcripts encoding IFN-β are implicated in rapid mRNA turnover [71, 72].

1.5
Evolution of the Type I IFNs

1.5.1
Vertebrate IFN Genes

IFN genes are present in all vertebrates and form an evolutionarily conserved critical component of the host defense system [73]. Even though downstream mediators of IFN signaling such as STAT (signal transducers and activators of transcription) proteins have been identified in invertebrates [74], no IFN genes have been identified in these organisms. A report stating the ability of recombinant feline IFN-ω to mediate antiviral effects on Japanese pearl oysters (*Pinctada fucata martensii*) infected with akoya virus [75] and bind to receptors on hemocytes from these oysters presents interesting possibilities of IFN-like systems existing in invertebrates [76].

Comparison of human and mouse IFN genes and their promoters show large regions of conservation, indicating some expansion of the type I IFN gene cluster occurred before the divergence of mouse and human from a common mammalian

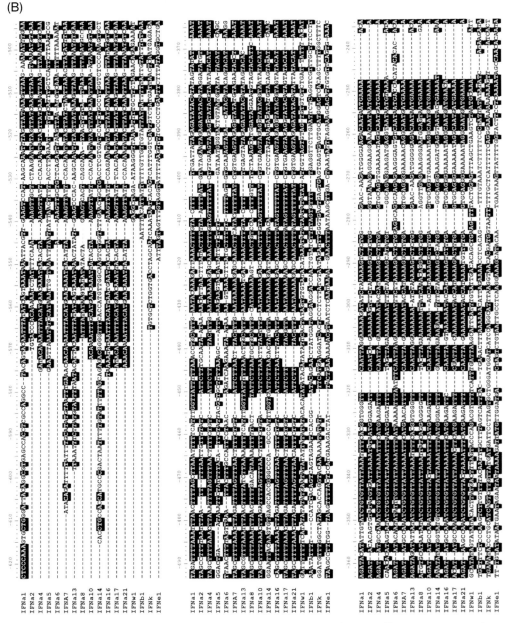

Fig. 1.2. (B) Nucleotide sequence alignment of human IFN promoter. Alignment was performed using ClustalW algorithm [79] of 500 bp 5′-flanking sequences containing the promoter elements of human IFN genes. The regulatory elements present in this region are indicated by boxed regions. (This figure also appears with the color plates.)

1.5 Evolution of the Type I IFNs | 15

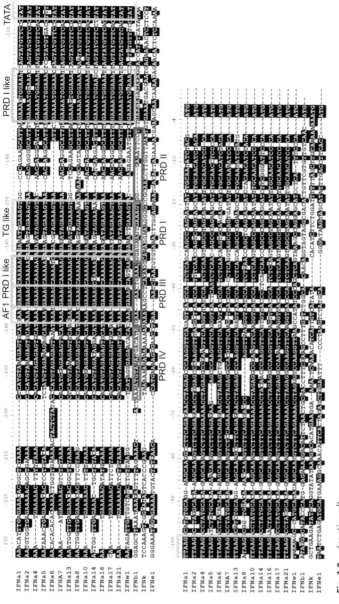

Fig. 1.2. (continued)

ancestor [12]. The type I IFN systems of different mammalian species show distinct similarities, with most mammals having multiple genes encoding IFN-α, and a single gene encoding IFN-β and most other type I IFNs. A small number of *Ifnb* gene duplications have been observed in ruminants. The three main type I IFN subfamilies, IFN-α, -β and -ω, are believed to have diverged after the mammalian–avian separation, but before the radiation of the eutherian mammals [77]. Phylogenetic analysis of mammalian, bird and fish IFNs demonstrates the clustering of similar subtypes from different species (Fig. 1.3). Phylogenetic analysis also suggests that genes encoding IFN-κ, -β and -ε may have diverged first from the ancestral IFN gene [57, 78, 79]. IFN-ω is believed to have diverged about 120 MYA from IFN-α, and is represented as multiple genes in some species and appears to have been lost in others [57]. The genes encoding IFN-τ have unique physiological functions involved in pregnancy and are believed to have originated from IFN-ω, with multiple copies of IFN-τ seen in some ungulate ruminants [55]. Only a single functional IFN-δ gene is present in porcine genomes and its progenitor may have pre-dated mammals.

Multiple subtypes of IFN-α appear to diverge after speciation, and it is notable that particular subtypes such as IFN-δ, -ξ, -τ and -ν found in pigs, rodents, ruminants and cats do not have individual subtypes in other species (except very closely related species, like IFN-τ within ruminants). Thus, once a species forms, it appears to be important to evolve multiple subtypes of IFN-αs. The reason for this is likely to be due to both different functions carried out by individual subtypes and to ensure some are produced in response to a broad range of pathogens. Numerous mammalian IFNs have been cloned and sequenced, and demonstrate the homology of these proteins in diverse species. A recent report of cloning and expression of five feline type I IFN genes demonstrates the similarity of these proteins even among distantly related mammals. Feline IFN-α1, -α2, -α3 and -α6 have very high sequence identity to each other, and these IFN sequences show 60% identity to human IFN-α2 and approximately 50% identity to human IFN-ω. Feline IFN-α5 is similar to feline IFN-ω due to the presence of an additional 5 amino acids inserted at position 139 of the sequence [80, 81]. Similarly, equine IFN-α subtypes showed 71–77% identity to human IFN-αs, while the equine IFN-β subtype showed 55% identity to the human ortholog [44].

Genomic information from marsupial and monotreme species, together with that from eutherian mammals, allows comparison of molecular evolution and IFN gene function in mammals with a relatively recent common ancestry. The type I IFNs of marsupial and monotreme species are classed into group I and II based on sequence identity. Analysis of the tammar wallaby *Macropus eugenii* genome demonstrated that both IFN-α and -β are present in this species [82]. Marsupial and monotreme IFN sequences also demonstrated that tammar wallaby group II and echidna (*Tachyglossus aculeatus*) group I IFNs contain the conserved Cys99 residue seen in eutherian (placental) mammals [82–84]. The wallaby group II IFNs are similar to the eutherian IFN-α and these IFNs also contain a complete set of conserved cysteine residues. Analysis of monotreme IFN genes in echidna resulted in the identification of two distinct groups of IFNs – three group I genes

1.5 Evolution of the Type I IFNs | 17

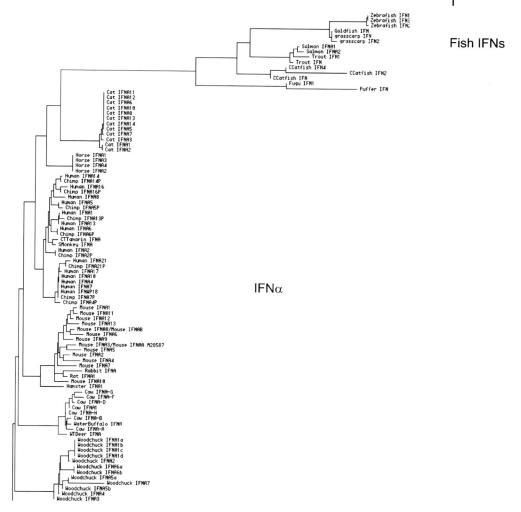

Fig. 1.3. Phylogenetic tree of vertebrate type I IFNs. The IFN amino acid sequences were aligned using the ClustalW algorithm [79], and the phylogenetic tree was constructed by calculating the distances between sequences using nongapped positions and corrected for multiple substitutions using the Kimura correction.

and a single group II gene [84]. Interestingly, NF-κB sites are present in the wallaby group I IFNs and human/murine IFN-β promoters, and similar to promoters of eutherian IFN-α genes wallaby IFN group II promoters lack NF-κB-binding sites. These putative, wallaby IFN promoter regions also contain GAAA-rich elements that are core binding sites for IRFs and related regulatory proteins. These regulatory regions are similar to elements seen in promoters of eutherian IFNs. Full-length IFN sequences, cysteine profiles and 5′ promoter sequences from

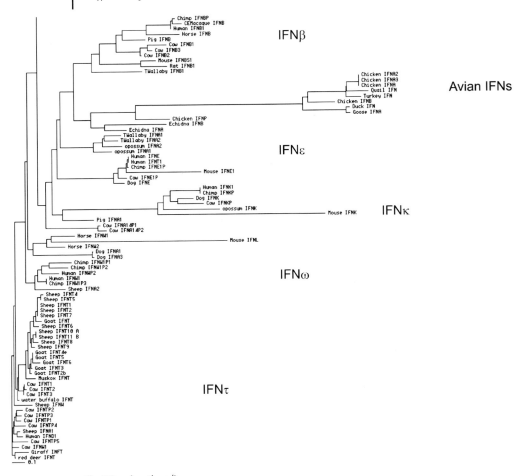

Fig. 1.3. (continued)

echidna also demonstrate that echidna group I IFN genes are more similar to those encoding primate IFN-α and the group II gene is similar to IFN-β [84]. Thus, placental mammals, marsupials and monotremes have evolved similar IFN systems.

Cloning nonmammalian vertebrate IFN genes was initially difficult because of weak sequence homology with their mammalian counterparts. However, the sequencing of multiple vertebrate genomes has yielded a large amount of sequence information on nonmammalian IFNs during the last few years. Recent outbreaks of avian flu and swine flu, and the impact on the poultry industry and farming, together with viral infections in aquaculture farming and implications of viral diseases of avian origin such as SARS (severe acute respiratory syndrome) and influenza on human health have rekindled interest in avian, porcine and fish IFNs.

The first nonmammalian IFN was cloned from virus-infected chick embryo cDNA [85], and chicken IFN was purified from embryonated chicken eggs and chick embryo cells [86–88]. This protein was shown to be a glycosylated, acid- and heat-stable protein of 20–30 kDa. Avian type I IFNs are classified as IFN I (α-like) and IFN II (β-like) genes [89]. Unlike mammals, which have their IFN genes located on autosomes, birds have type I IFN genes on their sex chromosomes. Comparison of the completed genome sequences revealed extensive conserved synteny between parts of human chromosome 9 (which contain the human IFN gene cluster) and chicken sex-determining chromosome Z, with 17 of the 24 known chicken Z chromosomal genes having orthologs in chromosome 9 [12]. Interestingly, the chromosomal region close to the IFN gene cluster on human chromosome 9 contains genes involved in testicular differentiation and hemizygous deletion of this region results in human XY sex reversal, suggesting parts of chromosome 9 may have been derived from sex chromosomes of a common ancestor of mammals and birds in the evolutionary past [12]. Cluster of about 10 IFN-α-like genes and a single IFN-β-like gene, all of which are intronless, were identified on the short arms of chicken and long arm of duck Z chromosomes [90, 91]. Induction of IFN-like activity in other avian species such as turkey, pheasant, bobwhite quail, guinea fowl and duck has been reported, and IFN genes from numerous bird species have been cloned [92–94], indicating the presence of these genes in many avian species. Even though there is no detectable sequence homology between mammalian and avian IFN promoter sequences [78], genome analysis identified specific similarities between avian IFN-α-like gene promoters, with promoters of mammalian IFN-α subtypes. These IFN-α-like genes lack the NF-κB transcription factor binding elements that are found in mammalian IFN-β and avian IFN-β-like gene promoter regions [94, 95]. However, phylogenetic analysis supports the hypothesis that the gene duplication event that gave rise to mammalian IFN-α and -β may have occurred after the last common ancestor of birds and mammals diverged. Therefore, the avian and mammalian IFN-α and -β subtypes are not orthologs, and may have evolved independently from a common ancestor in mammals and birds to give rise to similar subtype functionality [78].

For many years IFN-like antiviral activity has been observed in virally infected fish and fish cell lines [96]. The existence of IFN-stimulated genes (ISGs) and other downstream mediators of IFN activity were identified in a number of fish species. Homologs of *Mx*, *Vig1*, *Vig2*, *IRFs*, *PKR* and *ISG15* genes were identified in fish, and a transcriptome analysis of rainbow trout (*Oncorhynchus mykiss*) revealed that IFN stimulated genes make a major contribution to response against a rhabdovirus infection [97, 98]. Interestingly, the promoter of the single fugu *Mx* gene can be induced by human IFN, when transfected into human cells [99]. The first fish IFN gene was identified by searching a zebrafish (*Danio rerio*) expressed sequence tag (EST) database [98]. This fish IFN cDNA encoded a 185-amino-acid protein with a hydrophobic 22-amino-acid signal peptide. These proteins display only about 15% sequence identity to mammalian and avian IFNs, and unlike the acidic mammalian and avian IFN proteins the fish proteins are highly basic [98]. Later this zebrafish IFN sequence was used to uncover an IFN gene in puffer fish (*Fugu rubripes*) [98], and similar genes have been cloned from channel catfish (*Icta-*

lurus punctatus) [100] and Atlantic salmon (*Salmio salar*) [101]. Unlike in mammals, the presence of introns is characteristic of fish IFNs, with both Atlantic salmon and zebrafish IFNs having a gene structure of five exons and four introns [101]. Another IFN gene (*TnIFN*), which is structurally related to type I and λ IFNs, was cloned from *Tetraodon nigroviridis*. This *TnIFN* has four introns similar to the λ IFNs, and was shown to induce *Mx* and *PKR* genes [102]. Information on IFN genes in amphibian species is extremely rare. However, *xIRF6*, which is an IRF-related gene, expressed during early developmental stage of *Xenopus laevis* has been cloned [103]. Existence of reptilian IFNs has been inferred from antiviral activity mediated by acid-stable, heat-resistant 33-kDa proteins produced by turtle and tortoise (*Testudo graeca*) cells infected with virus or stimulated with IFN inducers such as poly(I:C) and LPS [104, 105]. Similar IFN-like activity was seen in snake (*Vipera russelli*) VH2, VSW and lizard (*Gekko gecko*) Gl-1 cell lines infected with the rabies virus [106].

1.5.2
The Expansion and Divergence of the IFN Genes

The type I IFN genes were believed to have originated from a common ancestor and expanded by successive duplications. Figure 1.3 represents a phylogenetic tree of type I IFNs, demonstrating the evolutionary relatedness of the IFN subtypes from different species. Due to the importance of the IFN system in host defense, expansion of the IFN cluster may have occurred initially by selection for redundancy. Once this selection pressure abated they may have diverged and acquired specialized functions or degenerated into pseudogenes [107–109]. Analysis of genomes reveal that newly created genes appear to be governed by positive Darwinian selection, with rapid changes in amino acid sequence and gene structure occurring in very short periods of evolution, particularly within IFN-α subtypes (e.g. feline IFN-α in Fig. 1.3). This positive selection is important in the interaction between structure, function, genotype and phenotype [100, 110]. Some viruses encode proteins such as soluble receptors that are IFN antagonists or other proteins that can disrupt IFN signaling pathways, enabling unhindered viral replication. The presence of multiple IFNs may prevent a viral- or other pathogen-produced IFN antagonist molecule disabling the IFN system completely. Coevolution of the IFN genes and the IFNAR receptors may have also resulted in selection for IFNs with different binding affinities and signaling responses. Thus, the lack of introns may have been another factor in selective pressure towards gene duplication since alternative splicing without introns was not an option to generate multiple functional gene products. In addition to differences in protein-coding sequences, differences between elements in the IFN promoter regions may have evolved a level of control of tissue- or pathogen-specific and temporal regulation of IFN expression, and aid in mounting appropriate host defense responses against different microorganisms. Differences in promoters may also enable IFNs such as trophoblast IFN (IFN-τ) to play physiological roles. Thus, the development of multiple IFN subtypes may account for the diverse range of biological roles these cytokines play.

1.6
Natural and Induced Mutations in IFN Genes

Mutations in type I IFN genes, whether naturally occurring or induced in model organisms and recombinant experimental systems, have provided important insights into the function of these cytokines.

Molecular genetic analysis aimed at linking disease to chromosomal regions identified a highly significant association between serum triglyceride levels and the IFN-α locus on HSA 9p21 in a communal founder population. Alleles at nearby loci are believed to protect against high triglyceride levels, as homozygocity at this locus is associated with low triglyceride levels [111].

Mutations within this gene cluster have revealed the tumor suppressor potential of the IFNs. A number of breakpoints have been identified in the IFN gene cluster, and IFN gene deletions have been associated with gliomas and leukemias, non-small cell lung cancers and others [10, 112–114]. A large number of scaffold/matrix associated regions (S/MARs) flanking functional IFN genes and pseudogenes were identified [115]. S/MARs are involved in shaping the chromatin of DNA into loop domains and in 9p21 these may be involved in organizing the IFN genes into a series of small 2- to 10-kb DNA loop domains which may predispose this region to breaks [115].

Functional genomic methods such as gene knockout studies in mice are invaluable in determining the function of IFNs in $vivo$. Due to the large number of IFN-α genes and their functional redundancies, no $Ifna$ gene knockouts have been generated; however, $Ifnar1$-deficient mice have been generated and these are deficient in their responses to all type I IFNs tested, including multiple IFN-α subtypes and IFN-β [116, 117]. IFN-β knockout mice have been produced [118, 119] and are more sensitive to viral infections [119, 120]. When these mice were analyzed, no abnormalities in their CD4 and CD8 T cell populations in peripheral blood, thymus and spleen was observed; however, activated splenic and lymph node T cells showed enhanced proliferation and decreased tumor necrosis factor-α production. A decrease in the number of circulating macrophages and granulocytes was also seen. Tumor growth was aggressive in these knockout mice, demonstrating the potent tumor inhibitory effects of this cytokine. This study demonstrated that IFN-β plays a fundamental nonredundant role in lymphoid development and myelopoiesis [121]. The IFN-β null mice also demonstrated that IFN-β production is necessary for generating an IFN-α response, implying the importance of IFN-β in IFN priming. These mice were also more resistant to septic shock induced by a high dose of Gram-negative bacterial LPS, suggesting the involvement of the type I IFN system in lethality due to septic shock [122]. The IFN-β knockout mice also demonstrated that IFN β plays an important role in the immunoregulation of allergic responses in mice [123] and in the central nervous system as they are more susceptible to chronic demyelinating experimental autoimmune encephalomyelitis [124].

Transgenic mice overexpressing murine IFN-α1 under the control of mouse metallothionein I promoter were generated [125]. In these mice, IFN-α1 was only expressed in the testis, and resulted in degeneration of spermatogenic cells and atrophy of seminiferous tubules [125]. Surprisingly, these transgenic mice were sterile or turned sterile over time. A similar effect was seen when IFN-β was overexpressed in transgenic mice, where degeneration of spermatids and sterility was observed [126]. These studies suggest that high levels of IFN expression in testis lead to male sterility. When transgenic mice expressing the murine active, hybrid human IFN-α A/D under the control of the human insulin promoter were generated, pancreatic IFN-α expression led to diabetes. Persistent hyperglycaemia with polyuria and polydipsia with inflammation centered on pancreatic islets was observed in these transgenic mice [127]. Neutralizing monoclonal antibodies against IFN-α were able to prevent this inflammation and diabetes [127]. Transgenic mice with targeted overexpression of murine IFN-κ in pancreatic islet β cells resulted in hyperglycaemia [128] and transgenic mice overexpressing IFN-β led to mild hyperglycaemia with 9% developing spontaneous diabetes. These mice were more sensitive to the diabetes-inducing agent streptozotocin, compared to normal mice [129]. These studies suggest that the *IFN* genes may be involved in the pathology of type I diabetes consistent with IFN-α expression in pancreatic islets seen in human patients with diabetes.

Mutagenesis studies of IFN genes have been important in the determination of functionally important domains and residues in the protein. Although all type I IFNs share five helical bundle structures, subtle structural differences may play a role in differential receptor interactions resulting in differential biological potencies or functions. Early studies demonstrated the importance of cysteine residues involved in disulfide bond formation and conserved tyrosine residues for optimal activity [130]. The role of specific amino acid residues in receptor interaction was shown by mutagenesis studies where a double substitution (N86E, Y92D) in the C helix of IFN-β abolished its capacity to induce this receptor association [131]. The ability of IFN-β and not IFN-α to induce the coimmunoprecipitation of both IFNAR-1 and -2 receptors suggested sequence differences between IFNs result in different interactions with the type I IFN receptor complex [132]. The ability of IFN-α and -β to interact with IFNAR-1 and -2 receptors differentially was further elucidated by receptor–ligand mutagenesis studies [133].

1.7
Secondary Structural Features of Type I IFNs

1.7.1
Conserved Amino Acid Residues

Conserved residues are critical for structure and function of IFNs, and analysis of these residues across different species has revealed patterns of evolution of these critical residues. Distinguishing features of mammalian IFNs include five highly conserved proline residues in IFN-αs, but the genes encoding IFN-ω and -τ contain only four of these conserved prolines, and lack the proline that immediately precedes a conserved cysteine at position 139. The feline IFN sequences all follow the human IFN-α proline pattern and are thus classified as feline IFN-αs [80]. Interestingly, IFN-α subtypes also contain conserved lysine residues near the N-terminus and tyrosine residues near the C-terminus that are critical to optimal activity [131]. Almost all IFN-α subtypes within a species have two conserved disulfide bonds (Cys1–Cys99 and Cys29–Cys139) [134, 135], while IFN-β forms a single intramolecular disulfide bridge equivalent to Cys29–Cys139 in IFN-α. Analysis of position, number and conservation of cysteine residues in different species has assisted in classifying IFN genes found in numerous avian and mammalian (eutherian, marsupial and monotreme) organisms. The cysteine pattern in IFN-α is conserved among mammals with four conserved disulfide bond-forming cysteines present. Only two of the conserved cysteine residues are present in IFN-β. With the exception of murine IFN-β and porcine IFN-δ, most eutherian IFNs conform to this characteristic cysteine pattern.

1.7.2
Post-translational Modifications of Type I IFNs

Glycosylation plays a critical role in protein folding, structure, targeting and pharmacokinetics. Glycosylation is rare among the human IFN-αs, with only two species glycosylated. IFN-α2b is O-glycosylated at Thr106 and IFN-α14c is N-glycosylated at Asn72 [136, 137]. IFN-α2b is the only IFN-α species with a threonine residue at position 106 and may be the only O-glycosylated human IFN-α protein [137, 138]. In virus-infected white blood or lymphoblastoid cells, two differentially glycosylated human IFN-α2 exists as a fully glycosylated form and a form with half the sugar content. In contrast to human IFNs, most murine IFN-α subtypes have a single putative N-glycosylation site at Asn78. Human IFN-β has three putative glycosylation sites at positions 29, 69 and 76, while none are present in the murine IFN-β. Interestingly, human IFN-β, unlike the murine protein, is glycosylated at Asn80 at the end of helix C [139]. These sugar residues project away from the core structure and interacts with Asn86 (helix A) and Gln23 (helix C) through hydrogen bond formation. This glycosylation has been shown to reduce the aggregation properties of IFN-β by shielding a hydrophobic patch implicated in oligomerization and is involved in providing temperature stability. When *Escherichia coli*-derived unglycosylated and glycosylated forms of recombinant IFN-β produced in CHO cells were compared, the glycosylated form was 10- to 15-fold more bioactive [139]. Similarly, the glycosylated form of feline IFN-α6 produced in yeast showed considerably higher antiviral activity than the *E. coli*-derived unglycosylated feline IFN-α6 [80]. Thus, glycosylation may contribute to the bioactivity of these proteins.

The two potential N-glycosylation sites present at Asn74 and Asn83 in the short BC loop seen in the mature sequence of IFN-ε are similar to those in human IFN-β and -ω, which have a single glycosylation site in this region [13]. The human IFN-ω protein is N-glycosylated at a position (Asn78) similar to human IFN-β

[46]. IFN-ζ also contains a single putative N-linked glycosylation site at Asn68 [50], while IFN-δ is N-glycosylated at Asn79 [140].

1.7.3
Conserved Cysteine Residues and Disulfide Bond Formation

The number and nature of disulfide bonds is critical for maintaining the full biological activity of IFN-α [141]. Curiously, human IFN-β lacks a second disulfide bond present in IFN-α structures, which might explain its different interaction with the receptor [109, 142, 143]. In contrast, murine IFN-β, although structurally similar to its human ortholog, has no intramolecular disulfide bonds [144]. Three cysteine residues are present in human and murine IFN-ε, with the predicted disulfide bond (Cys32–Cys140) linking the top of the AB loop to the top of the E helix. Unlike other IFNs, this protein lacks a disulfide bond connecting the A and C helices. However, secondary structure analysis indicates both human and mouse IFN-ε consist of six and five putative α helices, respectively, whilst the C-terminus helix of human IFN-ε is not present in the mouse protein [13]. Interestingly, IFN-κ in mouse and rat is characterized by the presence of a C-terminal cysteine residue in addition to the four cysteine residues conserved in other type I IFNs [48]. Unusually, IFN-ζ contains four cysteine residues that demonstrate no conservation to those in any other type I IFN [50]. When IFN-ζ was computationally modeled, it formed a structure of five long α helices and a short helix within the BC loop. The Cys52, Cys157 and Cys80, Cys130 residues of IFN-ξ are closer together on the structure and may form disulfide linkages. The unpaired cysteine residues may be involved in intramolecular bonding, resulting in dimer formation. Similarly, the cysteine content in porcine IFN-δ is also unusual, with five cysteine residues at positions Cys9, Cys56, Cys58, Cys107 and Cys145 that are not conserved with those in any other IFN. Four of those are clustered in the same structural region of the molecule, and may form possible disulfide linkages [140].

1.8
The Structure of Type I IFNs

Even though IFNs have only slightly different, closely related amino acid sequences, they display different biological potencies. The bioactivity of these cytokines is determined through specific high-affinity interactions with their cell surface receptors. The availability of large amounts of IFN proteins due to advances in recombinant DNA technology made crystallographic studies feasible in the early 1980s. Due to difficulty in crystallization of these proteins, more than a decade passed before the first IFN crystal structure was determined. The first reported crystals were for human recombinant IFN-α2a [145]. These were of insufficient quality for crystallography and no useful structural information was obtained. Sub-

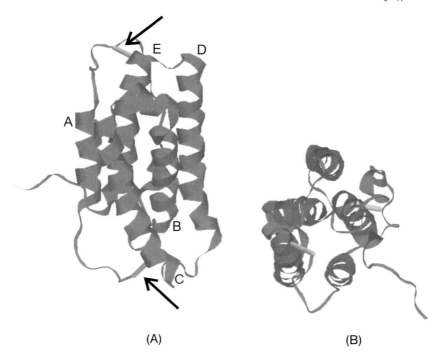

Fig. 1.4. Representative tertiary structure of human IFN-α. Tertiary structure of human IFN-α2 represented as a ribbon diagram, showing the five α-helices connected by loop regions. The disulfide bonds between cysteine residues are indicated by arrows. (A) Side view of the molecule. (B) View from above. (This figure also appears with the color plates.)

sequently, E. coli-derived recombinant murine IFN-β crystals were obtained [146, 147], and the crystal structure for murine IFN-β was solved at 2.6 and 2.15 Å resolution, paving the way for better understanding of type I IFN structure [144, 148].

The determination of tertiary structure through nuclear magnetic resonance (NMR) spectroscopy of human IFN-α2a [149], and X-ray crystallographic studies of human IFN-α2b [150], murine and human IFN-β [144, 151], and ovine IFN-τ [152] have determined that the type I IFNs consist of five α helices (labeled A–E) which are linked by a long loop (AB loop) and three shorter loops (BC, CD and DE) (Fig. 1.4). The AB loop contains short segments of 3_{10} helices and encircles helix E with which it is linked by a disulfide bond. In most IFN-α species a second disulfide bond, which is absent in human IFN-β, connects the N-terminus of the molecule to helix C. Even though the core structures of IFNs are similar, large structural differences occur in the AB loops, and at both the C-terminal ends of helix B and BC loops. Interestingly, the crystal lattice of human IFN-β consisted of dimers that contained a zinc atom at the dimer interface. Similar zinc-mediated dimerization of human IFN-α2b was also observed, although murine IFN-β and ovine IFN-τ were crystallized in the monomer form. The physiological relevance of dimer formation has not been determined. Human and murine IFN-β show only around 50% amino acid sequence identity to the murine form, and the murine protein lacks the conserved disulfide bond-forming cysteine residue pair present in human IFN-β. A 5-amino-acid deletion in the AB loop region is also seen in the murine IFN-β protein. These differences seem to have very little effect on tertiary structure, with the crystal structure of the two proteins being very similar [151]. Human IFN-β displays a tertiary structure typical of type I IFNs, with helix A and B parallel to each other, and antiparallel to C, D and E helices. According to the fold classification of Presnell and Cohen, human IFN-β is classified as a left-handed, type 2 helix bundle that is defined by antiparallel A, B, C and E helices [151, 153].

Mutagenesis studies have shown that the AB loop is critical for high-affinity binding and sequence differences in this region may be important for the difference seen among the IFN subtypes. Hybrid scanning, site-directed mutagenesis and other techniques have identified three functionally important segments on the IFN-β polypeptide sequence. These segments are spatially close to each other, indicating a receptor-binding interface [148]. The NMR structure of IFN-α2 bound to the IFNAR-2 extracellular domain was determined and demonstrated the presence of a predominantly aliphatic hydrophobic patch on the receptor that interacted with a matching hydrophobic surface of IFN-α2 [154]. In addition to this, an adjacent motif of alternating charged side-chains was involved in guiding the two proteins together into a complex [154, 155]. Comparison of the structure of ovine IFN-τ with the human IFN-α2b structure enabled the prediction of binding sites of different IFNs for the IFNAR receptors.

These structural studies, together with mutagenesis and receptor-binding studies, have provided a basis for understanding how structural differences between different subtypes affect biological activity. When the antiviral activities of murine type I IFN subtypes were measured relative to that of IFN-α1, most IFNs had similar antiviral activity (IFN-α2, -α5, -α6/8, -α6T, -α7/10, -α9, -α13 and -α14); however, IFN-α4, -α11, α12, -β and -ζ had up to 8- to 10-fold higher antiviral activity [38, 156–159]. The Arg58 and Asp59 residues have been associated with the high activity of IFN-α4 [38]. Interestingly, the antiproliferative activity of different murine IFN subtypes correlated with antiviral activity, with IFN-α11, -α12, -β and -ζ demonstrating up to 100-fold more potent antiproliferative activity relative to that of IFN-α1 [38]. In addition to differences in their biological potencies, different murine IFN subtypes have been shown to differentially activate STAT signaling [160]. Differences in biological activities and potencies of type I IFNs may therefore result from a combination of factors, such as subtle differences in sequence or structure and differences in receptor interaction or binding affinities [161–164]. Further studies elucidating the structural biology of the ligand–receptor complex will further enhance our understanding of structure–function relationships of IFNs and our ability to manipulate these proteins to improve therapeutic outcomes.

Thus structure–function studies together with evolutionary and phylogenetic analysis of both protein-coding genomic sequences and the regulatory sequences that dictate their patterns of expression have assisted in elucidating the biology of these therapeutically important cytokines.

References

1 A. Isaacs, J. Lindenmann, *Proc Roy Soc Lond B* **1957**, *147*, 258–267.
2 Y. H. Tan, R. P. Creagan, F. H. Ruddle, *Proc Natl Acad Sci USA* **1974**, *71*, 2251–2255.
3 D. L. Slate, F. H. Ruddle, *Cell* **1979**, *16*, 171–180.
4 C. Chany, C. Finaz, D. Weil, M. Vignal, N. Van Cong, J. de Grouchy, *Ann Genet* **1980**, *23*, 201–207.
5 S. Nagata, N. Mantei, C. Weissmann, *Nature* **1980**, *287*, 401–408.
6 M. Houghton, I. J. Jackson, A. C. Porter, S. M. Doel, G. H. Catlin, C. Barber, N. H. Carey, *Nucleic Acids Res* **1981**, *9*, 247–266.
7 D. Owerbach, W. J. Rutter, T. B. Shows, P. Gray, D. V. Goeddel, R. M. Lawn, *Proc Natl Acad Sci USA* **1981**, *78*, 3123–3127.
8 T. B. Shows, A. Y. Sakaguchi, S. L. Naylor, D. V. Goedell, R. M. Lawn, *Science* **1982**, *218*, 373–374.
9 O. I. O. Olopade, S. K. Bohlander, H. Pomykala, E. Maltepe, E. Van Melle, M. M. Le Beau, M. O. Diaz, *Genomics* **1992**, *14*, 437–443.
10 M. O. Diaz, H. M. Pomykala, S. K. Bohlander, E. Maltepe, K. Malik, B. Brownstein, O. Olopade, *Genomics* **1994**, *22*, 540–552.
11 M. Rubinstein, W. P. Levy, J. A. Moschera, C. Y. Lai, R. D. Hershberg, R. T. Bartlett, S. Pestka, *Arch Biochem Biophys* **1981**, *210*, 307–318.
12 S. J. Humphray, K. Oliver, A. R. Hunt, R. W. Plumb, J. E. Loveland, K. L. Howe, T. D. Andrews, S. Searle, S. E. Hunt, C. E. Scott, M. C. Jones, R. Ainscough, J. P. Almeida, K. D. Ambrose, R. I. Ashwell, A. K. Babbage, S. Babbage, C. L. Bagguley, J. Bailey, R. Banerjee, D. J. Barker, K. F. Barlow, K. Bates, H. Beasley, O. Beasley, C. P. Bird, S. Bray-Allen, A. J. Brown, J. Y. Brown, D. Burford, W. Burrill, J. Burton, C. Carder, N. P. Carter, J. C. Chapman, Y. Chen, G. Clarke, S. Y. Clark, C. M. Clee, S. Clegg, R. E. Collier, N. Corby, M. Crosier, A. T. Cummings, J. Davies, P. Dhami, M. Dunn, I. Dutta, L. W. Dyer, M. E. Earthrowl, L. Faulkner, C. J. Fleming, A. Frankish, J. A. Frankland, L. French, D. G. Fricker, P. Garner, J. Garnett, J. Ghori, J. G. Gilbert, C. Glison, D. V. Grafham, S. Gribble, C. Griffiths, S. Griffiths-Jones, R. Grocock, J. Guy, R. E. Hall, S. Hammond, J. L. Harley, E. S. Harrison, E. A. Hart, P. D. Heath, C. D. Henderson, B. L. Hopkins, P. J. Howard, P. J. Howden, E. Huckle, C. Johnson, D. Johnson, A. A. Joy, M. Kay, S. Keenan, J. K.

Kershaw, A. M. Kimberley, A. King, A. Knights, G. K. Laird, C. Langford, S. Lawlor, D. A. Leongamornlert, M. Leversha, C. Lloyd, D. M. Lloyd, J. Lovell, S. Martin, M. Mashreghi-Mohammadi, L. Matthews, S. McLaren, K. E. McLay, A. McMurray, S. Milne, T. Nickerson, J. Nisbett, G. Nordsiek, A. V. Pearce, A. I. Peck, K. M. Porter, R. Pandian, S. Pelan, B. Phillimore, S. Povey, Y. Ramsey, V. Rand, M. Scharfe, H. K. Sehra, R. Shownkeen, S. K. Sims, C. D. Skuce, M. Smith, C. A. Steward, D. Swarbreck, N. Sycamore, J. Tester, A. Thorpe, A. Tracey, A. Tromans, D. W. Thomas, M. Wall, J. M. Wallis, A. P. West, S. L. Whitehead, D. L. Willey, S. A. Williams, L. Wilming, P. W. Wray, L. Young, J. L. Ashurst, A. Coulson, H. Blocker, R. Durbin, J. E. Sulston, T. Hubbard, M. J. Jackson, D. R. Bentley, S. Beck, J. Rogers, I. Dunham, *Nature* **2004**, *429*, 369–374.
13 M. P. Hardy, C. M. Owczarek, L. S. Jermiin, M. Ejdeback, P. J. Hertzog, *Genomics* **2004**, *84*, 331–345.
14 E. De Maeyer, J. De Maeyer-Guignard, *Int Rev Immunol* **1998**, *17*, 53–73.
15 B. Cederblad, A. E. Gobl, G. V. Alm, *J Interferon Res* **1991**, *11*, 371–377.
16 S. B. Feldman, M. Ferraro, H. M. Zheng, N. Patel, S. Gould-Fogerite, P. Fitzgerald-Bocarsly, *Virology* **1994**, *204*, 1–7.
17 P. Fitzgerald-Bocarsly, *Biotechniques* **2002**, *22 (Suppl 16–20)*, 24–19.
18 S. Pestka, J. A. Langer, K. C. Zoon, C. E. Samuel, *Annu Rev Biochem* **1987**, *56*, 727–777.
19 N. Kadowaki, S. Antonenko, Y. J. Liu, *J Immunol* **2001**, *166*, 2291–2295.
20 P. Kontsek, G. Karayianni-Vasconcelos, E. Kontsekova, *Acta Virol* **2003**, *47*, 201–215.
21 C. D. Krause, S. Pestka, *Pharmacol Ther* **2005**, *106*, 299–346.
22 R. M. Roberts, *Endocine Rev* **1992**, *13*, 432–452.
23 R. M. Roberts, *Nature* **1993**, *362*, 583.
24 F. Lefevre, V. Boulay, *J Biol Chem* **1993**, *268*, 19760–19768.
25 J. M. Trent, S. Olson, R. M. Lawn, *Proc Natl Acad Sci USA* **1982**, *79*, 7809–7813.
26 C. Weissmann, H. Weber, *Prog Nucleic Acid Res Mol Biol* **1986**, *33*, 251–300.
27 Map viewer in NCBI Database. D. A. Benson, I. Karsch-Mizrachi, D. J. Liyman, J. Ostell, D. L. Wheeler, *Nucleic Acid Res* **2005**, *33*, **D34–38**.
28 ENSEMBL genome browser. J. Stalker, B. Gibbins, P. Meidl, J. Smith, W. Syooner, H. R. Hotz, A. V. Cox, *Genome Res* **2004**, *14*, 951–955.
29 J. Xu, S. Gu, S. Wang, J. Dai, C. Ji, Y. Jin, J. Qian, L. Wang, X. Ye, Y. Xie, Y. Mao, *Mol Biol Rep* **2003**, *30*, 239–242.
30 K. Yoshida, *Oncol Rep* **2005**, *13*, 1133–1137.
31 J. A. van der Korput, J. Hilkens, V. Kroezen, E. C. Zwarthoff, J. Trapman, *J Gen Virol* **1985**, *66*, 493–502.

32 V. van Pesch, H. Lanaya, J. C. Renauld, T. Michiels, *J Virol* **2004**, *78*, 8219–8228.
33 A. J. Gentles, S. Karlin, *Trends Genet* **1999**, *15*, 47–49.
34 M. K. Sakharkar, P. Kangueane, D. A. Petrov, A. S. Kolaskar, S. Subbiah, *Bioinformatics* **2002**, *18*, 1266–1267.
35 M. K. Sakharkar, V. T. Chow, I. Chaturvedi, V. S. Mathura, P. Shapshak, P. Kangueane, *Front Biosci* **2004**, *9*, 3262–3267.
36 M. K. Sakharkar, V. T. Chow, P. Kangueane, *In Silico Biol* **2004**, *4*, 0032.
37 M. K. Sakharkar, P. Kangueane, *BMC Bioinformatics* **2004**, *5*, 67.
38 D. N. N. Lee, R. Brissette, M. Chou, M. Hussain, D. S. Gill, M. J. Liao, D. Testa, *J Interferon Cytokine Res* **1995**, *15*, 341–349.
39 M. Ohlsson, J. Feder, L. L. Cavalli-Sforza, A. von Gabain, *Proc Natl Acad Sci USA* **1985**, *82*, 4473–4476.
40 P. J. Hertzog, L. A. O'Neill, J. A. Hamilton, *Trends Immunol* **2003**, *24*, 534–539.
41 R. Hauptmann, P. Swetly, *Nucleic Acids Res* **1985**, *13*, 4739–4749.
42 G. R. Adolf, I. Maurer-Fogy, I. Kalsner, K. Cantell, *J Biol Chem* **1990**, *265*, 9290–9295.
43 D. J. Capon, H. M. Shepard, D. V. Goeddel, *Mol Cell Biol* **1985**, *5*, 768–779.
44 A. Himmler, R. Hauptmann, G. R. Adolf, P. Swetly, *DNA* **1986**, *5*, 345–356.
45 I. Flores, T. M. Mariano, S. Pestka, *J Biol Chem* **1991**, *266*, 19875–19877.
46 P. Kontsek, *Acta Virol* **1994**, *38*, 345–360.
47 G. R. Adolf, *Mult Scler* **1995**, *1 (Suppl 1)*, S44–S47.
48 D. W. LaFleur, B. Nardelli, T. Tsareva, D. Mather, P. Feng, M. Semenuk, K. Taylor, M. Buergin, D. Chinchilla, V. Roshke, G. Chen, S. M. Ruben, P. M. Pitha, T. A. Coleman, P. A. Moore, *J Biol Chem* **2001**, *276*, 39765–39771.
49 B. Nardelli, L. Zaritskaya, M. Semenuk, Y. H. Cho, D. W. LaFleur, D. Shah, S. Ullrich, G. Girolomoni, C. Albanesi, P. A. Moore, *J Immunol* **2002**, *169*, 4822–4830.
50 K. Oritani, K. L. Medina, Y. Tomiyama, J. Ishikawa, Y. Okajima, M. Ogawa, T. Yokota, K. Aoyama, I. Takahashi, P. W. Kincade, Y. Matsuzawa, *Nat Med* **2000**, *6*, 659–666.
51 K. Oritani, P. W. Kincade, Y. Tomiyama, *J Mol Med* **2001**, *79*, 168–174.
52 K. Oritani, Y. Tomiyama, *Int J Hematol* **2004**, *80*, 325–331.
53 K. Oritani, S. Hirota, T. Nakagawa, I. Takahashi, S. Kawamoto, M. Yamada, N. Ishida, T. Kadoya, Y. Tomiyama, P. W. Kincade, Y. Matsuzawa, *Blood* **2003**, *101*, 178–185.
54 S. Kawamoto, K. Oritani, E. Asakura, J. Ishikawa, M. Koyama, K. Miyano, M. Iwamoto, S. Yasuda, H. Nakakubo, F. Hirayama, N. Ishida, H. Ujiie, H. Masaie, Y. Tomiyama, *Exp Hematol* **2004**, *32*, 797–805.

55 R. M. Roberts, T. Ezashi, C. S. Rosenfeld, A. D. Ealy, H. M. Kubisch, *Reprod Suppl* **2003**, *61*, 239–251.
56 M. D. Meyer, P. J. Jansen, W. W. Thatcher, M. Drost, L. Badinga, R. M. Roberts, J. Li, T. L. Ott, F. W. Bazer, *J Dairy Sci* **1995**, *78*, 1921–1931.
57 R. M. Roberts, L. Liu, Q. Guo, D. Leaman, J. Bixby, *J Interferon Cytokine Res* **1998**, *18*, 805–816.
58 K. A. Frazer, L. Pachter, A. Poliakov, E. M. Rubin, I. Dubchak, *Nucleic Acids Res* **2004**, *32*, W273–279.
59 N. Bray, I. Dubchak, L. Pachter, *Genome Res* **2003**, *13*, 97–102.
60 S. Goodbourn, *Semin Cancer Biol* **1990**, *1*, 89–95.
61 T. Maniatis, J. V. Falvo, T. H. Kim, T. K. Kim, C. H. Lin, B. S. Parekh, M. G. Wathelet, *Cold Spring Harb Symp Quant Biol* **1998**, *63*, 609–620.
62 N. Munshi, Y. Yie, M. Merika, K. Senger, S. Lomvardas, T. Agalioti, D. Thanos, *Cold Spring Harb Symp Quant Biol* **1999**, *64*, 149–159.
63 T. Agalioti, S. Lomvardas, B. Parekh, J. Yie, T. Maniatis, D. Thanos, *Cell* **2000**, *103*, 667–678.
64 S. Goodbourn, T. Maniatis, *Proc Natl Acad Sci USA* **1988**, *85*, 1447–1451.
65 C. M. Fan, T. Maniatis, *EMBO J* **1989**, *8*, 101–110.
66 T. Fujita, M. Miyamoto, Y. Kimura, J. Hammer, T. Taniguchi, *Nucleic Acids Res* **1989**, *17*, 3335–3346.
67 T. K. Kim, T. Maniatis, *Mol Cell* **1997**, *1*, 119–129.
68 S. L. Schafer, R. Lin, P. A. Moore, J. Hiscott, P. M. Pitha, *J Biol Chem* **1998**, *273*, 2714–2720.
69 D. Thanos, T. Maniatis, *Cell* **1992**, *71*, 777–789.
70 W. Du, D. Thanos, T. Maniatis, *Cell* **1993**, *74*, 887–898.
71 L. A. Whittemore, T. Maniatis, *Mol Cell Biol* **1990**, *10*, 1329–1337.
72 M. Paste, G. Huez, V. Kruys, *Eur J Biochem* **2003**, *270*, 1590–1597.
73 U. Schultz, B. Kaspers, P. Staeheli, *Dev Comp Immunol* **2004**, *28*, 499–508.
74 C. R. Dearolf, *Cell Mol Life Sci* **1999**, *55*, 1578–1584.
75 T. Miyazaki, N. Nozawa, T. Kobayashi, *Dis Aquat Organ* **2000**, *43*, 15–26.
76 T. Miyazaki, T. Taniguchi, J. Hirayama, N. Nozawa, *Dis Aquat Organ* **2002**, *51*, 135–138.
77 A. L. Hughes, *J Mol Evol* **1995**, *41*, 539–548.
78 A. L. Hughes, R. M. Roberts, *J Interferon Cytokine Res* **2000**, *20*, 737–739.
79 J. D. Thompson, D. G. Higgins, T. J. Gibson, *Nucleic Acids Res* **1994**, *22*, 4673–4680.
80 R. Wonderling, T. Powell, S. Baldwin, T. Morales, S. Snyder, K. Keiser, S. Hunter, E. Best, M. J. McDermott, M. Milhausen, *Vet Immunol Immunopathol* **2002**, *89*, 13–27.
81 N. Nakamura, T. Sudo, S. Matsuda, A. Yanai, *Biosci Biotechnol Biochem* **1992**, *56*, 211–214.
82 G. A. Harrison, L. J. Young, C. M. Watson, K. B. Miska, R. D. Miller, E. M. Deane, *Cytokine* **2003**, *21*, 105–119.

83 G. A. Harrison, K. A. McNicol, E. M. Deane, *Dev Comp Immunol* **2004**, *28*, 927–940.
84 G. A. Harrison, K. A. McNicol, E. M. Deane, *Immunol Cell Biol* **2004**, *82*, 112–118.
85 M. J. Sekellick, A. F. Ferrandino, D. A. Hopkins, P. I. Marcus, *J Interferon Res* **1994**, *14*, 71–79.
86 U. Krempien, I. Redmann, C. Jungwirth, *J Interferon Res* **1985**, *5*, 209–214.
87 M. Kohase, H. Moriya, T. A. Sato, S. Kohno, S. Yamazaki, *J Gen Virol* **1986**, *67*, 215–218.
88 M. J. Sekellick, P. I. Marcus, *Methods Enzymol* **1986**, *119*, 115–125.
89 J. W. Lowenthal, P. Staeheli, U. Schultz, M. J. Sekellick, P. I. Marcus, *J Interferon Cytokine Res* **2001**, *21*, 547–549.
90 C. Sick, U. Schultz, P. Staeheli, *J Biol Chem* **1996**, *271*, 7635–7639.
91 I. Nanda, C. Sick, U. Munster, B. Kaspers, M. Schartl, P. Staeheli, M. Schmid, *Chromosoma* **1998**, *107*, 204–210.
92 J. M. Moehring, W. R. Stinebring, *Nature* **1970**, *226*, 360–361.
93 R. E. Ziegler, W. K. Joklik, *Methods Enzymol* **1981**, *78*, 563–570.
94 M. Suresh, K. Karaca, D. Foster, J. M. Sharma, *J Virol* **1995**, *69*, 8159–8163.
95 C. Sick, U. Schultz, U. Munster, J. Meier, B. Kaspers, P. Staeheli, *J Biol Chem* **1998**, *273*, 9749–9754.
96 R. Nygaard, S. Husgard, A. I. Sommer, J. A. Leong, B. Robertsen, *Fish Shellfish Immunol* **2000**, *10*, 435–450.
97 C. O'Farrell, N. Vaghefi, M. Cantonnet, B. Buteau, P. Boudinot, A. Benmansour, *J Virol* **2002**, *76*, 8040–8049.
98 S. M. Altmann, M. T. Mellon, D. L. Distel, C. H. Kim, *J Virol* **2003**, *77*, 1992–2002.
99 W. H. Yap, A. Tay, S. Brenner, B. Venkatesh, *Immunogenetics* **2003**, *54*, 705–713.
100 M. Long, M. Deutsch, W. Wang, E. Betran, F. G. Brunet, J. Zhang, *Genetica* **2003**, *118*, 171–182.
101 B. Robertsen, V. Bergan, T. Rokenes, R. Larsen, A. Albuquerque, *J Interferon Cytokine Res* **2003**, *23*, 601–612.
102 G. Lutfalla, H. R. Crollius, N. Stange-Thomann, O. Jaillon, K. Mogensen, D. Monneron, *BMC Genomics* **2003**, *4*, 29.
103 S. Hatada, M. Kinoshita, S. Takahashi, R. Nishihara, H. Sakumoto, A. Fukui, M. Noda, M. Asashima, *Gene* **1997**, *203*, 183–188.
104 A. S. Galabov, *Methods Enzymol* **1981**, *78*, 196–208.
105 J. H. Mathews, A. V. Vorndam, *J Gen Virol* **1982**, *61*, 177–186.
106 T. J. Wiktor, H. F. Clark, *Infect Immun* **1972**, *6*, 988–995.
107 M. O. Diaz, *Semin Virol* **1995**, *6*, 143–149.
108 M. O. Diaz, D. Testa, *Biotherapy* **1996**, *8*, 157–162.
109 G. R. Foster, N. B. Finter, *J Viral Hepat* **1998**, *5*, 143–152.
110 D. Liberles, M. Wayne, *Genome Biol* **2002**, *3*, 1018.1011–1018.1014.

111 D. L. Newman, M. Abney, H. Dytch, R. Parry, M. S. McPeek, C. Ober, *Hum Mol Genet* **2003**, *12*, 137–144.
112 K. Ichimura, E. E. Schmidt, N. Yamaguchi, C. D. James, V. P. Collins, *Cancer Res* **1994**, *54*, 3127–3130.
113 G. Sternik, M. G. Pittis, M. Gutierrez, R. A. Diez, L. Sen, *Medicina (B Aires)* **1998**, *58*, 463–468.
114 C. D. James, J. He, E. Carlbom, M. Nordenskjold, W. K. Cavenee, V. P. Collins, *Cancer Res* **1991**, *51*, 1684–1688.
115 P. L. Strissel, H. A. Dann, H. M. Pomykala, M. O. Diaz, J. D. Rowley, O. I. Olopade, *Genomics* **1998**, *47*, 217–229.
116 S. Y. Hwang, P. J. Hertzog, K. A. Holland, S. H. Sumarsono, M. J. Tymms, J. A. Hamilton, G. Whitty, I. Bertoncello, I. Kola, *Proc Natl Acad Sci USA* **1995**, *92*, 11284–11288.
117 U. Muller, U. Steinhoff, L. F. Reis, S. Hemmi, J. Pavlovic, R. M. Zinkernagel, M. Aguet, *Science* **1994**, *264*, 1918–1921.
118 L. Erlandsson, R. Blumenthal, M. L. Eloranta, H. Engel, G. Alm, S. Weiss, T. Leanderson, *Curr Biol* **1998**, *8*, 223–226.
119 R. Deonarain, A. Alcami, M. Alexiou, M. J. Dallman, D. R. Gewert, A. C. Porter, *J Virol* **2000**, *74*, 3404–3409.
120 R. Deonarain, D. Cerullo, K. Fuse, P. P. Liu, E. N. Fish, *Circulation* **2004**, *110*, 3540–3543.
121 R. Deonarain, A. Verma, A. C. Porter, D. R. Gewert, L. C. Platanias, E. N. Fish, *Proc Natl Acad Sci USA* **2003**, *100*, 13453–13458.
122 M. Karaghiosoff, R. Steinborn, P. Kovarik, G. Kriegshauser, M. Baccarini, B. Donabauer, U. Reichart, T. Kolbe, C. Bogdan, T. Leanderson, D. Levy, T. Decker, M. Muller, *Nat Immunol* **2003**, *4*, 471–477.
123 V. Matheu, A. Treschow, V. Navikas, S. Issazadeh-Navikas, *J Allergy Clin Immunol* **2003**, *111*, 550–557.
124 I. Teige, A. Treschow, A. Teige, R. Mattsson, V. Navikas, T. Leanderson, R. Holmdahl, S. Issazadeh-Navikas, *J Immunol* **2003**, *170*, 4776–4784.
125 A. C. Hekman, J. Trapman, A. H. Mulder, J. L. van Gaalen, E. C. Zwarthoff, *J Biol Chem* **1988**, *263*, 12151–12155.
126 Y. Iwakura, M. Asano, Y. Nishimune, Y. Kawade, *EMBO J* **1988**, *7*, 3757–3762.
127 T. A. Stewart, B. Hultgren, X. Huang, S. Pitts-Meek, J. Hully, N. J. MacLachlan, *Science* **1993**, *260*, 1942–1946.
128 G. Vassileva, S. C. Chen, M. Zeng, S. Abbondanzo, K. Jensen, D. Gorman, B. M. Baroudy, Y. Jiang, N. Murgolo, S. A. Lira, *J Immunol* **2003**, *170*, 5748–5755.
129 M. Pelegrin, J. C. Devedjian, C. Costa, J. Visa, G. Solanes, A. Pujol, G. Asins, A. Valera, F. Bosch, *J Biol Chem* **1998**, *273*, 12332–12340.
130 M. W. Beilharz, I. T. Nisbet, M. J. Tymms, P. J. Hertzog, A. W. Linnane, *J Interferon Res* **1986**, *6*, 677–685.
131 L. Runkel, L. Pfeffer, M. Lewerenz, D. Monneron, C. H. Yang, A. Murti, S. Pellegrini, S. Goelz, G. Uze, K. Mogensen, *J Biol Chem* **1998**, *273*, 8003–8008.
132 E. Croze, D. Russell-Harde, T. C. Wagner, H. Pu, L. M. Pfeffer, H. D. Perez, *J Biol Chem* **1996**, *271*, 33165–33168.

133 M. LEWERENZ, K. E. MOGENSEN, G. UZE, *J Mol Biol* **1998**, *282*, 585–599.
134 R. WETZEL, *Nature* **1981**, *289*, 606–607.
135 R. WETZEL, L. J. PERRY, D. A. ESTELL, N. LIN, H. L. LEVINE, B. SLINKER, F. FIELDS, M. J. ROSS, J. SHIVELY, *J Interferon Res* **1981**, *1*, 381–390.
136 T. A. NYMAN, N. KALKKINEN, H. TOLO, J. HELIN, *Eur J Biochem* **1998**, *253*, 485–493.
137 G. R. ADOLF, I. KALSNER, H. AHORN, I. MAURER-FOGY, K. CANTELL, *Biochem J* **1991**, 276 (Pt 2), 511–518.
138 T. A. NYMAN, H. TOLO, J. PARKKINEN, N. KALKKINEN, *Biochem J* **1998**, *329*, 295–302.
139 L. RUNKEL, W. MEIER, R. B. PEPINSKY, M. KARPUSAS, A. WHITTY, K. KIMBALL, M. BRICKELMAIER, C. MULDOWNEY, W. JONES, S. E. GOELZ, *Pharm Res* **1998**, *15*, 641–649.
140 F. LEFEVRE, M. GUILLOMOT, S. D'ANDREA, S. BATTEGAY, C. LA BONNARDIERE, *Biochimie* **1998**, *80*, 779–788.
141 M. W. BEILHARZ, I. T. NISBET, M. J. TYMMS, P. J. HERTZOG, A. W. LINNANE, *J Interferon Res* **1986**, *6*, 677–685.
142 P. LAMKEN, M. GAVUTIS, I. PETERS, J. VAN DER HEYDEN, G. UZE, J. PIEHLER, *J Mol Biol* **2005**, *350*, 476–488.
143 L. C. ROISMAN, D. A. JAITIN, D. P. BAKER, G. SCHREIBER, *J Mol Biol* **2005**, *353*, 271–281.
144 T. SENDA, S. SAITOH, Y. MITSUI, *J Mol Biol* **1995**, *253*, 187–207.
145 D. L. MILLER, H. F. KUNG, S. PESTKA, *Science* **1982**, *215*, 689–690.
146 S. MATSUDA, G. KAWANO, S. ITOH, Y. MITSUI, Y. IITAKA, *J Biol Chem* **1986**, *261*, 16207–16209.
147 S. MATSUDA, T. SENDA, S. ITOH, G. KAWANO, H. MIZUNO, Y. MITSUI, *J Biol Chem* **1989**, *264*, 13381–13382.
148 T. SENDA, T. SHIMAZU, S. MATSUDA, G. KAWANO, H. SHIMIZU, K. T. NAKAMURA, Y. MITSUI, *EMBO J* **1992**, *11*, 3193–3201.
149 W. KLAUS, B. GSELL, A. M. LABHARDT, B. WIPF, H. SENN, *J Mol Biol* **1997**, *274*, 661–675.
150 R. RADHAKRISHNAN, L. J. WALTER, A. HRUZA, P. REICHERT, P. P. TROTTA, T. L. NAGABHUSHAN, M. R. WALTER, *Structure* **1996**, *4*, 1453–1463.
151 M. KARPUSAS, M. NOLTE, C. B. BENTON, W. MEIER, W. N. LIPSCOMB, S. GOELZ, *Proc Natl Acad Sci USA* **1997**, *94*, 11813–11818.
152 R. RADHAKRISHNAN, L. J. WALTER, P. S. SUBRAMANIAM, H. M. JOHNSON, M. R. WALTER, *J Mol Biol* **1999**, *286*, 151–162.
153 S. R. PRESNELL, F. E. COHEN, *Proc Natl Acad Sci USA* **1989**, *86*, 6592–6596.
154 J. H. CHILL, R. NIVASCH, R. LEVY, S. ALBECK, G. SCHREIBER, J. ANGLISTER, *Biochemistry* **2002**, *41*, 3575–3585.
155 J. H. CHILL, S. R. QUADT, R. LEVY, G. SCHREIBER, J. ANGLISTER, *Structure* **2003**, *11*, 791–802.
156 M. VAN HEUVEL, I. J. BOSVELD, P. KLAASSEN, E. C. ZWARTHOFF, J. TRAPMAN, *J Interferon Res* **1988**, *8*, 5–14.
157 J. TRAPMAN, M. VAN HEUVEL, P. DE JONGE, I. J. BOSVELD, P. KLAASSEN, E. C. ZWARTHOFF, *J Gen Virol* **1988**, *69*, 67–75.

158 M. Van Heuvel, I. J. Bosveld, A. A. Mooren, J. Trapman, E. C. Zwarthoff, *J Gen Virol* **1986**, *67*, 2215–2222.

159 P. Harle, V. Cull, L. Guo, J. Papin, C. Lawson, D. J. Carr, *Antiviral Res* **2002**, *56*, 39–49.

160 V. S. Cull, P. A. Tilbrook, E. J. Bartlett, N. L. Brekalo, C. M. James, *Blood* **2003**, *101*, 2727–2735.

161 M. Aguet, M. Grobke, P. Dreiding, *Virology* **1984**, *132*, 211–216.

162 A. Meister, G. Uze, K. E. Mogensen, I. Gresser, M. G. Tovey, M. Grutter, F. Meyer, *J Gen Virol* **1986**, *67*, 1633–1643.

163 T. Yamaoka, S. Kojima, S. Ichi, Y. Kashiwazaki, T. Koide, Y. Sokawa, *J Interferon Cytokine Res* **1999**, *19*, 1343–1349.

164 J. Piehler, G. Schreiber, *J Mol Biol* **1999**, *294*, 223–237.

2
Activation of Interferon Gene Expression Through Toll-like Receptor-dependent and -independent Pathways

Peyman Nakhaei, Suzanne Paz and John Hiscott

2.1
Introduction

Upon recognition of a specific molecular component of viruses or other pathogens, the host cell activates multiple signaling cascades, culminating in the production of cytokines and chemokines that disrupt virus replication, and initiate innate and adaptive immune responses [1–3]. Antiviral and antimicrobial signaling results in the activation of host transcription factors such as nuclear factor (NF)-κB, the interferon (IFN)-regulatory factors (IRF) and activating protein 1 (AP-1) via post-translational modifications – primarily phosphorylation events – in the absence of *de novo* protein synthesis. Once activated via virus-specific signaling mechanisms, these cellular proteins are recruited to the promoters of type I IFN and other immunomodulatory genes in a precise and coordinated manner to establish the innate antiviral state.

The IFN family is composed of transcriptionally activated and secreted proteins with pleiotropic biological effects on the host. IFNs play a central role in the resistance of mammalian hosts to pathogens, and modulate antiviral and immune response [4, 5]. The IFN family is classified into two subgroups: type I IFNs (IFN-α, -β, -ω and -λ) and type II IFN (IFN-γ), characterized mainly by the receptor complex used for signaling, the cell type from which they are secreted and their intrinsic biological properties (Tab. 2.1).

Type I IFNs (IFN-β and multiple types of IFN-α), also referred to as viral IFNs, play an essential role in the host immune response against viruses. IFN-α, previously referred to as leukocyte IFN, is comprised of at least (1) 13 IFN-α species, whereas IFN-β, also known as fibroblast IFN, is a single species. Each subtype of IFN-α is encoded by its own gene and is regulated by its own promoter sequence, all lacking introns [6]. Type II IFN, also known as IFN-γ, is encoded by one gene that contains four exons and three introns; expression is controlled by lymphoid restricted transcription factors such as nuclear factor of activated T cells (NF-AT),

The first two authors contributed equally to the work

Tab. 2.1. Biological properties of IFNs

	Type I (viral IFNs)		Type II (immune IFNs)
	IFN-α	IFN-β	IFN-λ (IFN-λ1, -λ2, -λ3)	IFN-γ
Principal producing cells	pDCs epithelial cells fibroblasts leukocytes macrophages	pDCs epithelial cells fibroblasts leukocytes macrophages	pDCs epithelial cells	macrophages T Lymphocytes
Inducing agents	dsRNA viruses TLRs	dsRNA viruses TLRs	Viruses TLRs	antigens mitogens IL-2 and -12
Major function	Inhibit viral replication Increase expression of class I and class II MHC molecules Increase natural killer, T_h1 response Increase antigen-presenting cell functions Increase Apoptosis	Inhibit viral replication Stimulate class I MHC molecules Stimulate natural killer activation	Antiviral protection	Inhibit viral replication enhance activity of macrophages Increase expression of class I and class II MHC molecules Increase T_h1 response Stimulate natural killer activation, delayed

Molecular weight (kDa)	16–27	28–35	20–25
Receptor	IFN-αR1 and -αR2	IFN-αR1 and -αR2	IFN-λ1: 20–33 IFN-λ2 and -λ3: 22 CRF2–12 (IFN-γR1) CRF2–4 (IL-10R2)	IFN-γR1 and -γR2
Chromosomal location	IFN: 9p21 Receptor: 21q21	IFN: 9p21 Receptor: 21q21	IFN: 19q13	IFN: 12q24 Receptor: 6q

Type I and II IFNs are described with respect to their principal producing cells, inducing agents and major functions. Molecular weights, corresponding receptors and chromosomal locations are also described.

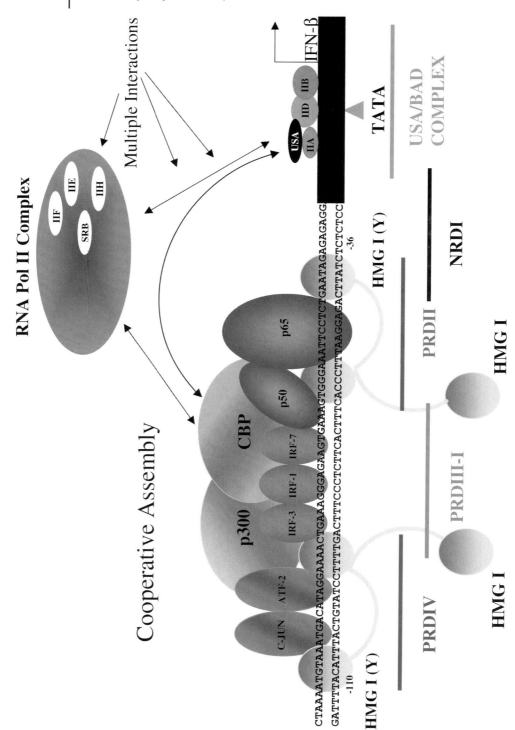

signal transducer and activator of transcription (STAT), T-bet, cAMP response element modulation protein (CREM)/activating transcription factor (ATF), GATA-3 and yin-yang-1 [7–12]. The transcriptional activation of IFN-γ is also regulated by a consensus GATA motif and an NFIL-2-like element, located between amino acids −108 and −40 of the human IFN-γ gene [10].

Once secreted, type I IFNs act in auto- or paracrine fashion through associated IFN receptors, through Janus kinase (JAK)–STAT signaling to induce a set of IFN-stimulated genes (ISGs) that require *de novo* protein synthesis. ISG transcription is achieved via the JAK–STAT signaling pathway and involves an IFN-α/β-activated transcription factor complex termed ISGF-3, composed of IRF-9, and STAT-1 and -2 [13–15]. This complex will subsequently bind to a common enhancer element referred to as the IFN-stimulated responsive element (ISRE), resulting in the induction of IFNs and other cytokines such as RANTES and interleukin (IL)-15 [16–24]. Another type I IFN involved in the mediation of antiviral protection is IFN-λ (IFN-λ1, -λ2 and -λ3) that utilizes a different set of IFN receptors IFN-λR1 (CRF2-12) and IL-10R2 (CRF2-4) [25], receptors that are also shared with the IL-10 and -22 receptor complex.

IFN-γ binds as a dimer to a specific tetrameric receptor composed of IFNGR-1 and two IFNGR-2 chains [26, 27]. Following oligomerization of the receptors, Janus tyrosine kinase JAK-1 and -2 are activated followed by STAT-1 recruitment, phosphorylation and dimerization [termed IFN-γ-activated factor (GAF)]. The GAF complex then translocates to the nucleus and binds to GAS motifs, which leads to transcription of multiple genes. The main route to transcription by IFN-γ is the GAF complex, but ISGF-3 and other transcription factors are also utilized. Reciprocally, IFN-α/β can also utilize GAF for signaling.

2.2
IFN-β Gene Transcription

In order for gene activation to occur following appropriate physiological stimulation, a multicomponent transcription enhancer complex composed of transcriptional activators, architectural proteins and coactivators [28, 29] must first assemble. The best characterized model of stimuli-dependent induction of transcription is the virus-inducible induction of the hIFN-β gene (Fig. 2.1). The virus inducible

Fig. 2.1. Schematic representation of the IFN-β promoter enhanceosome. This diagram shows the composition and organization of the human IFN-β enhanceosome. The nucleotide sequence (−110 to −36) is situated upstream from the starting transcription site. The positive regulatory sites as well as the negative regulatory domain are named PRDs and NRD, respectively. The transcriptional proteins binding to the PRDs are also shown (c-Jun, ATF-2, IRFs and NF-κB), as well as the architectural proteins HMGI(Y) and the transcriptional coactivators proteins p300 and CBP. The multiple protein–protein interactions required for transcriptional synergy of the IFN-β promoter derived from the cooperative assembly of the enhanceosome, cooperative assembly between the enhanceosome and the USA/BAD complex, and interaction with the RNA polymerase II complex [32]. (This figure also appears with the color plates.)

enhancer is a 60-bp region immediately upstream of the TATA box, located between −110 and −36 relative to the transcription start site of the hIFN-β gene. The promoter enhancer region contains four positive (PRDI–IV) and one negative regulatory domains (NRDI); PRDI and III contain the binding sequences for IRF-3 and -7, respectively, as well as for other IRF members; PRDII is recognized by NF-κB heterodimers, and PRDIV by ATF-2 and c-Jun heterodimers (Fig. 2.1). Virus infection leads to the recruitment of coactivators [GCN5 and CREB-binding protein (CBP/p300)], as well as the high-mobility-group protein [HMGI(Y)], which binds to the minor groove of DNA at four sites within the enhancer and contributes to the formation of a stable nucleoprotein complex called the enhanceosome.

This virus-inducible enhancer is silent in uninfected cells in part through the inhibitory effect of an NF-κB-regulating factor (NRF) that overlaps the PRDII site [30, 31], but is quickly induced to high levels upon viral infection [32, 33]. The positioning of nucleosomes upstream of the IFN-β gene is also critical for gene silencing [34–37]. Viral infection results in hyperacetylation of histones H3 and H4 localized in the IFN-β promoter [38]. Hyperacetylation of histones is known to play a crucial role in inducible gene expression. Enhanceosome assembly following infection requires precise spacing between the factor binding sites to ensure that each of the enhanceosome components simultaneously contact one another and DNA [39]. Another complex, composed of transcription factors TFIIB, A and D, along with the coactivator complex upstream stimulatory activity (USA), is necessary for enhanceosome-dependent transcriptional synergy [32, 40]. Figure 2.1 illustrates the various protein–protein interactions of the enhanceosome complex, as well as interactions with the RNA polymerase II complex and the USA/BAD complex.

The transient transcriptional activation of the different members of the IFN-α multigene family is a primary cellular response to viral infection and, just as for IFN-β, virus-induced IFN-α expression is mediated by regulatory sequences within 200 bp of the transcriptional start site which includes a 46-bp fragment called the virus response element (VRE-A1) [41, 42]. IRF-3 and -7 have been shown to differentially regulate the activation of immediate early and delayed IFN-α gene transcription by binding to the IRF-E elements of the IFN-α promoter [43–46].

2.3
IRF Family Members

Nine human IRFs have been identified (IRF-1, -2, -3, -4/Pip/ISCAT, -5, -6, -7, -8/ICSBP and -9/ISGF-3γ/p48); all share homology in the N-terminal DNA-binding domain (DBD), and contain characteristic five tryptophan repeat elements located within the first 150 amino acids, each separated by 10–18 amino acids [47]. The IRF DNA-binding domain mediates specific binding to GAAANN and AANNN-GAA sequences, termed IRF-E [48]. In addition to their role in immune regulation, IRFs are also involved in regulation of cell cycle, apoptosis and tumor suppression (Tab. 2.2). The ISRE-containing genes are regulated by various IRFs through the JAK–STAT pathway by stimulation of type I or type II IFN stimulation, although certain genes can also be regulated independently of the IFN-triggered, JAK–STAT pathway through virus infection.

The C-terminal region of all IRFs contains a unique domain involved in protein–protein association, termed the IRF-association domain (IAD), which is necessary for homo- and heterodimerization interactions among IRF family members or other modulators. Biochemical and structural studies have determined that the transactivation functions of IRF-3, -4, -5 and -7, which reside in the IAD, are suppressed by autoinhibitory structures [20, 21, 49–51].

2.4
Role of IRFs in Virus-mediated IFN Activation

2.4.1
IRF-3

IRF-3 is a constitutively expressed protein of 55 kDa (427 amino acids) present in most cell types [52]. In addition to the N-terminal DNA-binding domain, IRF-3 contains a transactivation domain (amino acids 134–394) and two autoinhibitory domains (IDs), found within the proline-rich sequence (amino acids 134–197) and at the extreme C-terminal end (amino acids 407–414). These IDs interact to generate a closed conformation that is likely to mask the C-terminal IAD, the DBD and the nuclear localization sequence (NLS) (Fig. 2.2). IRF-3 is maintained in an inactive form in the cytoplasm by its autoinhibitory interactions. Upon virus infection or double-stranded (ds) RNA treatment of cells, C-terminal phosphorylation by TBK1 [tumor necrosis factor (TNF) receptor-associated factor (TRAF)-family-member-associated NF-κB activator (TANK)]-binding kinase or IKKε/IKKi [(IκB) inhibitor of NF-κB kinase ε] occurs at specific serine residues in the C-terminus of IRF-3 [53–56], resulting in a conformational change that allows dimerization, translocation to the nucleus, association with CBP/p300 coactivators and transcriptional activation of IRF-E-containing promoters [17, 18, 57–59].

IRF-3 phosphorylation was initially detected as posttranslational modification following Sendai virus infection [17, 18, 60]. Interestingly, two alterations of the IRF-3 protein were observed: a slower migration form of IRF-3, and an overall reduction in the amount of IRF-3 in transfected and control cells [17]. The slower migration was shown to reflect virus-dependent phosphorylation of the C-terminal domain of IRF-3 protein [17, 18, 58–60]. Yoneyama et al. demonstrated that virus-induced phosphorylation occurred at residues Ser385 and Ser386 [17]. The C-terminal region comprises a cluster of phosphoacceptor sites ^{382}GGASSLENTVDLHISNSHPLSLTSDQYKAYLQD^{414} (Tab. 2.3). Using various deletion and point mutations, C-terminal phosphorylation was also localized between amino acids 395 and 407, a region adjacent to the Ser385/Ser386 residues.

Tab. 2.2. Biological properties of IRF proteins

IRF member	Expression pattern	Inducers	Activator or repressor	Knockout phenotype	Basic function
IRF-1 (ISGF-2)	most cell types constitutive and inducible	IFNs, IL-1, IL-12, TNF virus infection and LPS prolactin concanavalin A	activator, phosphorylation required	decreased number of CD8$^+$ cells and increased susceptibility to pathogens due to abnormal type I IFN gene induction Il-12 dysregulated in macrophages	immunity/defense and regulates apoptosis and tumor repression promotes T$_h$1 responses and natural killer cell development
IRF-2	most cell types constitutive	type I IFN virus infection	activator or repressor, depending on corepressors present	increased susceptibility to viruses as well as skin disorders IL-12 dysregulated in macrophages	repressor of IRF-1 and -9 oncogene and antiviral defense/immune regulator Promotes T$_h$1 responses and natural killer cell development may regulate NO
IRF-3	most cell types constitutive	virus infection dsRNA TLR-3 and -4	activator, phosphorylation required	increased susceptibility to virus infection impaired IFN response	innate antiviral immunity apoptosis protection
IRF-4 (Pip/LSIRF/ ICSAT	activated T cells, B cells constitutive and inducible	PMA HTLV-I Tax CD3-specific antibody (T cells) IL-4, antibody specific for IgM or DC40 (splenic B cells)	activator or repressor, depending on coactivators present	profound defects in function of B and T cells Impaired T$_h$2 response Impaired CD4$^+$ DCs	promotes B cell proliferation promotes differentiation of mitogen-activated T cells

IRF-5	B cells and DCs (constitutive)	viruses (NDV, VSV, HSV-1) Type I IFNs TLR-7 and -9	activator, phosphorylation required	impaired secretion levels of IL-6, IL-12 and TNF-α by TLR activation resistance to lethal shock induced by unmethylated DNA or LPS	induction of pro-inflammatory cytokines
IRF-6	mesodermal in early development epithelial, muscular skeletal cells	ND	activator or repressor?	ND	development of lip, palate, skin (formation of cleft lip and/or palate; Van der Woude's syndrome)
IRF-7	B cells, monocytes (moderate) and pDCs (high) other cell types (inducible)	IFN virus infection CpG-containing oligodeoxynucleotides TLR-7, -8 and -9	activator, regulated by phosphorylation	profound defect in IFN-α	transcriptional activation of virus-inducible cellular genes regulator of IFN-α
IRF-8 (ICSBP)	hematopoietic cells and immune cells constitutive and inducible	type II IFN and LPS (macrophages) TCR engagement (T cells)	activator or repressor, depending on coactivators present	lacks IRF-2, IL-2β and reactive oxygen intermediates impaired IFN-γ	regulating expression of IFN-α and -β negative regulator role
IRF-9 (ISG3γ/p48)	constitutive	type II IFN	activator, depending on coactivators present	ND	act as DNA recognition subunit activating transcription in response to IFN-α

This table represents a detailed description of all the members of the IRF family with respect to their expression patterns, inducing agents and knockout phenotypes. In addition, the basic functions and the roles as an activator or repressor of the different IRF members are described.

44 | *2 Activation of Interferon Gene Expression*

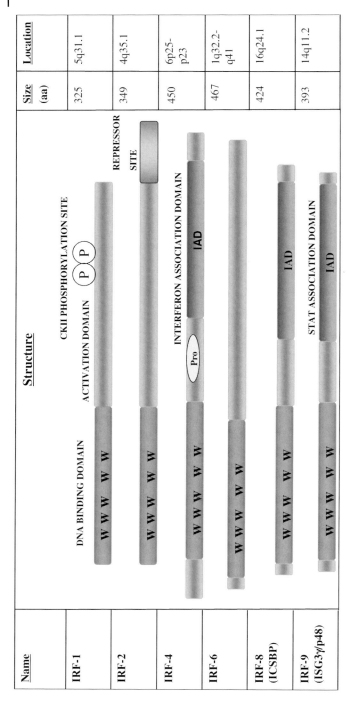

Fig. 2.2. Schematic representation of IRF-1, -2, -4, -6, -8 and -9 transcription factors. This diagram demonstrates the general structural homology amongst the IRF family. Different domains are shown: DBD, CKII phosphorylation site, repressor site and IAD. The size of each IRF as well as their chromosomal location is also described. (This figure also appears with the color plates.)

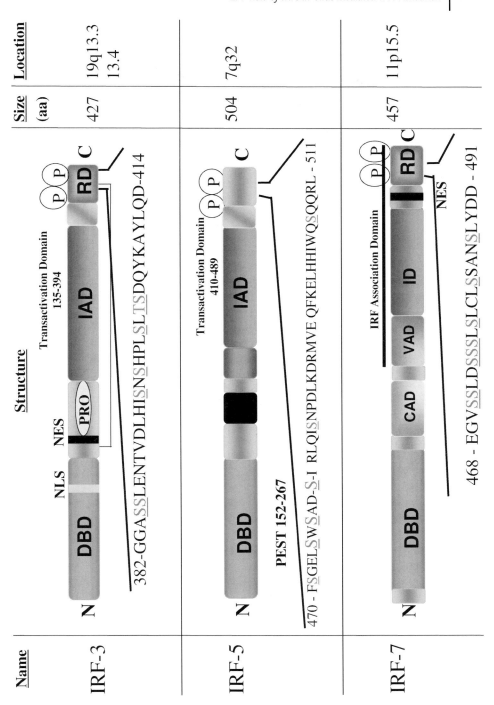

Name	Structure	Size (aa)	Location
IRF-3	Transactivation Domain 135–394 N — DBD — NLS NES PRO — IAD — P P — RD — C 382-GGASSLENTVDLHISNSHPLSLTSDQYKAYLQD-414	427	19q13.3–13.4
IRF-5	Transactivation Domain 410–489 N — DBD — IAD — P P — C PEST 152–267 470 - FSGELSWSAD-S-I RLQISNPDLKDRMVE QFKELHHIWQSQQRL - 511	504	7q32
IRF-7	IRF Association Domain N — DBD — CAD — VAD — ID — P P — RD — C NES 468 - EGVSSLDSSSLSLCLSSANSLYDD - 491	457	11p15.5

The generation of point mutations also demonstrated that Ser396 and Ser398 were also phosphorylated following virus infection [18]. Servant et al. demonstrated for the first time that IRF-3 is phosphorylated *in vivo* on Ser396 following Sendai virus infection, expression of viral nucleocapsid (N) or dsRNA treatment [53]. This site was identified as the minimal phosphoacceptor site since a single point mutation of Ser396 to the phosphomimetic Asp was sufficient to generate a constitutively active form of IRF-3 [53]. Ser396 and Ser398 were also identified as specific targets of TBK1 and IKKε [54, 61].

Analysis of the crystal structure of the IRF-3 C-terminal domain by Qin et al. [62] provided insight into the mechanism of IRF-3 activation by phosphorylation. Interestingly, most of the serine and threonine residues within the C-terminus are partially buried by hydrophobic residues, and are located adjacent to acidic residues, indicating that phosphorylation may perturb the autoinhibitory interactions due to charge repulsion [62–64]. The sites of phosphorylation may be grouped into three clusters. First, the Ser385/Ser386 cluster is located in the H4-β12 turn [62, 64, 65], connecting the IAD to the C-terminal ID, in which phosphorylation of Ser385 is speculated to lead to local structural destabilization with a profound effect on the autoinhibitory structure. The second cluster is composed of Ser396 and Ser398, located within the L6 loop, and this location is involved in the interaction with the N-terminal autoinhibitory helix H1. The phosphorylation of Ser396 and Ser398 most likely disrupts the interaction between the N- and C-terminus of the autoinhibitory structure [18, 62, 65]. The last cluster contains Ser402, Thr404 and Ser405, where Ser402 is located on the β13 strand. Both Thr404 and Ser405 are located on the β13-H5 turn. The β13 strand and the β13-H5 turn form a portion of the hydrophobic core that stabilizes the autoinhibitory structure; phosphorylation of this cluster likely contributes to the unfolding of the structure. These observations generally support the biochemical studies indicating that IRF-3 phosphorylation leads to structural rearrangement of the autoinhibitory structures. Nonetheless, the mechanisms of activation and accessibility of the virus-activated kinases to the sites of phosphorylation remains unclear.

Other interacting molecules may also contribute significantly to the conformational changes in IRF-3. A novel interaction of cyclophilin B (CypB) with the ID of IRF-3 was recently described [66]; CypB possesses a *cis–trans* peptidyl–prolyl isomerase (PPIase) activity [67], which facilitates protein folding through its PPIase activity. Knocking down CypB using small interfering RNA (siRNA) *in vivo* resulted in the inhibition of virus-induced IRF-3 activation and subsequent IFN-β induction. Overall, CypB interaction with IRF-3 is reminiscent of the functional interaction described earlier between IRF-4 and FKBP52 [68].

The importance of IRF-3 in regulating the early and late phase of IFN expression was demonstrated by Sato et al. through the generation of IRF-3 knockout mice.

Fig. 2.3. Schematic representation of IRF-3, -5 and -7 transcription factors. Different domains are shown: NLS, NES, DBD, CAD, VAD, IAD and the signal response domain (RD). The sequence of amino acids 382–414 (IRF-3) and the sequence of amino acids 468–491 (IRF-7) are amplified below the schematic. Important amino acids are shown in larger letters with the respective position number. (This figure also appears with the color plates.)

IRF-3$^{-/-}$ mice were more susceptible to encephalomyocarditis virus (ECMV) infection than normal mice and serum IFN levels from EMCV-infected mice were significantly lower in the IRF-3$^{-/-}$ mice than wild-type mice [22]. Additionally, cells defective in the expression of both IRF-3 and -7 completely failed to induce type I IFNs in response to infection by many viruses; the lack of IFN responsiveness was recovered by coexpressing both proteins, clearly demonstrating that IRF-3 and -7 have essential and distinct roles in the transcriptional efficiency of diverse IFN-α/β genes [22]. Sato et al. found that IRF-3 function was important for early-phase activation of IFN-β in an IFN-independent manner, but was also important for the late-phase potentiation of overall IFN-α/β mRNA induction in cooperation with IRF-7 [22].

2.4.2
IRF-5

IRF-5 is another recently characterized participant in the IFN activation pathway; IRF-5 is constitutively active in B cells as well as dendritic cells (DCs), and is inducible in other cell types [69, 70]. The 57-kDa IRF-5 protein is localized to the cytoplasm of unstimulated cells and accumulates in the nucleus following virus infection. Two NLS elements, necessary for virus translocation to the nucleus, are found in the IRF-5 protein [51, 71]. Lin et al. also demonstrated that IRF-5 possesses a nuclear export sequence (NES) element that controls the dynamic shuttling between the cytoplasm and the nucleus [72]. Although both NLS and NES are active in unstimulated cells, the NES element is dominant, resulting in the predominant cytoplasmic localization of IRF-5. Interestingly, coexpression of IKKε or TBK1 resulted in phosphorylation and dimerization of IRF-5, but nuclear accumulation was not observed, suggesting that aspects of IRF-5 activation may be distinct from the virus-induced mechanisms used by other IRFs. In contrast, Mancl et al. argue that TBK1- and IKKε-mediated phosphorylation of IRF-5 is functional, and that isoforms V3/4 and V5, and not V1 and V2 used by Lin et al., are targets for functional phosphorylation by TBK1/IKKε [73, 74]. Whether these kinases truly phosphorylate IRF-5 *in vivo* remains to be clarified.

In contrast to IRF-3 and -7, which are activated following viral infection, dsRNA or LPS, the activation of IRF-5 has only been observed with viruses such as Newcastle disease virus (NDV), vesicular somatitis virus (VSV), human simplex virus (HSV) type I [51], while Sendai virus, dsRNA [poly(I:C)] or LPS – activators of IRF-3 and -7 – do not activate IRF-5 [51], suggesting that the pathway involved in IRF-5 activation is distinct from that regulating IRF-3 and -7. In keeping with these distinct functions, IRF-5 mainly induces IFN-α subspecies [69]; in IRF-5-expressing cells, IFN-α8 was preferentially produced following viral infection, but when IRF-7 was also expressed, IFN-α1 was induced.

Another study demonstrated that IRF-5 not only activated IFN-α gene induction, but also repressed IFN gene activity depending on the IRF-interacting partners [71]. Interaction between IRF-5 and -7 (which was dependent on IRF-7 phosphorylation) did not lead to activation, but rather to the repression of IFN-α gene transcription. This observation was explained by the formation of IRF-5/7 heterodimers, which altered the composition of the IFN-α enhanceosome complex [71]. In contrast, IFN-α upregulation occurred with the cooperation with IRF-3. Overexpression of IRF5 also upregulated other type I IFNs, such as IFN-α2, -α14 and -β.

Recently, Takaoka et al. assessed the regulation of gene expression by IRF-5 using IRF-5$^{-/-}$ hematopoietic cells in the context of Toll-like receptor (TLR)-7 and myeloid differentiation factor 88 (MyD88) signaling (see below). Induction of IL-6, IL-12p40 and TNF-α were severely impaired, while the induction of IFN-α remained unaffected in the knockout background [75]. It was also determined that IRF-5 interacted with and was activated by MyD88 and TRAF6; and TLR-4 and -9 engagement resulted in IRF-5 translocation and cytokine gene activation [75]. Thus, IRF5 appears to be selectively involved in the induction of pro-inflammatory cytokines by TLRs.

2.4.3
IRF-7

IRF-7 was first described to bind and repress the Epstein–Bar virus (EBV) Qp promoter which regulates expression of the EBV nuclear antigen 1 (EBVA1) [76] and the importance of IRF-7 in IFN regulation was recognized shortly after its discovery [77–79]. IRF-7 is an IFN- and virus-inducible protein in most cells with transcriptional activity that depends on C-terminal phosphorylation; constitutive expression of IRF-7 is restricted to B cells and DCs. The IRF-7 protein contains multiple regulatory domains: a constitutive activation domain (CAD) between amino acids 150 and 246 of human IRF-7A, located adjacent to the conserved DBD [77, 80]; an ID that silences transactivation activity; and a region that increases basal and virus inducible activity termed the virus-activated domain (VAD) located between amino acids 278 and 305 [80]. Interestingly, removal of the amino acids 248–467 region creates a highly active form of IRF-7 (Fig. 2.3).

The serine-rich C-terminal region between amino acids 471 and 487 (Fig. 2.2) is the target of virus-induced phosphorylation [44, 76, 78] by TBK1 and IKKε [54, 81]. Ser477 and Ser479 appear to be critical targets for TBK1 and IKKε, since their substitution to Ala resulted in an IRF-7 that was unable to respond to virus challenge [21]. Substitution of Ser471 and Ser472 with the phosphomimetic amino acid Asp did not lead to the expected increased transactivation activity, but rather decreased activity, whereas the same phosphomimetic substitution of Ser477 and Ser479 led to a constitutively active IRF-7 [20, 81]. IRF-7 protein has a short half-life of approximately 30 min, which may represent a mechanism that ensures transient IFN induction [22].

A comprehensive structure–activity analysis of murine IRF-7 phosphorylation sites using mutant proteins, two-dimensional analysis and transcriptional readout assays was recently reported [82]. Essential phosphorylation events were mapped to amino acids 437–438 and a redundant set of sites at either amino acids 429–431 or 441. IRF-7 was heterogeneously phosphorylated and greater phosphorylation correlated with increased transactivation. Interestingly, a distinct serine cluster at amino acids 425 and 426 was also essential for IRF-7 activation, although the essential role of this motif did not involve phosphorylation. Rather, these serine residues appear to represent a conformational element required for IRF-7 function or a recognition motif for the virally activated kinases.

Using IRF-7$^{-/-}$ mice, Honda et al. demonstrated that IRF-7 is essential for the induction of type I IFN via virus-mediated, MyD88-independent and -dependent TLR signaling pathways. The IRF-7$^{-/-}$ mice developed normally with no overt differences in the hematopoietic cell populations [83]. However, IFN-α mRNA induction was completely inhibited and IFN-β levels were significantly reduced in IRF7$^{-/-}$ cells. Also, serum IFN levels were significantly lower in IRF-7$^{-/-}$ MEFs [83]. In the double IRF-3/7 knockout, IFN-β levels were completely abrogated. In contrast, MyD88$^{-/-}$ mouse embryonic fibroblasts (MEFs) induced IFN-α/β mRNA to similar levels as wild-type MEFs in response to virus, suggesting that IFN induction was MyD88 independent, but IRF-7-dependent [83].

Plasmacytoid DCs (pDCs) are major IFN-producing cells and stand out amongst other cells in their ability to produce high amounts of IFN-α/β following engagement of TLR-9 by unmethylated DNA or stimulation of human TLR-7/murine TLR-8 by single-stranded (ss) viral RNA [84–88]. pDCs utilize a MyD88-dependent pathway of IFN-α/β induction, which is also dependent on IRF-7 [83]. Furthermore, induction of IFN-α by TLR-9 in pDCs as well as the induction of CD8$^+$ T cell responses was completely dependent on IRF-7. Therefore, IRF-7 is not only important in the development of innate immunity, but also clearly plays a central role in adaptive immunity. Comparison between IRF-3$^{-/-}$ and -7$^{-/-}$ pDCs revealed that IFN induction through TLR-7/8 or -9 was normal in IRF-3$^{-/-}$ cells, but completely ablated in IRF-7$^{-/-}$ cells, thus demonstrating that IRF-7 is essential and IRF-3 dispensable for the MyD88-dependent induction of IFN-α/β genes via the nucleic acid-recognizing TLR subfamily [83].

2.5
IFN Signaling Pathways

2.5.1
TLR-dependent Signaling to IFN Activation

2.5.1.1 TLR Overview

The primary role of the innate immune response is to limit the spread of invading pathogens such as viruses and bacteria. Invading pathogens are recognized by specific motifs or pathogen-associated molecular patterns (PAMPs) through different TLRs [89–91]. The Toll receptor was originally identified as an essential receptor for the establishment of a dorso-ventral pattern in *Drosophila* [92]. Subsequently, multiple homologs of the Toll receptor were identified in mammalian cells and have been designated TLRs. The TLR family now consists of 11 members; of these,

Tab. 2.3. TLRs and their ligands

Receptor	Ligand	Origin of Ligand
TLR-1	triacyl lipopetides Soluble factors	bacteria and mycobacteria *Neisseria meningitidis*
TLR-2	heat shock protein 70 peptidoglycan lipoprotein/lipopeptides HCV core and nonstructural 3 protein	host Gram-positive bacteria various pathogens hepatitis C Virus
TLR-3	dsRNA	viruses
TLR-4	LPS envelope protein taxol	Gram-negative bacteria mouse mammary tumor virus plants
TLR-5	flagellin	bacteria
TLR-6	zymosan lipoteichoic acid diacyl lipopetides	fungi Gram-positive bacteria mycoplasma
TLR-7	ssRNA imidazoquinoline	viruses synthetic compounds
TLR-8	ssRNA imidazoquinoline	viruses synthetic compounds
TLR-9	CpG-containing DNA	bacteria and viruses
TLR-10	not determined	not determined
TLR-11	profilin-like molecule	*Toxoplasma gondii*

TLR-2, -3, -4, -7 (human)/-8 (murine) and -9 have been implicated in the recognition of different viral nucleic acid and/or protein motifs. In addition, TLR family members are expressed differentially among immune cells and respond to different components of invading pathogens [93] (Tab. 2.3). The cytoplasmic intracellular tail of TLRs shows high homology with that of the IL-1 receptor family, although the leucine-rich repeat (LRR)-containing extracellular domains are unrelated. Within the intracellular Toll-interacting region (TIR) domain, there are three well conserved regions which are essential for TLR signaling [94, 95]. The surface expression of TLRs on immature DCs is low, with only a few hundred molecules per cell, whereas TLR expression levels are more prominent on monocytes and can number up to a thousand molecules per cell. Upon stimulation with their appropriate ligand, TLRs in general form either homodimers or heterodimers in order to induce an effective signaling cascade [61]. Over the past several years, it has become evident that TLR signaling plays an important role in the induction of IRFs and TLR-responsive genes (see figure 2.4) [90].

2.5.1.2 TLR-3 Signaling

TLR-3 is a receptor for dsRNA – long considered a functional byproduct of intracellular virus replication – and TLR-3 engagement transmits signals that activate IFN and inflammatory cytokines through IRF and NF-κB signaling pathways. Early after infection, incoming virus particles or ribonucleoprotein complexes may be recognized within the endosomal compartment, while late after infection following replication and cell lysis, viral dsRNA is released into the extracellular space where it is available to bind the 904-amino-acid TLR-3. TLR-3 is expressed in intracellular vesicular compartments in DCs and on the cell surface in intestinal epithelial cells, but not in monocytes, polymorphonuclear leukocytes, or B, T and natural killer cells [96–98].

IRF-3 phosphorylation is mediated by the TLR-3-associated molecule TRIF/ TICAM-1 and functions independently of the MyD88 pathway. TRIF consists of an N-terminal proline-rich domain, a TIR domain and C-terminal proline-rich domain [99]. The N-terminal region of TRIF directly associates with TBK1 [61, 100]; resulting in IRF-3 phosphorylation. TRAF6, a ubiquitin ligase also interacts with the N-terminal region of TRIF [101]. Following virus infection, this association leads to the activation of IKKα/β and NF-κB, which upregulates the transcription of pro-inflammatory genes such as IL-6, IL-1β and TNF-α (Fig. 2.6). The recruitment of TBK1 to the N-terminal of TRIF initiates a signaling cascade that culminates in IRF-3 activation and induction of IFN-β, RANTES and IP-10. In addition, the phosphatidylinositol-3-kinase pathway (PI3K) also contributes to dsRNA- and TLR-3-dependent IRF-3 phosphorylation. Specific mutations of the tyrosine residues of TLR-3 Tyr759 and Tyr858 inhibit the recruitment of PI3K to the receptor and TBK1 activation, respectively [102]. As a result, partial IRF3 phosphorylation, dimerization and nuclear translocation occur, but activation of the IFN-β promoter is inhibited, suggesting that the PI3K/Akt pathway is essential for full dsRNA signaling to IRF-3. This blockade is reversed with TRIF overexpression, indicating that TRIF is downstream and/or independent of the PI3K pathway. Assembly of a multiprotein complex containing TLR-3, TBK1, IRF-3 and PI3K may be essential for complete TLR-3 signaling pathway [102, 103].

For NF-κB activation, both the C- and N-terminal of TRIF independently activate an NF-κB response. The N-terminal region of TRIF contains three TRAF6-binding motifs that associate with the TRAF-C domain of TRAF6, leading to NF-κB induction [101, 104]. Mutation of the TRAF6-binding motifs of TRIF abolished binding between TRIF and TRAF6, and partially reduced NF-κB promoter activity [100]. The C-terminal region of TRIF also recruits the kinase receptor-interacting protein (RIP-1) through its RIP homotypic interaction motif and induces the NF-κB pathway, whereas RIP-3 inhibited this pathway [105–107].

As mentioned before, TLR-3 localizes to the intracellular vesicular compartment in DCs and is not present on the cell surface [108]. Additionally, DC populations

52 *2 Activation of Interferon Gene Expression*

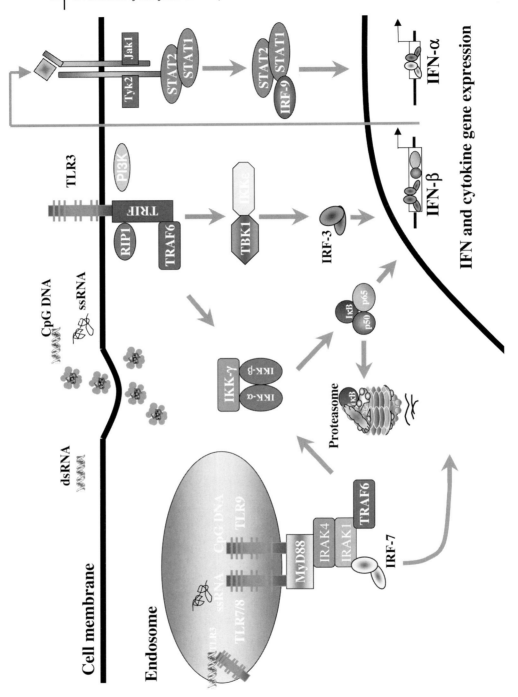

differentially express TLR-3 along with -7 (human)/-8 (murine) and -9. TLR-3 is not expressed in pDCs, but is highly expressed in human monocyte-derived DCs (Tab. 2.3). Upon TLR-7 and -9 stimulation with their respective ligand, pDCs produce a high level of Type I IFN, mainly IFN-α, whereas myeloid DCs mainly produce IL-12 and IFN-β upon TLR-3 stimulation, suggesting a differential response in distinct DC subtypes [109, 110]. Furthermore, upon poly(I:C) stimulation of TLR-3, DCs produce IFN-β and IL-12p70, and also upregulate costimulatory molecules such as CD80, CD83 and CD86 by a mechanism dependent on TRIF [108, 111].

2.5.1.3 TLR-4 Signaling

LPSs are major components of the outer membrane of Gram-negative bacteria. The host defense response to LPS includes the production of a variety of pro-inflammatory cytokines, such as TNF-α, IFN-β, as well as inducible NO (iNOS). TLR-4, the receptor for LPS, was the first mammalian homolog of *Drosophila Toll* discovered [112] and is a type I transmembrane protein; the cytoplasmic domain contains a TIR domain which is shared by all TLRs. Upon bacterial infection, a lipid-binding protein (LBP), an acute-phase protein that circulates in the liver, binds to the lipid A moiety of LPS [113]. Soluble CD14 binds and concentrates LPS present outside the cell. LBP-bound LPS forms a ternary complex with CD14, enabling the transfer of LPS to the TLR-4–MD2 complex [114]. MD2 is a secreted glycoprotein that acts as an extracellular adaptor protein that binds LPS and is essential for TLR-4 signaling to occur [115]. Upon binding of LPS, the TLR-4–MD2 complex homodimerizes and initiates the ensuing signaling cascade which bifurcates into two distinct pathways: MyD88 dependent and independent (fig. 2.5).

MyD88 is an adaptor protein which contains a C-terminal TIR domain and an N-terminal death domain [116]. MyD88$^{-/-}$ mice have revealed that the activation of NF-κB and mitogen-activated protein kinase (MAPK) still occurred in response to

Fig. 2.4. Overview of the TLR dependent signaling of the IFN response. TLR3 stimulation with dsRNA on the cell membrane or the endosomal compartment leads to the recruitment of the TRIF adaptor molecule and the activation of TBK1 and IKKε kinases that phosphorylate IRF-3 and induce the transcription of IFN-β. TRAF6 association with the TRIF complex results in NF-κB induction via IKKα/β/γ. IFN-β is released from infected cells, recognized by specific IFN receptors on adjacent cells and triggers the JAK/STAT pathway. Among the genes activated by the JAK/STAT pathway, IRF7 is induced and activated by phosphorylation. IRF7 expression results in the production of multiple IFN-α subtypes and amplification of the IFN response. pDCs express on their endosomes hTLR7/mTLR8 and TLR9 that recognize ssRNA and CpG DNA, respectively. The recognition of specific PAMPs results in the recruitment and formation of a complex composed of MyD88, IRAK4, IRAK1 leading to IRF7 phosphorylation and induction of type I IFN in a MyD88-dependent fashion. TRAF6 association with the MyD88 complex leads to the activation of the signalosome complex (IKKα, IKKβ, IKKγ), which initiates the NF-κB signaling cascade and subsequent inflammatory response. (This figure also appears with the color plates.)

2 Activation of Interferon Gene Expression

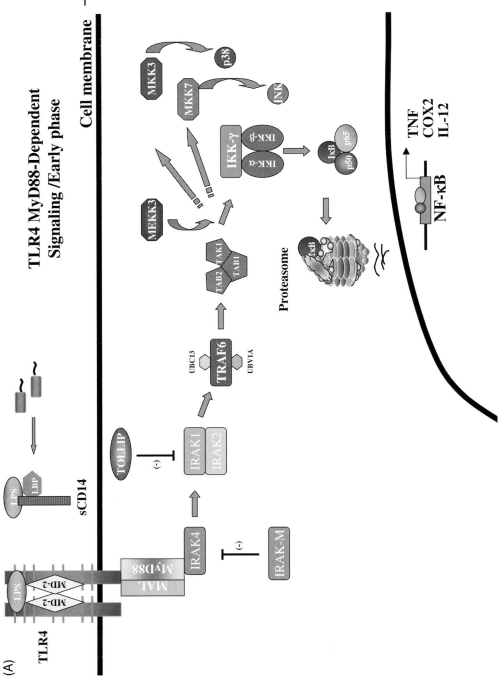

Fig. 2.5. (a) TLR-4 MyD88-dependent pathway – early phase. LPS from the outer membrane of Gram-negative bacteria is presented by the CD14–LBP complex to the TLR-4 receptor complex to initiate a MyD88-dependent signaling pathway. LPS signaling leads to the early activation of NF-κB. After activation and phosphorylation of IRAK, TRAF6 becomes activated and gives rise to the production of pro-inflammatory cytokines such as TNF, COX2 and IL-12.

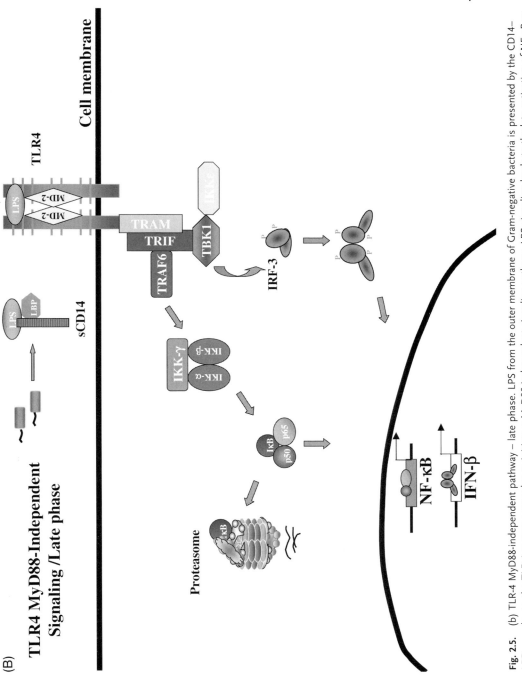

Fig. 2.5. (b) TLR-4 MyD88-independent pathway – late phase. LPS from the outer membrane of Gram-negative bacteria is presented by the CD14–LBP complex to the TLR-4 receptor complex to initiate a MyD88-independent signaling pathway. LPS signaling leads to the late activation of NF-κB as well as IFN-β induction. Following recruitment of TRIF to the receptor complex, TBK1 associates and initiates phosphorylation of IRF3, whereas TRAF6 associates with TRIF leading to the activation of NF-κB. (This figure also appears with the color plates.)

LPS, but in a delayed manner. In addition, it was shown that IRF-3 phosphorylation, as well as IFN-β induction were all unaffected in MyD88$^{-/-}$ mice, thus suggesting that MyD88 is an adaptor protein involved in the early host response to infection [103] (fig. 2.5a). A second adaptor protein, MyD88 adaptor-like protein (MAL), also known as TIRAP, was identified and found to be indispensable along with MyD88 in TLR-4 signaling for NF-κB activation [117]. Initially, IRAK4, a principle mediator in TLR-4 signaling in the MyD88-dependent signaling pathway, phosphorylates IRAK1 forming a complex which interacts with TRAF6 [118]; autophosphorylation then leads to the disassociation of the negative regulator of IRAK1, Tollip. Hyperphosphorylated IRAK1 dissociates from the receptor complex to form a new complex with IRAK2 and TRAF6 [119]. Subsequently, TRAF6 physically interacts with the ubiquitin conjugating enzyme complex Ubc13/Uev1A to catalyze the formation of a unique Lys63-linked polyubiquitin chain that positively regulates the NF-κB signaling pathway [120]. TRAF6 then becomes activated, associates with TAK1-binding protein (TAB2), which in turn activates MAPK kinase TAK1 (transforming growth factor-β activated kinase) which is constitutively associated with its adaptor protein TAB1 [121, 122]. This leads to the activation of MAPKs, such as extracellular signal-regulated kinases (ERKs), p38, and c-Jun N-terminal kinase (JNK). In addition, TRAF6 activates the IκBα kinase complex (IKK), leading to the phosphorylation and degradation of IκBα, and finally the activation of NF-κB (Fig. 2.5a) [123, 124].

In the MyD88-independent signaling pathway, TRIF-related adaptor molecule (TRAM) along with TRIF recruit the TBK1/IKKε complex to activate IRF-3, leading to induction of IFN-α/β [125] (fig. 2.5b). Following the discovery of TRIF, TRAM was also identified as an essential adaptor protein in the MyD88 independent signaling cascade. TRAM was shown to activate IRF-3 and -7, and NF-κB independently of MyD88. In addition, TRAM is upstream of TRIF in the signaling cascade, as TRAM cannot restore IFN-β induction in response to LPS stimulation when overexpressed in TRIF$^{-/-}$ cells [126]. TRIF also binds to TRAF6 via N-terminal TRAF6-binding domains leading to the activation of the signalosome, followed by ubquitination and degradation of IκB, culminating in late phase NF-κB activation. Furthermore, TRIF recruits and binds the heterodimer TBK1 and IKKε [61]. This complex phosphorylates IRF-3, which dimerzies and translocates to the nucleus to bind the ISRE to induce IFN-β production (Fig. 2.5b).

LPS mediates the maturation of DCs via a MyD88-independent pathway [127]. When MyD88-deficient bone marrow-derived DCs were cultured with LPS, cell surface expression of costimulatory molecules, such as CD40, CD80 and CD86 was upregulated, as well as T cell proliferation. By comparison, TLR-4 knockout DCs were unable to upregulate costimulatory molecules expression when cocultured with LPS. However, when wild-type DCs were stimulated with TLR-4 (MyD88 dependent) and TLR2 ligands (MyD88 independent), costimulatory molecule expression was upregulated by both MyD88-dependent and -independent pathways [127]. In sum, the innate immune response to LPS is highly complex and controlled, and includes two types of responses that are characterized by the early activation of pro-inflammatory cytokines as well as a late activation of IFN-β.

2.5.1.4 TLR-7 Signaling

While most cell types are able to produce type I IFN, pDCs are particularly adept at secreting very high IFN levels [86]. pDCs survey their environment for viruses by endocytic uptake, and TLR-7 is required in the recognition of ssRNA viruses such as VSV and influenza [128]. Single-stranded oligoribonucleotides introduced into the endosomal, but not cytoplasmic, compartment trigger TLR-7 activation [87, 88]. In TLR-7 signaling, type I IFN secretion by pDCs is MyD88 dependent [128], and both MyD88 and TRAF6 are essential to induce IFN-α production.

Upon stimulation of TLR-7 by ssRNA, IRAK1 is recruited to the complex with MyD88, TRAF6 and IRAK4 (fig. 2.4). IRAK4 phosphorylates IRAK1, triggering autophosphorylation of IRAK1 and increasing its affinity for TRAF6 [129]. *In vitro* studies demonstrate that IRAK1 binds and phosphorylates IRF-7, although to date endogenous IRF-7 phosphorylation by IRAK1 has not been shown. In IRAK1 knockout mice, both TLR-7- and -9-mediated IFN-α production was abolished. In addition, TLR-9 stimulation by A/D-type CpG oligodeoxynucleotides (ODNs) activated IRF-7 in pDCs [129]. Hence, IRAK1 is a key regulator for TLR-7- and -9-mediated IFN-α production. Furthermore, while MyD88 interacts with and activates IRF-7, it fails to activate IRF-3, thus indicating that IRF-7 is a key regulator of type I IFN in pDCs. IRF-7$^{-/-}$ mouse-derived pDCs demonstrated complete IFN nonresponsiveness upon TLR-7 and -9 stimulation, whereas pDCs from IRF-3$^{-/-}$ mice showed normal induction. Hence, IRF-3 appears dispensable for the induction of type I IFN in pDCs [83].

TLR-7 can also induce type I IFN via IRF-5 [74]. Both IRF-5 and -7 regulate the expression of distinct and overlapping IFN-α subsets. Upon TLR-7 stimulation, IRAK1 and TRAF6 are required for IRF-5 activation. Recently, our laboratory demonstrated that IRF-5 is phosphorylated by TBK1, yet this does not lead to nuclear translocation and activation [72]. IRF-5 was not only important for IFN-α, but also for IFN-β induction. In contrast, IRF-3 is not activated in the TLR-7 signaling pathway. Thus, IRF-5 and IRF-7 may function as heterodimers to regulate IFN-α gene transcription [74]. Takaoka et al. showed that in IRF5$^{-/-}$ mice had a marked decrease in the levels of pro-inflammatory cytokines in serum when induced with either unmethylated DNA or LPS. Specifically, the induction of IL-6, IL-12p40 and TNF-α were severely impaired, while the induction of IFN-α remained unaffected in IRF5$^{-/-}$ mice. IRF-5 interacts with and is activated by MyD88 and TRAF6, and TLR-4 and -9 activation result in the IRF-5 translocation and cytokine gene activation [75].

2.5.1.5 TLR-9 Signaling

Bacterially derived or synthetic DNA that contains unmethylated CpG is highly immunostimulatory, and binds TLR-9 in the endosomal compartment [130, 131]. CpG-containing DNA activates TLR-9, which in turn recruits MyD88 and IRAK family members through homophilic interactions between their death domains [132] (fig. 2.4). IRAK4 phosphorylates IRAK1, which enables IRAK1 to interact with TRAF6, leading to the phosphorylation of IRF-7 and subsequent production of type I IFN [129]. *In vitro* studies demonstrate that TBK1 and IKKε phosphorylate IRF-7 [54]. However, TBK$^{-/-}$ or IKKε$^{-/-}$ pDCs stimulated with CpG ODNs have normal IFN-α production, and thus other kinases appear to be involved in IRF-7 phosphorylation in TLR-7 and -9 signaling. *In vivo* studies in IRAK$^{-/-}$ pDCs stimulated with A/D-type CpG ODNs demonstrate that IRAK1 is dispensable for TLR-9-mediated induction of NF-κB. In sum, IRAK1 seems to play an important role in the induction of IFN-α in pDCs but is not a factor in the induction of pro-inflammatory cytokines such as TNF-α, IL-6 and IL-12p40. In contrast, the kinase activity of IRAK4 is essential for the optimal induction of pro-inflammatory cytokines and NF-κB [133]. In DCs, IRF-8 is essential in NF-κB activation in TLR-9 signaling [134]. IRF-8 is an important factor for the development and activation of DCs, and IRF-8$^{-/-}$ mice are completely unresponsive to CpG and fail to induce NF-κB. However, this type of regulation is restricted to TLR-9 signaling in DCs [134].

2.5.2
TLR-independent Signaling

2.5.2.1 Retinoic Acid Inducible Gene (RIG)-I Signaling

Recently, Yonemaya et al. discovered a separate pathway utilizing the retinoic acid inducible gene I (RIG-I) that activates virus and dsRNA-induced innate antiviral responses. RIG-I contains two caspase activation and recruitment domains (CARD) at its N-terminus and a DExD/H box RNA helicase activity in the C-terminal part of the molecule. CARD motifs were first identified in caspases, and are known to be involved in homophilic protein-protein interactions found in many signaling pathways. The discovery of RIG-I demonstrated the importance of this RNA helicase in the production of IFN-β upon infection with Newcastle disease virus (NDV) [135]. NDV enters the cell by membrane fusion, releasing its nucleocapsid into the cytoplasm and initiating a signaling cascade that leads to the production of type I IFN. Using a deletion construct that encompasses only the CARD domain, Yoneyama et al demonstrated that the truncated form of RIG-I induced IRF-3 phosphorylation, dimerization, nuclear translocation and transactivation of IFN genes [135]. The helicase domain expressed alone prevented IRF-3 activation upon viral infection thus functioning as a dominant negative mutant that was capable of binding synthetic RNA and 5′ or 3′ nontranslated region of Hepatitis C Virus (HCV) RNA [136, 137]. RIG-I recognition of dsRNA in the cytoplasm causes a structural change that enables the CARD domain to interact with other adaptor protein(s), initiating downstream events that lead to IRF-3 activation and IFN-β production (Fig 2.6).

The importance of the RIG-I pathway in antiviral immunity was demonstrated with the generation of RIG-I-deficient mice, revealing that this sensor, and not the TLR system, is required for development of the antiviral response in most cells, with the exception of plasmacytoid dendritic cells (pDCs). Reciprocally, the TLR system, but not RIG-I, is required for IFN secretion in pDCs [142]. Whether RIG-I and TLR represent complementary pathways required for full antiviral immunity remains to be investigated.

The adaptor molecule that links RIG-I sensing of incoming viral RNA and downstream activation events was recently elucidated by four independent groups (Fig. 2.6) [142–146]. MAVS/IPS/VISA/Cardif contains an amino-terminal CARD domain and a carboxyl-terminal mitochondrial transmembrane sequence that localizes this protein to the mitochondrial membrane, thus suggesting a novel role for mitochondrial signaling in the cellular innate response [142–146]. The fact that MAVS functionality requires mitochondrial association suggests a linkage between recognition of viral infection, the development of innate immunity and mitochondrial function.

In the context of hepatitis C virus infection, the RIG-I pathway is a target for viral evasion by the viral NS3/4A protease [136–138]. Significantly, the study by Meylan et al. demonstrated that Cardif is cleaved at its C-terminal end – adjacent to the mitochondrial targeting domain – by the NS3/4A protease of Hepatitis C virus [145]. An important prediction from these studies is that NS3/4A cleavage of MAVS/Cardif at cysteine- 508 results in its dissociation from the mitochondrial membrane and disruption of signaling to the antiviral immune response. Using a combination of biochemical analysis, subcellular fractionation and confocal microscopy, Li et al. demonstrated that NS3/4A results in the cleavage of MAVS/Cardif and causes relocation from the mitochondrial membrane to the cytosolic fraction of the cytoplasm. These results provide an example of host-pathogen interaction in which the virus evades innate immunity by dislodging a pivotal signaling adaptor from the mitochondria [146].

2.5.2.2 Melanoma Differentiation-associated Gene-5 (*mda-5*)

mda-5 was first discovered by subtraction hybridization as a gene induced during differentiation, cancer reversion and programmed cell death [139]. Like the RNA helicase RIG-I, this gene contains CARD and DExH group RNA helicase domains. IFN-β induces *mda-5* expression in a dose-dependent manner and, based on the absence of prior protein synthesis, *mda-5* is an early type I IFN-inducible gene which may contribute to the induction of apoptosis during IFN treatment [140]. Both *mda-5* and RIG-I are IFN inducible, and result in a more robust IFN output upon recognition of viral dsRNA in cells pretreated with IFN. *mda-5* stimulates the basal activity of the IFN-β promoter and *mda-5* overexpression enhances the activation of the IFN-β promoter in response to intracellular viral dsRNA [141]. It is unclear at this point whether *mda-5* represents a redundant or parallel pathway of CARD-mediated signaling to IFN activation.

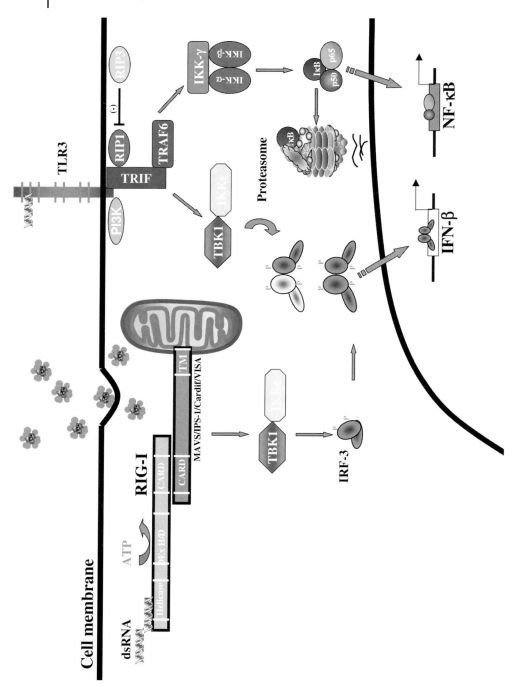

2.6
Conclusions

IFN activation in response to viral pathogens is a tightly regulated response that induces the production of IFNs to limit virus multiplication and reduce the severity of infection. The IFN family is meticulously regulated by postranscriptionally modified proteins such as the IRF family. IRF-3, -5 and -7 play critical roles in the virus-mediated regulation of IFNs. TLR-dependent and -independent signaling play an important role in the induction of IRFs and IFN-responsive genes. Recent evidence suggests that the TLR-independent molecule RIG-I is the primary sensor of viral dsRNA that induces the transcription of IFN-β by activating the TBK1 and IKKε kinases that phosphorylate IRF-3. It is likely that the secretion of these primary IFNs "prime" neighboring cells to induce a more robust IFN response by TLR-dependent pathways. TLR expression in human and mouse pDCs is limited to TLR-7/8 and -9. As a consequence, pDCs may represent the major sentinel in the production of type I IFN.

References

1 SAMUEL, C. E. 2001. Antiviral actions of interferons. *Clin Microbiol Rev* 14, 778–809.
2 SEN, G. C. 2001. Viruses and interferons. *Annu Rev Microbiol* 55, 255–281.
3 MALMGAARD, L. 2004. Induction and regulation of IFNs during viral infections. *J Interferon Cytokine Res* 24, 439–454.
4 TANIGUCHI, T., K. OGASAWARA, A. TAKAOKA, N. TANAKA. 2001. IRF family of transcription factors as regulators of host defense. *Annu Rev Immunol* 19, 623–655.
5 KATZE, M. G. 2002. Viruses and interferons: a fight for supremacy. *Nat Rev Immunol* 2, 675–687.
6 CIVAS, A., M. L. ISLAND, P. GENIN, P. MORIN, S. NAVARRO. 2002. Regulation of virus-induced interferon-A genes. *Biochimie* 84, 643–654.
7 TSUJI-TAKAYAMA, K., Y. AIZAWA, I. OKAMOTO, H. KOJIMA, K. KOIDE, M. TAKEUCHI, H. IKEGAMI, T. OHTA, M. KURIMOTO. 1999. Interleukin-18 induces interferon-gamma production through NF-kappaB and NFAT activation in murine T helper type 1 cells. *Cell Immunol* 196, 41–50.

Fig. 2.6. Recognition of dsRNA and/or incoming cytoplasmic ribonucleoprotein complexes by RIG-I leads to the recruitment of the MAVS/IPS-1/VISA/Cardif. CARD domain interactions between RIG-I and MAVS lead to the activation of NF-κB and IRF dependent pathways through IKKα/β and TBK1/IKKε kinases respectively. The TLR3 dependent pathway also leads to the induction of NF-κB and IRF dependent pathways activating the same molecules involved in RIG-I pathway with the exception of recruitment of the TRIF, TRAF6, and RIP1 molecule to the TIR domain of TLR3 for NF-κB activation. PI3K is required for the full activation of IRF-3 via TLR3. (This figure also appears with the color plates.)

8 Sica, A., T.-H. Tan, N. Rice, M. Kretzschmar, P. Ghosh, H. A. Young. 1992. The c-*rel* protooncogene product c-rel but not NF-κB binds to the intronic region of the human interferon-γ gene at a site related to an interferon-stimulable response element. *Proc Natl Acad Sci USA* 89, 1740–1744.

9 Sica, A., L. Dorman, V. Viggiano, M. Cippitelli, P. Ghosh, N. Rice, H. A. Young. 1997. Interaction of NF-kappaB and NFAT with the interferon-gamma promoter. *J Biol Chem* 272, 30412–30420.

10 Penix, L., W. M. Weaver, Y. Pang, H. A. Young, C. B. Wilson. 1993. Two essential regulatory elements in the human interferon gamma promoter confer activation specific expression in T cells. *J Exp Med* 178, 1483–1496.

11 Penix, L. A., M. T. Sweetser, W. M. Weaver, J. P. Hoeffler, T. K. Kerppola, C. B. Wilson. 1996. The proximal regulatory element of the interferon-gamma promoter mediates selective expression in T cells. *J Biol Chem* 271, 31964–31972.

12 Ye, R. D., Z. Pan, V. V. Kravchenko, D. D. Browning, E. R. Prossnitz. 1996. Gene transcription through activation of G-protein-coupled chemoattractant receptors. *Gene Expression* 5, 205–215.

13 Darnell, Jr., J. E., I. M. Kerr, G. R. Stark. 1994. Jak–STAT pathways and transcriptional activation in response to IFNs and other extracellular signaling proteins. *Science* 264, 1415–1421.

14 Darnell, Jr., J. E. 1997. STATs and gene regulation. *Science* 277, 1630–1635.

15 Caraglia, M., M. Marra, G. Pelaia, R. Maselli, M. Caputi, S. A. Marsico, A. Abbruzzese. 2005. Alpha-interferon and its effects on signal transduction pathways. *J Cell Physiol* 202, 323–335.

16 Juang, Y. T., W. Lowther, M. Kellum, W. C. Au, R. Lin, J. Hiscott, P. M. Pitha. 1998. Primary activation of interferon A and interferon B gene transcription by interferon regulatory factory-3. *Proc Natl Acad Sci USA* 95, 9837–9842.

17 Yoneyama, M., W. Suhara, Y. Fukuhara, M. Fukada, E. Nishida, T. Fujita. 1998. Direct triggering of the type I interferon system by virus infection: activation of a transcription factor complex containing IRF-3 and CBP/p300. *EMBO J* 17, 1087–1095.

18 Lin, R., C. Heylbroeck, P. M. Pitha, J. Hiscott. 1998. Virus-dependent phosphorylation of the IRF-3 transcription factor regulates nuclear translocation, transactivation potential, and proteasome-mediated degradation. *Mol Cell Biol* 18, 2986–2996.

19 Lin, R., C. Heylbroeck, P. Genin, P. M. Pitha, J. Hiscott. 1999. Essential role of interferon regulatory factor 3 in direct activation of RANTES chemokine transcription. *Mol Cell Biol* 19, 959–966.

20 Lin, R., Y. Mamane, J. Hiscott. 1999. Structural and functional analysis of interferon regulatory factor 3: localization of the transactivation and autoinhibitory domains. *Mol. Cell. Biol.* 19, 2465–2474.

21 LIN, R., Y. MAMANE, J. HISCOTT. **2000**. Multiple regulatory domains control IRF-7 activity in response to virus infection. *J Biol Chem* 275, 34320–34327.
22 SATO, M., H. SUEMORI, N. HATA, M. ASAGIRI, K. OGASAWARA, K. NAKAO, T. NAKAYA, M. KATSUKI, S. NOGUCHI, N. TANAKA, T. TANIGUCHI. **2000**. Distinct and essential roles of transcription factors IRF-3 and IRF-7 in response to viruses for IFN-alpha/beta gene induction. *Immunity* 13, 539–548.
23 AZIMI, N., Y. TAGAYA, J. MARINER, T. A. WALDMANN. **2000**. Viral activation of interleukin-15 (IL-15): characterization of a virus-inducible element in the IL-15 promoter region. *J Virol* 74, 7338–7348.
24 NAKAYA, T., M. SATO, N. HATA, M. ASAGIRI, H. SUEMORI, S. NOGUCHI, N. TANAKA, T. TANIGUCHI. **2001**. Gene induction pathways mediated by distinct irfs during viral infection. *Biochem Biophys Res Commun* 283, 1150–1156.
25 KOTENKO, S. V., G. GALLAGHER, V. V. BAURIN, A. LEWIS-ANTES, M. SHEN, R. P. DONELLY. **2003**. IFN-lambdas mediate antiviral protection through a distinct class II cytokine receptor complex. *Nat Immunol* 4, 69–77.
26 STARK, G. R., I. M. KERR, B. R. WILLIAMS, R. H. SILVERMAN, R. D. SCHREIBER. **1998**. How cells respond to interferons. *Annu Rev Biochem* 67, 227–264.
27 SCHRODER, K., P. J. HERTZOG, T. RAVASI, D. A. HUME. **2004**. Interferon-gamma: an overview of signals, mechanisms and functions. *J Leukoc Biol* 75, 163–189.
28 GROSSCHEDL, R. **1995**. Higher-order nucleoprotein complexes in transcription: analogies with site-specific recombination. *Curr Opin Cell Biol* 7, 362–370.
29 WERNER, M. H., S. K. BURLEY. **1997**. Architectural transcription factors: proteins that remodel DNA. *Cell* 88, 733.
30 NOURBAKHSH, M., H. HAUSER. **1997**. The transcriptional silencer protein NRF: a repressor of NF-kappa B enhancers. *Immunobiology* 198, 65–72.
31 NOURBAKHSH, M., H. HAUSER. **1999**. Constitutive silencing of IFN-beta promoter is mediated by NRF (NF-kappaB-repressing factor), a nuclear inhibitor of NF-kappaB. *EMBO J* 18, 6415–6425.
32 MANIATIS, T., J. V. FALVO, T. H. KIM, T. K. KIM, C. H. LIN, B. S. PAREKH, M. G. WATHELET. **1998**. Structure and function of the interferon-beta enhanceosome. *Cold Spring Harbor Symp Quant Biol* 63, 609–620.
33 MUNSHI, N., J. YIE, M. MERIKA, K. SENGER, S. LOMVARDAS, T. AGALIOTI, D. THANOS. **1999**. The IFN-beta enhancer: a paradigm for understanding activation and repression of inducible gene expression. *Cold Spring Harb Symp Quant Biol* 64, 149–159.
34 THANOS, D., T. MANIATIS. **1995**. Virus induction of human IFN-beta gene expression requires the assembly of an enhanceosome. *Cell* 83, 1091–1100.
35 SENGER, K., M. MERIKA, T. AGALIOTI, J. YIE, C. R. ESCALANTE, G. CHEN, A. K. AGGARWALL, D. THANOS. **2000**. Gene repression by co-activator repulsion. *Mol Cell* 4, 931–937.

36 MUNSHI, N., T. AGALIOTI, S. LOMVARDAS, M. MERIKA, G. CHEN, D. THANOS. **2001**. Coordination of a transcriptional switch by HMGI(Y) acetylation. *Science* 293, 1133–1136.

37 LOMVARDAS, S., D. THANOS. **2001**. Nucleosome sliding via TBP DNA binding *in vivo*. *Cell* 106, 685–696.

38 PAREKH, B. S., T. MANIATIS. **1999**. Virus infection leads to localized hyperacetylation of histones H3 and H4 at the IFN-β promoter. *Mol Cell* 3, 125–129.

39 MERIKA, M., D. THANOS. **2001**. Enhanceosomes. *Curr Opin Genet Dev* 11, 205–208.

40 CHEN-PARK, F. E., D. B. HUANG, B. NORO, D. THANOS, G. GHOSH. **2002**. The kappa B DNA sequence from the HIV long terminal repeat functions as an allosteric regulator of HIV transcription. *J Biol Chem* 277, 24701–24708.

41 RYALS, J., P. DIERKS, H. RAGG, C. WEISSMANN. **1985**. A 46-nucleotide promoter segment from an IFN-alpha gene renders an unrelated promoter inducible by virus. *Cell* 41, 497–507.

42 KUHL, D., J. DE LA FUENTE, M. CHATURVEDI, S. PARIMOO, J. RYALS, F. MEYER, C. WEISSMANN. **1987**. Reversible silencing of enhancers by sequences derived from the human IFN-alpha promoter. *Cell* 50, 1057–1069.

43 MARIE, I., J. E. DURBIN, D. E. LEVY. **1998**. Differential viral induction of distinct interferon-alpha genes by positive feedback through interferon regulatory factor-7. *EMBO J* 17, 6660–6669.

44 MARIE, I., E. SMITH, A. PRAKASH, D. E. LEVY. **2000**. Phosphorylation-induced dimerization of interferon regulatory factor 7 unmasks DNA binding and a bipartite transactivation domain. *Mol Cell Biol* 20, 8803–8814.

45 LEVY, D. E., I. MARIE, E. SMITH, A. PRAKASH. **2002**. Enhancement and diversification of IFN induction by IRF-7-mediated positive feedback. *J Interferon Cytokine Res* 22, 87–93.

46 LEVY, D. E. **2002**. Whence interferon? Variety in the production of interferon in response to viral infection. *J Exp Med* 195, F15–18.

47 HARADA, H., T. FUJITA, M. MIYAMOTO, Y. KIMURA, M. MARUYAMA, A. FURIA, T. MIYATA, T. TANIGUCHI. **1989**. Structurally similar but functionally distinct factors, IRF-1 and IRF-2, bind to the same regulatory elements of IFN and IFN-inducible genes. *Cell* 58, 729–739.

48 ESCALANTE, C. R., J. YIE, D. THANOS, A. K. AGGARWAL. **1998**. Structure of IRF-1 with bound DNA reveals determinants of interferon regulation. *Nature* 391, 103–106.

49 AU, W. C., W. S. YEOW, P. M. PITHA. **2001**. Analysis of functional domains of interferon regulatory factor 7 and its association with IRF-3. *Virology* 280, 273–282.

50 BRASS, A., E. KEHRLI, C. EISENBEIS, U. STORB, H. SINGH. **1996**. Pip, a lymphoid restricted IRF, contains a regulatory domain that is important for autoinhibition and ternary complex formation with the Ets factor PU.1. *Genes Dev* 10, 2335–2347.

51 BARNES, B. J., M. J. KELLUM, A. E. FIELD, P. M. PITHA. **2002**.

Multiple regulatory domains of IRF-5 control activation, cellular localization, induction of chemokines that mediate recruitment of T lymphocytes. *Mol Cell Biol* 22, 5721–5740.

52 AU, W.-C., P. A. MOORE, W. LOWTHER, Y.-T. JUANG, P. M. PITHA. **1995**. Identification of a member of the interferon regulatory factor family that binds to the interferon-stimulated response element and activates expression of interferon-induced genes. *Proc Natl Acad Sci USA* 92, 11657–11661.

53 SERVANT, M. J., N. GRANDVAUX, B. R. tenOEVER, D. DUGUAY, R. LIN, J. HISCOTT. **2003**. Identification of the minimal phosphoacceptor site required for *in vivo* activation of interferon regulatory factor 3 in response to virus and double-stranded RNA. *J Biol Chem* 278, 9441–9447.

54 SHARMA, S., B. R. tenOEVER, N. GRANDVAUX, G. P. ZHOU, R. LIN, J. HISCOTT. **2003**. Triggering the interferon antiviral response through an IKK-related pathway. *Science* 300, 1148–1151.

55 FITZGERALD, K. A., S. M. McWHIRTER, K. L. FAIA, D. C. ROWE, E. LATZ, D. T. GOLENBOCK, A. J. COYLE, S. M. LIAO, T. MANIATIS. **2003**. IKKepsilon and TBK1 are essential components of the IRF3 signaling pathway. *Nat Immunol* 4, 491–496.

56 McWHIRTER, S. M., K. A. FITZGERALD, J. ROSAINS, D. C. ROWE, D. T. GOLENBOCK, T. MANIATIS. **2004**. IFN-regulatory factor 3-dependent gene expression is defective in Tbk1-deficient mouse embryonic fibroblasts. *Proc Natl Acad Sci USA* 101, 233–238.

57 NAVARRO, L., K. MOWEN, S. RODEMS, B. WEAVER, N. REICH, D. SPECTOR, M. DAVID. **1998**. Cytomegalovirus activates interferon immediate–early response gene expression and an interferon regulatory factor 3-containing interferon-stimulated response element-binding complex. *Mol Cell Biol* 18, 3796–3802.

58 SATO, M., N. TANAKA, N. HATA, E. ODA, T. TANIGUCHI. **1998**. Involvement of the IRF family transcription factor IRF-3 in virus-induced activation of the IFN-beta gene. *FEBS Lett* 425, 112–116.

59 WEAVER, B. K., K. P. KUMAR, N. C. REICH. **1998**. Interferon regulatory factor 3 and CREB-binding protein/p300 are subunits of double-stranded RNA-activated transcription factor DRAF1. *Mol Cell Biol* 18, 1359–1368.

60 WATHELET, M. G., C. H. LIN, B. S. PARAKH, L. V. RONCO, P. M. HOWLEY, T. MANIATIS. **1998**. Virus infection induces the assembly of coordinately activated transcription factors on the IFN-beta enhancer *in vivo*. *Mol Cell* 1, 507–518.

61 FITZGERALD, K. A., D. C. ROWE, B. J. BARNES, D. R. CAFFREY, A. VISINTIN, E. LATZ, B. MONKS, P. M. PITHA, D. T. GOLENBOCK. **2003**. LPS–TLR-4 signaling to IRF-3/7 and NF-kappaB involves the toll adapters TRAM and TRIF. *J Exp Med* 198, 1043–1055.

62 QIN, B. Y., C. LIU, S. S. LAM, H. SRINATH, R. DELSTON, J. J. CORREIA, R. DERYNCK, K. LIN. **2003**. Crystal structure of IRF-3 reveals mechanism of autoinhibitory and virus-induced phosphoactivation. *Nat Struct Biol* 10, 913–921.

63 SERVANT, M. J., N. GRANDVAUX, J. HISCOTT. **2002**.

Multiple signaling pathways leading to the activation of interferon regulatory factor 3. *Biochem Pharmacol 64*, 985–992.

64 YONEYAMA, M., W. SUHARA, T. FUJITA. **2002**. Control of IRF-3 activation by phosphorylation. *J Interferon Cytokine Res 22*, 73–76.

65 MOUSTAKAS, A., C. H. HELDIN. **2003**. The nuts and bolts of IRF structure. *Nat Struct Biol 10*, 874–876.

66 OBATA, Y., K. YAMAMOTO, M. MIYAZAKI, K. SHIMOTOHNO, S. KOHNO, T. MATSUYAMA. **2005**. Role of cyclophilin B in activation of interferon regulatory factor-3. *J Biol Chem 18*, 18355–18360.

67 PRICE, E. R., L. D. ZYDOWSKY, M. J. JIN, C. H. BAKER, F. D. MCKEON, C. T. WALSH. **1991**. Human cyclophilin B: a second cyclophilin gene encodes a peptidyl-prolyl isomerase with a signal sequence. *Proc Natl Acad Sci USA 88*, 1903–1907.

68 MAMANE, Y., S. SHARMA, L. PETROPOULOS, R. LIN, J. HISCOTT. **2000**. Posttranslational regulation of IRF-4 activity by the immunophilin FKBP52. *Immunity 12*, 129–140.

69 BARNES, B. J., P. A. MOORE, P. M. PITHA. **2001**. Virus-specific activation of a novel interferon regulatory factor, IRF-5, results in the induction of distinct interferon alpha genes. *J Biol Chem 276*, 23382–23390.

70 IZAGUIRRE, A., B. J. BARNES, S. AMRUTE, W. S. YEOW, N. MEGJUGORAC, J. DAI, D. FENG, E. CHUNG, P. M. PITHA. **2003**. Comparative analysis of IRf and IFN-alpha expression in human plasmocytoid and monocyte-derived dendritic cells. *J Leukoc Biol 74*, 1125–1138.

71 BARNES, B. J., A. E. FIELD, P. M. PITHA. **2003**. Virus-Induced heterodimer formation between IRF-5 and IRF-7 modulates assembly of the IFNA enhanceosome *in vivo* and transcriptional activity of IFNA genes. *J Biol Chem 278*, 16630–16641.

72 LIN, R., L. YANG, M. ARGUELLO, C. PENAFUERTE, J. HISCOTT. **2005**. A CRM1-dependent nuclear export pathway is involved in the regulation of IRF-5 subcellular localization. *J Biol Chem 280*, 3088–3095.

73 MANCL, M. E., G. HU, N. SANGSTER-GUITY, S. L. OLSHALSKY, K. HOOPS, P. M. PITHA, B. J. BARNES. **2005**. Two discrete promoters regulate the alternative-spliced human interferon regulatory factor-5 isoforms: multiple isoforms with distinct cell type-specific expression, localization, regulation and function. *J Biol Chem 280*, 21078–21090.

74 SCHOENEMEYER, A., B. J. BARNES, M. E. MANCL, E. LATZ, N. GOUTAGNY, P. M. PITHA, K. A. FITZGERALD, D. T. GOLENBOCK. **2005**. The interferon regulatory factor, IRF5, is a central mediator of TLR7 signaling. *J Biol Chem 280*, 17005–17012.

75 TAKAOKA, A., H. YANAI, S. KONDO, G. DUNCAN, H. NEGISHI, T. MIZUTANI, S. I. KANO, K. HONDA, Y. OHBA, T. W. MAK, T. TANIGUCHI. **2005**. Integral role of IRF-5 in the gene induction programme activated by Toll-like receptors. *Nature 434*, 243–249.

76 ZHANG, L., J. S. PAGANO. **1997**. IRF-7, a new interferon regulatory factor associated with Epstein Barr Virus latency. *Mol Cell Biol 17*, 5748–5757.

77 Au, W. C., P. A. Moore, D. W. LaFleur, B. Tombal, P. M. Pitha. 1998. Characterization of the interferon regulatory factor-7 and its potential role in the transcription activation of interferon A genes. *J Biol Chem 273*, 29210–29217.
78 Sato, M., N. Hata, M. Asagiri, T. Nakaya, T. Taniguchi, N. Tanaka. 1998. Positive feedback regulation of type I IFN genes by the IFN-inducible transcription factor IRF-7. *FEBS Lett 441*, 106–110.
79 Yeow, W. S., W. C. Au, Y. T. Juang, C. D. Fields, C. L. Dent, D. R. Gewert, P. M. Pitha. 2000. Reconstitution of virus-mediated expression of interferon alpha genes in human fibroblast cells by ectopic interferon regulatory factor-7. *J Biol Chem 275*, 6313–6320.
80 Putzke, J. D., M. A. Williams, J. F. Daniel, B. A. Foley, J. K. Kirklin, T. J. Boll. 2000. Neuropsychological functioning among heart transplant candidates: a case control study. *J Clin Exp Neuropsychol 22*, 95–103.
81 tenOever, B. R., S. Sharma, W. Zou, Q. Sun, N. Grandvaux, I. Julkunen, H. Hemmi, M. Yamamoto, S. Akira, W. C. Yeh, R. Lin, J. Hiscott. 2004. Activation of TBK1 and IKKvarepsilon kinases by vesicular stomatitis virus infection and the role of viral ribonucleoprotein in the development of interferon antiviral immunity. *J Virol 78*, 10636–10649.
82 Caillaud, A., A. G. Hovanessian, D. E. Levy, I. J. Marie. 2005. Regulatory serine residues mediate phosphorylation-dependent and phosphorylation-independent activation of interferon regulatory factor 7. *J Biol Chem 280*, 17671–17677.
83 Honda, K., H. Yanai, H. Negishi, M. Asagiri, M. Sato, T. Mizutani, N. Shimada, Y. Ohba, A. Takaoka, N. Yoshiba, T. Taniguchi. 2005. IRF-7 is the master regulator of type-1 interferon-dependent immune responses. *Nature 434*, 772–777.
84 Hemmi, H., T. Kaisho, K. Takeda, S. Akira. 2003. The roles of Toll-like receptor 9, MyD88, and DNA-dependent protein kinase catalytic subunit in the effects of two distinct CpG DNAs on dendritic cell subsets. *J Immunol 170*, 3059–3064.
85 Wagner, M., H. Poeck, B. Jahrsdoerfer, S. Rothenfusser, D. Prell, B. Bohle, E. Tuma, T. Giese, J. W. Ellwart, S. Endres, G. Hartmann. 2004. IL-12p70-dependent T_h1 induction by human B cells requires combined activation with CD40 ligand and CpG DNA. *J Immunol 172*, 954–963.
86 Colonna, M., G. Trinchieri, Y. J. Liu. 2004. Plasmacytoid dendritic cells in immunity. *Nat Immunol 5*, 1219–1226.
87 Diebold, S. S., T. Kaisho, H. Hemmi, S. Akira, C. Reis e Sousa. 2004. Innate antiviral responses by means of TLR7-mediated recognition of single-stranded RNA. *Science 303*, 1529–1531.
88 Heil, F., H. Hemmi, H. Hochrein, F. Ampenberger, C. Kirschning, S. Akira, G. Lipford, H. Wagner, S. Bauer. 2004. Species-specific recognition of single-stranded RNA via toll-like receptor 7 and 8. *Science 303*, 1526–1529.
89 Kaisho, T., S. Akira. 2004. Pleiotropic function of Toll-like receptors. *Microb Infect 6*, 1388–1394.

90 TAKEDA, K., S. AKIRA. 2005. Toll-like receptors in innate immunity. *Int Immunol* 17, 1–14.

91 IWASAKI, A., R. MEDZHITOV. 2004. Toll-like receptor control of the adaptive immune responses. *Nat Immunol* 5, 987–995.

92 LeMOSY, E. K., D. KEMLER, C. HASHIMOTO. 1998. Role of Nudel protease activation in triggering dorsoventral polarization of the *Drosophila* embryo. *Development* 125, 4045–4053.

93 ULEVITCH, R. J. 2000. Molecular mechanisms of innate immunity. *Immunol Res* 21, 49–54.

94 XU, Y., X. TAO, B. SHEN, T. HORNG, R. MEDZHITOV, J. L. MANLEY, L. TONG. 2000. Structural basis for signal transduction by the Toll/interleukin-1 receptor domains. *Nature* 408, 111–115.

95 RADONS, J., S. GABLER, H. WESCHE, C. KORHERR, R. HOFMEISTER, W. FALK. 2002. Identification of essential regions in the cytoplasmic tail of interleukin-1 receptor accessory protein critical for interleukin-1 signaling. *J Biol Chem* 277, 16456–16463.

96 STURM, A., C. SCHULTE, R. SCHATTON, A. BECKER, E. CARIO, H. GOEBELL, A. U. DIGNASS. 2000. Transforming growth factor-beta and hepatocyte growth factor plasma levels in patients with inflammatory bowel disease. *Eur J Gastroenterol Hepatol* 12, 445–450.

97 MUZIO, M., N. POLENTARUTTI, D. BOSISIO, M. K. PRAHLADAN, A. MANTOVANI. 2000. Toll-like receptors: a growing family of immune receptors that are differentially expressed and regulated by different leukocytes. *J Leukoc Biol* 67, 450–456.

98 VISINTIN, A., A. MAZZONI, J. H. SPITZER, D. H. WYLLIE, S. K. DOWER, D. M. SEGAL. 2001. Regulation of Toll-like receptors in human monocytes and dendritic cells. *J Immunol* 166, 249–255.

99 OSHIUMI, H., T. TSUJITA, K. SHIDA, M. MATSUMOTO, K. IKEO, T. SEYA. 2003. Prediction of the prototype of the human Toll-like receptor gene family from the pufferfish, *Fugu rubripes*, genome. *Immunogenetics* 54, 791–800.

100 JIANG, Z., T. W. MAK, G. SEN, X. LI. 2004. Toll-like receptor 3 (TLR3)-mediated activation of NFkappa B and IRF-3 diverges at Toll-Il-1 receptor domain containing adaptor inducing IFN-beta. *Proc Natl Acad Sci USA* 101, 3533–3538.

101 SATO, S., M. SUGIYAMA, M. YAMAMOTO, Y. WATANABE, T. KAWAI, K. TAKEDA, S. AKIRA. 2003. Toll/IL-1 receptor domain-containing adaptor inducing IFN-beta (TRIF) associates with TNF receptor-associated factor 6 and TANK-binding kinase 1, activates two distinct transcription factors, NF-kappa B and IFN-regulatory factor-3, in the Toll-like receptor signaling. *J Immunol* 171, 4304–4310.

102 SARKAR, S. N., K. L. PETERS, C. P. ELCO, S. SAKAMOTO, S. PAL, G. C. SEN. 2004. Novel roles of TLR3 tyrosine phosphorylation and PI3 kinase in double-stranded RNA signaling. *Nat Struct Mol Biol* 11, 1060–1067.

103 JIANG, Z., M. ZAMANIAN-DARYOUSH, H. NIE, A. M. SILVA, B. R. WILLIAMS, X. LI. 2003. Poly(I–C)-induced Toll-like receptor 3 (TLR3)-mediated activation of NFkappa B and MAP kinase is

through an interleukin-1 receptor-associated kinase (IRAK)-independent pathway employing the signaling components TLR3–TRAF6–TAK1–TAB2–PKR. *J Biol Chem 278*, 16713–16719.

104 YE, H., J. R. ARRON, B. LAMOTHE, M. CIRILLI, T. KOBAYASHI, N. K. SHEVDE, D. SEGAL, O. K. DZIVENU, M. VOLOGODSKAIA, M. YIM, K. DU, S. SINGH, J. W. PIKE, B. G. DARNAY, Y. CHOI, H. WU. **2002**. Distinct molecular mechanism for initiating TRAF6 signalling. *Nature 418*, 443–447.

105 MEYLAN, E., K. BURNS, K. HOFMANN, V. BLANCHETEAU, F. MARTINON, M. KELLIHER, J. TSCHOPP. **2004**. RIP1 is an essential mediator of Toll-like receptor 3-induced NF-kappa B activation. *Nat Immunol 5*, 503–507.

106 CUSSON-HERMANCE, N., S. KHURANA, T. H. LEE, K. A. FITZGERALD, M. A. KELLIHER. **2005**. Rip1 mediates the Trif-dependent toll-like receptor 3- and 4-induced NF-{kappa}B activation but does not contribute to interferon regulatory factor 3 activation. *J Biol Chem 280*, 36560–36566.

107 SUN, X., J. YIN, M. A. STAROVASNIK, W. J. FAIRBROTHER, V. M. DIXIT. **2002**. Identification of a novel homotypic interaction motif required for the phosphorylation of receptor-interacting protein (RIP) by RIP3. *J Biol Chem 277*, 9505–9511.

108 MATSUMOTO, M., K. FUNAMI, M. TANABE, H. OSHIUMI, M. SHINGAI, Y. SETO, A. YAMAMOTO, T. SEYA. **2003**. Subcellular localization of Toll-like receptor 3 in human dendritic cells. *J Immunol 171*, 3154–3162.

109 DEGLI-ESPOSTI, M. A., M. J. SMYTH. **2005**. Close encounters of different kinds: dendritic cells and NK cells take centre stage. *Nat Rev Immunol 5*, 112–124.

110 REIS E SOUSA, C. **2004**. Toll-like receptors and dendritic cells: for whom the bug tolls. *Semin Immunol 16*, 27–34.

111 CELLA, M., M. SALIO, Y. SAKAKIBARA, H. LANGEN, I. JULKUNEN, A. LANZAVECCHIA. **1999**. Maturation, activation, protection of dendritic cells induced by double-stranded RNA. *J Exp Med 189*, 821–829.

112 MEDZHITOV, R., C. A. JANEWAY, JR. **1997**. Innate immunity: the virtues of a nonclonal system of recognition. *Cell 91*, 295–298.

113 SCHUMANN, R. R., S. R. LEONG, G. W. FLAGGS, P. W. GRAY, S. D. WRIGHT, J. C. MATHISON, P. S. TOBIAS, R. J. ULEVITCH. **1990**. Structure and function of lipopolysaccharide binding protein. *Science 249*, 1429–1431.

114 TOBIAS, P. S., K. SOLDAU, J. A. GEGNER, D. MINTZ, R. J. ULEVITCH. **1995**. Lipopolysaccharide binding protein-mediated complexation of lipopolysaccharide with soluble CD14. *J Biol Chem 270*, 10482–10488.

115 VISINTIN, A., E. LATZ, B. G. MONKS, T. ESPEVIK, D. T. GOLENBOCK. **2003**. Lysines 128 and 132 enable lipopolysaccharide binding to MD-2, leading to Toll-like receptor-4 aggregation and signal transduction. *J Biol Chem 278*, 48313–48320.

116 BURNS, K., F. MARTINON, C. ESSLINGER, H. PAHL, P. SCHNEIDER, J. L. BODMER, F. DI MARCO, L. FRENCH, J. TSCHOPP. **1998**. MyD88, an adapter protein involved in interleukin-1 signaling. *J Biol Chem 273*, 12203–12209.

117 Horng, T., G. M. Barton, R. Medzhitov. **2001**. TIRAP: an adapter molecule in the Toll signaling pathway. *Nat Immunol* 2, 835–841.

118 Li, S., A. Strelow, E. J. Fontana, H. Wesche. **2002**. IRAK-4: a novel member of the IRAK family with the properties of an IRAK-kinase. *Proc Natl Acad Sci USA* 99, 5567–5572.

119 Cao, Z., J. Xiong, M. Takeuchi, T. Kurama, D. V. Goeddel. **1996**. TRAF6 is a signal transducer for interleukin-1. *Nature* 383, 443–446.

120 Deng, L., C. Wang, E. Spencer, L. Yang, A. Braun, J. You, C. Slaughter, C. Pickart, Z. J. Chen. **2000**. Activation of the IkappaB kinase complex by TRAF6 requires a dimeric ubiquitin-conjugating enzyme complex and a unique polyubiquitin chain. *Cell* 103, 351–361.

121 Ninomiya-Tsuji, J., K. Kishimoto, A. Hiyama, J. Inoue, Z. Cao, K. Matsumoto. **1999**. The kinase TAK1 can activate the NIK-IkB as well as the MAP kinase cascade in the IL-1 signaling pathway. *Nature* 398, 252–256.

122 Yamaguchi, K., K. Shirakabe, H. Shibuya, K. Irie, I. Oishi, N. Ueno, T. Taniguchi, E. Nishida, K. Matsumoto. **1995**. Identification of a member of the MAPKKK family as a potential mediator of TGF-beta signal transduction. *Science* 270, 2008–2011.

123 Akira, S., K. Takeda, T. Kaisho. **2001**. Toll-like receptors: critical proteins linking innate and acquired immunity. *Nat Immunol* 2, 675–680.

124 Hemmi, H., O. Takeuchi, S. Sato, M. Yamamoto, T. Kaisho, H. Sanjo, T. Kawai, K. Hoshino, K. Takeda, S. Akira. **2004**. The roles of two IkappaB kinase-related kinases in lipopolysaccharide and double stranded RNA signaling and viral infection. *J Exp Med* 199, 1641–1650.

125 Yamamoto, M., S. Sato, K. Mori, K. Hoshino, O. Takeuchi, K. Takeda, et al. **2002**. Cutting edge: a novel Toll/IL-1 receptor domain-containing adapter that preferentially activates the IFN-beta promoter in the Toll-like receptor signaling. *J Immunol* 169, 6668–6672.

126 Yamamoto, M., S. Sato, H. Hemmi, S. Uematsu, K. Hoshino, T. Kaisho, O. Takeuchi, K. Takeda, S. Akira. **2003**. TRAM is specifically involved in the Toll-like receptor 4-mediated MyD88-independent signaling pathway. *Nat Immunol* 4, 1144–1150.

127 Kaisho, T., S. Akira. **2001**. Dendritic-cell function in Toll-like receptor- and MyD88-knockout mice. *Trends Immunol* 22, 78–83.

128 Lund, J. M., L. Alexopoulou, A. Sato, M. Karow, N. C. Adams, N. W. Gale, A. Iwasaki, R. A. Flavell. **2004**. Recognition of single-stranded RNA viruses by Toll-like receptor 7. *Proc Natl Acad Sci USA* 101, 5598–5603.

129 Uematsu, S., S. Sato, M. Yamamoto, T. Hirotani, H. Kato, F. Takeshita, M. Matsuda, C. Coban, K. J. Ishii, T. Kawai, O. Takeuchi, S. Akira. **2005**. Interleukin-1 receptor-associated kinase-1 plays an essential role for Toll-like receptor (TLR)7- and TLR9-mediated interferon-α induction. *J Exp Med* 201, 915–923.

130 HEMMI, H., O. TAKEUCHI, T. KAWAI, T. KAISHO, S. SATO, H. SANJO, M. MATSUMOTO, K. HOSHINO, H. WAGNER, K. TAKEDA, S. AKIRA. 2000. A Toll-like receptor recognizes bacterial DNA. *Nature 408*, 740–745.

131 LATZ, E., A. VISINTIN, T. ESPEVIK, D. T. GOLENBOCK. 2004. Mechanisms of TLR9 activation. *J Endotoxin Res 10*, 406–412.

132 DUNNE, A., L. A. O'NEILL. 2003. The interleukin-1 receptor/Toll-like receptor superfamily: signal transduction during inflammation and host defense. *Sci STKE 2003*, re3.

133 LYE, E., C. MIRTSOS, N. SUZUKI, S. SUZUKI, W. C. YEH. 2004. The role of interleukin 1 receptor-associated kinase-4 (IRAK-4) kinase activity in IRAK-4-mediated signaling. *J Biol Chem 279*, 40653–40658.

134 TSUJIMURA, H., T. TAMURA, H. J. KONG, A. NISHIYAMA, K. J. ISHII, D. M. KLINMAN, K. OZATO. 2004. Toll-like receptor 9 signaling activates NF-kappaB through IFN regulatory factor-8/IFN consensus sequence binding protein in dendritic cells. *J Immunol 172*, 6820–6827.

135 YONEYAMA, M., M. KIKUCHI, T. NATSUKAWA, N. SHINOBU, T. IMAIZUMI, M. MIYAGISHI, K. TAIRA, S. AKIRA, T. FUJITA. 2004. The RNA helicase RIG-I has an essential function in double-stranded RNA-induced innate antiviral responses. *Nat Immunol 5*, 730–737.

136 FOY, E., K. LI, R. SUMPTER, JR., Y. M. LOO, C. L. JOHNSON, C. WANG, P. M. FISH, M. YONEYAMA, T. FUJITA, S. M. LEMON, M. GALE, JR. 2005. Control of antiviral defenses through hepatitis C virus disruption of retinoic acid-inducible gene-I signaling. *Proc Natl Acad Sci USA 79*, 2689–2699.

137 SUMPTER, JR., R., Y. M. LOO, E. FOY, K. LI, M. YONEYAMA, T. FUJITA, S. M. LEMON, M. GALE, JR. 2005. Regulating intracellular antiviral defense and permissiveness to hepatitis C virus RNA replication through a cellular RNA helicase, RIG-I.I. *J Virol 79*, 2689–2699.

138 BREIMAN, A., N. GRANDVAUX, R. LIN, C. OTTONE, S. AKIRA, M. YONEYAMA, T. FUJITA, J. HISCOTT, E. F. MEURS. 2005. Inhibition of RIG-I-dependent signaling to the interferon pathway during hepatitis C virus expression and restoration of signaling by IKKepsilon. *J Virol 79*, 3969–3978.

139 KANG, D. C., R. V. GOPALKRISHNAN, Q. WU, E. JANKOWSKY, A. M. PYLE, P. B. FISHER. 2002. mda-5: an interferon-inducible putative RNA helicase with double-stranded RNA-dependent ATPase activity and melanoma growth-suppressive properties. *Proc Natl Acad Sci USA 99*, 637–642.

140 KANG, D. C., R. V. GOPALKRISHNAN, L. LIN, A. RANDOLPH, K. VALERIE, S. PESTKA, P. B. FISHER. 2004. Expression analysis and genomic characterization of human melanoma differentiation associated gene-5, *mda-5*: a novel type I interferon-responsive apoptosis-inducing gene. *Oncogene 23*, 1789–1800.

141 ANDREJEVA, J., K. S. CHILDS, D. F. YOUNG, T. S. CARLOS, N. STOCK, S. GOODBOURN, R. E. RANDALL. 2004. The V proteins of paramyxoviruses bind the IFN-inducible RNA helicase, *mda-5*, and inhibit its activation of the IFN-beta promoter. *Proc Natl Acad Sci USA 101*, 17264–17269.

142 Kawai, T., K. Takahashi, S. Sato, C. Coban, H. Kumar, H. Kato, K. J. Ishii, O. Takeuchi, S. Akira. **2005**. IPS-1, an adaptor triggering RIG-I- and Mda5-mediated type I interferon induction. *Nat Immunol* 6, 981–988.

143 Seth, R. B., L. Sun, C. K. Ea, Z. J. Chen. **2005**. Identification and characterization of MAVS, a mitochondrial antiviral signaling protein that activates NF-κB and IRF 3. *Cell* 122, 669–682.

144 Xu, L. G., Y. Y. Wang, K. J. Han, L. Y. Li, Z. Zhai, H. B. Shu. **2005**. VISA is an adaptor protein required for virus-triggered IFN-beta signaling. *Mol Cell* 19, 727–740.

145 Meylan, E., J. Curran, K. Hofmann, D. Moradpour, M. Binder, R. Bartenschlager, J. Tsachopp. **2005**. Cardif is an adaptor protein in the RIG-I antiviral pathway and is targeted by hepatitis C virus. *Nature* 437, 1167–1172.

146 Xiao-Dong, L., S. Lijun, R. B. Seth. G. Pineda, Z. J. Chen. **2005**. Hepatitis C virus protease NS3/4A cleaves mitochondrial antiviral signaling protein off the mitochondria to evade innate immunity. *PNAS* 102, 17717–17722.

3
Interferon Proteins: Structure, Production and Purification

Dimitris Platis and Graham R. Foster

3.1
Introduction

The type I interferons (IFNs) are a family of closely related cytokines consisting of 12 different IFN-α subtypes, one IFN-β subtype and one IFN-ω subtype [1, 2]. Of the many different type I IFNs that are produced naturally, only IFN-α2 and -β are currently in widespread clinical use. In recent years, modified artificial IFNs have been produced and one of these, consensus IFN, is currently in clinical use [3]. Type I IFNs, chiefly IFN-α2, are used clinically to treat a variety of hematological malignancies and viral hepatitis, and a great deal of effort and ingenuity has been spent in determining optimal production and purification approaches. In this chapter, we will briefly review the structure of the type I IFNs before describing the historical development of production systems for the manufacture and purification of type I IFNs.

3.2
The Structure of Type I IFNs

All of the naturally occurring IFN-α subtypes have a similar amino acid sequence (they are over 70% similar) and Fig. 3.1 illustrates the amino acid sequence of the different subtypes along with the sequence of IFN-β. It is probable that the structure of the other type I IFNs will be very similar, although this has not yet been formally proven.

The three-dimensional structure of the type I IFNs was first resolved for murine IFN-β [4] in 1992. Subsequent studies on crystallized human IFN-α2b in the form of a zinc-mediated dimer allowed the structure of the human type I IFN to be determined to a resolution of 2.9 Å [5], and this structure has now been confirmed and extended using high-resolution nuclear magnetic resonance [6]. Based on the sequence similarity of the other type I IFNs (see Fig. 3.1) it is probable that the

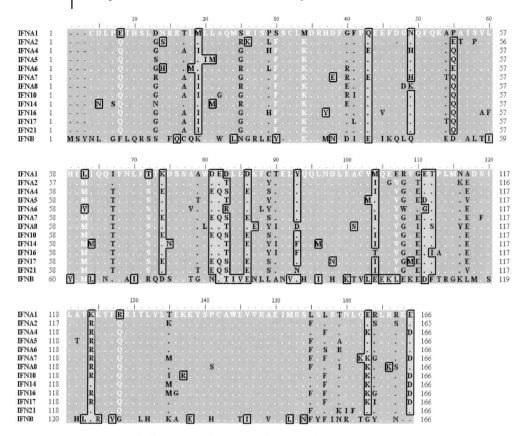

Fig. 3.1. Multiple amino acid sequence alignment of all the natural IFN subtypes. All the amino acid sequences of the natural IFN subtypes (including IFN-β, 1AU1) have been aligned using ClustalW multiple alignment (Bioedit sequence analysis package).

structures of other human type I IFNs will be similar. The type I IFNs form a cuboical structure composed of five α helices (termed A–E) with their associated linking loops. The cuboidal IFN binds to the heterodimeric receptor that consists of receptor components termed IFNAR-1 and -2. Based on studies of modified and mutated IFNs, Uze et al. [7] have put forward a model for the receptor–ligand interaction in which the A helix initially binds to IFNAR-1 and the D helix binds to IFNAR-2. Rotation of the IFN and, possibly, changes in the receptor lead to an "active complex" in which helices A and C bind to IFNAR-1, and the AB loop and D helix bind to IFNAR-2. This model remains hypothetical and further crystallographic studies will be required to define the events that take place when the cuboidal IFNs interact with their receptor.

3.3
Production and Purification of Type I IFNs

3.3.1
Leukocyte-derived IFN – First Steps in Producing Commercial IFN

The type I IFNs were first clearly identified in 1957 when Isaacs and Lindenmann [8] demonstrated that an "interfering factor" which inhibited viral replication could be produced by cells that were exposed to inactivated virus. Early attempts to produce large quantities of the "interfering protein" focused on exposing cells to active or inactive virus in the hope of producing large quantities of this "interfering factor". Initial attempts were met with very limited success, but persistence eventually paid off and it was found that infecting human leukocytes with Sendai or Newcastle disease virus led to the production of substantial amounts of IFN [9–11]. In collaboration with the Finish blood transfusion service, Kari Cantell was able to develop conditions that maximized the production of type I IFNs from Sendai-treated human leukocytes and this allowed the production of sufficient material to perform early clinical studies [12] that confirmed the therapeutic potential of this exciting new molecule.

Early efforts to purify IFN from crude extracts relied upon the remarkable stability of IFN at low pH. IFN was initially dissolved into a solution of acidic ethanol and, by slowly raising the pH of the solution, early workers managed to gradually remove any impurities by precipitation. By repeating the process several times, considerably pure IFN could be obtained (P-IF) compared to the initial crude extracts or slightly processed samples containing concentrated preparations of IFN (C-IF). By taking advantage of the robustness of the IFN molecule under extreme conditions of pH, detergents and temperature, crude protocols were formulated which allowed partial purification of IFN from leukocytes [13]. However, the purity did not exceed 0.1–1% by weight and, in fact, the IFN used in the early clinical trials in the 1970s consisted of IFN of extremely low purity. As well as antiviral properties, these early crude preparations exhibited additional antiprotozoal and antibacterial properties, and inhibited cellular proliferation. However, these properties could not be definitively ascribed to IFN since contaminating proteins were present. Although partial purification of IFN on sodium dodecylsulfate polyacrylamide gels was achieved in the 1970s [14], it was not until 1978 when type I IFN was isolated in homogeneity and in sufficient quantities that complete chemical, functional and physical characterization of molecule was possible [15–17]. This was achieved by the introduction of chromatographic methods, such as high-pressure liquid chromatography (HPLC) [18–20].

The production of leukocyte-derived IFN continues to the present day. The type I IFN that is produced is a mixture of different IFN-α subtypes; there are some who claim that this "natural" IFN-α has beneficial properties and *in vitro* studies suggest that this may indeed be the case [21]. However, no controlled clinical trials have compared leukocyte-derived IFN-α with the recombinant form and therefore it is unclear whether leukocyte-derived IFN-α really does have beneficial properties or not.

3.3.2
Lymphoblastoid IFN – Towards more Reliable Supplies of IFN

Large-scale production of human IFN from leukocytes clearly requires the use of large numbers of white blood cells derived from volunteer blood donors. A major advance in the production of type I IFN involved a switch from leukocytes to lymphoblastoid lines [22–25] which markedly increased the potential for commercial production. Initially, there were hesitations about the use of human "immortal" cell lines because of fears that any product derived from these cells might cause cancer to its recipient. However, these fears subsided and a lymphoblastoid cell line (Namalwa) was identified that produced large amounts of IFN after Sendai virus induction. The IFN produced from these cultured cells could be purified relatively easily by employing standard chromatographic methods, such as gel filtration or anion/cation-exchange chromatography already developed for leukocyte IFN. The resulting product, lymphoblastoid IFN, was produced commercially for many years and marked under the trade name Wellferon. Production continues today in Japan where the product is marketed under the brand name Sumiferon.

3.3.3
Cloned Type I IFNs – An Inexhaustible Supply of Therapeutic Material

The development of gene cloning technology in the 1970s led to the possibility of producing human proteins in bacteria. Introduction of the IFN genes initially in bacteria [26, 27], and later in a variety of expression systems including yeast [28], insect cells [29] mammalian cells [30] and even in transgenic plants [31], has allowed for commercial production of IFN in almost limitless amounts. The relatively small size and compactness of the IFN protein combined with the lack of any functional glycosylation (unglycosylated IFNs are functionally identical to their glycosylated counterparts in so far as studies have been performed) certainly contributes to high yield and bioactivity in a variety of production platforms. More importantly, more economically viable production approaches can be used since proper glycosylation is not an issue. Bioactive IFN has been obtained from several expression systems using a combination of different chromatographic techniques [32, 33]. In addition to chromatographic purification of the type I IFNs, the proteins can be purified to a greater or lesser extent by the use of affinity-binding techniques. As well as the natural IFN-binding protein ligand, the IFN receptor, several compounds have been shown to bind IFN and can, potentially, be used as an affinity chromatography ligand [34–37]. In addition, monoclonal antibodies raised against partially purified IFN recovered from induced leukocytes and immobilized on cross-linked agarose can be used to separate and purify type I IFNs [38]. In the laboratory a wide variety of different approaches have been used to prepare small amounts of IFN for experimental use and some of these are listed in Tab. 3.1.

A necessary prerequisite for the use of IFN in clinical practice is its production in large quantities, i.e. at a medium and large scale. Significant progress has been made towards the scaling-up of the IFN production process [9, 24, 44–47], and ad-

Tab. 3.1. Different protocols utilized for the purification of IFN-αs under laboratory-based conditions

Activity [(U/mg) × 10^8]	Purification steps
3.2	cation exchange, metal chelation, gel filtration [33]
2.5	immunoaffinity, cation exchange, gel filtration [32]
2.8	mimetic ligand-affinity chromatography [37]
1	metal chelation, reverse-phase HPLC [39]
1.7	gel filtration [40]
1.8	metal chelation [41]
1.1	metal chelation [42]
30	ion exchange [43]

vances in laboratory-scale production and purification of IFN have been adopted by the pharmaceutical industry in an effort to achieve more economic production and downstream processing. Industrial-scale production and purification of IFN is often hampered by economic factors and patents restricting its choice of production and purification methods; however, in general, a common approach is followed and this is illustrated in Fig. 3.2. Crude material derived, usually, from genetically engineered bacteria acts as the starting point for the purification process. Unfortunately the crude IFN derived from bacteria is usually present in insoluble aggregated particles (inclusion bodies) within the bacteria. This crude material is inactive and has to be renatured to acquire its full biological activity [48]. A variety of tricks have been used to renature and refold the crude IFN, and these essentially involve the use of compounds that reduce protein aggregation and assist refolding [49] followed by refolding in appropriate buffers. Optimized, commercially sensitive, approaches are now in widespread use. The crude, refolded IFN can be isolated using a variety of immunoadsorbants. However, even pure IFN extracts are not totally homogeneous and are comprised of IFN oligomers, monomers of different molecular forms or even fragments of the molecule [50] and final purification usually involves a combination of the chromatographic techniques mentioned above to obtain a homogeneous well-characterized product.

This multiple-step approach has been used from the 1980s and has certainly been refined in recent years. Its main drawback is that product loss increases geometrically with each step, resulting in reduced final yield. However, significant advances in our knowledge of protein expression (vectors, protein secretion) and refinement of chromatography techniques have minimized product losses or have even allowed for the purification of IFN in a single step.

In all cases, a basic production protocol is followed which is composed of the following steps: fermentation, cell separation, extraction, purification and formulation.

Industrial-scale production of IFN utilizes a protocol comprised of series of back-to-back clarification steps designed to achieve not only higher purity, but also the production of properly formulated product through a totally controlled process. Be-

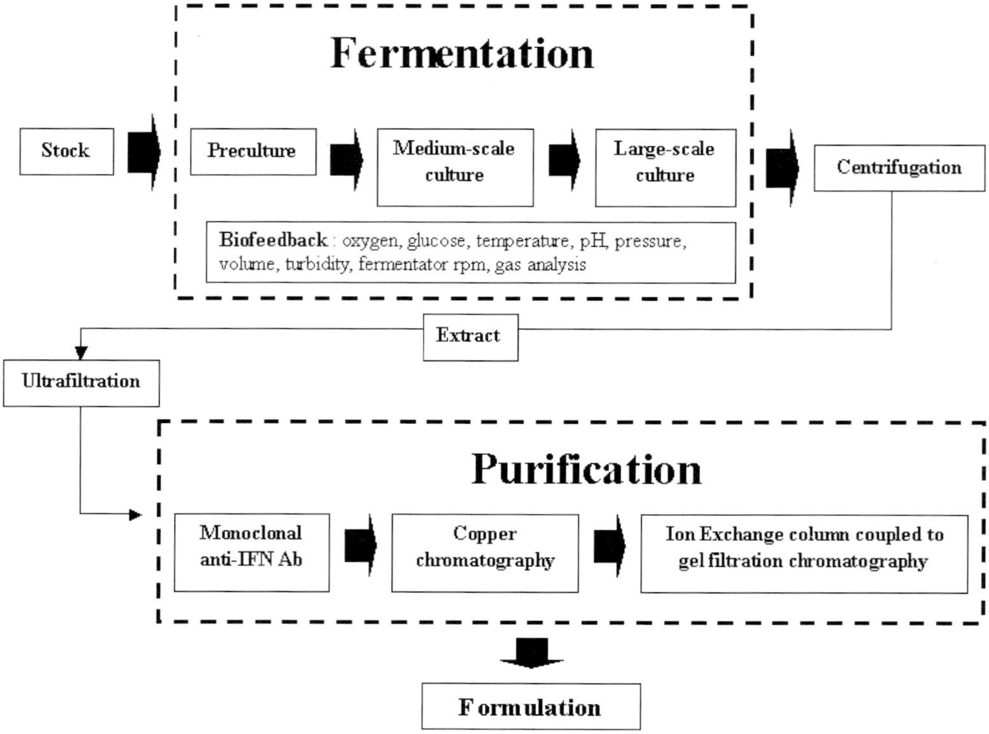

Fig. 3.2. Schematic diagram of the IFN production process.

cause of the clinical use of the final product, strict Good Clinical Practice protocols have to be applied during the entire production and purification procedure, and the purified IFN has to be free of toxins, chemicals or other biological pathogens, such as viruses and bacteria (formulated product). Such exhaustive downstream processing protocols are certainly followed by considerable losses in the amount of the final product. Nevertheless, the development of computer-assisted fermentation processes coupled to high-density cell cultivation techniques have considerably contributed to the production of pure IFN at an industrial scale by offering an unprecedented level of control of the fermentation and the subsequent downstream processes, ensuring high product quality and process reproducibility.

3.4
Long-acting IFNs

The natural type I IFNs are rapidly degraded by proteolytic breakdown in the kidneys and the resulting cleavage products are excreted. This rapid metabolism leads to a very short half-life for unmodified IFNs which requires frequent dosing intervals. To overcome the problems associated with their short half-lifes a number of different approaches have been developed to reduce the excretion of the type I IFNs and increase their serum half-life.

Polyethylene glycol (PEG) is an inert molecule that can form long, covalent chains. Using a variety of different approaches, PEG chains of various lengths can be attached to biological proteins. The resulting proteins retain their biological activity, but have greatly reduced rates of degradation and hence much longer half-lives. To improve the biostability of the type I IFNs, pegylated versions have recently been developed and these are now widely licensed for the treatment of hepatitis C [51]. Two different pegylated IFNs are currently available. One consists of a 12-kDa PEG chain linked by a cleavable bond to an IFN-α2b. This cleavable bond breaks down in solution liberating unmodified IFN-α2b. Hence, this PEG IFN is probable best regarded as a slow-release preparation that liberates active drug in solution. The alternative pegylated IFN consists of a long, branched PEG chain that is covalently linked to the IFN-α2a molecule. This pegylated IFN has a long half-life and circulates intact [51]. The pegylated IFNs that are currently available are derived from standard IFN-α2a and -α2b (produced as outlined above) that has been chemically linked to the PEG chain and then purified for clinical use.

The currently available pegylated IFNs are used in once-weekly dosing and attempts are currently being made to produce active type I IFNs that can be administered less frequently. The natural protein albumin can be linked to IFN to generate a very long-acting IFN which requires dosing as infrequently as once a week. This novel compound (Albuferon) is currently undergoing clinical trials [52].

3.5
Summary

The type I IFNs are a family of structurally related, cuboidal proteins that can be produced by viral stimulation of mammalian cells or generated from appropriately modified, genetically engineered, bacteria. Early commercial IFNs were produced from stimulated cells, but refinements in recombinant protein production techniques have led to bacterially derived IFNs dominating the therapeutic market place. The bacterially derived IFNs are purified to homogeneity using an iterative process involving a variety of purification steps and the resulting products have now been in widespread clinical use for over a decade with striking success. Recent advances in our understanding of the limitations of short-acting IFNs have led to the development of modified IFNs, chiefly pegylated IFNs, that have enhanced pharmacokinetic properties and are clinically beneficial. It seems likely that further refinements to both the structure and production of the type I IFNs will lead to further clinically beneficial advances in the foreseeable future.

References

1 Pestka S, Langer T, Zoon K, Samuel C. Interferons and their actions. *Annu Rev Biochem* **1987**, *56*, 727–777.
2 Adolf G. Monoclonal antibodies and enzyme immunoassays specific for human interferon omega: evidence that interferon omega is a component of human leucocyte interferon. *Virology* **1990**, *175*, 410–417.
3 Heathcote J. Consensus interferon: a novel interferon for the treatment of hepatitis C. *J Viral Hepat* **1998**, *5 (Suppl 1)*, 13–18.
4 Senda T, Saitoh S, Mitsui Y. Refined crystal structure of recombinant murine interferon-beta at 2.15 A resolution. *J Mol Biol* **1995**, *253*, 187–207.
5 Radhakrishnan R, Walter LJ, Hruza A, Reichert P, Trotta PP, Nagabhushan TL, et al. Zinc mediated dimer of human interferon-alpha 2b revealed by X-ray crystallography. *Structure* **1996**, *4*, 1453–1463.
6 Klaus W, Gsell B, Labhardt A, Wipf B, Senn H. The three dimensional high resolution structure of human interferon alpha 2a determined by heteronuclear NMR spectroscopy in solution. *J Mol Biol* **1997**, *274*, 661–675.
7 Uze G, Lutfalla G, Mogensen K. Alpha and beta interferons and their receptor and their friends and relatives. *J Interferon Cytokine Res* **1995**, *15*, 3–26.
8 Isaacs A, Lindenmann J. Virus interference: the interferon. *Proc Roy Soc Med* **1957**, *147*, 258–267.
9 Cantell K, Hirvonen S. Large-scale production of human leukocyte interferon containing 10^8 units per ml. *J Gen Virol* **1978**, *39*, 541–543.
10 Familletti PC, McCandliss R, Pestka S. Production of high levels of human leukocyte interferon from a continuous human myeloblast cell culture. *Antimicrob Agents Chemother* **1981**, *20*, 5–9.
11 Lee S, vanRooyen C, Ozere R. Additional studies of interferon production by human leukemic leukocytes *in vitro*. *Cancer Res* **1969**, *29*, 645–652.
12 Jones BR, Coster DJ, Falcon MG, Cantell K. Clinical trials of topical interferon therapy of ulcerative viral keratitis. *J Infect Dis* **1976**, *133 (Suppl)*, A169–A172.
13 Cantell K, Hirvonen S, Mogensen KE, Pyhala L. Human leukocyte interferon: production, purification, stability, and animal experiments. *In Vitro Monogr* **1974**, 35–38.
14 Berg K. Purification and characterization of murine and human interferons. A review of the literature of the 1970s. *Acta Pathol Microbiol Immunol Scand* **1982**, *279*, S1–S136.
15 Rubinstein M, Levy WP, Moschera JA, Lai CY, Hershberg RD, Bartlett RT, et al. Human leukocyte interferon: isolation and characterization of several molecular forms. *Arch Biochem Biophys* **1981**, *210*, 307–318.
16 Rubinstein M, Rubinstein S, Familletti PC, Gross MS, Miller RS, Waldman AA, et al. Human leukocyte interferon purified to homogeneity. *Science* **1978**, *202*, 1289–1290.

17 Friesen HJ, Stein S, Pestka S. Purification of human fibroblast interferon by high-performance liquid chromatography. *Methods Enzymol* **1981**, *78(A)*, 430–435.
18 Friesen HJ, Stein S, Evinger M, Familletti PC, Moschera J, Meienhofer J, et al. Purification and molecular characterization of human fibroblast interferon. *Arch Biochem Biophys* **1981**, *206*, 432–450.
19 Rubinstein M, Rubinstein S, Familletti PC, Miller RS, Waldman AA, Pestka S. Human leukocyte interferon: production, purification to homogeneity, and initial characterization. *Proc Natl Acad Sci USA* **1979**, *76*, 640–644.
20 Stein S, Kenny C, Friesen HJ, Shively J, Del Valle U, Pestka S. NH_2-terminal amino acid sequence of human fibroblast interferon. *Proc Natl Acad Sci USA* **1980**, *77*, 5716–5719.
21 Fan SX, Skillman DR, Liao MJ, Testa D, Meltzer MS. Increased efficacy of human natural interferon alpha (IFN-alpha n3) versus human recombinant IFN-alpha 2 for inhibition of HIV-1 replication in primary human monocytes. *AIDS Res Hum Retroviruses* **1993**, *9*, 1115–1122.
22 Zoon KC, Bridgen PJ, Smith ME. Production of human lymphoblastoid interferon by Namalwa cells cultured in serum-free media. *J Gen Virol* **1979**, *44*, 227–229.
23 Zoon KC, Smith ME, Bridgen PJ, zur Nedden D, Anfinsen CB. Purification and partial characterization of human lymphoblast interferon. *Proc Natl Acad Sci USA* **1979**, *76*, 5601–5605.
24 Finter N. Large scale production of human interferon from lymphoblastoid cells. *Tex Rep Biol Med* **1981**, *41*, 175–178.
25 Finter N, Fantes K, Johnston M. Human lymphoblastoid cells as a source of interferon. *Dev Biol Stand* **1977**, *38*, 343–348.
26 Weismann C, Nagata S, Boll M, Fountoulakis M, Fujisawa A, Fujisawa J, et al. Structure and expression of human alpha-interferon gene. *Princess Takamatsu Symp* **1982**, *12*, 1–22.
27 Pestka S. The human interferons – from protein purification and sequence to cloning and expression in bacteria: before, between, and beyond. *Arch Biochem Biophys* **1983**, *221*, 1–37.
28 Singh A, Lugovoy J, Kohr W, Perry L. Synthesis, secretion and processing of alpha-factor-interferon fusion proteins in yeast. *Nucleic Acids Res* **1984**, *12*, 8927–8938.
29 Ruttanapumma R, Nakamura M, Takehara, K. High level expression of recombinant chicken interferon-alpha using baculovirus. *J Vet Med Sci* **2005**, *67*, 25–28.
30 Rossmann C, Sharp N, Allen G, Gewert D. Expression and purification of recombinant, glycosylated human interferon alpha 2b in murine myeloma NSo cells. *Protein Expr Purif* **1996**, *7*, 335–342.
31 Daniell H, Chebolu S, Kumar S, Middleton M, Falconer R. Chloroplast-derived vaccine antigens and other therapeutic proteins. *Vaccine* **2005**, *23*, 1779–1783.
32 Tarnowski SJ, Roy SK, Liptak RA, Lee DK, Ning RY. Large-scale purification of recombinant human leukocyte interferons. *Methods Enzymol* **1986**, *119*, 153–165.

33 Thatcher DR. Purification of recombinant human IFN-alpha 2. *Methods Enzymol* **1986**, *119*, 166–177.
34 Chadha KC, Grob PM, Mikulski AJ, Davis LR, Jr., Sulkowski E. Copper chelate affinity chromatography of human fibroblast and leucocyte interferons. *J Gen Virol* **1979**, *43*, 701–706.
35 Grob PM, Chadha KC. Separation of human leukocyte interferon components by concanavalin A–agarose affinity chromatography and their characterization. *Biochemistry* **1979**, *18*, 5782–5786.
36 Jankowski WJ, von Muenchhausen W, Sulkowski E, Carter WA. Binding of human interferons to immobilized Cibacron Blue F3GA: the nature of molecular interaction. *Biochemistry* **1976**, *15*, 5182–5187.
37 Swaminathan S, Khanna N. Affinity purification of recombinant interferon-alpha on a mimetic ligand adsorbent. *Protein Expr Purif* **1999**, *15*, 236–242.
38 Staehelin T, Hobbs DS, Kung H, Lai CY, Pestka S. Purification and characterization of recombinant human leukocyte interferon (IFLrA) with monoclonal antibodies. *J Biol Chem* **1981**, *256*, 9750–9754.
39 Beldarrain A, Cruz Y, Cruz O, Navarro M, Gil M. Purification and conformational properties of a human interferon alpha2b produced in *Escherichia coli*. *Biotechnol Appl Biochem* **2001**, *33*, 173–182.
40 Liu PT, Ta T, Villarete L. High-yield expression and purification of human interferon alpha-1 in *Pichia pastoris*. *Protein Expr Purif* **2001**, *22*, 381–387.
41 Platis D, Foster GR. High yield expression, refolding, and characterization of recombinant interferon alpha2/alpha8 hybrids in *Escherichia coli*. *Protein Expr Purif* **2003**, *31*, 222–230.
42 Neves FO, Ho P, Raw I, Pereira C, Moreira C, Nascimento A. Overexpression of a synthetic gene encoding human alpha interferon in *Escherichia coli*. *Protein Expr Purif* **2004**, *35*, 353–359.
43 Srivastava P, Bhattacharaya P, Pandey G, Mukherjee K. Overexpression and purification of recombinant human interferon alpha2b in *Escherichia coli*. *Protein Expr Purif* **2005**, *41*, 313–322.
44 Riesenberg D, Menzel K, Schulz V, Schumann K, Veith G, Zuber G, et al. High cell density fermentation of recombinant *Escherichia coli* expressing human interferon alpha 1. *Appl Microbiol Biotechnol* **1990**, *34*, 77–82.
45 Babu KR, Swaminathan S, Marten S, Khanna N, Rinas U. Production of interferon-alpha in high cell density cultures of recombinant *Escherichia coli* and its single step purification from refolded inclusion body proteins. *Appl Microbiol Biotechnol* **2000**, *53*, 655–660.
46 Yang XM, Xu L, Eppstein L. Production of recombinant human interferon-alpha 1 by *Escherichia coli* using a computer-controlled cultivation process. *J Biotechnol* **1992**, *23*, 291–301.
47 Bridgen PJ, Anfinsen CB, Corley L, Bose S, Zoon KC, Ruegg UT, et al. Human lymphoblastoid interferon. Large

scale production and partial purification. *J Biol Chem* **1977**, *252*, 6585–6587.

48 MARSTON FA. The purification of eukaryotic polypeptides synthesized in *Escherichia coli*. *Biochem J* **1986**, *240*, 1–12.

49 SORENSEN HP, KK M. Soluble expression of recombinant proteins in the cytoplasm of *Escherichia coli*. *Microb Cell Fact* **2005**, *4*, 1.

50 HOCHULI E, GILLESSEN D, KOCHER H. Specificity of the immunoadsorbent used for large-scale recovery of interferon alpha-2a. *J Chromatogr* **1987**, *411*, 371–378.

51 FOSTER GR. Review article: pegylated interferons: chemical and clinical differences. *Aliment Pharmacol Ther* **2004**, *20*, 825–30.

52 SUNG C, NARDELLI B, LAFLEUR DW, BLATTER E, CORCORAN M, OLSEN HS, et al. An IFN-beta-albumin fusion protein that displays improved pharmacokinetic and pharmacodynamic properties in nonhuman primates. *J Interferon Cytokine Res* **2003**, *23*, 25–36.

4
Interferon-γ: Gene and Protein Structure, Transcription Regulation, and Actions

Ana M. Gamero, Deborah L. Hodge, David M. Reynolds, Maria Cecilia Rodriguez-Galan, Mansour Mohamadzadeh and Howard A. Young

4.1
Introduction

In 1965, E. F. Wheelock reported in the journal *Science* [1] on an interferon (IFN)-like virus inhibitor induced in human leukocytes by phytohemagglutinin. This report is generally accepted as being the first description of IFN-γ. Also designated as "immune" or type II IFN, due to its production by leukocytes (in contrast to the cell sources for IFN-α and -β) and its immunoregulatory effects, IFN-γ has since been reported to have many additional properties beyond its antiviral activity. Designated as IFN-γ in 1980 by an international committee [2], one of the first activities attributed to this protein that was distinct from its antiviral activity was that of macrophage-activating factor (MAF) [3], but it was not until 1983 that IFN-γ was determined to be the molecule exhibiting this biological activity [4].

IFN-γ genes have been identified in all mammals, birds and even the fugu fish. However, no such distantly related gene has been reported in lower eukaryotes. The cDNA is not known to undergo alternate splicing and is approximately 1.2–1.3 kb in length. Although not definitively demonstrated, the gene is thought to be transcribed from both alleles, which puts it in contrast to interleukin (IL)-2 and -4. IFN-γ mRNA is induced by many extracellular signals, including soluble mediators (e.g. IL-2, -12, -15 and -18) and by crosslinking of cell surface receptors (e.g. CD3, CD16 and LY49 activating receptors), and when multiple signals are combined a synergistic induction of the mRNA is observed – a response seen with few other cytokines. While the major cell types that express IFN-γ are T cells (both CD4 and CD8), natural killer (NK) cells and NKT cells, numerous other cell types have been reported to express this mRNA, including macrophages, B cells and neutrophils. However, other than a report demonstrating that autocrine secretion of IFN-γ negatively regulates immature B cell homing [5], the physiological significance of IFN-γ expression by alternate cell types has remained elusive.

The human protein is 166 amino acids, of which 23 amino acids represent a hydrophobic signal sequence. While a viable and fertile IFN-γ null mouse has been created [6], as yet there have been no reports of humans born without an intact IFN-γ gene. In contrast, there are a number of reports of humans born with a defective IFN-γ receptor (IFN-γR) chain, and these defects sensitize these individuals to *Mycobacterium* and *Salmonella* infections (for a review, see [7]).

In this chapter, we shall cover a number of aspects of IFN-γ biology, including the epigenetic and molecular regulation of IFN-γ gene transcription and protein expression, the pathways involved in IFN-γ signaling, its role in innate and cellular immune maturation and function, and the effects of IFN-γ on tumor development and growth.

4.2
IFN-γ Gene Structure and Regulation

IFN-γ is a single-copy gene that has been identified in mammals, birds and fish. The human IFN-γ gene and cDNA sequences were initially reported in the early 1980s [8–10]. The human gene is located on chromosome 12 and the mouse gene on chromosome 10. Although sequence analysis of the IFN-γ gene from many species revealed that the gene structure is highly conserved, and consists of four exons and three introns (Fig. 4.1), the nucleotide sequence is less conserved and there is only 40% similarity at the amino acid level among human, mouse and rat. In humans, the IFN-γ coding sequence is invariant [11]; however, several single nucleotide polymorphisms (SNPs) exist in the promoter, intron 1 and 3′-untranslated regions (UTRs) [12–15]. Of these, the IFN-γ promoter SNP, which consists of a G to T transition at position −179, influences IFN-γ transcription through the creation of a potential AP-1 site and increased IFN-γ transcription in response to tumor necrosis factor (TNF)-α [12]. Another highly polymorphic CA microsatellite nucleotide repeat is located in the first intron, and is linked to altered IFN-γ production and the development of a variety of diseases, including rheumatoid arthritis [16] lung transplant allograft fibrosis [17] and acute graft-versus-host disease [18].

4.2.1
Transcriptional Regulation

Careful examination of the proximal promoter region (−300 to −1) of the mouse and human genes reveals an approximate 80% identity between species, suggesting that this region is significant for the regulation of IFN-γ gene transcription. Recently, two highly homologous and evolutionarily conserved noncoding sequence (CNS) elements were identified within the IFN-γ locus [19, 20]. The first CNS site is situated approximately 5 kb upstream of the IFN-γ transcriptional start site, and

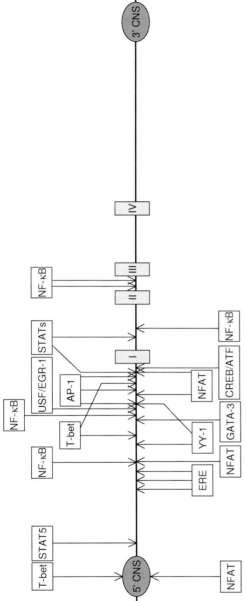

Fig. 4.1. Structure of the IFN-γ gene. The diagram depicts the IFN-γ gene structure. Exons I–IV are shown as boxed regions, while the 5′- and 3′-CNSs are represented as ovals. Putative and defined *trans*-acting factor binding sites are depicted by arrows along the promoter and intronic regions of the IFN-γ gene.

binds NFAT and T-bet, two transcription factors that are essential for IFN-γ expression and T helper (T$_h$) 1 cell development [21]. The second CNS is located approximately 18 kb downstream of the initiation codon. DNA–protein interactions in this region have not been extensively characterized; however, T-bet-dependent histone modifications have been mapped to this region [20]. Regulatory elements in the proximal IFN-γ promoter and intronic regions enhance IFN-γ expression, but do not influence its cell-specific expression. Thus, newly described distal CNSs may exist to facilitate high levels of cell- and tissue-specific expression of IFN-γ.

The proximal IFN-γ promoter and intervening sequences contain binding sites for many *trans*-activating regulatory factors that in part include T-bet, GATA-3, nuclear factor (NF)-κB, NF-AT, yin-yang-1 (YY-1), signal transducers and activators of transcription (STATs), Jun, AP-2, OCT-1, peroxisome proliferator-activated receptor (PPAR)-γ, and cAMP response element-binding protein (CREB)/activating transcription factor (ATF)-2. These factors act as enhancers with the exception of the negative regulatory factors YY-1 and PPAR-γ. YY-1 represses IFN-γ transcription by binding to several sites in the IFN-γ promoter and by competing with AP-1 for binding to overlapping sequence elements [22]. PPAR-γ repression activity localizes to the IFN-γ proximal promoter and interferes with c-Jun activation of the IFN-γ gene [23]. T-bet acts as an enhancer and regulates lineage specific expression of IFN-γ as demonstrated by its requirement for IFN-γ expression in CD4$^+$ derived T$_h$1 cells [21]. In contrast, CD8$^+$ T cells require both T-bet and Eomesodermin (Eomes), a T-bet paralog, for elevated cell-specific expression of IFN-γ. This is demonstrated by a severe reduction in IFN-γ expression in CD8$^+$ T cells from T-bet$^{-/-}$ mice that can be restored by overexpression of Eomes in the CD8$^+$ T cells from these animals [24, 25]. A number of putative T-bet response elements have been identified within the first 2300 nucleotides of the IFN-γ promoter [21, 26, 27]. Of these, several T-bet half sites within the first 300 bp upstream of the transcriptional start site appear to be most critical for T-bet-specific activation of IFN-γ expression [26] GATA-3 is a complex regulatory factor in that it suppresses IFN-γ expression in T$_h$ cells but enhances expression in natural killer (NK) cells [28, 29]. The mechanism by which GATA-3 suppresses IFN-γ expression does not depend on direct binding of GATA-3 to the IFN-γ promoter but instead acts through a GATA-3-mediated downregulation of STAT-4. An alternative pathway for GATA-3 upregulation of IFN-γ is unknown.

4.2.2
Epigenetic Regulation

Epigenetic changes that include chromatin remodeling, DNA methylation and histone acetylation provide stable and long-term transcriptional control of the IFN-γ gene locus. To identify regulatory regions of the IFN-γ gene, early chromatin remodeling studies examined the accessibility of DNase I to the IFN-γ gene. Regions of highly condensed chromatin containing nontranscribed DNA are not readily accessible to limited DNase I digestion. However, DNase I cleavage will occur at so-called hypersensitive sites (HSs) where the chromatin unwraps and DNA becomes accessible. Stimulated T cells, nonstimulated T cells and non-T cells have distinct patterns of IFN-γ DNase I hypersensitivity, suggesting that alterations in chromatin structure do influence a cells ability to produce IFN-γ [30, 31]. In addition, Rao et al. have demonstrated that new and permanent DNase I HSs develop in the IFN-γ gene upon T cell differentiation of naïve T_h cells into IFN-γ-producing $T_h 1$ cells. These HSs do not appear during T cell differentiation into $T_h 2$ cells that produce interleukin 4 (IL-4) *in lieu* of IFN-γ [32].

Recently, a chromatin conformation capture assay (3C assay) was used to examine *in vivo* alterations in the IFN-γ gene during T cell differentiation. It was found that after 5 days in either neutral, $T_h 1$ or $T_h 2$ culture conditions, the IFN-γ locus from the differentiated $T_h 1$ and $T_h 2$ cells had undergone conformational changes that were distinct not only from one another, but also from undifferentiated cells. Moreover, these changes were enhanced following a secondary restimulation by anti-CD3 [33]. Ultimately, the conformational changes reflected an IFN-γ gene structure that is more open in effector $T_h 1$ cells and highly compact in $T_h 2$ cells.

DNA methylation is a robust regulator of gene expression where the nucleotide cytosine is modified to 5-methylcytosine at CpG residues resulting in reduced or silenced gene expression directly by blocking DNA–protein interactions or indirectly by recruiting corepressors such as histone deacetylases. In contrast, DNA demethylation allows access of regulatory proteins to chromatin sequences to upregulate gene transcription. Several reports show that hypomethylation of the IFN-γ promoter between positions -200 and $+1$ contributes to IFN-γ gene transcription [34–37]. Furthermore, a *Sna*BI recognition site (TACGTA), which is highly conserved among all species, is a critical methylation target. Hypomethylation at this site correlates with high levels of IFN-γ expression in both T_h cell lines and in newly activated primary T cells [34, 35, 37, 38]. Dissection of the IFN-γ locus has revealed this region to be highly methylated in nonactivated T cells. However, antigen activation, cellular differentiation and proliferation of T cells result in a rapid demethylation of the IFN-γ locus [39]. Unlike T cells, NK cells produce IFN-γ without prior cellular preactivation. Direct cytokine stimulation of NK cells results in rapid transcriptional activation of the IFN-γ gene that coincides with a constitutively demethylated IFN-γ locus [40]. Thus, epigenetic modifications, in particular CpG methylation, identify fundamental molecular differences between T and NK cells with regard to their ability to produce IFN-γ.

Histone acetylation is a reversible modification that is associated with increased transcriptional gene activation. Using a T_h cell differentiation model, it was shown that T cell receptor stimulation of naïve T cells results in rapid acetylation of both IFN-γ and IL-4 loci irrespective of the $T_h 1$ and $T_h 2$ culture conditions [41]. Yet, continued culturing in $T_h 1$- or $T_h 2$-polarizing conditions leads to selective, specific and distinct acetylation patterns for the IFN-γ and IL-4 genes. An in-depth analysis of the IFN-γ locus shows that $T_h 2$ polarization results in hyperacetylation of at least 50 kb of DNA sequence upstream and downstream of the IFN-γ gene. Moreover, this extended pattern of histone acetylation is highly dependent on the $T_h 1$-polarizing cytokine IL-12 [42].

4.2.3
Post-transcriptional Regulation

Post-transcriptional regulatory control of IFN-γ expression is the least understood. Stabilization of IFN-γ mRNA expression [43] induced by the combination of IL-2 and -12 indicates that regulation of this gene is controlled by both transcriptional and post-transcriptional mechanisms. One proposed mechanism by which IL-12 controls IFN-γ mRNA expression is by nuclear sequestration of pre-existing IFN-γ transcripts that is relieved in response to secondary stimulation resulting in rapid nucleocytoplasmic shuttling of the IFN-γ mRNA with subsequent protein synthesis [44]. A different study found that IFN-γ mRNA can autoregulate its own translation through a pseudoknot located in its 5'-UTR. The RNA pseudoknot activates protein kinase PKR that in turn phosphorylates initiation factor eIF-2, which results in a strong reduction in mRNA translation [45]. Supporting evidence for IFN-γ post-transcriptional control is derived from IFN-γ reporter mice where the IFN-γ 3'-UTR sequence is replaced with a Yellow Fluorescent Protein (YFP) sequence, and NK and NKT cells from these mice are spontaneously fluorescent. Unlike wild-type mice, these mice show basal IFN-γ protein expression that in parallel displays constitutive IFN-γ mRNA accumulation [46].

The signaling pathways responsible for post-transcriptional control of IFN-γ are not entirely elucidated. Several studies demonstrate IL-12 and -18 signal through the mitogen-activated protein kinase (MAPK) p38 pathway to regulate IFN-γ expression at the transcriptional and post-transcriptional levels, with the post-transcriptional regulation mediated through sequence elements found within the 3'-UTR of the IFN-γ mRNA [47, 48]. Overall, post-transcriptional control of IFN-γ expression is not due to a single mRNA conformation or a single sequence element, but instead requires the involvement of both 5'- and 3'-UTRs to confer stimulus-specific control of IFN-γ at multiple locations within the cell.

IFN-γ is an important immunomodulator, and thus it is not surprising that the IFN-γ gene is highly conserved and regulated through complex and multifaceted mechanisms that include DNA/histone modifications, tissue-specific protein–DNA interactions and mRNA control. Much progress has been made in recent years to delineate the mechanisms underlying the regulation of IFN-γ expression. Future *in vivo* studies using a combination of transgenic, targeted DNA deletion and knockout approaches may allow a more complete understanding of how expression of this complex gene is regulated.

4.3
IFN-γ Signal Transduction

The antiviral, antigrowth and immunoregulatory effects IFN-γ exerts on cells were believed to be mediated by a specific set of genes induced by this cytokine. However, how IFN-γ delivered an intracellular signal from engaging the receptor at the

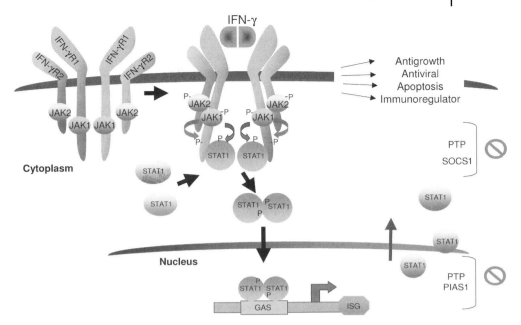

Fig. 4.2. Signal transduction of IFN-γ. IFN-γ binding to its receptor causes JAK-1 and -2 auto and *trans* tyrosine phosphorylation, respectively. Activated JAK-1 phosphorylates IFN-γR1 and STAT-1 becomes recruited to the receptor to be phosphorylated by JAK-1. Activated STAT-1 dissociates from the receptor, dimerizes with another STAT-1 and translocates to the nucleus to drive ISG expression. Inactivation of the JAK–STAT pathway is controlled by PTPs, SOCS-1 and PIAS-1. (This figure also appears with the color plates.)

cell surface to the nucleus to activate gene transcription was initially not readily understood. The primary signaling pathway activated by IFN-γ was uncovered in the early 1990s and since then has been known as the Janus kinase (JAK)–STAT pathway, as depicted in Fig. 4.2. In this section, we will present a description of our current understanding of the JAK–STAT signal transduction pathway and the significance of additional signaling cascades that lead to the biological effects of IFN-γ.

4.3.1
The JAK–STAT Signaling Pathway

IFN-γ signals primarily via the JAK–STAT transduction pathway to induce most of its biological activities. This involves the participation of a preassembled IFN-γR which consists of two ligand-binding IFN-γR1 chains each associated with one IFN-γR2 chain. IFN-γRs belong to the class II cytokine receptor family and are constitutively expressed on virtually all cells. The IFN-γR1 gene is located on human chromosome 6 and murine chromosome 10, and encodes a 90-kDa protein. The IFN-γR2 gene is located on human chromosome 21 and murine chromosome 16, and encodes a 62-kDa protein [49]. As both IFNγ-R chains lack intrinsic kinase and phosphatase activity, additional molecules are recruited for signal transduction. Two members of the family of Janus kinases, JAK-1 and -2, are constitutively associated with IFN-γR1 and -γR2 respectively. Biologically active IFN-γ is a noncovalent homodimer that upon binding to its receptor causes the intracellular domains of both receptors to open, recruit signaling molecules and then induce JAK-2 autophosphorylation and the subsequent JAK-1 transphosphorylation by JAK-2. Activated JAK-1 phosphorylates tyrosine residues 440 and 419 of each IFN-γR1 subunit in human and mouse, respectively, that serves as docking sites to recruit STAT-1. JAK-1 phosphorylates STAT-1 on Tyr701 that causes STAT-1 to dissociate from the receptor and form dimers via SH2 domain–phosphotyrosyl interactions. STAT-1 dimers translocate to the nucleus to bind DNA at γ-activated sequence (GAS) elements with the consensus sequence $TTCN_{(2-4)}GAA$ that is found in the promoters of IFN-inducible genes (ISG). Biochemical and genetic analysis support the significance of the JAK–STAT pathway in mediating IFN-γ signaling. Cells lines lacking IFN-γR, STAT-1, or JAK-1 or -2, when complemented with the missing gene, show restoration of IFN-γ mediated biological responses [49, 50]. Deficiency of either JAK-1 or -2 genes in mice is embryonic lethal [51, 52]. Although mice deficient in STAT-1 or IFN-γR are viable, they remain highly susceptible to viral, bacterial and parasitic infections, and are more prone to form tumors [53–56]. In addition, naturally occurring, heterozygous loss-of-function mutations in IFN-γR as well as in STAT-1 have been found in individuals who are susceptible to mycobacterial infection [57–63].

4.3.2
Activation of Alternate Signaling Pathways

Several signaling pathways, in addition to the JAK–STAT pathway, are activated by IFN-γ. Serine phosphorylation of STAT-1 at residue 727 is essential for maximal gene expression and occurs independently from STAT-1 tyrosine phosphorylation [64, 65]. This allows for the interaction of STAT-1 with MCM5 and BRCA1 to augment IFN-γ-mediated gene transcription [66]. Other studies have shown that stress, lipopolysaccharide (LPS), and inflammatory cytokines like TNF-β and IL-1 induce serine phosphorylation of STAT-1, and synergize with IFN-γ to enhance STAT-1-mediated gene activation [67]. Evidence to support the biological impact of STAT-1 serine phosphorylation in IFN-γ signaling comes from mice expressing a Ser727 to Ala STAT-1 mutation. These mice not only show defective STAT-1-mediated gene expression, but most importantly they fail to clear bacterial infections [68]. Identification of the kinase responsible for the serine phosphorylation of STAT-1 has remained elusive. Several studies indicate the involvement of the MAPK pathway in serine phosphorylation of STAT-1 [64]. IFN-γ activation of serine/threonine kinase p42/ERK-2 mediates serine phosphorylation of STAT-1 that when inhibited, impairs gene transcription [69, 70]. Raf-1, the upstream kinase responsible for the activation of ERK-2 not only is activated by IFN-γ, but also requires STAT-1 and JAK-1 to properly function [71, 72]. In addition, p38MAPK is also activated by IFN-γ, yet it remains unclear whether this kinase is directly involved in the serine phosphorylation of STAT-1 [67, 73, 74]. Work from several groups has also demonstrated that phosphatidyliositol-3-kinase (PI3K) and its effector kinase AKT [75, 76], protein kinase C δ [77] and calmodulin-dependent kinase (CAMKII) [78] are activated by IFN-γ, and play a role in the serine phosphorylation of STAT-1. Furthermore, IFN-γ also promotes the phosphorylation of the p65 subunit of NF-κB, thus increasing the transactivation potential of NF-κB [79]. More recently, the inhibitor of κB kinase (IKK) was reported to be required for the transcription of a subset of ISGs, that occurs independently of NF-κB activation and is downstream of STAT-1 tyrosine phosphorylation [80].

IFN-γ also activates a STAT-1-independent signaling pathway. Cells deficient in STAT-1 remain responsive to IFN-γ as these cells proliferate and are protected from apoptosis, even though they remain more susceptible to viral infection. In the absence of STAT-1, IFN-γ upregulates the expression of c-*myc* and c-*jun* – genes that are implicated in promoting cell growth and survival [81, 82]. Furthermore, IFN-γ also weakly induces the activation of STAT-3, which antagonizes the activities of STAT-1 and competes for binding to Tyr419 of IFN-γR1. However, in the absence of STAT-1, STAT-3 activation is stronger and prolonged in response to IFN-γ and drives the expression of genes such as suppressor of cytokine signaling (SOCS)-3 and C/EBP-δ whose promoters contain GAS elements [83].

4.3.3
Regulation of IFN-γ Signaling

Distinct mechanisms exist that regulate IFN-γ activation of the JAK–STAT pathway. For instance, IFN-γ contains a nuclear localization signal (NLS) in the C-terminus [84, 85], and a model of nuclear translocation of IFN-γ has been proposed wherein IFN-γ assists STAT-1 localization to the nucleus and binding to DNA to activate gene expression [86, 87]. As described, IFN-γ forms a complex with IFN-γR. Recruitment of STAT-1 to the complex occurs next via association with IFN-γR1, resulting in endocytosis of the receptor and translocation of IFN-γ:IFN-γR1:STAT-1 to the nucleus while IFN-γR2 is maintained on the cell surface. In contrast, formation of an IFN-γ:IFN-γR1 complex can lead to IFN-γR desensitization. IFN-γ binding to its receptor promotes the internalization of IFN-γR1 resulting in downregulation of surface receptor expression. The internalized receptor can enter the endosomal pathway where IFN-γR1 either dissociates from IFN-γ and is recycled to the cell surface, while IFN-γ is degraded [88, 89] or IFN-γR1 is targeted for degradation [90].

Another key regulatory control involves the IFN-γ inducible gene, SOCS-1 [91].

IFN-γ creates its own negative feedback mechanism with SOCS-1, as it associates with activated JAK-2 to inhibit its kinase activity [92]. For instance, overexpression of SOCS-1 in cells results in loss of IFN-γ responsiveness, and mice deficient in SOCS-1 develop a complex fatal neonatal complex and show hypersensitivity to IFN-γ [93–95]. The mechanism by which SOCS-1 regulates STAT-1 activation was recently elucidated. In addition to phosphorylation of Tyr419 of the IFN-γR1 in mice, IFN-γ also phosphorylates Tyr441. Mutation of this residue inhibits SOCS-1 from binding to IFN-γR1 and JAK-2, and prolongs STAT-1 activation after IFN-γ stimulation [96].

Protein tyrosine phosphatases (PTP) also play a crucial role in downregulating IFN-γ signaling. At the receptor level, SHP-2 is a PTP that acts as a negative regulator of the JAK–STAT pathway by inducing JAK dephosphorylation, thus inhibiting further STAT-1 phosphorylation [97]. Cells deficient in SHP-2 display enhanced STAT-1 binding to DNA, augmented antigrowth activity to IFN-γ and increased caspase-1 expression [97]. In contrast, inactivation of STAT-1 by distinct PTPs takes place in the nucleus [98, 99]. Recently, TC-45, the nuclear form of TC-PTP, was identified as a nuclear PTP responsible for the dephosphorylation of STAT-1 [100]. TC-PTP directly dephosphorylates STAT-1 and, to a lesser extent, STAT-3. As demonstrated in cell lines deficient in TC-PTP, dephosphorylation of IFN-γ induced tyrosine phosphorylated STAT-1 is defective, but can be restored when cells are complemented with TC-PTP [100].

Protein inhibitor of activated STAT (PIAS)-1 also negatively regulates IFN-γ-mediated STAT-1 signaling. Support for this observation and its significance in IFN-γ-mediated immune responses come from mice deficient in PIAS-1 wherein mice show enhanced antimicrobial activity [101]. It was proposed that PIAS-1 selectively regulates the expression of a subset of ISGs by interfering with STAT-1 binding to the gene promoter. Furthermore, PIAS proteins can also function as regulators of small ubiquitin-like modifier (SUMO) modification. Like ubiquitination, SUMO modification targets proteins for degradation. PIAS1 was recently reported to enhance the SUMO modification of STAT-1. Although Lys703 of STAT-1 was mapped as the site for SUMO modification, mutation of this residue nevertheless abolished STAT-1 SUMO modifications, but did not abrogate the inhibitory activity of PIAS-1 [102].

Most recently, arginine methylation of STAT-1 was discovered as another post-translational modification required for IFN-mediated gene activation [103]. STAT-1 arginine methylation appears to control the rate of STAT-1 dephosphorylation. Inhibition of STAT-1 arginine methylation not only leads to defective STAT-1 tyrosine dephosphorylation, but also abrogates gene transcription. This seems to occur by the enhanced association of PIAS1 with activated STAT-1, that in turn prevents TC-PTP from binding to STAT-1 to be inactivated [104].

In summary, IFN-γ activates and orchestrates several signaling cascades that together lead to the manifestation of the biological effects of this cytokine. Perturbation in any of these signaling events leads to cytokine signaling de-regulation, resulting in either a gain or loss of function that can be deleterious to the host.

4.4
IFN-γ in T_h Cell Development

IFN-γ is an important mediator in the development of host defense where both innate and adaptive immunity become integrated. Adaptive immunity is heavily dependent on the generation of antigen-specific $CD4^+$ and $CD8^+$ T cells to fight disease. $CD4^+$ T cells act as a bridge that helps shape the development of an immune response and have been designated T_h cells. Mosmann et al. classified T_h cells into two subgroups, T_h1 and T_h2, based on their cytokine production profiles [105]. T_h1 cells secrete IFN-γ, IL-2, TNF-α and TNF-β, all of which affect cellular immunity, whereas T_h2 cells produce IL-4, -5, -6 and -13 that in turn regulate B cell function affecting humoral immunity. Naïve $CD4^+$ T cells, the T_h precursor, can differentiate into either T_h1 or T_h2 effector cells. This decision is not only determined by T cell receptor (TCR) engagement of antigen bound to major histocompatibility complex (MHC) class II molecules on the surface of antigen-presenting cells (APCs), but is also strongly influenced by concentration of antigen, the nature of the APC and the existent cytokine milieu. Importantly, numerous laboratories have demonstrated that IFN-γ and IL-12 primarily determine T_h1 development, whereas IL-4 promotes T_h2 development. Furthermore, IFN-γ secreted by APCs, NK, $CD8^+$ and $CD4^+$ T cells can directly influence naïve $CD4^+$ T cells toward a T_h1 phenotype by establishing a positive feedback mechanism to amplify the T_h1 response and inhibiting production of IL-4 by T_h2 cells. Moreover, among the repertoire of APCs, dendritic cells (DCs) have been implicated as the primary type of APC involved in presenting MHC class II antigenic peptides to the naïve $CD4^+$ T cells, based on their potent antigen-presenting capabilities, high MHC class II expression and are localized to sites where they can encounter naïve $CD4^+$ T cells [106]. Specific subsets of DCs produce factors that stimulate production of cytokines such as IL-12 (important for T_h1) and IL-6 (T_h2), and recently it has been shown that APCs (DCs and macrophages) can produce IFN-γ – a potent signal for T_h1 development [107].

4.4.1
Signaling Pathways Involved in T Cell Development

Signaling pathways that promote T_h differentiation are activated by specific cytokines or upon APCs interaction with naïve T cells. One of these pathways involves members of Tec family of nonreceptor tyrosine kinases. Txk in humans and Rlk, the mouse homolog, are implicated in T_h1 development. Txk expression was shown to be restricted to T_h0 and T_h1 cells, and transfection of Txk transactivated an IFN-γ promoter construct along with elevating endogenous IFN-γ levels [108, 109]. Conversely, Itk has been implicated in T_h2 development; in particular, the regulation of IL-4 with no visible effects on IFN-γ [110]. This role for the Tec family in T cell differentiation is supported by *in vivo* studies of infectious disease models [111]. Studies looking at JNK-1 and -2, p38MAPK, and GADD45 have implicated the MAPK pathways in T_h1 signaling (specifically IL-12 mediated) and IFN-γ production with little to no effects on T_h2 cytokine production [112].

The significance of IFN-γ in T_h1-mediated immune responses is exemplified in IFN-γ, -γR1 and -γR2, and STAT-1 knockout mice. These mice have impaired immune responses and increased susceptibility to microbial pathogens and viruses [6, 53–55], also seen in humans with natural occurring mutations in the IFN-γR signaling pathway [61–63]. One of the major roles for IFN-γ in the T_h1 response is activation of macrophages resulting in increased phagocytosis, upregulation of MHC class I and II, and production of IL-12, nitric oxide (NO) and superoxides to eliminate intracellular pathogens [113]. Mice infected with *Leishmania major* mount a T_h1 response; however, administration of anti-IFN-γ antibodies antagonized this response and instead generated a T_h2 response that failed to clear the infection [114, 115]. To further support the role of IFN-γ in T_h1 development, *in vitro* studies showed IFN-γ-stimulated naïve CD4$^+$ T cells differentiated along the T_h1 pathway in the absence of IL-12 [116]. Furthermore, CD4$^+$ T cells deficient in IFN-γR2 show impaired T_h1 responses to anti-CD3/CD28 stimulation *in vitro* and *in vivo* [117]. Through these studies, IFN-γ has been shown to be essential for IL-12 induced T_h1 development, partly through the maintenance of IL-12Rβ2 expression [118–120]. The absence of IFN-γ stabilizes T_h2 development, in part through the downregulation of the IL-12Rβ2 chain, and hence the loss of IL-12 responsiveness, and addition of IFN-γ into developing T_h2 cultures restores IL-12Rβ2 expression and IFN-γ secretion by these cells [119].

IFN-γ signaling also promotes the activation of the transcription factor T-bet, a member of the T-Box family of transcription factors. T-bet regulation in T_h1 development is controlled primarily by IFN-γ signaling through STAT-1 which causes chromatin remodeling and activation of IFN-γ and IL-12Rβ2 gene expression, thus establishing a positive feedback loop and commitment to T_h1 development [21, 121, 122]. Simultaneously, a negative feedback loop is also created in part by T-bet resulting in inhibition of T_h2 development/cytokine production and downregulation of IL-4R activation through the inhibition of the T_h2 transcription factor GATA-3 [123–126]. T-bet inhibition of GATA-3 is due to Itk kinase-mediated interaction of T-bet with GATA3 interfering with GATA-3s ability to bind its target DNA [127]. Conversely, activation of GATA-3 during T_h2 lineage commitment through IL-4/STAT-6 activation can directly or indirectly establish a negative feedback loop to inhibit IFN-γ secretion and subsequently IL-12R 2 expression [123, 128]. Overall, after TCR engagement, IFN-γ in the cellular environment appears to be a vital trigger to polarize naïve CD4$^+$ T cells toward T_h1 development. IFN-γ pathway activation upregulates T-bet, which in turn establishes IFN-γ production in the CD4$^+$ T cell and creates a positive feedback loop, where IFN-γ up regulates IL-12Rβ2 expression, establishes IL-12 responsiveness and the cell becomes committed to the T_h1 lineage pathway. Finally, IFN-γ establishes a negative feedback loop to inhibit T_h2 development stabilizing the T_h1 lineage.

Similar to T cells, human NK cells undergo a linear development passing through stages that resemble T_h1 and T_h2 cytokine phenotypes [129]. Furthermore, it has recently been shown that IFN-γ directly affects the IL-4-dependent proliferation of these NK cells [130]. In summary, IFN-γ, through its expression and direct effects on key lymphoid cell populations, represents a critical bridge between the innate and adaptive host immune response.

4.5 IFN-γ and DCs

As discussed in this chapter, IFN-γ is involved in the modulation of MHC class I and II cell surface expression, enhancement of cytotoxic killing by antigen-specific T cells, and activation of macrophages. In this section, we discuss the role of IFN-γ following T and NK cell activation after their encounter with professional APCs, specifically DC subsets.

DCs, widely distributed in lymphoid and nonlymphoid tissue, are a complex, heterogeneous group of multifunctional APCs that comprise an essential component of the immune system [131]. DC differentiation, governed by several cytokines, results in the development of myeloid DCs (MDCs), lymphoid DCs (LDC) and Langerhans cells (LCs). In addition, plasmacytoid DCs reside in various organs that are specialized to secrete high levels of type I IFN in response to viral infection [132]. Thus DCs have become recognized as being an essential component of the both the host innate and adaptive immune response.

4.5.1 IFN-γ and T Cell–DC Crosstalk

DCs, by way of cytokine secretion, shape the phenotype and functional properties of T_h cells; in contrast, T cells profoundly affect the function of DCs. For example, DCs present processed immunogenic peptides on MHC class II to precursor T_h cells [133], which are activated by costimulatory signals (e.g. CD40, CD86) delivered from DCs in lymphoid organs. The T cell/DC dialog via CD40/CD40 ligand [134] triggers DC maturation [135] followed by IL-12 and -18 secretion, which promotes IFN-γ production in pre-T_h1 cells [134, 136]. T_h1-derived IFN-γ then signals through STAT-1, which induces the activation of the T-box transcription factor T-bet, which is a critical T_h1/IFN-γ inducer (described above). Even though T-bet acts before IL-12 signaling, DC-derived IL-12 activates STAT-4 and stabilizes IFN-γ expression in terminally differentiated T_h1 cells. Furthermore, T cell-derived IFN-γ upregulates the expression of MHC class I and II, induces the production of IL-6, and arrests IL-8 production in DCs and macrophages [137]. Moreover, induced IFN-γ plays an essential role in enabling the functional priming of already committed T_h1 cells. IFN-γ-primed T_h1 cells preferentially express several unique molecules such as IL-12Rβ2 chain, IL-18 receptors, P-selectin glycoprotein ligand-1, Chandra, CXCR3 and the CXCR5 chemokine receptors that ultimately enable them to migrate to B cell follicles [138–140] and interact with naïve antigen-

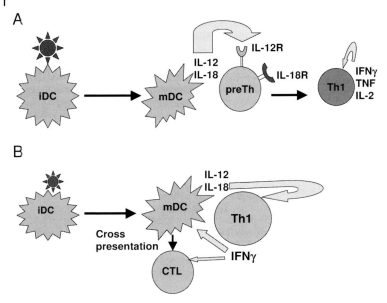

Fig. 4.3. (A) DCs capture, internalize and process pathogens in endosomal compartments. Their primary sequence is then presented as small peptides via MHC class II molecules to naive T cells. In parallel, DCs can be induced to secrete IL-12 and -18, which skew precursor T cells toward T_h1 polarization. (B) Immature DCs also internalize immunogenic antigens, but this cannot be cross-presented via MHC class I to CD8$^+$ T cells. IFN-γ secreted by T cells can significantly improve this process where mature DCs can then cross-present immunogenic peptides to rare CD8$^+$ T cells. In addition, IFN-γ upregulates the expression of IL-2 receptors that facilitate the expansion of CTL. (This figure also appears with the color plates.)

presenting B cells [141]. As summarized in Fig. 4.3(A), IFN-γ production is indirectly influenced by crosstalk between DCs and T cells. DC presentation of antigen and release of IL-12 and IL-18 to T cell subsets induce a T_h1 stimulatory response with IFN-γ production. The release of this T cell-derived IFN-γ then positively activates DCs and fuels a robust T_h1 immune response.

Generally, cross-presentation by professional APCs requires that immunogenic antigens gain access to the machinery that is strongly involved in endogenous MHC class I processing. Studies have focused on the unique capacity of DCs to process soluble exogenous immunogens and their subsequent presentation to killer T lymphocytes as a method to enhance the antitumor response in the host. Recently it has been demonstrated that immature LCs are completely deficient in cross-presentation of immunogens [142]. Exogenous IFN-γ compensates for this deficiency in LCs by allowing LCs exposed to peptides derived from exogenous tumor antigen to cross present to CD8$^+$ T cells. Thus, when antigen-specific CD8$^+$ T cells are activated by DCs, the ensuing release of IFN-γ, by CD4$^+$ and CD8$^+$ T cells promotes antigen-specific T cell proliferation and recruitment of other APCs to the site of infection as summarized in Fig. 4.3(B). This feature of IFN-γ may be important to limit other cell types in the microenvironment where DCs can induce specific cell-mediated immune responses.

4.5.2
Signals through Toll-like Receptors (TLRs) Activate DCs and Influence IFN-γ Expression

Sensing and recognition of microbial subunits by innate molecular components directly triggers various immune responses that regulate adaptive immune response. Pattern recognition receptors (PRR) known collectively as TLRs signal through myeloid differentiation factor 88 (MyD88) in DC subsets. Signaling through one or a combination of certain TLRs activates and biases antigen-presenting DCs to prime either T_h1 or T_h2 responses as depicted in Fig. 4.4. It leads to differential expression of phenotypic markers and cytokines that ultimately modulate early immune responses. The ultimate release of IFN-γ by DC-stimulated T cells is influenced by both the type of pathogen and the type of engaged DC. Accordingly, it has recently been shown that DC subsets express distinct repertoires of TLRs that enable them, when engaged by a pathogen, to become activated and subsequently secrete cytokines. For example, studies show that *Bordetella pertussis* binds TLR-4 and induces DC maturation with IL-12 secretion, leading to IFN-γ expression by T cells, NKT cells and NK cells [143]. In contrast, a TLR-4 deficiency enhanced IL-10 release and established a microenvironment that was favorable for proliferation of T regulatory (T_r) and T_h2 cells [144]. In another set of experiments, activation of TLR-3, -4, -7 and -9, but not -2, on DCs, through the use of TLR agonists, activated

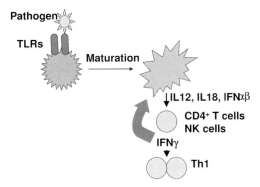

Fig. 4.4. Recognition of conserved subunits of infectious pathogens by TLRs induces DC maturation resulting in critical pro-inflammatory cytokine production (e.g. IL-12 or -18, or IFN-α/β). Such cytokines induce the expression of IFN-γ in T and NK cells. This IFN-γ can affect further DC and T activation. (This figure also appears with the color plates.)

naive and memory T cells and this T cell activation was directly mediated by IFN-α/β, and IL-12 and -18 secreted by DCs and IFN-γ produced by NK cells [145]. Furthermore IFN-α/β was found to be critical for T cell bystander activation and production of IFN-γ by NK cells *in vivo* [145].

Taken together, these data indicate that triggering of multiple receptors on DCs associated with the innate immune system influences the type and magnitude of immune response to a pathogen, and the subsequent triggering of IFN-γ gene expression.

4.6
IFN-γ – Role in Tumor Development and Growth

IFN-γ was approved for clinical use in 1986 as an antitumor and antiviral therapeutic agent. Despite the remarkable antitumor responses reported in various animal studies [146–151], data obtained from clinical trials of IFN-γ have been thus far disappointing. As described above, IFN-γ signals primarily through the JAK–STAT pathway and experimental studies performed in mice deficient in IFN-γR α subunit (IFN-γR1), IFN-γ or STAT-1 demonstrate the crucial role of IFN-γ in mediating rejection of transplantable tumors and in preventing primary tumor development [149, 151].

The diversity of mechanisms through which IFN-γ exerts its antitumor activity is quite broad, and includes a direct toxic effect on tumor cells, inhibition of angiogenesis during tumor development, and induction of a potent innate and adaptive immune response to the tumor cells. In this section, we review the most important mechanisms wherein IFN-γ participates as an essential mediator of the antitumor immune response.

4.6.1
IFN-γ in Tumor Growth and Survival

STAT-1, a component of the IFN-γ signaling cascade, has been proposed to act as a tumor suppressor gene and is required for the direct apoptotic effects that IFN-γ exerts on a wide variety of tumor cell lines [147, 152–155]. For instance, while a number of human tumor cell lines express Fas receptor, most of them are resistant to Fas-mediated cell death. However, IFN-γ can sensitize tumor cells to rapidly undergo apoptosis via activation of the Fas signaling pathway [156, 157]. This effect has been proposed to occur by the synergistic actions of caspase-1 activation and the induction of the IFN-γ-inducible gene IFN consensus sequence-binding protein (ICSBP) [156]. Another group has reported that cleavage of the antiapoptotic protein Bcl-2 can be detected in apoptotic melanoma cells due to IFN-γ mediated up-regulation of Fas expression [158]. IFN-γ can also restore TNF-related apoptosis-inducing ligand (TRAIL)-induced apoptosis in resistant neuroblastoma (NB) cells by upregulating caspase-8 expression [159]. IFN-γ-mediated tumor apoptosis is not the only mechanism by which this cytokine restricts tumor growth. It has also been postulated that IFN-γ can exert direct antiproliferative effects on a wide variety of tumor cells [147, 151, 152, 155]. Two molecular models have been proposed for this antiproliferative effect. One model is based on the STAT-1-dependent activation of the cyclin-dependent kinase inhibitors, CDKIs (e.g. p21$^{\text{WAF1/CIP1}}$) [152], whereas the other demonstrates CDKI independence [155]. Furthermore, it has been demonstrated that a continuous interaction of IFN-γ with its receptor is necessary for sustained expression of p21 and the transcription factor IRF-1, whose expression is associated with growth inhibition in ovarian cancer cells [160].

4.6.2
Inhibition of Angiogenesis by IFN-γ

Establishment of solid tumors *in vivo* requires the formation of new blood vessels in order to provide a continuous blood supply and this process, known as angiogenesis, has become an important target for the treatment of solid tumors. In this context, IFN-γ induced antiangiogenesis represents another mechanism of tumor rejection [161–163]. It remains unresolved, however, whether IFN-γ inhibits angiogenesis by acting directly on endothelial cells or indirectly through activity on other nonhematopoietic cells in the tumor stroma. Another controversial issue is the cell source of IFN-γ in the antiangiogenic response. Qin et al. showed that in the J558L myeloma tumor model, depletion of CD8$^+$, but not CD4$^+$, T cells in immunized mice restored blood vessel formation in the tumor [163], while other studies have revealed the importance of CD4$^+$ T cells in the IFN-γ-dependent inhibition of tumor angiogenesis [161]. IFN-γ produced by NK and NKT cells can inhibit murine endothelial cell proliferation *in vitro*, and depletion of NK cells caused a significant, but partial, inhibition of tumor growth and angiogenesis *in vivo* [162]. Interestingly, the *in vivo* efficacy of CD8$^+$ cells in tumor rejection seems to correlate with IFN-γ production rather than cytolytic activity as demonstrated in perforin and Fas ligand null mice [164, 165].

The timely recruitment of T cells to the tumor site and their IFN-γ production appears to be crucial for the inhibition of angiogenesis in the tumor microenvironment. In naïve mice, tumor-specific immunity develops concomitant with tumor growth. Before T cells are recruited to the primary tumor site, they must be primed with tumor antigen and this process does not always develop early enough to prevent rapid tumor growth; therefore, angiostasis might be ineffective. Strong support for this model came from experiments conducted in mice bearing a 1- to 2-week-old tumors. Mice rechallenged with the same tumor in a distant site showed continuous growth of the primary tumor, while rejection of the secondary tumor and this rejection occurred in a T cell-dependent manner. Overall, these observations support the model of IFN-γ-induced angiostasis as a general mechanism involved in tumor rejection, where T and/or NK cells can participate, but do not act exclusively, as tumor cell killers.

4.5.3
Role of IFN-γ in Promoting Immune Responses against Tumors

IFN-γ is recognized as an important mediator of immunosurveillance. The concept of "immunosurveillance" represents only one phase of the broader term better known as "cancer immunoediting". This concept is based on the notion that the immune system continuously defends the host against tumor development. This process consists of three phases: elimination, equilibrium and escape. In some instances, the immune system of a normal individual can promote tumor development because tumor cells with reduced immunogenicity manage to escape immune recognition and death [166]. For example, tumors overexpressing a dominant-negative form of the IFN-γR1 display enhanced tumorigenicity and reduced immunogenicity when transplanted into naïve syngeneic hosts [167]. In addition, mice deficient in IFN-γR1 and STAT-1, in contrast to their wild-type counterparts, are far more sensitive to the tumor-inducing capacity of methylcholanthrene (MCA) as these mice develop more tumors and have a shortened tumor latency period [149]. In this context, although IFN-γ contributes to the prevention of tumor development, immunoediting can sometimes be deleterious to the host because it selects for tumor variants that have adapted to survive in an immunocompetent host.

NK and NKT cells are the earliest producers of IFN-γ [168, 169]. These cells proliferate, produce cytokines and acquire killing competency from exposure to IFN-γ and IL-12 [170]. Acting as early suppliers of IFN-γ, they bridge the initiation of a general antitumor innate immune response to the specific activation of cytolytic T lymphocytes (CTL) [169]. IFN-γ is a powerful activator of macrophages [168, 171, 172] and these activated macrophages provide activation signals to T cells. As described earlier, T cell activation is dependent on the engagement of the TCR by antigen presentation in complex with MHC class II molecules displayed on macrophages. Expression of MHC class II molecules is low on resting macrophages, but is upregulated after exposure to IFN-γ [172]. Incubation of macrophages with IFN-γ also enhances respiratory burst and NO secretion, and upregulates expression of cytotoxic ligands of the tumor necrosis family, including TNF-α, Fas ligand and TRAIL [171, 172]. IFN-γ-mediated recruitment and activation of macrophages to the tumor site has been demonstrated in a rat glioma model [173], and is also proposed to occur in humans as one mechanism in the host antitumor response [174, 175]. IFN-γ facilitates the trafficking of leukocytes to the tumor site by inducing the expression of adhesion molecules, including intercellular adhesion molecule (ICAM)-1, vascular cell adhesion molecule (VCAM)-1 and the lymphocyte function-associated antigen (LFA)-1 on the surface of the vascular endothelium, and inflammatory leukocytes. Additionally, IFN-γ directs gene induction of chemoattractants including MIG and IP-10, chemokines that participate in the recruitment of activated T cells to sites of tumor growth and inflammation [113, 176].

The importance of the IFN-γ signaling pathway in establishing a protective adaptive antitumor immune response has been addressed by different laboratories

[177–179]. Shankaran et al. have shown a correlation between lack of expression of antigens in IFN-γ-insensitive tumor cells lines (RAD.gR28 and 30) and poor induction of a protective adaptive antitumor immune response. When RAD.gR38 cells were transfected with molecules that participate in the antigen presentation process (i.e. TAP-1 and H-2Kb) and then injected into immunocompetent mice, tumor rejection was complete. This process required the participation of both CD4$^+$ and CD8$^+$ T cells since tumor rejection was not evident in RAG-2$^{-/-}$ mice that are devoid of T cells [179]. Furthermore, it has been demonstrated that T cells derived from immunized STAT-1-deficient mice produced half the levels of IFN-γ and lacked cytolytic activity when compared with wild-type mice [177]. Other groups have also confirmed these data using STAT-6-deficient mice where T$_h$2 development is blocked and the T$_h$1 immune response is exacerbated [178]. In this model, tumor rejection was accompanied by enhanced tumor-specific IFN-γ production and CTL activity [178]. These results suggest that IFN-γ plays an important role in directing the appropriate T$_h$1/T$_h$2 balance during the development of an effective anti-tumor immune response.

As summarized in this section, IFN-γ can mediate antitumor immunity in mouse models by diverse mechanisms; however, immunotherapeutic strategies utilizing IFN-γ in human cancer therapy remain elusive. Thus, the role of IFN-γ as a critical player in preventing tumor growth and mediating tumor destruction needs to be more thoroughly analyzed in order to facilitate its future therapeutic applications.

4.7 Summary

As discussed in this chapter, IFN-γ has multiple biological features that demonstrate its importance in host innate and adaptive immune responses. The expression of IFN-γ can affect all host cells and as prolonged expression is deleterious, mechanisms are in place for the transcriptional and post-transcriptional regulation of this gene. The width breath of IFN-γ biological activity is perhaps best highlighted by the fact that IFN-γ gene expression can be induced in NK and T cells by many diverse signals, and this induction is controlled by regulatory elements that have been highly conserved through evolution. The fact that Mother Nature has seen fit to conserve the nucleic acid regulatory elements that control its expression more strongly than the primary amino acid sequence suggest that maintenance of the regulation of IFN-γ gene expression is essential to the development of host immunity to prevent cancer and infectious disease. Thus, IFN-γ has evolved to represent a critical element in uniting the innate and adaptive immune response.

References

1 E. F. Wheelock, *Science* **1965**, *149*, 310.
2 W. E. I. Stewart, *Nature* **1980**, *286*, 110.
3 B. R. Bloom, B. Bennett, *Science* **1966**, *153*, 80–82.
4 C. F. Nathan, H. W. Murray, M. E. Wiebe, B. Y. Rubin, *J Exp Med* **1983**, *158*, 670–689.
5 L. Flaishon, R. Hershkoviz, F. Lantner, O. Lider, R. Alon, Y. Levo, R. A. Flavell, I. Shachar, *J Exp Med* **2000**, *192*, 1381–1388.
6 D. K. Dalton, S. Pitts-Meek, S. Keshav, I. S. Figari, A. Bradley, T. A. Stewart, *Science* **1993**, *259*, 1739–1742.
7 S. D. Rosenzweig, S. M. Holland, *Curr Opin Pediatr* **2004**, *16*, 3–8.
8 R. Devos, H. Cheroutre, Y. Taya, W. Fiers, *J Interferon Res* **1982**, *2*, 409–420.
9 P. W. Gray, D. V. Goeddel, *Nature* **1982**, *298*, 859–863.
10 Y. Taya, R. Devos, J. Tavernier, H. Cheroutre, G. Engler, W. Fiers, *EMBO J* **1982**, *1*, 953–958.
11 C. Hayden, E. Pereira, P. Rye, L. Palmer, N. Gibson, M. Palenque, I. Hagel, N. Lynch, J. Goldblatt, P. Lesouef, *Clin Exp Allergy* **1997**, *27*, 1412–1416.
12 J. H. Bream, A. Ping, X. Zhang, C. Winkler, H. A. Young, *Genes Immun* **2002**, *3*, 165–169.
13 C. Chevillard, S. Henri, F. Stefani, D. Parzy, A. Dessein, *Eur J Immunogenet* **2002**, *29*, 53–56.
14 S. Henri, F. Stefani, D. Parzy, C. Eboumbou, A. Dessein, C. Chevillard, *Genes Immun* **2002**, *3*, 1–4.
15 V. Pravica, C. Perrey, A. Stevens, J. H. Lee, I. V. Hutchinson, *Hum Immunol* **2000**, *61*, 863–866.
16 A. Khani-Hanjani, D. Lacaille, D. Hoar, A. Chalmers, D. Horsman, M. Anderson, R. Balshaw, P. A. Keown, *Lancet* **2000**, *356*, 820–825.
17 M. Awad, V. Pravica, C. Perrey, A. El Gamel, N. Yonan, P. J. Sinnott, I. V. Hutchinson, *Hum Immunol* **1999**, *60*, 343–346.
18 J. Cavet, A. M. Dickinson, J. Norden, P. R. Taylor, G. H. Jackson, P. G. Middleton, *Blood* **2001**, *98*, 1594–1600.
19 D. U. Lee, O. Avni, L. Chen, A. Rao, *J Biol Chem* **2004**, *279*, 4802–4810.
20 M. Shnyreva, W. M. Weaver, M. Blanchette, S. L. Taylor, M. Tompa, D. R. Fitzpatrick, C. B. Wilson, *Proc Natl Acad Sci USA* **2004**, *101*, 12622–12627.
21 S. J. Szabo, S. T. Kim, G. L. Costa, X. Zhang, C. G. Fathman, L. H. Glimcher, *Cell* **2000**, *100*, 655–669.
22 J. Ye, M. Cippitelli, L. Dorman, J. R. Ortaldo, H. A. Young, *Mol Cell Biol* **1996**, *16*, 4744–4753.
23 R. Cunard, Y. Eto, J. T. Muljadi, C. K. Glass, C. J. Kelly, M. Ricote, *J Immunol* **2004**, *172*, 7530–7536.
24 B. M. Sullivan, A. Juedes, S. J. Szabo, M. von Herrath, L. H. Glimcher, *Proc Natl Acad Sci USA* **2003**, *100*, 15818–15823.
25 E. L. Pearce, A. C. Mullen, G. A. Martins, C. M. Krawczyk,

A. S. Hutchins, V. P. Zediak, M. Banica, C. B. DiCioccio, D. A. Gross, C. A. Mao, H. Shen, N. Cereb, S. Y. Yang, T. Lindsten, J. Rossant, C. A. Hunter, S. L. Reiner, *Science* **2003**, *302*, 1041–1043.

26 J. Y. Cho, V. Grigura, T. L. Murphy, K. Murphy, *Int Immunol* **2003**, *15*, 1149–1160.

27 M. Soutto, F. Zhang, B. Enerson, Y. Tong, M. Boothby, T. M. Aune, *J Immunol* **2002**, *169*, 4205–4212.

28 O. Kaminuma, F. Kitamura, N. Kitamura, M. Miyagishi, K. Taira, K. Yamamoto, O. Miura, S. Miyatake, *FEBS Lett* **2004**, *570*, 63–68.

29 S. I. Samson, O. Richard, M. Tavian, T. Ranson, C. A. Vosshenrich, F. Colucci, J. Buer, F. Grosveld, I. Godin, J. P. Di Santo, *Immunity* **2003**, *19*, 701–711.

30 K. J. Hardy, B. M. Peterlin, R. E. Atchison, J. D. Stobo, *Proc Natl Acad Sci USA* **1985**, *82*, 8173–8177.

31 K. J. Hardy, B. Manger, M. Newton, J. D. Stobo, *J Immunol* **1987**, *138*, 2353–2358.

32 S. Agarwal, A. Rao, *Immunity* **1998**, *9*, 765–775.

33 E. R. Eivazova, T. M. Aune, *Proc Natl Acad Sci USA* **2004**, *101*, 251–256.

34 P. R. Falek, S. Z. Ben Sasson, M. Ariel, *Cytokine* **2000**, *12*, 198–206.

35 D. R. Fitzpatrick, K. M. Shirley, L. E. McDonald, H. Bielefeldt-Ohmann, G. F. Kay, A. Kelso, *J Exp Med* **1998**, *188*, 103–117.

36 C. M. Tato, G. A. Martins, F. A. High, C. B. DiCioccio, S. L. Reiner, C. A. Hunter, *J Immunol* **2004**, *173*, 1514–1517.

37 H. A. Young, P. Ghosh, J. Ye, J. Lederer, A. Lichtman, J. R. Gerard, L. Penix, C. B. Wilson, A. J. Melvin, M. E. McGurn, *J Immunol* **1994**, *153*, 3603–3610.

38 J. A. Mikovits, H. A. Young, P. Vertino, J. P. Issa, P. M. Pitha, S. Turcoski-Corrales, D. D. Taub, C. L. Petrow, S. B. Baylin, F. W. Ruscetti, *Mol Cell Biol* **1998**, *18*, 5166–5177.

39 J. J. Bird, D. R. Brown, A. C. Mullen, N. H. Moskowitz, M. A. Mahowald, J. R. Sider, T. F. Gajewski, C. R. Wang, S. L. Reiner, *Immunity* **1998**, *9*, 229–237.

40 D. R. Fitzpatrick, K. M. Shirley, A. Kelso, *J Immunol* **1999**, *162*, 5053–5057.

41 O. Avni, D. Lee, F. Macian, S. J. Szabo, L. H. Glimcher, A. Rao, *Nat Immunol* **2002**, *3*, 643–651.

42 W. Zhou, S. Chang, T. M. Aune, *Proc Natl Acad Sci USA* **2004**, *101*, 2440–2445.

43 J. Ye, J. R. Ortaldo, K. Conlon, R. Winkler-Pickett, H. A. Young, *J Leukoc Biol* **1995**, *58*, 225–233.

44 D. L. Hodge, A. Martinez, J. G. Julias, L. S. Taylor, H. A. Young, *Mol Cell Biol* **2002**, *22*, 1742–1753.

45 Y. Ben Asouli, Y. Banai, Y. Pel-Or, A. Shir, R. Kaempfer, *Cell* **2002**, *108*, 221–232.

46 D. B. Stetson, M. Mohrs, R. L. Reinhardt, J. L. Baron, Z. E. Wang, L. Gapin, M. Kronenberg, R. M. Locksley, *J Exp Med* **2003**, *198*, 1069–1076.

47 A. Mavropoulos, G. Sully, A. P. Cope, A. R. Clark, *Blood* **2005**, *105*, 282–288.

48 S. Zhang, M. H. Kaplan, *J Immunol* **2000**, *165*, 1374–1380.
49 G. R. Stark, I. M. Kerr, B. R. Williams, R. H. Silverman, R. D. Schreiber, *Annu Rev Biochem* **1998**, *67*, 227–264.
50 K. Schroder, P. J. Hertzog, T. Ravasi, D. A. Hume, *J Leukoc Biol* **2004**, *75*, 163–189.
51 E. Parganas, D. Wang, D. Stravopodis, D. J. Topham, J. C. Marine, S. Teglund, E. F. Vanin, S. Bodner, O. R. Colamonici, J. M. van Deursen, G. Grosveld, J. N. Ihle, *Cell* **1998**, *93*, 385–395.
52 S. J. Rodig, M. A. Meraz, J. M. White, P. A. Lampe, J. K. Riley, C. D. Arthur, K. L. King, K. C. Sheehan, L. Yin, D. Pennica, E. M. Johnson, Jr., R. D. Schreiber, *Cell* **1998**, *93*, 373–383.
53 J. E. Durbin, R. Hackenmiller, M. C. Simon, D. E. Levy, *Cell* **1996**, *84*, 443–450.
54 S. Huang, W. Hendriks, A. Althage, S. Hemmi, H. Bluethmann, R. Kamijo, J. Vilcek, R. M. Zinkernagel, M. Aguet, *Science* **1993**, *259*, 1742–1745.
55 R. Kamijo, J. Le, D. Shapiro, E. A. Havell, S. Huang, M. Aguet, M. Bosland, J. Vilcek, *J Exp Med* **1993**, *178*, 1435–1440.
56 M. A. Meraz, J. M. White, K. C. Sheehan, E. A. Bach, S. J. Rodig, A. S. Dighe, D. H. Kaplan, J. K. Riley, A. C. Greenlund, D. Campbell, K. Carver-Moore, R. N. DuBois, R. Clark, M. Aguet, R. D. Schreiber, *Cell* **1996**, *84*, 431–442.
57 S. E. Dorman, S. M. Holland, *J Clin Invest* **1998**, *101*, 2364–2369.
58 S. E. Dorman, C. Picard, D. Lammas, K. Heyne, J. T. van Dissel, R. Baretto, S. D. Rosenzweig, M. Newport, M. Levin, J. Roesler, D. Kumararatne, J. L. Casanova, S. M. Holland, *Lancet* **2004**, *364*, 2113–2121.
59 S. Dupuis, C. Dargemont, C. Fieschi, N. Thomassin, S. Rosenzweig, J. Harris, S. M. Holland, R. D. Schreiber, J. L. Casanova, *Science* **2001**, *293*, 300–303.
60 S. Dupuis, E. Jouanguy, S. Al-Hajjar, C. Fieschi, I. Z. Al-Mohsen, S. Al-Jumaah, K. Yang, A. Chapgier, C. Eidenschenk, P. Eid, A. Al Ghonaium, H. Tufenkeji, H. Frayha, S. Al-Gazlan, H. Al-Rayes, R. D. Schreiber, I. Gresser, J. L. Casanova, *Nat Genet* **2003**, *33*, 388–391.
61 E. Jouanguy, F. Altare, S. Lamhamedi, P. Revy, J. F. Emile, M. Newport, M. Levin, S. Blanche, E. Seboun, A. Fischer, J. L. Casanova, *N Engl J Med* **1996**, *335*, 1956–1961.
62 E. Jouanguy, S. Lamhamedi-Cherradi, D. Lammas, S. E. Dorman, M. C. Fondaneche, S. Dupuis, R. Doffinger, F. Altare, J. Girdlestone, J. F. Emile, H. Ducoulombier, D. Edgar, J. Clarke, V. A. Oxelius, M. Brai, V. Novelli, K. Heyne, A. Fischer, S. M. Holland, D. S. Kumararatne, R. D. Schreiber, J. L. Casanova, *Nat Genet* **1999**, *21*, 370–378.
63 M. J. Newport, C. M. Huxley, S. Huston, C. M. Hawrylowicz, B. A. Oostra, R. Williamson, M. Levin, *N Engl J Med* **1996**, *335*, 1941–1949.
64 Z. Wen, Z. Zhong, J. E. Darnell, Jr., *Cell* **1995**, *82*, 241–250.

65 X. Zhu, Z. Wen, L. Z. Xu, J. E. Darnell, Jr., *Mol Cell Biol* **1997**, *17*, 6618–6623.
66 J. J. Zhang, Y. Zhao, B. T. Chait, W. W. Lathem, M. Ritzi, R. Knippers, J. E. Darnell, Jr., *EMBO J* **1998**, *17*, 6963–6971.
67 P. Kovarik, D. Stoiber, P. A. Eyers, R. Menghini, A. Neininger, M. Gaestel, P. Cohen, T. Decker, *Proc Natl Acad Sci USA* **1999**, *96*, 13956–13961.
68 L. Varinou, K. Ramsauer, M. Karaghiosoff, T. Kolbe, K. Pfeffer, M. Muller, T. Decker, *Immunity* **2003**, *19*, 793–802.
69 M. David, E. 3. Petricoin, C. Benjamin, R. Pine, M. J. Weber, A. C. Larner, *Science* **1995**, *269*, 1721–1723.
70 V. A. Nguyen, J. Chen, F. Hong, E. J. Ishac, B. Gao, *Biochem J* **2000**, *349*, 427–434.
71 M. Sakatsume, L. F. Stancato, M. David, O. Silvennoinen, P. Saharinen, J. Pierce, A. C. Larner, D. S. Finbloom, *J Biol Chem* **1998**, *273*, 3021–3026.
72 L. F. Stancato, C. R. Yu, E. F. Petricoin, A. C. Larner, *J Biol Chem* **1998**, *273*, 18701–18704.
73 K. C. Goh, S. J. Haque, B. R. Williams, *EMBO J* **1999**, *18*, 5601–5608.
74 K. Ramsauer, I. Sadzak, A. Porras, A. Pilz, A. R. Nebreda, T. Decker, P. Kovarik, *Proc Natl Acad Sci USA* **2002**, *99*, 12859–12864.
75 A. Navarro, B. Anand-Apte, Y. Tanabe, G. Feldman, A. C. Larner, *J Leukoc Biol* **2003**, *73*, 540–545.
76 H. Nguyen, C. V. Ramana, J. Bayes, G. R. Stark, *J Biol Chem* **2001**, *276*, 33361–33368.
77 D. K. Deb, A. Sassano, F. Lekmine, B. Majchrzak, A. Verma, S. Kambhampati, S. Uddin, A. Rahman, E. N. Fish, L. C. Platanias, *J Immunol* **2003**, *171*, 267–273.
78 J. S. Nair, C. J. DaFonseca, A. Tjernberg, W. Sun, J. E. Darnell, Jr., B. T. Chait, J. J. Zhang, *Proc Natl Acad Sci USA* **2002**, *99*, 5971–5976.
79 M. R. Rani, A. R. Asthagiri, A. Singh, N. Sizemore, S. S. Sathe, X. Li, J. D. DiDonato, G. R. Stark, R. M. Ransohoff, *J Biol Chem* **2001**, *276*, 44365–44368.
80 N. Sizemore, A. Agarwal, K. Das, N. Lerner, M. Sulak, S. Rani, R. Ransohoff, D. Shultz, G. R. Stark, *Proc Natl Acad Sci USA* **2004**, *101*, 7994–7998.
81 C. V. Ramana, N. Grammatikakis, M. Chernov, H. Nguyen, K. C. Goh, B. R. Williams, G. R. Stark, *EMBO J* **2000**, *19*, 263–272.
82 C. V. Ramana, M. P. Gil, Y. Han, R. M. Ransohoff, R. D. Schreiber, G. R. Stark, *Proc Natl Acad Sci USA* **2001**, *98*, 6674–6679.
83 Y. Qing, G. R. Stark, *J Biol Chem* **2004**, *279*, 41679–41685.
84 P. S. Subramaniam, M. G. Mujtaba, M. R. Paddy, H. M. Johnson, *J Biol Chem* **1999**, *274*, 403–407.
85 P. S. Subramaniam, M. M. Green, J. 3. Larkin, B. A. Torres, H. M. Johnson, *J Interferon Cytokine Res* **2001**, *21*, 951–959.
86 C. M. Ahmed, M. A. Burkhart, M. G. Mujtaba, P. S. Subramaniam, H. M. Johnson, *J Cell Sci* **2003**, *116*, 3089–3098.

87 P. S. Subramaniam, H. M. Johnson, *J Immunol* **2002**, *169*, 1959–1969.
88 P. Anderson, Y. K. Yip, J. Vilcek, *J Biol Chem* **1983**, *258*, 6497–6502.
89 A. Celada, R. D. Schreiber, *J Immunol* **1987**, *139*, 147–153.
90 D. G. Fischer, D. Novick, P. Orchansky, M. Rubinstein, *J Biol Chem* **1988**, *263*, 2632–2637.
91 M. M. Song, K. Shuai, *J Biol Chem* **1998**, *273*, 35056–35062.
92 T. A. Endo, M. Masuhara, M. Yokouchi, R. Suzuki, H. Sakamoto, K. Mitsui, A. Matsumoto, S. Tanimura, M. Ohtsubo, H. Misawa, T. Miyazaki, N. Leonor, T. Taniguchi, T. Fujita, Y. Kanakura, S. Komiya, A. Yoshimura, *Nature* **1997**, *387*, 921–924.
93 W. S. Alexander, R. Starr, J. E. Fenner, C. L. Scott, E. Handman, N. S. Sprigg, J. E. Corbin, A. L. Cornish, R. Darwiche, C. M. Owczarek, T. W. Kay, N. A. Nicola, P. J. Hertzog, D. Metcalf, D. J. Hilton, *Cell* **1999**, *98*, 597–608.
94 H. Sakamoto, H. Yasukawa, M. Masuhara, S. Tanimura, A. Sasaki, K. Yuge, M. Ohtsubo, A. Ohtsuka, T. Fujita, T. Ohta, Y. Furukawa, S. Iwase, H. Yamada, A. Yoshimura, *Blood* **1998**, *92*, 1668–1676.
95 H. Sakamoto, I. Kinjyo, A. Yoshimura, *Leuk Lymphoma* **2000**, *38*, 49–58.
96 Y. Qing, A. P. Costa-Pereira, D. Watling, G. R. Stark, *J Biol Chem* **2005**, *280*, 1849–1853.
97 M. You, D. H. Yu, G. S. Feng, *Mol Cell Biol* **1999**, *19*, 2416–2424.
98 M. David, P. M. Grimley, D. S. Finbloom, A. C. Larner, *Mol Cell Biol* **1993**, *13*, 7515–7521.
99 R. L. Haspel, M. Salditt-Georgieff, J. E. Darnell, Jr., *EMBO J* **1996**, *15*, 6262–6268.
100 J. ten Hoeve, M. de Jesus Ibarra-Sanchez, Y. Fu, W. Zhu, M. Tremblay, M. David, K. Shuai, *Mol Cell Biol* **2002**, *22*, 5662–5668.
101 B. Liu, S. Mink, K. A. Wong, N. Stein, C. Getman, P. W. Dempsey, H. Wu, K. Shuai, *Nat Immunol* **2004**, *5*, 891–898.
102 R. S. Rogers, C. M. Horvath, M. J. Matunis, *J Biol Chem* **2003**, *278*, 30091–30097.
103 K. A. Mowen, J. Tang, W. Zhu, B. T. Schurter, K. Shuai, H. R. Herschman, M. David, *Cell* **2001**, *104*, 731–741.
104 W. Zhu, T. Mustelin, M. David, *J Biol Chem* **2002**, *277*, 35787–35790.
105 T. R. Mosmann, H. Cherwinski, M. W. Bond, M. A. Giedlin, R. L. Coffman, *J Immunol* **1986**, *136*, 2348–2357.
106 M. K. Jenkins, A. Khoruts, E. Ingulli, D. L. Mueller, S. J. McSorley, R. L. Reinhardt, A. Itano, K. A. Pape, *Annu Rev Immunol* **2001**, *19*, 23–45.
107 D. M. Frucht, T. Fukao, C. Bogdan, H. Schindler, J. J. O'Shea, S. Koyasu, *Trends Immunol* **2001**, *22*, 556–560.
108 J. Kashiwakura, N. Suzuki, H. Nagafuchi, M. Takeno, Y. Takeba, Y. Shimoyama, T. Sakane, *J Exp Med* **1999**, *190*, 1147–1154.
109 Y. Takeba, H. Nagafuchi, M. Takeno, J. Kashiwakura, N. Suzuki, *J Immunol* **2002**, *168*, 2365–2370.

110 D. J. Fowell, K. Shinkai, X. C. Liao, A. M. Beebe, R. L. Coffman, D. R. Littman, R. M. Locksley, *Immunity* **1999**, *11*, 399–409.

111 E. M. Schaeffer, G. S. Yap, C. M. Lewis, M. J. Czar, D. W. McVicar, A. W. Cheever, A. Sher, P. L. Schwartzberg, *Nat Immunol* **2001**, *2*, 1183–1188.

112 C. Dong, R. J. Davis, R. A. Flavell, *Annu Rev Immunol* **2002**, *20*, 55–72.

113 U. Boehm, T. Klamp, M. Groot, J. C. Howard, *Annu Rev Immunol* **1997**, *15*, 749–795.

114 M. Belosevic, D. S. Finbloom, P. H. Van Der Meide, M. V. Slayter, C. A. Nacy, *J Immunol* **1989**, *143*, 266–274.

115 Z. E. Wang, S. L. Reiner, S. Zheng, D. K. Dalton, R. M. Locksley, *J Exp Med* **1994**, *179*, 1367–1371.

116 L. M. Bradley, D. K. Dalton, M. Croft, *J Immunol* **1996**, *157*, 1350–1358.

117 B. Lu, C. Ebensperger, Z. Dembic, Y. Wang, M. Kvatyuk, T. Lu, R. L. Coffman, S. Pestka, P. B. Rothman, *Proc Natl Acad Sci USA* **1998**, *95*, 8233–8238.

118 S. E. Macatonia, C. S. Hsieh, K. M. Murphy, A. O'Garra, *Int Immunol* **1993**, *5*, 1119–1128.

119 S. J. Szabo, A. S. Dighe, U. Gubler, K. M. Murphy, *J Exp Med* **1997**, *185*, 817–824.

120 C. A. Wenner, M. L. Guler, S. E. Macatonia, A. O'Garra, K. M. Murphy, *J Immunol* **1996**, *156*, 1442–1447.

121 M. Afkarian, J. R. Sedy, J. Yang, N. G. Jacobson, N. Cereb, S. Y. Yang, T. L. Murphy, K. M. Murphy, *Nat Immunol* **2002**, *3*, 549–557.

122 A. C. Mullen, F. A. High, A. S. Hutchins, H. W. Lee, A. V. Villarino, D. M. Livingston, A. L. Kung, N. Cereb, T. P. Yao, S. Y. Yang, S. L. Reiner, *Science* **2001**, *292*, 1907–1910.

123 J. L. Grogan, M. Mohrs, B. Harmon, D. A. Lacy, J. W. Sedat, R. M. Locksley, *Immunity* **2001**, *14*, 205–215.

124 H. Huang, W. E. Paul, *J Exp Med* **1998**, *187*, 1305–1313.

125 A. P. Mountford, P. S. Coulson, A. W. Cheever, A. Sher, R. A. Wilson, T. A. Wynn, *Immunology* **1999**, *97*, 588–594.

126 W. Ouyang, M. Lohning, Z. Gao, M. Assenmacher, S. Ranganath, A. Radbruch, K. M. Murphy, *Immunity* **2000**, *12*, 27–37.

127 E. S. Hwang, S. J. Szabo, P. L. Schwartzberg, L. H. Glimcher, *Science* **2005**, *307*, 430–433.

128 W. Ouyang, S. H. Ranganath, K. Weindel, D. Bhattacharya, T. L. Murphy, W. C. Sha, K. M. Murphy, *Immunity* **1998**, *9*, 745–755.

129 B. Perussia, M. J. Loza, *Trends Immunol* **2003**, *24*, 235–241.

130 M. J. Loza, B. Perussia, *Int Immunol* **2004**, *16*, 23–32.

131 J. Banchereau, R. M. Steinman, *Nature* **1998**, *392*, 245–252.

132 M. Colonna, G. Trinchieri, Y. J. Liu, *Nat Immunol* **2004**, *5*, 1219–1226.

133 B. Dubois, B. Vanbervliet, J. Fayette, C. Massacrier, C. Van Kooten, F. Briere, J. Banchereau, C. Caux, *J Exp Med* **1997**, *185*, 941–951.

134 M. Cella, D. Scheidegger, K. Palmer-Lehmann, P. Lane, A. Lanzavecchia, G. Alber, *J Exp Med* **1996**, *184*, 747–752.

135 I. S. Grewal, R. A. Flavell, *Immunol Today* **1996**, *17*, 410–414.
136 M. F. Mackey, J. R. Gunn, C. Maliszewsky, H. Kikutani, R. J. Noelle, R. J. Barth, Jr., *J Immunol* **1998**, *161*, 2094–2098.
137 B. J. Czerniecki, P. A. Cohen, M. Faries, S. Xu, J. G. Roros, I. Bedrosian, *Crit Rev Immunol* **2001**, *21*, 157–178.
138 A. Kelso, *Immunol Today* **1995**, *16*, 374–379.
139 G. Kelsoe, B. Zheng, *Curr Opin Immunol* **1993**, *5*, 418–422.
140 J. M. Weiss, J. Sleeman, A. C. Renkl, H. Dittmar, C. C. Termeer, S. Taxis, N. Howells, M. Hofmann, G. Kohler, E. Schopf, H. Ponta, P. Herrlich, J. C. Simon, *J Cell Biol* **1997**, *137*, 1137–1147.
141 B. Pulendran, M. Karvelas, G. J. Nossal, *Proc Natl Acad Sci USA* **1994**, *91*, 2639–2643.
142 M. Matsuo, Y. Nagata, E. Sato, D. Atanackovic, D. Valmori, Y. T. Chen, G. Ritter, I. Mellman, L. J. Old, S. Gnjatic, *Proc Natl Acad Sci USA* **2004**, *101*, 14467–14472.
143 S. C. Higgins, E. C. Lavelle, C. McCann, B. Keogh, E. McNeela, P. Byrne, B. O'Gorman, A. Jarnicki, P. McGuirk, K. H. Mills, *J Immunol* **2003**, *171*, 3119–3127.
144 F. Re, J. L. Strominger, *J Immunol* **2004**, *173*, 7548–7555.
145 F. Granucci, I. Zanoni, N. Pavelka, S. L. Van Dommelen, C. E. Andoniou, F. Belardelli, M. A. Degli Esposti, P. Ricciardi-Castagnoli, *J Exp Med* **2004**, *200*, 287–295.
146 M. P. Colombo, G. Trinchieri, *Cytokine Growth Factor Rev* **2002**, *13*, 155–168.
147 K. M. Detjen, K. Farwig, M. Welzel, B. Wiedenmann, S. Rosewicz, *Gut* **2001**, *49*, 251–262.
148 H. Ikeda, L. J. Old, R. D. Schreiber, *Cytokine Growth Factor Rev* **2002**, *13*, 95–109.
149 D. H. Kaplan, V. Shankaran, A. S. Dighe, E. Stockert, M. Aguet, L. J. Old, R. D. Schreiber, *Proc Natl Acad Sci USA* **1998**, *95*, 7556–7561.
150 Z. Qin, T. Blankenstein, *Immunity* **2000**, *12*, 677–686.
151 S. E. Street, E. Cretney, M. J. Smyth, *Blood* **2001**, *97*, 192–197.
152 B. Chen, L. He, V. H. Savell, J. J. Jenkins, D. M. Parham, *Cancer Res* **2000**, *60*, 3290–3298.
153 S. Fulda, K. M. Debatin, *Oncogene* **2002**, *21*, 2295–2308.
154 J. J. Sironi, T. Ouchi, *J Biol Chem* **2004**, *279*, 4066–4074.
155 C. Vivo, F. Levy, Y. Pilatte, J. Fleury-Feith, P. Chretien, I. Monnet, L. Kheuang, M. C. Jaurand, *Oncogene* **2001**, *20*, 1085–1093.
156 K. Liu, S. I. Abrams, *J Immunol* **2003**, *170*, 6329–6337.
157 W. A. Selleck, S. E. Canfield, W. A. Hassen, M. Meseck, A. I. Kuzmin, R. C. Eisensmith, S. H. Chen, S. J. Hall, *Mol Ther* **2003**, *7*, 185–192.
158 T. Kamei, M. Inui, S. Nakamura, K. Okumura, A. Goto, T. Tagawa, *Melanoma Res* **2003**, *13*, 153–159.
159 X. Yang, M. S. Merchant, M. E. Romero, M. Tsokos, L. H. Wexler, U. Kontny, C. L. Mackall, C. J. Thiele, *Cancer Res* **2003**, *63*, 1122–1129.

160 F. Burke, P. D. Smith, M. R. Crompton, C. Upton, F. R. Balkwill, *Br J Cancer* **1999**, *80*, 1236–1244.
161 G. Beatty, Y. Paterson, *J Immunol* **2001**, *166*, 2276–2282.
162 Y. Hayakawa, K. Takeda, H. Yagita, M. J. Smyth, L. Van Kaer, K. Okumura, I. Saiki, *Blood* **2002**, *100*, 1728–1733.
163 Z. Qin, J. Schwartzkopff, F. Pradera, T. Kammertoens, B. Seliger, H. Pircher, T. Blankenstein, *Cancer Res* **2003**, *63*, 4095–4100.
164 H. Winter, H. M. Hu, W. J. Urba, B. A. Fox, *J Immunol* **1999**, *163*, 4462–4472.
165 C. Becker, H. Pohla, B. Frankenberger, T. Schuler, M. Assenmacher, D. J. Schendel, T. Blankenstein, *Nat Med* **2001**, *7*, 1159–1162.
166 G. P. Dunn, A. T. Bruce, H. Ikeda, L. J. Old, R. D. Schreiber, *Nat Immunol* **2002**, *3*, 991–998.
167 A. S. Dighe, E. Richards, L. J. Old, R. D. Schreiber, *Immunity* **1994**, *1*, 447–456.
168 G. J. Bancroft, R. D. Schreiber, E. R. Unanue, *Curr Top Microbiol Immunol* **1989**, *152*, 235–242.
169 N. Y. Crowe, M. J. Smyth, D. I. Godfrey, *J Exp Med* **2002**, *196*, 119–127.
170 M. J. Smyth, N. Y. Crowe, D. G. Pellicci, K. Kyparissoudis, J. M. Kelly, K. Takeda, H. Yagita, D. I. Godfrey, *Blood* **2002**, *99*, 1259–1266.
171 M. Duff, P. P. Stapleton, J. R. Mestre, S. Maddali, G. P. Smyth, Z. Yan, T. A. Freeman, J. M. Daly, *Ann Surg Oncol* **2003**, *10*, 305–313.
172 A. H. Klimp, E. G. de Vries, G. L. Scherphof, T. Daemen, *Crit Rev Oncol Hematol* **2002**, *44*, 143–161.
173 S. Frewert, F. Stockhammer, G. Warschewske, A. C. Zenclussen, S. Rupprecht, H. D. Volk, C. Woiciechowsky, *Neurosci Lett* **2004**, *364*, 145–148.
174 J. C. Eymard, M. Lopez, A. Cattan, O. Bouche, J. C. Adjizian, J. Bernard, *Eur J Cancer* **1996**, *32A*, 1905–1911.
175 T. Lesimple, A. Moisan, F. Guille, C. Leberre, R. Audran, B. Drenou, L. Toujas, *J Immunother* **2000**, *23*, 675–679.
176 C. S. Tannenbaum, N. Wicker, D. Armstrong, R. Tubbs, J. Finke, R. M. Bukowski, T. A. Hamilton, *J Immunol* **1996**, *156*, 693–699.
177 F. Fallarino, T. F. Gajewski, *J Immunol* **1999**, *163*, 4109–4113.
178 A. K. Kacha, F. Fallarino, M. A. Markiewicz, T. F. Gajewski, *J Immunol* **2000**, *165*, 6024–6028.
179 V. Shankaran, H. Ikeda, A. T. Bruce, J. M. White, P. E. Swanson, L. J. Old, R. D. Schreiber, *Nature* **2001**, *410*, 1107–1111.

5
Interferon and Related Receptors

Sidney Pestka and Christopher D. Krause

5.1
Introduction

The mammalian interferons (IFN-α, -β, -ε, -κ, -ω, -δ, -τ, -ν and -γ) and IFN-like molecules [limitin, and interleukin (IL)-28A, -28B and -29] are a subset of the class 2 α-helical cytokines. Their corresponding receptors consist of three different groups that are activated by three groups of ligands: IFN-α, -β, -ε, -κ, -ω, -δ, -τ, -ν and limitin, IFN-γ, and IL-28A, -28B and -29. This chapter summarizes the current understanding of the receptors for these ligands.

5.2
IFNs and IFN-like Molecules in Brief

Type I IFNs consist of eight classes [1–11]: IFN-α, -β, -ε, -κ, -ω, -δ, -τ and -ν. Type II IFN consists of IFN-γ only. Four IFN-like cytokines – limitin (found only in mice), IL-28A, -28B and -29 – found in human and other mammals are functionally related to the Type I IFNs (Tab. 5.1). IFN-α, -β, -ε, -κ, -ω and -ν, and IL-28A, -28B and -29 are found in humans [11–13], whereas IFN-δ [14] and -τ [15], and limitin [8] are not. IFN-τ was described first as ovine trophoblast protein-1 and is found in ungulates where it is required for implantation of the ovum [15], but there is no direct human homolog. Human IFN-κ, although it exhibits low specific antiviral activity, is expressed in human keratinocytes [10]. IL-28A, -28B and IL-29 are found in humans, and in other mammals, birds and amphibians [11, 16, 17], and function like Type I IFNs. Human IFN-ε [18] has not been characterized in significant detail, but an implication in reproductive function is possible [19]. IFNs were the first cytokines discovered and the first to be used therapeutically [4, 9, 20, 21].

The Interferons: Characterization and Application. Edited by Anthony Meager
Copyright © 2006 WILEY-VCH Verlag GmbH & Co. KGaA, Weinheim
ISBN: 3-527-31180-7

Tab. 5.1. IFNs and IFN-like molecules

Ligand	Common alternate names/members	Gene locus	Receptor complex chain 1	Receptor complex chain 2	Signal transduction pathways employed
Type I IFN	IFN-α, -β, -δ, -ε, -κ, -τ, -ω, -ν, limitin	9p21 + 3 (T) (α, ω), 9p21 + 3 (T) (β), 9p21 + 1 (T) (κ), 9p21 + 3 (T) (ε), 9p21 + 3 (T) (ν) [11]	IFN-αR1	IFN-αR2	JAK-1 + TYK-2, STAT 2 + 1 + 3 + 5, mitogen-activated protein kinase (MAPK), PRMT1 [148]
Type II IFN	IFN-γ	12q14 + 3 (C)	IFN-γR1	IFN-γR2	JAK-1 + JAK-2, STAT 1 + 3 + 5, Akt, MAPK [148]
IL-28A, -28B, -29	IFN-λ2, -λ3, -λ1 [17]	19q13 + 2 (T, C, T)	IL-28R1	IL-10R2	JAK-1 + TYK-2, STAT 2 + 1 + 3 + 5 [17, 18, 146]

This table delineates the genome-encoded ligands (or their close orthologs). IL-28A, -28B and -29 are clustered together. Also, Type I IFN members have been clustered together. The columns from left to right indicate standard names for the ligand or ligand family, alternate names for the ligand (or major subgroups of the Type I IFN family), the location of the genes in the human genome, the components of its receptor complex (the receptor chain predicted or known to bind JAK-1 is listed first) and signal transduction initiated by the ligand receptor. Only locations of the human genes are indicated. "(C)" indicates transcription occurs toward the centromere, while "(T)" indicates transcription occurs away from the centromere. Although most IFN-α genes are transcribed toward the telomere, some are transcribed toward the centromere. Data were obtained from: http://www.ncbi.nlm.nih.gov/mapview/map_search.cgi?taxid=9606.

5.3
The Receptors

This chapter focuses on three groups of receptors that are activated by three classes of ligands: the Type I IFN receptor complex that is activated by numerous ligands (IFN-α, -β, -ε, -κ, -ω, -δ, -τ, -ν and limitin); the Type II receptor complex activated only by IFN-γ; and the IL-28R1/10R2 receptor complex activated by IL-28A, -28B and -29. Similar to most cytokines and growth factors, the actions of IFNs are mediated by an interaction with specific cell surface receptors and the Janus kinase (JAK)–signal transducers and activators of transcription (STAT) signal transduction pathway [4, 22–27]. Competition binding studies demonstrated that Type I ligands share the same receptor complex, whereas Type II IFN (IFN-γ) binds to a distinct receptor [4, 24, 28–31].

5.3.1
Receptor Nomenclature

The designations of the IFN receptor components are given in Tab. 5.2. These receptor complexes consist of two or more components. It also appears that the indi-

Tab. 5.2. The receptors for IFNs and IFN-like molecules

Standard name	Common alternate names/members	Gene locus[a]	Ligands
IFN-γR1	IFN-γR [106], IFN-γRα [152]	6q23 + 2 (C)	IFN-γ
IFN-γR2	AF-1 [87], IFN-γRβ [122]	21q22 + 11 (T)	IFN-γ
IFN-αR1	IFN-αR [72], IFN-αRα, IFNAR-1	21q22 + 11 (T)	IFN-α, -β, -ω, -δ, -ε, -κ, -τ, -ν, limitin
IFN-αR2	IFN-α/βR [93], IFN-αR2a [34], IFN-αR2b, IFN-αR2c, IFN-αR2$_L$ [95], IFN-αR2$_S$, IFN-αR2-1 [96], IFN-αR2-2, IFN-αR2-3, IFNAR-2	21q22 + 11 (T)	IFN-α, -β, -ω, -δ, -ε, -κ, -τ, -ν, limitin
IL-10R2	IL-10Rβ [147], CRFB4 [153], CRF2-4 [154]	21q22 + 11 (T)	cellular and viral IL-10, -22, -26, -28A, -28B, -29
IL28R1	CRF2-12 [155], IL-28Rα [18], IFN-λR1 [17], LICRII [146]	1p36 + 11 (T)	IL-28A, -28B, -29

The columns from left to right indicate the currently accepted name for the receptor, alternate names or physiologically relevant splice variants for the receptor, the location of its gene in the human genome and ligands which signal through that receptor chain when it is part of an intact receptor complex. Only locations of the human genes are indicated. "(C)" indicates transcription occurs toward the centromere, while "(T)" indicates transcription occurs away from the centromere. Data were obtained from: http://www.ncbi.nlm.nih.gov/mapview/map_search.cgi?taxid=9606.

vidual components may contribute to one extent or another to ligand binding so that designations such as an α subunit for the ligand binding component and the β subunit for the signal transduction subunit are not warranted. Particularly, in the case of the human Type I IFN receptor complex, the IFN-αR2 (also designated IFN-αRβ, IFNAR-2) chain is the major ligand-binding chain, not chain 1 (IFN-αR1; also designated IFN-αRα, IFNAR-1). The subunits were named in the order in which they were cloned and discovered as distinct entities. Alternative designations for the subunits that have been used are also given (Tab. 5.2).

5.4
The Type I IFN Receptor

The receptor complex for the Type I IFNs consists of two chains – IFN-αR1 and -αR2 (Tab. 5.2 and Fig. 5.1). The IFNs and IFN-like molecules signal through the JAK–STAT pathway [21, 32, 33]. IFN-α, -β, -ε, -κ, -ω, -δ and -τ, and limitin signal through the Type I receptors IFN-αR1 and -αR2c, and very likely do IFN-ε and -ν, although the pathway of IFN-ε and -ν has not yet been defined (Tabs. 5.1 and 5.2). All the Type I IFNs activate STAT-1 and -2, TYK-2, and JAK-1, and induce genes that have the IFN-stimulated response element (ISRE) in the promoter [34, 35]. IFN-γ uses the unique receptor chains IFN-γR1 and -γR2 (Tabs. 5.1 and 5.2, and Fig. 5.3), and activates JAK-1 and -2, and STAT-1 that in turn induces genes containing the γ activation sequence (GAS) in the promoter [35, 36]. In addition to the JAK–STAT pathway, the Type I IFNs utilize other pathways, some of which are noted in Tab. 5.1. Other pathways independent of the JAK–STAT pathway can initiate signal transduction by the Type I IFNs (Tab. 5.1). For example, cells from mice lacking STAT-1 can still respond to Type I IFNs: IFN-α and -β can modulate proliferative responses in phagocytes from STAT-1-deficient mice, and STAT-1 is not necessary for some functions of the IFN-γ [37–39]. Understanding of these events is further complicated by the multiplicity of the Type I IFNs [12] that exhibit different activities, although they interact with the same receptor. This is an area that needs to be explored in great detail to understand this family of proteins and their mechanisms of action.

The Type I IFNs exhibit a wide breadth of biological activities: antiviral, antiproliferative, stimulation of cytotoxic activity [40–46] of a variety of cells of the immune system [T cells, natural killer (NK) cells, monocytes, macrophages, dendritic cells (DCs)], increasing expression of tumor-associated surface antigens [46–49], stimulation of other surface molecules such as MHC Class I antigens [47, 50, 51], induction and/or activation of proapoptotic genes and proteins [e.g. tumor necrosis factor (TNF)-related apoptosis-inducing ligand (TRAIL), caspases, Bak and Bax] [52], repression of antiapoptotic genes [e.g. Bcl-2, inhibitor of apoptosis protein (IAP)] [52], modulation of differentiation [53–55] and antiangiogenic activity [56, 57]. All these actions make IFN a most promising agent to treat various diseases. The challenge is to be able to use this enormous potential of the IFNs without the de-

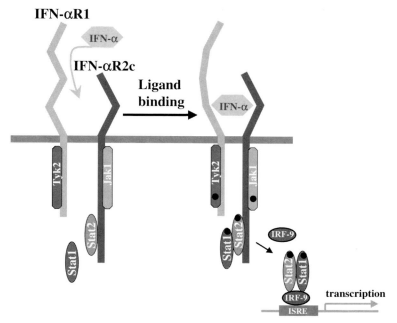

Fig. 5.1. Model of the Type I IFN receptor complex. The IFN-α receptor complex consists of two different chains, IFN-αR1 and -αR2c [67, 70, 88, 93, 94]. The ligand IFN-α is a monomer that binds to the two-chain complex [153]. Upon entry of IFN-α into the complex, STAT-2 binds to IFN-αR2c and recruits STAT-1. STAT-1 and -2 are phosphorylated, then released, associate with IRF-9, translocated to the nucleus and activate genes that contain the ISRE as described [26]. It is reported that limitin uses the Type I IFN receptor complex [154, 155]; however, although limitin requires TYK-2, it does not require STAT-1 for B cell growth inhibition where Daxx may replace the need for STAT-1 [155, 156]. Based on homology with the IFN-γ receptor complex [143], we believe that the Type I IFN receptor complex is preassembled and that a conformational change accompanies the activation of the receptor complex by Type I IFN. (This figure also appears with the color plates.)

bilitating side-effects. Appropriate technology to deliver locally the IFNs in tumors could overcome the problem of systemic side-effects. Overall, it is highly likely that IFNs will play a major role in the next generation of novel antitumor and antiviral therapies.

5.4.1
Discovery of the Type I Receptor Complex

Somatic cell genetic studies with human × rodent hybrid cells containing various combinations of human chromosomes provided evidence that the presence of human chromosome 21 confers sensitivity of the rodent cells to human Type I IFNs

[58–63], and the region was later localized to a 3-Mb segment of chromosome 21q around 21q22.1 [64–67]. Mouse cells with functional human IFN Type I IFN receptors were first isolated by Jung and Pestka [68] and by Revel et al. [69]. The mouse cells transfected with total human DNA exhibited the properties of a Type I IFN receptor in that they responded to both Type I IFNs tested: human IFN-α and -β. Although primary transformants were obtained by these groups, no molecular clone was obtained that provided Type I IFN receptor activity. Employing the procedures of Jung and Pestka [68], Uzé et al. [27], by switching the selection procedure from human IFN-αA and -β to human IFN-α8 (human IFN-αB2), were able to obtain a molecular clone that they designated the Type I IFN receptor. However, this cDNA clone (now known as human IFN-αR1) did not significantly bind Type I IFNs other than human IFN-α8 and yielded very little response even to human IFN-α8 when expressed in cells [70–72]. Results with polyclonal and monoclonal antibodies began to suggest that two separate subunits might be involved [73–77]. During this period, however, a number of groups obtained antibodies suggesting that the Type I receptor consisted of multiple components [63, 67, 73–81] – an hypothesis consistent with the inability of Jung and Pestka [68] to obtain stable secondary transformants.

Since the IFN-αR1 chain required an additional component to complete a functional Type I IFN-α complex, a number of groups began to search for the additional chain. With the use of the yeast artificial chromosome F136C5 (αYAC) containing a segment of human chromosome 21 introduced into Chinese hamster ovary (CHO) cells [82–87], it was demonstrated that cells containing this αYAC responded to all human Type I IFNs tested: human IFN-αA (IFN-α2), -α8B2, -ω, -αA/D (*Bgl*) and -β. With cells containing this αYAC clone [67, 83, 85], the Type I IFNs exhibited antiviral activity, MHC class I induction and binding of the IFNs to the cells.

Direct proof of the requirement for the IFN-αR1 chain for Type I IFN receptor function was demonstrated by experiments disrupting the *IFNAR1* (IFN-αR1) gene of the αYAC clone [88]. The resultant ΔαYAC with a deletion of the segment expressing the IFN-αR1 chain was transferred to CHO cells, and eliminated the ability of Type I IFNs to induce MHC class I antigens and exhibit antiviral activity [67, 88]. By subsequent transfection of the cDNA for human IFN-αR1 into the cells containing the ΔαYAC, activity of Type I IFNs was reconstituted. Furthermore, homozygous deletions of *IFNAR1* in mice (IFN-αR1$^{0/0}$) were reported to cause enhanced susceptibility to several viruses, and eliminate antiproliferative activity of IFN-α and IFN-β [89, 90]. Thus, the human IFN-αR1 subunit plays a critical role in the functional human Type I IFN receptor complex and the additional component necessary for activity together with IFN-αR1 gene was encoded in this αYAC.

The results indicated that all the genes necessary to reconstitute a biologically active Type I human IFN receptor complex are located within the human DNA insert of this αYAC clone. Although the human IFN-αR1 subunit plays a critical role in the functional human Type I IFN receptor complex whose components are encoded on this αYAC, because binding of ligands was retained in the cells containing the ΔαYAC with the deletion, it was clear that one or more additional subunits encoded on the αYAC were primarily responsible for the ligand binding.

Shortly after the studies describing the αYAC, a second chain, IFN-αR2b, was reported, but it exhibited no intrinsic activity alone or together with the human IFN-αR1 chain [91]. The human gene *IFNAR2* encoding this chain is located on the αYAC and also localized to the 3 × 1S region of chromosome 21. This human IFN-αR2b chain was reported to bind Type I IFNs and antibodies to this chain were able to coimmunoprecipitate JAK-1 [91]. However, expression of neither IFN-αR1 nor -αR2b chains, nor the combination of both, in hamster cells was able to reconstitute functional human receptor activity [85, 92]. In addition, the CHO cells containing the ΔαYAC with the disrupted *IFNAR1* gene retained the ability to bind human IFN-αA and human IFN-αB2 [88]. This paradox was partially resolved when U5 cells [26] that do not respond to Type I IFNs were able to be reconstituted to respond to the IFNs by introduction of the αYAC (Cook et al., unpublished). Unfortunately, no stable transformants were obtained with the αYAC in U5 cells. The data indicated, however, that another component encoded on the αYAC was necessary for response to Type I IFNs. This component was subsequently cloned [93, 94] and shown to be a long form of the human IFN-αR2b chain called human IFN-αR2c (Tab. 5.2). All three chains (human IFN-αR2a, -αR2b and -αR2c) are encoded by the same *IFNAR2* gene. The human IFN-αR2c chain contains a large intracellular domain compared to the short intracellular domain of the human IFN-αR2b chain. The human IFN-αR2c chain reconstitutes U5 cells for responses to Type I IFNs. The human IFN-αR2a protein is a soluble form of the receptor consisting of the extracellular domain of the human IFN-αR2b and -αR2c chains. Thus, it was established that the functional Type I receptor consists of the IFN-αR1 and -αR2c chains (Fig. 5.1).

The data show that hamster cells containing the αYAC exhibit the properties expected for a functional Type I human IFN receptor complex. The αYAC provides genes which are necessary and sufficient to encode this functional Type I IFN receptor complex as measured by three distinct biological assays. Also the specific binding of both ^{32}P-labeled IFN-αA and IFN-αB2 to hamster cells fused to this αYAC demonstrated that the biological functions induced by these IFNs were reflected in the interaction between the ligand and receptor complex. Since this αYAC contains the entire gene for the cloned human IFN-αR1 receptor subunit (∼30 kb) and for the human IFN-αR2 chains [67, 88], there may be other genes responsible for the formation of the receptor complex present in this αYAC. The results suggest that a high-affinity receptor is composed of the cloned human IFN-αR1 and -αR2c receptor subunits (Fig. 5.1).

5.4.2
Diversity of the Interaction of Type I IFNs with the Receptor

As noted above, hamster cells containing the yeast artificial chromosome αYAC respond to all Type I human IFNs including IFN-αA, -β and -ω [67, 88, 95]. The αYAC contains at least two genes encoding IFN-αR chains that are required for response to Type I human IFNs: human IFN-αR1 and -αR2c. A splice variant of the human

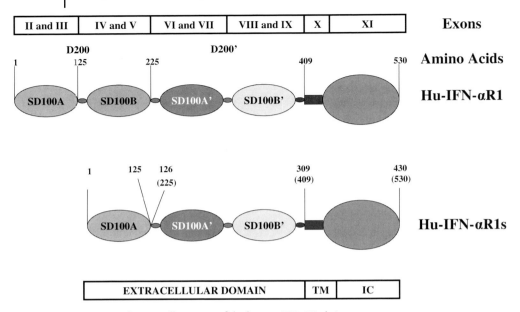

Fig. 5.2. Schematic Illustration of the human IFN-αR1 chain and the splice variant human IFN-αR1s. The domains of the human IFN-αR1 chain are shown. The splice variant chain, human IFN-αR1s, lacks exons IV and V. (This figure also appears with the color plates.)

IFN-αR1 chain designated human IFN-αR1s or -αR1b [95] (Fig. 5.2) was isolated. CHO cells containing the disrupted αYAC, which contains a deletion in the human IFN-αR1 gene, ΔαYAC, were transfected with expression vectors for the human IFN-αR1 and -αR1s chains yielding cells ΔαYAC/αR1 and ΔαYAC/αR1s, respectively. With these cells, two Type I IFNs were identified that can interact with the splice variant (human IFN-αR1s) and with the human IFN-αR1 chains: human IFN-αA (-α2) and -ω [95]. Two other Type I IFNs, human IFN-α8 (-αB2) and -α21 (-αF), are capable of signaling through the human IFN-αR1 chain only, but cannot utilize the splice variant human IFN-αR1s [95]. Human IFN-αR1 and -αR1s differ in that the latter is missing a single subdomain of the four subdomains of the receptor extracellular domain encoded by exons 4 and 5 of the *IFNAR1* gene expressing human IFN-αR1 (Fig. 5.2). Therefore, different Type I IFNs can interact differently with the human IFN-αR1 receptor chain and the splice variant chain (human IFN-αR1s) is functional. All the Type I IFNs tested bind to the hamster ΔαYAC/αR1s cells that contain the splice variant so that the differences in functional activity do not reflect gross differences in binding to the receptor components. It thus appears that some Type I IFNs (e.g. human IFN-B2, -αF) require the two subdomains absent in the splice variant (Fig. 5.2) and some can function without these domains (e.g. human IFN-αA and -ω) as described [95]. The results indicate that the various human IFN-α species and other Type I IFNs interact with the receptor differently, thereby accounting to some degree for their differential activities. The conservation of the multitude of Type I IFNs, but not their sequences, throughout evolution of the mammals is consistent with unique functional roles for each of the Type I IFNs [11]. However, the mechanisms by which the various IFNs exhibit different patterns of activity are poorly understood [42, 43, 46, 96, 97].

5.5
The Type II IFN (IFN-γ) Receptor

5.5.1
Chromosomal Localization of the IFN-γ Receptor Ligand-binding Chain and Discovery of Two Chains Required for Activity

It was shown by study of ligand-binding competition that the IFN-γ receptor was distinct from that of the receptor that bound IFN-α and -β [28, 98]. Through somatic cell genetic techniques, Rashidbaigi et al. [101] demonstrated that the gene for the ligand-binding chain of the human IFN-γ receptor was localized to human chromosome 6q. Nevertheless, although somatic cell hybrids that contain this region of or the entire chromosome 6 exhibited excellent binding of IFN-γ, the ligand was unable to initiate any biological activities. This study led Jung et al. [100, 101] to the discovery that an additional component located on human chromosome 21q was required for function together with the IFN-γR1 chain encoded on human chromosome 6. Thus, two species-specific components were involved as part of the functional IFN-γ receptor: the ligand-binding chain of the receptor (IFN-γR1) and the second chain of the receptor (IFN-γR2) we initially designated accessory factor-1 (AF-1) that is required for signal transduction. These somatic cell genetic experiments led to the discovery of the location of the ligand-binding chain (IFN-γR1) to chromosome 6 and the location of the second chain (IFN-γR2) to human chromosome 21, and set the stage to isolate and clone these components. The comparable mouse chains were localized to mouse chromosomes 10 and 16, respectively [102, 103]. Table 5.2 summarizes the nomenclature for these chains. Subsequent to the localization of the ligand-binding chain, both the human [104] and mouse [105–109] IFN-γR1 chains were cloned and shown to bind ligand. In addition, these studies confirmed the observations that the ligand-binding component was insufficient for generating a biological response. To generate a biological response, it was necessary to have the ligand-binding chain plus the second chain IFN-γR2 (AF-1). It should be noted that the biological response measured in these assays was induction of class I MHC antigen expression. By reconstituting functional activity with the cloned human IFN-γR1 and human chromosome 21, it was definitively demonstrated that the ligand-binding chain was the necessary and sufficient component contributed by human chromosome 6 [110]. It was thus

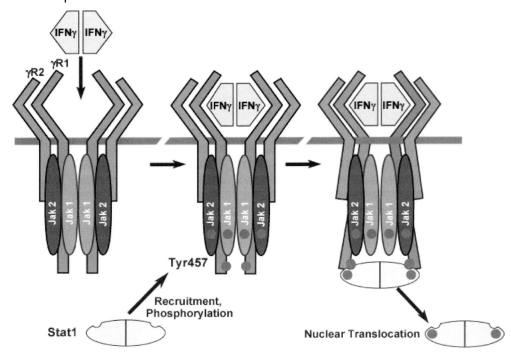

Fig. 5.3. Model of the IFN-γR complex. The IFN-γR complex consists of two different chains, IFN-γR1 and -γR2 [9, 33, 85, 99–101, 110, 120, 121, 152, 157–164]. The ligand IFN-γ is a dimer that binds to two IFN-γR1 chains, but does not directly bind to the IFN-γR2 chain in the absence of the IFN-γR1 chain. The two receptor chains are preassembled prior to binding of IFN-γ [143]. Upon ligand binding, the JAK kinases cross-phosphorylate each other (solid circles). The activated JAK kinases then phosphorylate Tyr457 of each IFN-γR1 chain that serves as the recruitment site for STAT-1α which, in turn, attaches to phosphotyrosine Tyr457 of IFN-γR1, moves the receptor chains apart [143] and is phosphorylated by the JAK kinases. Once phosphorylated, the phosphorylated STAT-1α proteins detach from each IFN-γR1 chain, forming the transcription factor that is translocated to the nucleus to activate IFN-γ regulated genes. (This figure also appears with the color plates.)

concluded that the functional receptor consisted of two chains that are now designated IFN-γR1 and -γR2 (Tab. 5.2 and Fig. 5.3).

Study of chimeric human and mouse IFN-γR1 chains demonstrated that the ability of mouse and human IFN-γ species to stimulate MHC class I antigen induction in various cells transfected with the chimeric receptors required the homologous extracellular domain of the receptor. For example, the chimeric receptor HMM (human extracellular, murine transmembrane and murine intracellular domains, respectively) was able to respond to human IFN-γ only in mouse cells or hamster cells containing human chromosome 21 or 21q [111–114].

5.5.2
The IFN-γR1 Chain

The IFN-γR1 chain binds the ligand IFN-γ. The functional architecture of the IFN-γR1 intracellular domain is concentrated in two specific regions. Distal from the membrane, a five-residue sequence ($Y^{457}DKPH^{461}$) is required for all signal transduction [115–117]. This sequence is completely conserved in mammals, birds and frogs receptors. Y^{457} is phosphorylated upon ligand binding, and serves as a recruitment site for STAT-1α [118]. The function of the other residues in the YDKPH motif are not known, but D^{458} and H^{461} are required [116, 117]. Specific mutations have shown that the C-terminal 29 amino acids have no apparent function [116]. A membrane-proximal $L^{266}PKS^{269}$ motif is required for receptor activity and JAK-1 binding [119].

5.5.3
The Second Receptor Chain (IFN-γR2)

The second chain of the human IFN-γ receptor was isolated by Soh et al. [83, 85] and Cook et al. [84], and the mouse chain by Hemmi et al. [120]. The cDNA clones encode the necessary species-specific factor and are able to substitute for human chromosome 21 to reconstitute the human IFN-γ receptor-mediated induction of class I HLA antigens. However, the factor encoded by the cDNA does not confer full antiviral protection against encephalomyocarditis virus (ECMV), suggesting that an additional factor encoded on human chromosome 21 may be required for reconstitution of full antiviral activity against EMCV [83–85, 121]. Similar observations were made with the mouse IFN-γR1 and -γR2 chains. Substitution of each of the tyrosine residues of the intracellular domain by phenylalanine did not alter the ability of the mouse IFN-γR2 chain to support signal transduction [122]. Kotenko et al. [92] showed that the intracellular domain of the human IFN-γR2 chain lacking the terminal 49 residues was totally inactive in MHC class I antigen induction, JAK-1 and -2 phosphorylation, STAT-1α activation, and tyrosine phosphorylation of the IFN-γR1 chain. Furthermore, cells expressing the human IFN-γR2 chain alone did not crosslink or bind ligand; however, when both human IFN-γR1 and -γR2 chains were coexpressed in hamster cells, the association constant for binding of ligand increased about 3-fold and a specific band of a crosslinked human IFN-γ:human IFN-γR2 complex was observed [92, 123, 124]. This indicated that the two chains of the receptor must be in close proximity as suggested by experiments with chimeric receptors noted above. JAK-2 was found to associate with the IFN-γR2 intracellular domain [92]. Residues 263–267 and 270–274 of the intracellular domain of IFN-γR2 act as a JAK-2-binding site [124]. The cytoplasmic domains of the IFN-γR2 subunits, like the cytoplasmic domains of the IFN-γR1 chains, can be interchanged between species with no loss of biological activity, confirming that the species-specific interaction of the IFN-γR1 and -γR2 chains involves only the extracellular domains of the two proteins [125].

5.5.4
The Functional IFN-γR Complex

A model for the functional IFN-γR is shown in Fig. 5.3. The IFN-γR1 chains bind the ligand, whereas the IFN-γR2 chains serve to complete the complex for signal transduction.

5.5.4.1 Specificity of Ligand Binding

There are two distinct interactions which underlie species specificity: the binding of ligand to the IFN-γR1 chain, and the interaction of the IFN-γR1 and -γR2 chains, independent of ligand. The interaction of IFN-γ with the IFN-γR1 chain is highly species specific. Human IFN-γ stimulates no activity on rodent cells whatsoever and rodent IFN-γ exhibits no activity on human cells. This effect is attributed to the interaction of the ligand with the IFN-γR1 chain. The presence of the second chain IFN-γR2 increases the overall affinity of the ligand to the receptor complex slightly [92, 123, 126].

5.5.4.2 Specificity of the Interactions of the Two Chains

Chimeric receptors were constructed between the mouse and human IFN-γR1 chains [111–114]. The extracellular, transmembrane and intracellular domains were swapped between the human and mouse IFN-γR1 chains. The chains were introduced into mouse–human and hamster–human somatic cell hybrids that contained human chromosome 21 which expressed the human IFN-γR2 chain. These results established that the extracellular domains of the two chains must be matched from the same species to enable the ligand to initiate signal transduction. In a complementary series of experiments with hamster–mouse somatic cell hybrid cells, it was shown that mouse IFN-γ could only activate the hybrid cells when the mouse chromosome 16 was present in the hybrids and the IFN-γR1 chain contained the extracellular mouse domain of this chain [103, 113].

5.5.5
Specificity of Signal Transduction

Each cytokine that utilizes the JAK–STAT signal transduction pathway activates a distinct combination of members of the JAK and STAT families. Thus, either the JAKs, the STATs or both could contribute to the specificity of ligand action. With the use of chimeric receptors involving the IFN-γR complex as a model system, it was demonstrated that JAK-2 activation is not an absolute requirement for IFN-γ signaling [127]. Other members of the JAK family can functionally substitute for JAK-2. IFN-γ can signal through the activation of JAK family members other than JAK-2 as measured by STAT-1α activation and MHC class I antigen expression. This indicates that JAKs are interchangeable in the JAK–STAT signal transduction pathway. The necessity for the activation of one particular kinase during signaling can be overcome by recruiting another kinase to the receptor complex. The results suggest that the JAKs do not contribute significantly to the specificity of signal transduction.

The STATs represent proteins containing SH2, SH3 and DNA-binding domains (for reviews, see [26, 128]). The highly selective and specific interaction between STAT SH2 domains and the phosphotyrosine-containing STAT recruitment sites on the intracellular domains of the cytokine receptors determines which STATs are to be recruited to a particular receptor complex [129, 130]. Thus, the major specificity of the pathway is likely due in large part to the specificity of the STAT recruitment sites on the receptor chains. Other molecules that interact with JAKs and STATs may contribute to the specificity and range of the interaction [131, 132].

Patients with a deficiency in the IFN-γR (-γR1 or -γR2 chains) are susceptible to mycobacterial infections [133–139]. Bacille Calmette-Guerin (BGC) vaccine is often fatal in children with deficiencies in the IFN-γR1 chain.

5.5.6
Receptor Structure

Crystallographic analysis of the IFN-γ:IFN-γR1 complex suggests that each monomer of the IFN-γ homodimer binds one IFN-γR1 and one IFN-γR2 subunit [140].

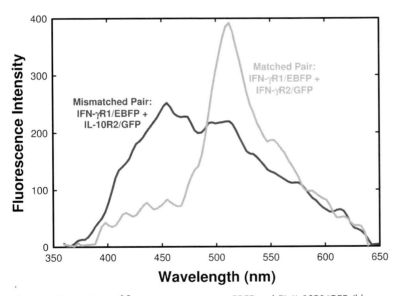

Fig. 5.4. Comparison of fluorescence emission spectra of cells expressing the matched and mismatched pair of receptor chains. The matched receptor chains are FL-IFN-γR2/GFP and IFN-γR1/EBFP (green curve); the mismatched receptor chains are IFN-γR1/EBFP and FL-IL-10R2/GFP (blue curve). The fluorescence emission spectra in response to two-photon excitation at 760 nm are shown. (Figure modified from data of Krause et al. [143].) (This figure also appears with the color plates.)

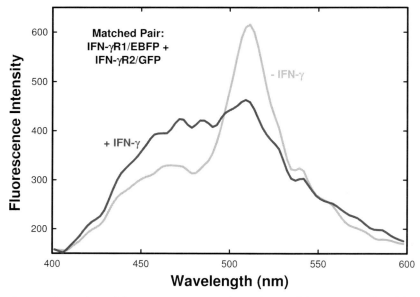

Fig. 5.5. Comparison of fluorescence spectra of cells expressing the matched pair of receptor chains in the presence and absence of IFN-γ. The matched receptor chains are FL-IFN-γR2/GFP and IFN-γR1/EBFP. The spectrum in green was taken in the absence of IFN-γ. IFN-γ (3500 U mL^{-1}) was then added to the medium and the spectrum taken (blue curve) of the same region in the same cell. The fluorescence emission spectra in response to two-photon excitation at 760 nm are shown. (Figure modified from data of Krause et al. [143].) (This figure also appears with the color plates.)

Biochemical analyses [141, 143] showed that one IFN-γ homodimer binds two IFN-γR1 chains. Thus the signal-transducing complex of IFN-γ consists of the IFN-γ homodimer bound to two IFN-γR1 and two IFN-γR2 chains (Fig. 5.2) which recruit JAK-1 and -2, respectively [92, 125, 127, 132], and JAK-2 phosphorylates JAK-1 following which the kinases phosphorylate Tyr457 of the IFN-γR1 chain. The phosphorylated segment of each IFN-γR1 chain recruits STAT-1α, which is then phosphorylated by JAK-1 or -2, then released to form the active STAT-1α transcription factor.

5.5.7
Preassembly of the Receptor Complex

A series of experiments were designed to test the hypothesis that the cell surface IFN-γR chains are preassembled rather than associated by ligand that has been the general belief. To evaluate this in live cells, fluorescence resonance energy transfer (FRET), a powerful spectroscopic technique that has been used to determine molecular interactions and distances between the donor and acceptor, was used [143]. The IFN-γR1 and -γR2 chains were labeled with Blue and Green Fluorescent Proteins (BFP and GFP) by fusing the BFP and GFP to the receptor proteins, then transfecting the chains into cells. In contrast to the prevailing view, the results demonstrated that the receptor chains are preassociated (Fig. 5.4) and that the intracellular domains move apart on binding the ligand IFN-γ (Fig. 5.5) [143].

5.6
The IL-28R1 and -10R2 Receptor Complex

IL-28A, -28B and -29 are induced by virus infection, exhibit antiviral activity, and use the JAK–STAT pathway and ISRE [16, 17, 144]. The IFN-like molecules IL-28A, -28B and -29 use the receptor complex consisting of IL-10R2 and -28R1 chains (Tabs. 5.1 and 5.2). Discoveries of the IFN and many cytokine receptors set the

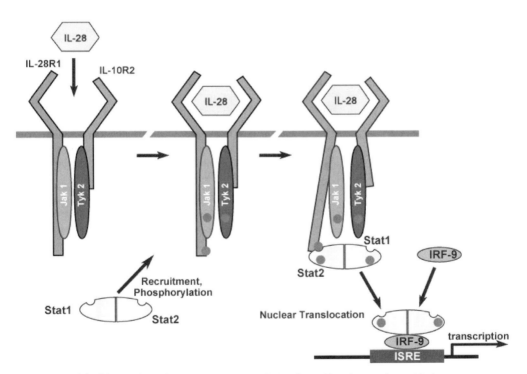

Fig. 5.6. Model of the IL-28R1/10R2 receptor complex. This receptor complex consists of two chains, IL-28R1 and -10R2, that are associated with JAK-1 and TYK-2, respectively, as shown. The STAT-1/2 heterodimer appears to function together with IRF-9 as the latent transcription factor. Based on analogy with the IFN-γR complex, we believe that this receptor complex is preassembled and undergoes a conformation change after its contact with ligand. (This figure also appears with the color plates.)

groundwork for discovery of other receptors [12], so that it was not necessary to use the difficult procedures that were required to identify the IFN receptor components. Thus, the discovery of the IL-28/29 cytokines and their receptor were reported at the same time [16, 17]. The IL-10R2 chain was discovered as the second chain of the IL-10 receptor complex [145, 146]. IL-10R2, the common chain of the class 2 cytokines, serves as the second chain of this complex. A summary of the IL-28R1/10R2 receptor complex and its signal transduction is shown in Fig. 5.6.

5.7
Overview of Multichain Receptors

The multichain cytokine class II receptors have two major chains exemplified by the IFN-γR complex (Fig. 5.3). The ligand-binding chain (IFN-γR1) and the accessory chain (AC; IFN-γR2) serve as a foundation for the functional IFN-γR complex [92, 125, 127, 132]. The question arises why should two separate chains have evolved when one in the correct configuration would suffice as for growth hormone. We postulate that the presence of two distinct chains provides for more effective control and fine tuning of responses to ligand. For example, the differences in response of T helper (T_h) 1 and T_h2 cells to IFN-γ results from the lack of expression of the IFN-γR2 chain in the T_h1 subset [147, 148], and allows exquisite fine-tuning of sensitivity to IFN-γ. Receptor preassembly [11] can allow multichain receptor complexes to function as a single entity and enable a rapid activation after contact with ligand. It is also possible that receptors with multiple chains could recruit additional factors into the complex to generate a wider variety of intracellular signals. This could explain how receptors with multiple subunits could activate a greater number of specific pathways and signals than those with fewer elements in the receptor complex.

5.8
Global Summary

The Type I and II IFN receptors have been substantially delineated so they have served as a base to understand other class 2 cytokine receptors. In particular, IFN-γ and the IFN-γR1 chain have been crystallized as singular entities and as a dual structure that has provided some insight into the interaction. Furthermore, fluorescent resonance energy measurements of the IFN-γ receptor chains have shown they are preassembled and are beginning to provide insight into the details of the events underlying the multiple interactions of the ligand, receptor and associated components. The IL-28R1/10R2 receptor complex is likely to fit reasonably well into the general structure of the IL-10 and IFN-γ receptors. Surprisingly, while the Type I IFN receptor has been generally outlined, there are many gaps that need to be filled to understand how this complex group of ligands produces so many different patterns of activity. It would be a major achievement to fill these gaps – an achievement that would have a significant impact on developing new generations of Type I IFNs.

References

1 PESTKA S. Methods in Enzymology. Interferon, Part. A. 1981, *79(A)*, 1–632.
2 PESTKA S. Cloning of human interferons. 599–601. In: Pestka, S., editor. *Methods Enzymol* Interferon, Part B. New York: Academic Press; **1981**, *79(B)*, 1–677.
3 PESTKA S. Interferon from 1981 to 1986. Pp 3–14. In: Pestka, S., editor. *Methods Enzymol* Interferon, Part C. **1986**, *119*, 1–845.
4 PESTKA S, LANGER JA, ZOON KC, SAMUEL CE. Interferons and their actions. *Annu Rev Biochem* **1987**, *56*, 727–777.
5 ROBERTS RM, CROSS JC, LEAMAN DW. Unique features of the trophoblast interferons. *Pharmacol Ther* **1991**, *51*, 329–345.
6 PESTKA S. The human interferon alpha species and hybrid proteins. *Semin Oncol* **1997**, *24 (Suppl 9)*, S9-4–S9-17.
7 STARK GR, KERR IM, WILLIAMS BR, SILVERMAN RH, SCHREIBER RD. How cells respond to interferons. *Annu Rev Biochem* **1998**, *67*, 227–264.
8 ORITANI K, MEDINA KL, TOMIYAMA Y, ISHIKAWA J, OKAJIMA Y, OGAWA M, et al. Limitin: an interferon-like cytokine that preferentially influences B-lymphocyte precursors. *Nat Med* **2000**, *6*, 659–666.
9 PESTKA S. The human interferon α species and receptors. *Biopolymers* **2000**, *55*, 254–287.
10 LAFLEUR DW, NARDELLI B, TSAREVA T, MATHER D, FENG P, SEMENUK M, et al. Interferon-κ, a novel Type I interferon expressed in human keratinocytes. *J Biol Chem* **2001**, *276*, 39765–39771.
11 KRAUSE CD, PESTKA S. Evolution of the class 2 cytokines and receptors, and discovery of new friends and relatives. *Pharmacol Ther* **2005**, *106*, 299–346.
12 PESTKA S, KRAUSE CD, SARKAR D, WALTER MR, SHI Y, FISHER PB. Interleukin-10 and related cytokines and receptors. *Annu Rev Immunol* **2004**, *22*, 929–979.
13 PESTKA S, KRAUSE CD, WALTER MR. Interferons, interferon-like cytokines, and their receptors. *Immunol Rev* **2004**, *202*, 8–32.
14 LEFEVRE F, GUILLOMOT M, D'ANDREA S, BATTEGAY S, LA BONNARDIERE C. Interferon-δ: the first member of a novel Type I interferon family. *Biochimie* **1998**, *80*, 779–788.
15 ROBERTS RM, CROSS JC, LEAMAN DW. Unique features of the trophoblast interferons. *Pharmacol Ther* **1991**, *51*, 329–345.
16 KOTENKO SV, GALLAGHER G, BAURIN VV, LEWIS-ANTES A, SHEN M, SHAH NK, et al. IFN-λs mediate antiviral protection through a distinct class II cytokine receptor complex. *Nat Immunol* **2003**, *4*, 69–77.

17 Sheppard P, Kindsvogel W, Xu W, Henderson K, Schlutsmeyer S, Whitmore TE, et al. IL-28, IL-29 and their class II cytokine receptor IL-28R. *Nat Immunol* **2003**, *4*, 63–68.

18 Chen J, Wood WI. Interferon PRO655. *US Patent 6,300,475*, **2003**.

19 Hardy MP, Owczarek CM, Jermiin LS, Ejdeback M, Hertzog PJ. Characterization of the Type I interferon locus and identification of novel genes. *Genomics* **2004**, *84*, 331–345.

20 Pestka S. The human interferons – from protein purification and sequence to cloning and expression in bacteria: before, between, and beyond. *Arch Biochem Biophys* **1983**, *221*, 1–37.

21 Stark GR, Kerr IM, Williams BR, Silverman RH, Schreiber RD. How cells respond to interferons. *Annu Rev Biochem* **1998**, *67*, 227–264.

22 Aguet M. High-affinity binding of ^{125}I-labeled mouse interferon to a specific cell surface receptor. *Nature* **1980**, *284*, 768–770.

23 Lengyel P. Biochemistry of interferons and their actions. *Annu Rev Biochem* **1982**, *51*, 251–282.

24 Langer JA, Pestka S. Interferon receptors. *Immunol Today* **1988**, *9*, 393–400.

25 Sen GC, Lengyel P. The interferon system. A bird's eye view of its biochemistry. *J Biol Chem* **1992**, *267*, 5017–5020.

26 Darnell JE, Jr., Kerr IM, Stark GR. Jak–Stat pathways and transcriptional activation in response to IFNs and other extracellular signaling proteins. *Science* **1994**, *264*, 1415–1421.

27 Uze G, Lutfalla G, Mogensen KE. Alpha and beta interferons and their receptor and their friends and relations. *J Interferon Cytokine Res* **1995**, *15*, 3–26.

28 Branca AA, Baglioni C. Evidence that Type I and II interferons have different receptors. *Nature* **1981**, *294*, 768–770.

29 Flores I, Mariano TM, Pestka S. Human interferon omega binds to the interferon-α/β receptor. *J Biol Chem* **1991**, *266*, 19875–19877.

30 Li J, Roberts RM. Structure–function relationships in the interferon-tau (IFN-tau). Changes in receptor binding and in antiviral and antiproliferative activities resulting from site-directed mutagenesis performed near the carboxyl terminus. *J Biol Chem* **1994**, *269*, 24826–24833.

31 Alexenko AP, Li J, Mathialagan N, Izotova L, Mariano TM, Pestka S, et al. Interaction of bovine interferon-τ with the Type I interferon receptor on daudi cells. *J Interferon Cytokine Res* **1995**, *15 (Suppl 1)*, S97.

32 Pestka S. The interferon receptors. *Semin Oncol* **1997**, *24 (Suppl 9)*, S9-18–S9-40.

33 Pestka S, Kotenko SV, Muthukumaran G, Izotova LS, Cook JR, Garotta G. The interferon γ (IFN-γ) receptor: a paradigm for the multichain cytokine receptor. *Cytokine Growth Factor Rev* **1997**, *8*, 189–206.

34 Fu XY, Schindler C, Improta T, Aebersold R, Darnell JE, Jr. The proteins of ISGF-3, the interferon α-induced transcriptional activator, define a gene family involved in

signal transduction. *Proc Natl Acad Sci USA* **1992**, *89*, 7840–7843.

35 DARNELL JE, JR. STATs and gene regulation. *Science* **1997**, *277*, 1630–1635.

36 SHUAI K, SCHINDLER C, PREZIOSO VR, DARNELL JE, JR. Activation of transcription by IFN-γ: tyrosine phosphorylation of a 91-kD DNA binding protein. *Science* **1992**, *258*, 1808–1812.

37 RAMANA CV, GRAMMATIKAKIS N, CHERNOV M, NGUYEN H, GOH KC, WILLIAMS BR, et al. Regulation of c-*myc* expression by IFN-γ through Stat1-dependent and -independent pathways. *EMBO J* **2000**, *19*, 263–272.

38 RAMANA CV, GIL MP, HAN Y, RANSOHOFF RM, SCHREIBER RD, STARK GR. Stat1-independent regulation of gene expression in response to IFN-γ. *Proc Natl Acad Sci USA* **2001**, *98*, 6674–6679.

39 GIL MP, BOHN E, O'GUIN AK, RAMANA CV, LEVINE B, STARK GR, et al. Biologic consequences of Stat1-independent IFN signaling. *Proc Natl Acad Sci USA* **2001**, *98*, 6680–6685.

40 HERBERMAN RB, ORTALDO JR, MANTOVANI A, HOBBS DS, KUNG H-F, PESTKA S. Effect of human recombinant interferon on cytotoxic activity of natural killer (NK) cells and monocytes. *Cell Immunol* **1982**, *67*, 160–167.

41 ORTALDO JR, HERBERMAN RB, PESTKA S. Augmentation of human natural killer cells with human leukocyte and human recombinant leukocyte interferon. In: HERBERMAN RB (Ed.), *NK Cells and Other Natural Effector Cells*. New York: Academic Press, **1982**, pp. 1279–1283.

42 REHBERG E, KELDER B, HOAL EG, PESTKA S. Specific molecular activities of recombinant and hybrid leukocyte interferons. *J Biol Chem* **1982**, *257*, 11497–11502.

43 ORTALDO JR, MASON A, REHBERG E, MOSCHERA J, KELDER B, PESTKA S, et al. Effects of recombinant and hybrid recombinant human leukocyte interferons on cytotoxic activity of natural killer cells. *J Biol Chem* **1983**, *258*, 15011–15015.

44 ORTALDO JR, MANTOVANI A, HOBBS D, RUBINSTEIN M, PESTKA S, HERBERMAN RB. Effects of several species of human leukocyte interferon on cytotoxic activity of nk cells and monocytes. *Int J Cancer* **1983**, *31*, 285–289.

45 ORTALDO JR, MASON A, REHBERG E, KELDER B, HARVEY C, OSHEROFF P, et al. Augmentation of NK activity with recombinant and hybrid recombinant human leukocyte interferons. In: DE MAEYER E, SCHELLEKENS H (Eds.), *The Biology of the Interferon System*. Amsterdam: Elsevier, **1983**, pp. 353–358.

46 ORTALDO JR, HERBERMAN RB, HARVEY C, OSHEROFF P, PAN YC, KELDER B, et al. A species of human α interferon that lacks the ability to boost human natural killer activity. *Proc Natl Acad Sci USA* **1984**, *81*, 4926–4929.

47 GREINER JW, HAND PH, NOGUCHI P, FISHER PB, PESTKA S, SCHLOM J. Enhanced expression of surface tumor-associated antigens on human breast and colon tumor cells after recombinant human leukocyte α-interferon treatment. *Cancer Res* **1984**, *44*, 3208–3214.

48 Greiner JW, Schlom J, Pestka S, Langer JA, Giacomini P, Kusama M, et al. Modulation of tumor associated antigen expression and shedding by recombinant human leukocyte and fibroblast interferons. *Pharmacol Ther* **1985**, *31*, 209–236.

49 Greiner JW, Guadagni F, Noguchi P, Pestka S, Colcher D, Fisher PB, et al. Recombinant interferon enhances monoclonal antibody-targeting of carcinoma lesions *in vivo*. *Science* **1987**, *235*, 895–898.

50 Basham TY, Bourgeade MF, Creasey AA, Merigan TC. Interferon increases HLA synthesis in melanoma cells: interferon-resistant and -sensitive cell lines. *Proc Natl Acad Sci USA* **1982**, *79*, 3265–3269.

51 Dolei A, Capobianchi MR, Ameglio F. Human interferon-γ enhances the expression of class I and class II major histocompatibility complex products in neoplastic cells more effectively than interferon-α and interferon-β. *Infect Immun* **1983**, *40*, 172–176.

52 Clemens MJ. Interferons and apoptosis. *J Interferon Cytokine Res* **2003**, *23*, 277–292.

53 Keay S, Grossberg SE. Interferon inhibits the conversion of 3T3-L1 mouse fibroblasts into adipocytes. *Proc Natl Acad Sci USA* **1980**, *77*, 4099–4103.

54 Fisher PB, Hermo H, Jr., Prignoli DR, Weinstein IB, Pestka S. Hybrid recombinant human leukocyte interferon inhibits differentiation in murine B-16 melanoma cells. *Biochem Biophys Res Commun* **1984**, *119*, 108–115.

55 Fisher PB, Prignoli DR, Hermo H, Jr., Weinstein IB, Pestka S. Effects of Combined treatment with interferon and mezerein on melanogenesis and growth in human melanoma cells. *J Interferon Res* **1985**, *5*, 11–22.

56 Fidler IJ. Regulation of neoplastic angiogenesis. *J Natl Cancer Inst Monogr* **2001**, 10–14.

57 Kerbel R, Folkman J. Clinical translation of angiogenesis inhibitors. *Nat Rev Cancer* **2002**, *2*, 727–739.

58 Tan YH, Tischfield J, Ruddle FH. The linkage of genes for the human interferon-induced antiviral protein and indophenol oxidase-B tarits to chromosome G-21. *J Exp Med* **1973**, *137*, 317–330.

59 Revel M, Bash D, Ruddle FH. Antibodies to a cell-surface component coded by human chromosome 21 inhibit action of interferon. *Nature* **1976**, *260*, 139–141.

60 Slate DL, Shulman L, Lawrence JB, Revel M, Ruddle FH. Presence of human chromosome 21 alone is sufficient for hybrid cell sensitivity to human interferon. *J Virol* **1978**, *25*, 319–325.

61 Epstein CJ, McManus NH, Epstein LB. Direct evidence that the gene product of the human chromosome 21 locus, IFRC, is the interferon-α receptor. *Biochem Biophys Res Commun* **1982**, *107*, 1060–1066.

62 Shulman LM, Kamarck ME, Slate DL, Ruddle FH, Branca AW, Baglioni C, et al. Antibodies to chromosome 21 coded cell surface components block binding of human alpha interferon but not gamma interferon to human cells. *Virology* **1984**, *137*, 422–427.

63 RAZIUDDIN A, GUPTA SL. Receptors for human interferon α: Two forms of interferon-receptor complexes identified by chemical cross linking. In: WILLIAMS BRG, SILVERMAN RH (Eds.), *The 2-5A System; Molecular and Clinical Aspects of the Interferon-Regulated Pathway*. New York: Alan R. Liss, **1985**, pp. 219–226.

64 LANGER JA, RASHIDBAIGI A, LAI LW, PATTERSON D, JONES C. Sublocalization on chromosome 21 of human interferon-alpha receptor gene and the gene for an interferon-gamma response protein. *Somat Cell Mol Genet* **1990**, *16*, 231–240.

65 LUTFALLA G, ROECKEL N, MOGENSEN KE, MATTEI M-G, UZÉ G. Assignment of human interferon-α receptor gene to chromosome 21q22.1 by *in situ* hybridization. *J Interferon Res* **1990**, *10*, 515–517.

66 JUNG V. *The Human Interferon Gamma Receptor and Signal Transduction*. Thesis. State University of New Jersey and Graduate School of Biomedical Sciences, Robert Wood Johnson Medical School, **1991**.

67 SOH J, MARIANO TM, LIM JK, IZOTOVA L, MIROCHNITCHENKO O, SCHWARTZ B, et al. Expression of a functional human Type I interferon receptor in hamster cells: application of functional yeast artificial chromosome (YAC) screening. *J Biol Chem* **1994**, *269*, 18102–18110.

68 JUNG V, PESTKA S. Selection and screening of transformed NIH3T3 cells for enhanced sensitivity to human interferons α and β. *Methods Enzymol* **1986**, *119*, 597–611.

69 REVEL M, COHEN B, ABRAMOVICH C, NOVICK D, RUBINSTEIN M, SHULMAN L. Components of the human Type I IFN receptor system. *J Interferon Res* **1991**, *11 (Suppl)*, S61.

70 UZE G, LUTFALLA G, GRESSER I. Genetic transfer of a functional human interferon alpha receptor into mouse cells: cloning and expression of its cDNA. *Cell* **1990**, *60*, 225–234.

71 UZÉ G, LUTFALLA G, BANDU MT, PROUDHON D, MOGENSEN KE. Behavior of a cloned murine interferon α/β receptor expressed in homospecific or heterospecific background. *Proc Natl Acad Sci USA* **1992**, *89*, 4774–4778.

72 LIM JK, XIONG J, CARRASCO N, LANGER JA. Intrinsic ligand binding properties of the human and bovine alpha-interferon receptors. *FEBS Lett* **1994**, *350*, 281–286.

73 COLAMONICI OR, D'ALESSANDRO F, DIAZ MO, GREGORY SA, NECKERS LM, NORDAN R. Characterization of three monoclonal antibodies that recognize the interferon alpha 2 receptor. *Proc Natl Acad Sci USA* **1990**, *87*, 7230–7234.

74 UZE G, LUTFALLA G, EID P, MAURY C, BANDU MT, GRESSER I, et al. Murine tumor cells expressing the gene for the human interferon alpha beta receptor elicit antibodies in syngeneic mice to the active form of the receptor. *Eur J Immunol* **1991**, *21*, 447–451.

75 COLAMONICI OR, PFEFFER LM, D'ALESSANDRO F, PLATANIAS LC, GREGORY SA, ROSOLEN A, et al. Multichain structure of the IFN-alpha receptor on hematopoietic cells. *J Immunol* **1992**, *148*, 2126–2132.

76 BENOIT P, MAGUIRE D, PLAVEC I, KOCHER H, TOVEY M, MEYER F. A monoclonal antibody to recombinant human

IFN-α receptor inhibits biologic activity of several species of human IFN-α, IFN-β and IFN-ω: Detection of heterogeneity of the cellular Type I IFN receptor. *J Immunol* **1993**, *150*, 707–716.

77 COLAMONICI OR, DOMANSKI P. Identification of a novel subunit of the Type I interferon receptor localized to human chromosome 21. *J Biol Chem* **1993**, *268*, 10895–10899.

78 HANNIGAN GE, LAU AS, WILLIAMS BR. Differential human interferon alpha receptor expression on proliferating and non-proliferating cells. *Eur J Biochem* **1986**, *157*, 187–193.

79 COLAMONICI OR, PFEFFER LM. Structure of the human interferon alpha receptor. *Pharmacol Ther* **1991**, *52*, 227–233.

80 MARIANO TM, DONNELLY R, SOH J, PESTKA S. Structure and Function of the Type I Interferon Receptor. In: BARON S, KLIMPEL GR, et al. (Eds.), Galveston, TX: University of Texas Medical Branch at Galveston, **1992**, pp. 129–138.

81 ABRAMOVICH C, SHULMAN LM, RATOVITSKI E, HARROCH S, TOVEY M, EID P, et al. Differential tyrosine phosphorylation of the IFNAR chain of the Type I interferon receptor and of an associated surface protein in response to IFN-alpha and IFN-beta. *EMBO J* **1994**, *13*, 5871–5877.

82 COOK JR, EMANUEL SL, PESTKA S. Yeast artificial chromosome fragmentation vectors that utilize URA3 selection. *Genet Anal Tech Appl* **1993**, *10*, 109–112.

83 SOH J, DONNELLY RJ, MARIANO TM, COOK JR, SCHWARTZ B, PESTKA S. Identification of a yeast artificial chromosome clone encoding an accessory factor for the human interferon gamma receptor: evidence for multiple accessory factors. *Proc Natl Acad Sci USA* **1993**, *90*, 8737–8741.

84 COOK JR, EMANUEL SL, DONNELLY RJ, SOH J, MARIANO TM, SCHWARTZ B, et al. Sublocalization of the human interferon-gamma receptor accessory factor gene and characterization of accessory factor activity by yeast artificial chromosomal fragmentation. *J Biol Chem* **1994**, *269*, 7013–7018.

85 SOH J, DONNELLY RJ, KOTENKO S, MARIANO TM, COOK JR, WANG N, et al. Identification and sequence of an accessory factor required for activation of the human interferon gamma receptor. *Cell* **1994**, *76*, 793–802.

86 SOH J, MARIANO TM, BRADSHAW G, DONNELLY RJ, PESTKA S. Generation of random internal deletion derivatives of YACs by homologous targeting to Alu sequences. *DNA Cell Biol* **1994**, *13*, 301–309.

87 EMANUEL SL, COOK JR, O'REAR J, ROTHSTEIN R, PESTKA S. New vectors for manipulation and selection of functional yeast artificial chromosomes (YACs) containing human DNA inserts. *Gene* **1995**, *155*, 167–174.

88 CLEARY CM, DONNELLY RJ, SOH J, MARIANO TM, PESTKA S. Knockout and reconstitution of a functional human Type I interferon receptor complex. *J Biol Chem* **1994**, *269*, 18747–18749.

89 MÜLLER U, STEINHOFF U, REIS LF, HEMMI S, PAVLOVIC J, ZINKERNAGEL RM, et al. Functional role of Type I and Type II interferons in antiviral defense. *Science* **1994**, *264*, 1918–1921.

90 Hwang SY, Hertzog PJ, Holland KA, Sumarsono SH, Tymms MJ, Hamilton JA, et al. A null mutation in the gene encoding a Type I interferon receptor component eliminates antiproliferative and antiviral responses to interferons alpha and beta and alters macrophage responses [published erratum appears in *Proc Natl Acad Sci USA* **1996**, *93*, 4519]. *Proc Natl Acad Sci USA* **1995**, *92*, 11284–11288.

91 Novick D, Cohen B, Rubinstein M. The human interferon alpha/beta receptor: characterization and molecular cloning. *Cell* **1994**, *77*, 391–400.

92 Kotenko SV, Izotova LS, Pollack BP, Mariano TM, Donnelly RJ, Muthukumaran G, et al. Interaction between the components of the interferon gamma receptor complex. *J Biol Chem* **1995**, *270*, 20915–20921.

93 Domanski P, Witte M, Kellum M, Rubinstein M, Hackett R, Pitha P, et al. Cloning and expression of a long form of the β subunit of the interferon α receptor that is required for signaling. *J Biol Chem* **1995**, *270*, 21606–21611.

94 Lutfalla G, Holland SJ, Cinato E, Monneron D, Reboul J, Rogers NC, et al. Mutant U5A cells complemented by an interferon-alpha/beta receptor subunit generated by alternative processing of a new member of the cytokine receptor gene cluster. *EMBO J* **1995**, *14*, 5100–5108.

95 Cook JR, Cleary CM, Mariano TM, Izotova L, Pestka S. Differential Responsiveness of a splice variant of the human Type I interferon receptor to interferons. *J Biol Chem* **1996**, *271*, 13448–13453.

96 Evinger M, Pestka S. Assay of growth inhibition in lymphoblastoid cell cultures. *Methods Enzymol* **1981**, *79(B)*, 362–368.

97 Evinger M, Rubinstein M, Pestka S. Antiproliferative and antiviral activities of human leukocyte interferons. *Arch Biochem Biophys* **1981**, *210*, 319–329.

98 Rashidbaigi A, Kung HF, Pestka S. Characterization of receptors for immune interferon in U937 cells with ^{32}P-labeled human recombinant immune interferon. *J Biol Chem* **1985**, *260*, 8514–8519.

99 Rashidbaigi A, Langer JA, Jung V, Jones C, Morse HG, Tischfield JA, et al. The gene for the human immune interferon receptor is located on chromosome 6. *Proc Natl Acad Sci USA* **1986**, *83*, 384–388.

100 Jung V, Rashidbaigi A, Jones C, Tischfield JA, Shows TB, Pestka S. Human chromosomes 6 and 21 are required for sensitivity to human interferon gamma. *Proc Natl Acad Sci USA* **1987**, *84*, 4151–4155.

101 Jung V, Jones C, Rashidbaigi A, Geyer DD, Morse HG, Wright RB, et al. Chromosome mapping of biological pathways by fluorescence-activated cell sorting and cell fusion: the human interferon gamma receptor as a model system. *Somat Cell Mol Genet* **1988**, *14*, 583–592.

102 Mariano TM, Kozak CA, Langer JA, Pestka S. The mouse immune interferon receptor gene is located on chromosome 10. *J Biol Chem* **1987**, *262*, 5812–5814.

103 Mariano TM, Muthukumaran G, Donnelly RJ, Wang N, Adamson MC, Pestka S, et al. Genetic mapping of the gene

for the mouse interferon-gamma receptor signaling subunit to the distal end of chromosome 16. *Mammal Genome* **1996**, *7*, 321–322.

104 AGUET M, DEMBIC Z, MERLIN G. Molecular cloning and expression of the human interferon-gamma receptor. *Cell* **1988**, *55*, 273–280.

105 KUMAR CS, MUTHUKUMARAN G, FROST LJ, NOE M, AHN Y-H, MARIANO TM, et al. Molecular characterization of the murine interferon gamma receptor cDNA. *J Biol Chem* **1989**, *264*, 17939–17946.

106 HEMMI S, PEGHINI P, METZLER M, MERLIN G, DEMBIC Z, AGUET M. Cloning of Murine interferon gamma receptor cDNAs expression in human cells mediates high-affinity binding but is not sufficient to confer sensitivity to murine interferon gamma. *Proc Natl Acad Sci USA* **1989**, *86*, 9901–9905.

107 GRAY PW, LEONG S, FENNIE EH, FARRAR MA, PINGEL JT, FERNANDEZ-LUNA J, et al. Cloning and expression of the cDNA for the murine interferon gamma receptor. *Proc Natl Acad Sci USA* **1989**, *86*, 8497–8501.

108 MUNRO S, MANIATIS T. Expression cloning of the murine interferon gamma receptor, cDNA. *Proc Natl Acad Sci USA* **1989**, *86*, 9248–9252.

109 COFANO F, MOORE SK, TANAKA S, YUHKI N, LANDOLFO S, APPELLA E. Affinity purification, peptide analysis and cDNA sequence of the mouse interferon-gamma receptor. *J Biol Chem* **1990**, *265*, 4064–4071.

110 JUNG V, JONES C, KUMAR CS, STEFANOS S, O'CONNELL S, PESTKA S. Expression and reconstitution of a biologically active human interferon gamma receptor in hamster cells. *J Biol Chem* **1990**, *265*, 1827–1830.

111 GIBBS VC, WILLIAMS SR, GRAY PW, SCHREIBER RD, PENNICA D, RICE G, et al. The extracellular domain of the human interferon gamma receptor interacts with a species-specific signal transducer. *Mol Cell Biol* **1991**, *11*, 5860–5866.

112 HEMMI S, MERLIN G, AGUET M. Functional characterization of a hybrid human–mouse interferon-gamma receptor: evidence for species-specific interaction of the extracellular receptor domain with a putative signal transducer. *Proc Natl Acad Sci USA* **1992**, *89*, 2737–2741.

113 HIBINO Y, KUMAR CS, MARIANO TM, LAI D, PESTKA S. Chimeric interferon gamma receptors demonstrate that an accessory factor required for activity interacts with the extracellular domain. *J Biol Chem* **1992**, *267*, 3741–3749.

114 KALINA U, OZMAN L, DADOVA KD, GENTZ R, GAROTTA G. The human gamma interferon receptor accessory factor encoded by chromosome 21 transduces the signal for the induction of 2′,5′-oligoadenylate-synthetase, resistance to virus cytopathic effect, and major histocompatibility complex class I antigens. *J Virology* **1993**, *67*, 1702–1706.

115 FARRAR MA, FERNANDEZ-LUNA J, SCHREIBER RD. Identification of two regions within the cytoplasmic domain of the human interferon-gamma receptor required for function. *J Biol Chem* **1991**, *266*, 19626–19635.

116 Cook JR, Jung V, Schwartz B, Wang P, Pestka S. Structural analysis of the human interferon gamma receptor: a small segment of the intracellular domain is specifically required for class I major histocompatibility complex antigen induction and antiviral activity. *Proc Natl Acad Sci USA* **1992**, *89*, 11317–11321.

117 Farrar MA, Campbell JD, Schreiber RD. Identification of a functionally important sequence in the C terminus of the interferon-gamma receptor. *Proc Natl Acad Sci USA* **1992**, *89*, 11706–11710.

118 Greenlund AC, Farrar MA, Viviano BL, Schreiber RD. Ligand-induced IFN gamma receptor tyrosine phosphorylation couples the receptor to its signal transduction system (p91). *EMBO J* **1994**, *13*, 1591–1600.

119 Kaplan DH, Greenlund AC, Tanner JW, Shaw AS, Schreiber RD. Identification of an interferon-gamma receptor alpha chain sequence required for JAK-1 binding. *J Biol Chem* **1996**, *271*, 9–12.

120 Hemmi S, Bohni R, Stark G, Di Marco F, Aguet M. A novel member of the interferon receptor family complements functionality of the murine interferon gamma receptor in human cells. *Cell* **1994**, *76*, 803–810.

121 Lembo D, Ricciardi-Castagnoli P, Alber G, Ozmen L, Landolfo S, Bluthmann H, et al. Mouse macrophages carrying both subunits of the human interferon-gamma (IFN-gamma) receptor respond to human IFN-gamma but do not acquire full protection against viral cytopathic effect. *J Biol Chem* **1996**, *271*, 32659–32666.

122 Hemmi S, Bohni R, Aguet M. *J Interferon Cytokine Res* **1994**, *14*, S94.

123 Marsters SA, Pennica D, Bach EA, Schreiber RD, Ashkenazi A. Interferon gamma signals via a high-affinity multisubunit receptor complex that contains two types of polypeptide chain. *Proc Natl Acad Sci USA* **1995**, *92*, 5401–5405.

124 Bach EA, Tanner JW, Marsters S, Ashkenazi A, Aguet M, Shaw AS, et al. Ligand-induced assembly and activation of the gamma interferon receptor in intact cells. *Mol Cell Biol* **1996**, *16*, 3214–3221.

125 Muthukumaran G, Donnelly RJ, Ebensperger C, Mariano TM, Garotta G, Dembic Z, et al. The intracellular domain of the second chain of the interferon-gamma receptor is interchangeable between species. *J Interferon and Cytokine Res* **1996**, *16*, 1039–1045.

126 Lai D. *The Mapping of the Murine and Human Interferon Gamma Receptor.* Thesis. State University of New Jersey and Graduate School of Biomedical Sciences, Robert Wood Johnson Medical School, **1994**.

127 Kotenko SV, Izotova LS, Pollack BP, Muthukumaran G, Paukku K, Silvennoinen O, et al. Other kinases can substitute for Jak2 in signal transduction by IFN-gamma. *J Biol Chem* **1996**, *271*, 17174–17182.

128 Fu X-Y. A direct signalling pathway through tyrosine kinases activation of SH2 domain-containing transcriptional factors. *J Leukoc Biol* **1995**, *57*, 529–535.

129 Heim MH, Kerr IM, Stark GR, Darnell JE, Jr. Contribution of STAT SH2 groups to specific interferon signaling by the Jak–STAT pathway. *Science* **1995**, *267*, 1347–1349.

130 Stahl N, Farruggella TJ, Boulton TG, Zhong Z, Darnell JE, Jr., Yancopoulos GD. Choice of STATs and other substrates specified by modular tyrosine-based motifs in cytokine receptors. *Science* **1995**, *267*, 1349–1353.

131 Pollack BP, Kotenko SV, Pestka S. Use of the yeast two-hybrid system to study interferon signal transduction. *J Interferon Cytokine Res* **1995**, *15 (Suppl 1)*, S67.

132 Muthukumaran G, Kotenko S, Donnelly R, Ihle JN, Pestka S. Chimeric erythropoietin–interferon gamma receptors reveal differences in functional architecture of intracellular domains for signal transduction. *J Biol Chem* **1997**, *272*, 4993–4999.

133 Jouanguy E, Altare F, Lamhamedi S, Revy P, Emile JF, Newport M, et al. Interferon-gamma-receptor deficiency in an infant with fatal bacille Calmette-Guerin infection. *N Engl J Med* **1996**, *335*, 1956–1961.

134 Newport MJ, Huxley CM, Huston S, Hawrylowicz CM, Oostra BA, Williamson R, et al. A mutation in the interferon-gamma-receptor gene and susceptibility to mycobacterial infection. *N Engl J Med* **1996**, *335*, 1941–1949.

135 Dorman SE, Holland SM. Mutation in the signal-transducing chain of the interferon-gamma receptor and susceptibility to mycobacterial infection. *J Clin Invest* **1998**, *101*, 2364–2369.

136 Jouanguy E, Lamhamedi-Cherradi S, Lammas D, Dorman SE, Fondaneche MC, Dupuis S, et al. A human IFNGR1 small deletion hotspot associated with dominant susceptibility to mycobacterial infection. *Nat Genet* **1999**, *21*, 370–378.

137 Doffinger R, Jouanguy E, Dupuis S, Fondaneche MC, Stephan JL, Emile JF, et al. Partial interferon-gamma receptor signaling chain deficiency in a patient with bacille Calmette-Guerin and *Mycobacterium abscessus* infection. *J Infect Dis* **2000**, *181*, 379–384.

138 Sasaki Y, Nomura A, Kusuhara K, Takada H, Ahmed S, Obinata K, et al. Genetic basis of patients with bacille Calmette-Guerin osteomyelitis in Japan: identification of dominant partial interferon-gamma receptor 1 deficiency as a predominant type. *J Infect Dis* **2002**, *185*, 706–709.

139 Dorman SE, Picard C, Lammas D, Heyne K, van Dissel JT, Baretto R, et al. Clinical features of dominant and recessive interferon gamma receptor 1 deficiencies. *Lancet* **2004**, *364*, 2113–2121.

140 Walter MR, Windsor WT, Nagabhushan TL, Lundell DJ, Lunn CA, Zauodny PJ, et al. Crystal structure of a complex between interferon-gamma and its soluble high-affinity receptor. *Nature* **1995**, *376*, 230–235.

141 Fountoulakis M, Zulauf M, Lustig A, Garotta G. Stoichiometry of interaction between interferon gamma and its receptor. *Eur J Biochem* **1992**, *208*, 781–787.

142 Fountoulakis M, Mesa C, Schmid G, Gentz R, Manneberg M, Zulauf M, et al. Interferon gamma receptor extracellular

domain expressed as IgG fusion protein in Chinese hamster ovary cells. Purification, biochemical characterization, and stoichiometry of binding. *J Biol Chem* **1995**, *270*, 3958–3964.

143 KRAUSE CD, MEI E, XIE J, JIA Y, BOPP MA, HOCHSTRASSER RM, et al. Seeing the light: preassembly and ligand-induced changes of the interferon gamma receptor complex in cells. *Mol Cell Proteomics* **2002**, *1*, 805–815.

144 DUMOUTIER L, LEJEUNE D, HOR S, FICKENSCHER H, RENAULD JC. Cloning of a new Type II cytokine receptor activating signal transducer and activator of transcription (STAT)1, STAT2 and STAT3. *Biochem J* **2003**, *370*, 391–396.

145 KOTENKO SV, KRAUSE CD, IZOTOVA LS, POLLACK BP, WU W, PESTKA S. Identification and functional characterization of a second chain of the interleukin-10 receptor complex. *EMBO J* **1997**, *16*, 5894–5903.

146 KOTENKO SV, PESTKA S. Jak–Stat signal transduction pathway through the eyes of cytokine class II receptor complexes. *Oncogene* **2000**, *19*, 2557–2565.

147 ADOLF GR, KALSNER I, AHORN H, MAURER-FOGY I, CANTELL K. Natural human interferon-alpha 2 is O-glycosylated. *Biochem J* **1991**, *276*, 511–518.

148 ADOLF GR, MAURER-FOGY I, KALSNER I, CANTELL K. Purification and characterization of natural human interferon omega 1. Two alternative cleavage sites for the signal peptidase. *J Biol Chem* **1990**, *265*, 9290–9295.

149 BACH EA, AGUET M, SCHREIBER RD. The IFNγ receptor: a paradigm for cytokine receptor signaling. *Annu Rev Immunol* **1997**, *15*, 563–591.

150 LUTFALLA G, MCINNIS MG, ANTONARAKIS SE, UZE G. Structure of the human CRFB4 gene: comparison with its IFNAR neighbor. *J Mol Evol* **1995**, *41*, 338–344.

151 LUTFALLA G, GARDINER K, UZE G. A new member of the cytokine receptor gene family maps on chromosome 21 at less than 35 kb from IFNAR. *Genomics* **1993**, *16*, 366–373.

152 KOTENKO SV. The family of IL-10-related cytokines and their receptors: related, but to what extent? *Cytokine Growth Factor Rev* **2002**, *13*, 223–240.

153 PESTKA S, KELDER B, FAMILLETTI PC, MOSCHERA JA, CROWL R, KEMPNER ES. Molecular weight of the functional unit of human leukocyte, fibroblast, and immune interferons. *J Biol Chem* **1983**, *258*, 9706–9709.

154 KAWAMOTO S, ORITANI K, ASADA H, TAKAHASHI I, ISHIKAWA J, YOSHIDA H, et al. Antiviral activity of limitin against encephalomyocarditis virus, herpes simplex virus, and mouse hepatitis virus: diverse requirements by limitin and alpha interferon for interferon regulatory factor 1. *J Virol* **2003**, *77*, 9622–9631.

155 AOKI K, SHIMODA K, ORITANI K, MATSUDA T, KAMEZAKI K, MUROMOTO R, et al. Limitin, an interferon-like cytokine, transduces inhibitory signals on B-cell growth through activation of Tyk2, but not Stat1, followed by induction and nuclear translocation of Daxx. *Exp Hematol* **2003**, *31*, 1317–1322.

156 Pestka S. Human leukocyte interferon Hu-IFN-α.001. *US Patent 5,789,551*, **1998**.

157 Llorente L, Zou W, Levy Y, Richaud-Patin Y, Wijdenes J, Alcocer-Varela J, et al. Role of interleukin 10 in the B lymphocyte hyperactivity and autoantibody production of human systemic lupus erythematosus. *J Exp Med* **1995**, *181*, 839–844.

158 Touitou R, Cochet C, Joab I. Transcriptional analysis of the Epstein–Barr virus interleukin-10 homologue during the lytic cycle. *J Gen Virol* **1996**, *77*, 1163–1168.

159 Park YB, Lee SK, Kim DS, Lee J, Lee CH, Song CH. Elevated interleukin-10 levels correlated with disease activity in systemic lupus erythematosus. *Clin Exp Rheumatol* **1998**, *16*, 283–288.

160 Liva SM, Voskuhl RR. Testosterone acts directly on CD4$^+$ T lymphocytes to increase IL-10 production. *J Immunol* **2001**, *167*, 2060–2067.

161 Takayama T, Morelli AE, Onai N, Hirao M, Matsushima K, Tahara H, et al. Mammalian and viral IL-10 enhance C–C chemokine receptor 5 but down-regulate C–C chemokine receptor 7 expression by myeloid dendritic cells: impact on chemotactic responses and *in vivo* homing ability. *J Immunol* **2001**, *166*, 7136–7143.

162 Tilg H, van Montfrans C, van den EA, Kaser A, van Deventer SJ, Schreiber S, et al. Treatment of Crohn's disease with recombinant human interleukin 10 induces the proinflammatory cytokine interferon gamma. *Gut* **2002**, *50*, 191–195.

163 Orange S, Horvath J, Hennessy A. Preeclampsia is associated with a reduced interleukin-10 production from peripheral blood mononuclear cells. *Hypertens Pregnancy* **2003**, *22*, 1–8.

164 Zou J, Clark MS, Secombes CJ. Characterisation, expression and promoter analysis of an interleukin 10 homologue in the puffer fish, *Fugu rubripes*. *Immunogenetics* **2003**, *55*, 325–335.

6
Type III Interferons: The Interferon-λ Family

Sergei V. Kotenko and Raymond P. Donnelly

6.1
Introduction

Hundreds of different viruses are able to infect humans and cause various diseases. Depending on the type of virus, the severity of the disease may range from an asymptomatic opportunistic viral infection to a life-threatening disease. Importantly, there are no absolute remedies to cure viral infections. In most cases the outcome of viral infections depends entirely on the ability of the immune system to recognize, restrain and, ultimately, to eliminate the virus. Therefore, the best therapeutic strategy to effectively deal with viral infections is to enhance the protective antiviral forces of the immune system, rather than to directly target the virus that is actually causing the disease.

The biology, life cycle and pathogenesis of different viruses are widely divergent. Nevertheless, based on current knowledge, it seems that the immune system employs essentially the same mechanisms of antiviral protection to battle many types of viruses. This mechanism relies entirely on the action of interferons (IFNs). These small secreted proteins are released by virus-infected cells to warn neighboring cells about viral presence and to force these cells to deploy various means of antiviral protection. The robust production and secretion of IFNs in response to viral infections is currently considered to be a key event for the establishment of a multifaceted antiviral response.

Until recently, our understanding of the rapid, innate response to viruses focused on the classical type I and II IFNs. Type I IFNs (IFN-α/β) are well known for their ability to induce potent antiviral protection in a wide variety of cells (see other chapters for details). However, recently, a new type of IFN, the type III IFNs (IFN-λs), has been identified nearly 50 years after the term "IFN" was first used to describe a group of proteins (factors) produced by virus-infected cells and capable of interfering with viral replication [1–3]. A family of proteins, distinct from type I IFNs, which are capable of inducing antiviral protection was discovered simultaneously by two research teams [2, 3]. Both groups found that three genes encode distinct, but highly homologous, proteins; based on their antiviral activity, we des-

The Interferons: Characterization and Application. Edited by Anthony Meager
Copyright © 2006 WILEY-VCH Verlag GmbH & Co. KGaA, Weinheim
ISBN: 3-527-31180-7

ignated these as a novel group of IFNs as IFN-λ1, -λ2 and -λ3 [3]. These proteins were also alternatively designated as interleukin (IL)-29, -28A and -28B, respectively, by the other group who identified them [2]. Consistent with their antiviral activity, the expression of IFN-λ mRNAs is inducible by viral infections in various cell lines. All three IFN-λ proteins utilize the same receptor complex for signaling and this receptor is distinct from the type I IFN receptor complex. Although IFN-λs do not use the IFN-α/β receptor complex for signaling, the engagement of either the IFN-λ or -α receptor complex results in activation of overlapping Janus kinase (JAK)–signal transducers and activators of transcription (STAT) signal transduction events, including formation of the IFN-stimulated gene factor 3 (ISGF-3) transcription factor complex. Therefore, the IFN-λ ligand–receptor system, upon engagement, leads to the establishment of an antiviral state by a mechanism similar to, but independent from, that used by type I IFNs. The discovery of this novel antiviral system represents a new direction in antiviral research and may lead to a better understanding of the complex nature of virus–host interactions. The discovery of the IFN-λ family may also lead to the development of novel antiviral therapeutic agents.

6.2
The Class II Cytokine Receptor Family (CRF2) and their Ligands

IFN-λs belong to a family of 27 related human cytokines which share common functional and structural features [4–14]. Because all of these cytokines use receptors from the class II CRF for signaling, these cytokines were collectively designated CRF2 cytokines. The CRF2 cytokines fall into four structurally related groups: the IL-10 family, the type I IFNs (IFN-α/β), type II IFN (IFN-γ) and, more recently, the type III IFNs (IFN-λ). The family of IL-10-related cytokines consists of six members of cellular origin: IL-10, -19, -20, -22, -24 and -26, as well as several viral cytokines. The type I IFN family consists of 13 IFN-α species, and a single species of IFN-β, -κ, -ω and -ε [6, 9, 15, 16]. There is also a mouse protein designated "limitin" with homology to type I IFN that apparently has no known human counterpart [17]. Type II IFN is represented by a single member, IFN-γ. IFN-γ is quite distinct in its primary structure from all other CRF2 ligands, although its tertiary structure is very similar to that of IL-10. The recently discovered IFN-λ family is comprised of three distinct, closely related proteins, IFN-λ1, -λ2 and -λ3 ([2, 3]; these cytokines are also referred to as IL-29, -28A and -28B, respectively, by some laboratories [2]).

Although their sequences and receptors are distinct from both type I IFN and the IL-10 family members, several observations indicate that the IFN-λ family is phylogenetically related to both the IL-10 and type I IFN families. IFN-λs demonstrate a clear, albeit very limited, sequence relationship (about 10–18% identity) to members of both the type I IFN and IL-10 families. IFN-λ genes are composed of five exons, thereby resembling the structural organization of genes encoding IL-10-related cytokines [7], while the genes encoding type I IFNs lack introns. Furthermore, the positions of the introns with respect to the protein reading frames are conserved for the IFN-λ genes and for the genes of the IL-10-related cytokines. As discussed below, IFN-λs signal through a receptor complex composed of the unique IFN-λR1 chain and the IL-10R2 chain – a protein that is also a component of the IL-10, -22 and -26 receptor complexes. These structural features make the IFN-λs seem more closely related to members of the IL-10 family. However, unlike the other cytokines in the IL-10 family, IFN-λs possess antiviral activity, which is considered to be a "unique family business" of the IFN family. Such "confused" identity of the IFN-λs is reflected in their nomenclature, being simultaneously designated as "IFNs" by our group [3] and as "ILs" by Sheppard et al. [2].

The functions of the CRF2 members and their ligands have been defined to differing degrees. All CRF2 cytokines mediate important biological activities associated with the induction and regulation of immune and inflammatory responses. While the anti-inflammatory and T cell-regulatory activities of IL-10 are well characterized [18], the functions of other IL-10-related cytokines are less well defined. IL-22 upregulates expression of several acute phase proteins in hepatocytes, induces expression of the pancreatitis-associated protein (PAP)-1 in pancreatic acinar cells and increases expression of β-defensin 2 and β-defensin 3 in keratinocytes [19–21]. Adenovirus-mediated delivery of IL-22 into mice induced systemic acute-phase response [22, 23]. IL-22 and -22R1 mRNAs were upregulated in diseased joints of animals with collagen-induced arthritis and IL-22 neutralizing antibody decreased severity of joint pathology [22]. Increasing evidence suggests a potential role for both IL-22 and -20 in the pathogenesis of psoriasis [21, 24, 25]. IL-20 and -22 receptor chains are expressed in keratinocytes from the lesional and non-lesional skin biopsies of patients with psoriasis, and IL-20 and -22 gene expression is elevated in the lesional psoriatic skin [21, 24, 25]. In addition, both IL-22 and -20 transgenic mice died within days after birth, sharing similar skin abnormalities comparable to those observed in psoriasis [24, 25]. However, the immune infiltrates commonly present in psoriatic skin were observed only in the skin of IL-22 transgenic mice [24, 25]. Subcutaneous administration of IL-22, but not -20, caused epidermal thickening and immune cell infiltration, which were neutralized by IL-22-binding protein (IL-22BP), a naturally occurring soluble receptor that acts as IL-22 antagonist [25–28]. The functional activities of IL-19 and -24 are not yet fully understood, but IL-19 expression is inducible by a variety of proinflammatory stimuli in human monocytes [29–31]. Because IL-19 and -24 use the same receptors that are used by IL-20 and -22, they may also share at least a subset of biological activities. The function of IL-26 is currently unknown. IFNs are major players in the establishment of a multi-faceted antiviral response. Type I IFNs induce potent antiviral activity. However, they also stimulate an increasing list of activities which link innate and acquired immunity, including dendritic cell (DC) maturation, T helper (T_h) cell biasing, B cell differentiation and natural killer (NK) activation. Although IFN-γ has weak intrinsic antiviral activity, it plays a stronger role in protection against intracellular parasites and is a major activator of antitumor defenses [32–34]. IFN-γ affects diverse aspects of innate immunity, such as the activation of macrophages. It also has strong effects on acquired responses, particularly in cell-mediated immunity, where it promotes the development of CD4$^+$ T$_h$1 cells and cytotoxic CD8$^+$ T cells, while suppressing CD4$^+$ T$_h$2 cells. Thus, IFN-γ is more commonly thought of as a T$_h$1 cytokine that functions in both innate and adaptive immunity.

Several structural and functional features unite the CRF2 cytokines. They all use CRF2 receptor proteins for signaling and these receptor proteins share homology in their extracellular domains. CRF2 cytokines also demonstrate limited primary and structural similarity. Although these cytokines regulate a broad array of biological activities, it is possible to ascribe a common function to these diverse cytokines. First of all, all of them are involved in the regulation of inflammatory responses. Some cytokines such as IFN-γ and type I IFNs promote inflammation, whereas IL-10 has strong anti-inflammatory activity. Second, the primary cellular source of all these cytokines is leukocytes such as lymphocytes and macrophages [29, 30]. A variety of microorganisms and their associated toxins are able to induce expression of these cytokines by leukocytes. For example, lipopolysaccharide (LPS), a component of all Gram-negative bacteria, induces coexpression of IL-10, -19, -20 and -24 mRNAs in monocytes [29–31]. Treatment of NK cells with formalin-fixed *Staphylococcus aureus* in the presence of IL-2 and -12 results in expression of IL-10, -22 and -26 mRNAs. T cells stimulated with anti-CD3 [T cell receptor (TCR) crosslinking] induce coexpression of IFN-γ, and IL-22 and -26 mRNA [29, 30]. Although IFNs (IFN-α/β and -λs) are produced by a wide variety of cells in response to viral infections, the most potent source of types I and III IFNs are plasmacytoid DCs (pDCs), which are also known as natural IFN-producing cells (NIPC) [35].

Thus, the CRF2 cytokines can be produced by various leukocyte populations in response to both antigen-specific (acquired) and nonspecific (innate) immune stimuli. In contrast, the primary targets of these cytokines are nonhematopoietic tissues and organs such as skin, lungs, gastrointestinal tract, liver and pancreas. The ability of the IL-10-related cytokines to activate gene expression in these target tissues is determined by the pattern of CRF2 receptor expression. Keratinocytes, hepatocytes and epithelial cells from various organs demonstrate a high level of expression of IL-20R1, -20R2 and 22R1, and IFN-λR1 mRNAs [29, 30]. However, immune cells are unresponsive to many of the CRF2 cytokines [29, 30]. A major exception to this paradigm is IL-10, which primarily acts on leukocytes such as monocytes and lymphocytes to inhibit inflammation [18]. IFN-γ is also produced primarily by T and NK cells. It promotes T$_h$1-type immune responses and has broad effects on a wide variety of nonhematopoietic cell types: upregulation of MHC class I expression and changes in proteosomal machinery induction of several angiostatic chemokines. Thus, a common functional scheme emerges for CRF2 cytokines: they are often produced by immune cells in response to nonspecific damage to the host, and they regulate inflammatory and native immune responses in damaged tissues.

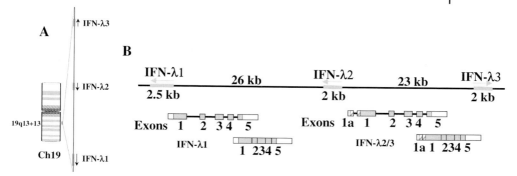

Fig. 6.1. Chromosomal localization and schematic structure of the genes for IFN-λs. The National Center for Biotechnology Information database was used to determine the chromosomal localization of the genes encoding IFN-λs and their receptor subunits (Fig. 6.2), and to generate the ideograms of chromosomes. (A) Chromosome 19 (Ch19) encodes three IFN-λ genes which are transcribed in the direction indicated by arrows. (B) Approximate sizes of the genes and intergenic distances are shown (in kb). Unspliced transcripts are schematically shown as strings of shaded/open boxes (exons) interrupted by lines (introns). Spliced transcripts are also shown as shaded/open boxes with vertical lines indicating relative positions of former introns. The coding regions of exons are shaded and the segments corresponding to 5′- and 3′-UTRs are left open. The IFN-λ2 and -λ3 genes have an additional exon (exon 1a) positioned upstream of their first coding exons (exon 1). Translation from the ATG codon within the exon 1a generates extra amino acids within the signal peptide of IFN-λ2 and -λ3 (hatched areas).

6.3
Genomic Structure

Genes encoding all three members of the novel IFN-λ cytokine family are clustered on human chromosome 19 (19q13.13 region; Fig. 6.1). The coding region of each of the genes is divided into five exons (exons 1–5 on Fig. 6.1). However, it appears that the IFN-λ2 and -λ3 genes have an additional exon (exon 1a) positioned upstream of their first coding exons (exon 1). Exon 1a contains an upstream ATG codon in-frame with the ATG codon encoded in exon 1. Translation from this upstream ATG codon generates four extra amino acids within the signal peptide of IFN-λ2 and -λ3. MetLysLeuAsp are the 4 amino acids which are added at the N-terminus of the IFN-λ2 and -λ3 sequences shown in Fig. 6.5 (below). Overall intron/exon organization of the genes encoding IFN-λs correlates well with the common conserved architecture of the genes encoding IL-10-related cytokines [7]. In contrast, the genes for the type I IFNs lack introns (Fig. 6.5 below).

The high degree of homology between human IFN-λs suggests that their genes derived from a common predecessor quite recently. It appears that, after the divergence of the IFN-λ1 and IFN-λ2 genes, there occurred a more recent duplication event in which a fragment containing the IFN-λ1 and -λ2 genes was copied and integrated back in to the genome in a head-to-head orientation with the IFN-λ1–IFN-λ2 segment. This duplication created the IFN-λ3 gene, which is almost identical to the IFN-λ2 gene not only in the coding region, but also in the upstream and downstream flanking sequences. Thus, promoters of the IFN-λ2 and -λ3 genes are very similar, and share several common elements with the IFN-λ1 promoter, suggesting that all three genes are likely to be regulated in a similar manner. In the duplicated fragment, a part which contained the IFN-λ1 gene was extensively mutated so that only separate pieces which do not encode a functional gene can be found in this region. Whereas the IFN-λ1 and -λ2 genes are transcribed towards the telomere, the IFN-λ3 gene is transcribed in the opposite direction. The IFN-λ1, -λ2 and -λ3 genes are separated from each other by 26 and 23 kb, respectively.

The genes encoding the IFN-λ receptor subunits, IFN-λR1 (also designated as IL-28R, LICR or CRF2-12 [2, 3, 36]) and IL-10R2 [2, 3, 37], are positioned on human chromosome 1 and 21, respectively (Fig. 6.2). The IFN-λR1 gene is positioned in close proximity to the IL-22R1 gene (1p36.11 chromosomal region). Both the IFN-λR1 gene and the adjacent IL-22R1 gene are transcribed towards the telomere and are positioned approximately 10 kb apart from one another. The IL-10R2 gene is clustered together with three genes encoding other members of the class II cytokine receptor family, IFN-αR1, -αR2 and -γR2, and all four genes are located on chromosome 21 (21q22.11 chromosomal region). All four genes are transcribed in the same direction toward the telomere. The IL-10R2 gene is surrounded by the IFN-αR2 gene 2 kb apart on one side and the IFN-αR1 gene 28 kb apart on another side. The IFN-λR1 and IL-10R2 genes have a similar intron/exon structure which is also shared by other genes encoding CRF2 proteins [7]. The coding regions of the receptor genes are composed of seven exons (exons 1 through 7, Fig. 6.2). Exon 1 encodes the 5′-untranslated region (UTR) and the signal peptide, the extracellular domain is encoded by exons 2, 3, 4, 5, and part of the exon 6. Exon 6 also encodes the transmembrane domain and the beginning of the intracellular domain. Exon 7 covers the rest of the intracellular domain and the 3′-UTR. The positions of the introns are highly conserved.

Interestingly, IFN-λR1 has three splice variants similar to those of IFN-αR2. Splice variants of both receptors may also be functionally similar [2, 3, 36, 38–40]. When all seven exons of the IFN-αR2 gene are present in the transcript, a signaling-competent IFN-αR2c splice variant is produced. Differential mRNA splicing generates two other variants, IFN-αR2a and -αR2b. A secreted soluble IFN-αR2a protein is a result of splicing out exon 6. A membrane-bound IFN-αR2b chain, with a shorter cytoplasmic domain than that in the IFN-αR2c chain, is encoded by the transcript lacking exon 7. The IFN-αR2b does not appear to stimulate intracellular activities and may function as a dominant-negative molecule. All three of these variant forms of IFN-αR2 have the same extracellular domain and all three can effectively bind IFN-α/β. IFN-λR1 appears to have analogous variants: one that encodes a full-length (seven exons) membrane-associated, functional receptor; a second that encodes a soluble receptor due to skipping exon 6; and a third that encodes a membrane-bound receptor with modified intracellular domain which is likely to be signaling-incompetent, that is generated by a distinct splicing event.

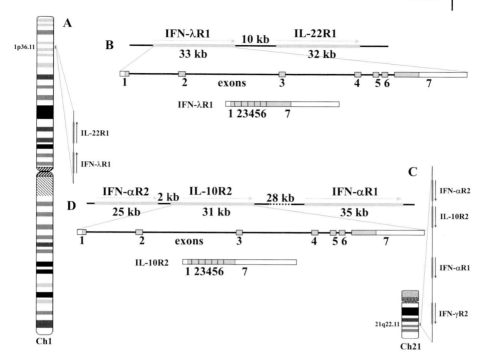

Fig. 6.2. Chromosomal localization and schematic structure of the genes encoding IFN-λR1 and IL-10R2. The chromosomal localization (Ch1) of the IFN-λR1 gene and its close neighbor, the IL-22R1 gene (A), and the chromosomal localization (Ch21) of the IL-10R2 gene and its neighbors, the genes encoding IFN-αR2 and IFN-αR1 (C), as well as the ideograms of chromosomes 1 and 21 are shown. The genes are transcribed in the direction indicated by arrows. Approximate sizes of the genes and intergenic distances are shown (in kb). (B and D) Schematic exon/intron structure of the gene is shown. Figure details are as described for Fig. 6.1.

IL-10R2 (CRF2-4), which is a part of the IL-10, -22 and -26, and IFN-λ receptor complexes, is ubiquitously expressed. Its mRNA is detectable by Northern blotting in most tissues albeit with very low level of expression in the brain [19, 29, 30, 41]. It appears that IFN-λR1 is constitutively expressed across a broad range of cell lines and tissues [2, 3]. The IFN-λR1 message is detectable in various normal tissues, including heart, kidney, skin, small intestine, lung, skeletal muscle and liver. The major IFN-λR1 transcript appears to be about 5 kb long. However, several IFN-λR1 transcripts of different sizes than the major transcript were observed in RNA derived from heart, skin and the MOLT-4 cell line, indicating alternative splicing and/or transcriptional termination variants. Among cell lines examined, hematopoietic (HL-60, K-562, MOLT-4 and Raji) and nonhematopoietic (HeLa S3, SW480, A549 and G-361) cell lines both express IFN-λR1 transcripts. Of these cell lines,

Raji cells expressed the highest levels of IFN-λR1. In general, all epithelial-like cell lines are responsive to IFN-λ. These include epitheloid carcinoma HeLa S3 cells, lung carcinoma A549 cells, keratinocyte HaCaT cells, hepatoma HuH7 cells and colorectal carcinoma HT-29 cells. However, several cell types such as fibroblast-like cell lines, primary fibroblasts and endothelial cells do not express IFN-λR1, and, therefore, are not responsive to IFN-λs.

6.4
Receptor Complex and Signaling

IFN-α/β and IFN-λs utilize distinct receptor complexes for signaling. Whereas type I IFNs exert their biological activities by signaling through a heterodimeric receptor complex composed of IFN-αR1 and -αR2c, IFN-λs bind and signal through a receptor complex formed by IFN-λR1 and IL-10R2 [2, 3] (Fig. 6.3).

IFN-λ ligand–receptor complex shares a common structure-functional architecture with other class II cytokine receptor complexes [4, 5, 7, 9, 42–44]. Ligands signaling through CRF2 activate predominantly the JAK–STAT signal transduction pathway, and require two distinct receptor subunits for signaling. The receptor chains within a given receptor complex can be divided into two types denoted R1 and R2. An R1 type subunit has a long intracellular domain which is associated with JAK-1 tyrosine kinase. The intracellular domain of an R1 subunit is phosphorylated on tyrosine residues after receptor engagement. This process facilitates recruitment of various SH2 domain-containing proteins, particularly STATs, to the receptor complex. Thus, the R1 chain defines the specificity of signaling. The STATs are then activated by JAK-mediated tyrosine phosphorylation, form homo- or heterodimers, dissociate from receptors and translocate to the nucleus. Various STAT combinations are activated in response to different cytokines and this in part determines the specificity of signaling. In the nucleus, the STAT dimers, in combination with other factors, modulate the finely tuned and well-orchestrated transcription of cytokine-inducible genes. In contrast, an R2-type subunit possesses a short intracellular domain which is associated with JAK-2 or TYK-2 tyrosine kinases and does not recruit STATs. It seems that the only function of the R2 subunit is to initiate signal transduction events by recruiting an additional tyrosine kinase to the receptor complex and, thus, allowing JAK cross-activation. Based on the length of their intracellular domains and participation in the STAT recruitment process, the receptor pairs within a given complex can be well defined according to the R1/R2 type division. IFN-αR1 is the only exception to the rule to a certain degree. It was reported that the IFN-αR1 intracellular domain can modulate type I IFN signaling: the deletion of a region within the IFN-αR1 intracellular domain created a receptor which produced an enhanced response [45].

The crosslinking pattern of radioactively labeled IFN-λ1 indicated that it is a monomer in solution [3]. IFN-λ2 is also a monomer (Zdanov, unpublished). Thus, binding of a monomeric IFN-λs is likely to engage one molecule of each

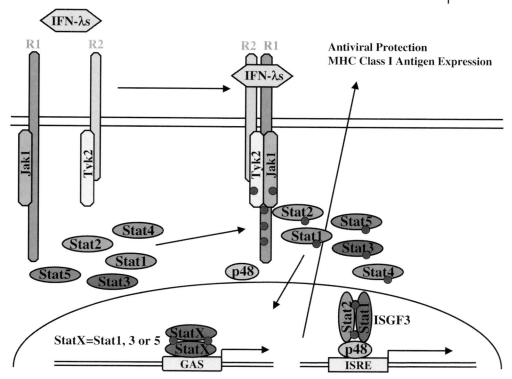

Fig. 6.3. Model of the IFN-λ receptor system. IFN-λs are likely to be monomers. The functional IFN-λ receptor complex consists of two receptor chains – the unique IFN-λR1 chain and the IL-10R2 chain. The IL-10R2 chain is a shared common chain in four receptor complexes, the IL-10, -22 and -26, and IFN-λ receptor complexes. Expression of both chains of the IFN-λ receptor complex is required for ligand binding and for assembling of the functional receptor complex. Ligand binding leads to the formation of the heterodimeric receptor complex, and to the initiation of a signal transduction cascade involving members of the JAK protein kinase family and the STAT family of transcription activators. The IL-10R2 chain is associated with TYK-2 [60] and the IFN-λR1 chain is likely to interact with JAK-1. Upon the ligand-induced heterodimerization of IFN-λ receptor chains receptor-associated JAKs crossactivate each other, phosphorylate the IFN-λR1 intracellular domain and, thus, initiate the cascade of signal transduction events. STAT-1, -2, -3, -4 and -5 are activated by IFN-λ leading to activation of biological activities, such as upregulation of MHC class I antigen expression and induction of antiviral protection. (This figure also appears with the color plates.)

IFN-λR1 and IL-10R2 subunits (Fig. 6.3). IFN-λs require both receptor chains for high-affinity binding and do not detectably interact with either single subunit expressed on the cell surface [3]. However, IFN-λ1 can form a complex with soluble IFN-λR1 in solution (Zdanov, unpublished results).

Despite signaling through distinct receptor complexes, type I IFNs and IFN-λs activate similar signaling events and biological activities, consistent with their common ability to induce an antiviral state in cells (Fig. 6.3). In both cases, receptor engagement leads *via* the activation of the JAK kinases JAK-1 and TYK-2, to the tyrosine phosphorylation of latent transcriptional factors of the STAT family: STAT-1, -2, -3, -4 and -5 [3, 46]. Phosphorylated STATs form various homo- and heterodimers and translocate to the nucleus, where they bind to specific DNA elements such as GAS (γ-activated sequence) in the promoters of IFN-responsive genes. Signaling through either type I or type III IFN receptor complexes also results in the activation of the ISGF-3 transcription complex, composed of STAT-1 and -2, and of the IFN-regulatory factor (IRF)-9 (ISGF-3 or p48). ISGF-3 regulates gene transcription by binding to an IFN-stimulated response element (ISRE). These IFN-activated transcriptional factors in combination with other enhanceosomal and transcriptosomal proteins modulate the transcription of IFN-responsive genes.

Although type I and type III IFNs act through receptors that do not share any apparent sequence homology in their intracellular domains, a very similar set of STATs is activated by both cytokine families. Analysis of the tyrosine-based motifs within the IFN-λR1 intracellular domain responsible for STAT activation revealed the molecular mechanism underlying the ability of both types of IFNs to induce the similar pattern of STAT activation [46]. Two tyrosines, Tyr343 and Tyr517, of IFN-λR1 can independently mediate STAT-2 activation by IFN-λs. In response to type I IFNs, one tyrosine of IFN-αR1 and two tyrosines of IFN-αR2 seem to play a role in STAT-2 recruitment and activation [47, 48]. Interestingly, the Tyr343-based motif of IFN-λR1 (YIEPPS) shows some similarities with that surrounding Tyr466 of IFN-αR1 (YVFFPS). In addition, the C-terminal amino acid sequence of IFN-λR1 containing Tyr517 (YMARstop) is very similar to the C-terminal amino acid sequence of IFN-αR2 containing Tyr512 (YIMRstop), which is also implicated in STAT-2 phosphorylation. Therefore, both type I and III IFN receptor subunits contain similar docking sites for STAT-2 recruitment and activation, YΨXXPS, where Ψ should be hydrophobic, and YXXR stop. Activation of STAT-2 requires the presence of either Tyr343 or Tyr517 of IFN-λR1, whereas activation of STAT-4 phosphorylation (and to some extent STAT-1 and -3 activation) was independent from IFN-λR1 tyrosine residues. The ability of IFN-λs to induce antiviral and antiproliferative activities is dependent on Tyr343 or Tyr517 of IFN-λR1, demonstrating that the activation of STAT-2 is pivotal for these biological activities, and neither STAT-4, -1 nor -3 activation, are sufficient to mediate the antiproliferative and antiviral activities of IFN-λs. Thus, Tyr343- and Tyr517-based motifs on IFN-λR1 are also likely to mediate ISGF-3 activation which is responsible for most of the IFN-λ-induced biological activities that are shared in common with IFN-α.

Since both the type I IFNs and IFN-λs activate the same transcriptional factors, particularly ISGF-3, it is not surprising that they induce similar biological activities. Cytokines from both families induce expression of several proteins which participate in antiviral responses, including double-stranded (ds) RNA-activated protein kinase (PKR), 2'–5'-oligoadenylate synthetase (2'–5'-OAS) and Mx proteins ([3] and unpublished data). These proteins have been shown to interfere with viral replication through various mechanisms. dsRNA-activated PKR phosphorylates

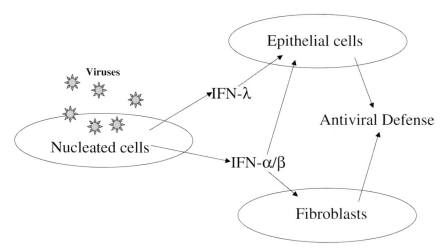

Fig. 6.4. Model of the antiviral response. Both type I and III IFNs are produced by a broad variety of nucleated cells in response to viral infections. Whereas all cell types express the type I IFN (IFN-α/β) receptor complex, the expression of the functional type III IFN (IFN-λ) receptor complex is limited primarily to the epithelial-like cells. Therefore, unlike type I IFNs, IFN-λs induce antiviral protection only in certain organs and tissues.

translation-initiation factor eIF-2α blocking protein synthesis. 2'–5'-OAS activates RNase L which cleaves mRNA and rRNA, thus inhibiting viral replication on both transcriptional and translational levels. Clearly, such drastic measures affect cellular viability. Indeed, both enzymes have been implicated in apoptosis. Mx proteins are GTPases and some of them possess antiviral activity, although the mechanism of their action is not completely understood. Mx proteins associate with ribonucleoprotein complexes of some viruses and can interfere with their transcriptional functioning and/or trafficking. Based on our current knowledge, it thus appears that both the IFN-α/β and -λ ligand–receptor systems can independently induce the establishment of an antiviral state by engaging similar signaling pathways and participants of the antiviral response (Fig. 6.4).

In general, IFN-λs do not inhibit proliferation of various cell lines including Daudi cells, which strongly respond to type I IFNs in an antiproliferative assay [2, 3, 46, 49]. However, IFN-λs do inhibit proliferation of certain tumor cell lines, such as LN319 human glioblastoma cell line [49] and in cells that are engineered to express high levels of IFN-λR1 from transfected plasmids [46] (Kotenko, unpublished). Therefore, it appears that the ability of IFN-λs to induce antiproliferative activity in cells depends on the level of IFN-λR1 expression.

A microarray experiment comparing the genes that are regulated by IFN-λ and by IFN-β demonstrated that both cytokines up- or downregulated the expression of the same set of genes, indicating that these cytokines should induce similar cellular responses [46]. Nevertheless, since the intracellular domains of IFN-αR2, -αR1 and -λR1 are very different, it is likely that some of the signaling molecules recruited and activated by IFN-α/β and -λs may differ. Therefore, it is possible that IFN-α/β and -λs each possess some unique biological activities.

Although it has not been documented, crosstalk between ligand-specific receptor complexes is likely to occur. The interaction between complexes may occur on several levels. Extensive receptor sharing may lead to sequestering of a shared chain by one complex which has higher affinity for the shared chain from another receptor complex with low affinity for this chain. For example, expression of IL-10R1/IFN-λR1 (extracellular/intracellular domain) in HT29 cells [3] which express endogenous functional IL-22 receptor complex noticeably weakened IL-22-induced signaling in these cells (unpublished). Competition for endogenous IL-10R2, a shared chain between the IL-10 and -22 receptor complexes, is a possible explanation for the observed inhibition. Competition for the pool of JAK kinases or STAT proteins could be another possible mechanism to account for the crosstalk between ligand–receptor complexes on the signaling level [50]. It is also possible that there are cell types that respond preferentially to either type I or III IFNs. Exposure to IFN-α/β or -γ may suppress responsiveness to IFN-λ by temporarily depleting the cytoplasmic pool of one or more of the ISGF-3 components: STAT-1 and -2, and IRF-9 (p48). This would decrease the availability of these transcription factor components for recruitment to the IFN-λ receptor complex when its ligand, IFN-λ, is present. Overexpression of the IFN-λR1 chain reduces responsiveness of the cells to type I IFNs [2].

6.5
Biological Activities

Antiviral protection is a complex process which involves several levels of defense. It begins with the efforts of an infected cell to prevent the replication of a virus and to signal other cells about viral presence, and finishes with the engagement of a broad array of immune protective forces to combat virus propagation in the body. One critical task of virus-infected cells is to release IFNs as signals to neighboring cells. Release of IFNs also alerts the immune system to the presence of an invading virus and induces antiviral protection prior to the spread of the virus. The IFNs can act in both an autocrine and paracrine manner to promote antiviral protection. Thus, the next wave of infected cells is better prepared to resist infection and, upon sensing the presence of a virus through various means, activates intracellular mechanisms of antiviral protection or even commits suicide (apoptosis) to prevent viral replication. IFNs can be produced by a wide variety of nucleated cells when infected by viruses or through other stimuli. In addition, there are specialized circulating cells that, while in relatively low abundance, produce high amounts of IFNs on a per-cell basis. These cells, sometimes called NIPCs, are a subset of DCs, pDCs, and can also respond to a virus through mechanisms independent of viral replication in the cells. A race between viral replication and the development of cellular and immune defense forces determines the severity, duration and outcome of a viral infection.

Type I IFNs (IFN-α/β) were viewed as the main players in many antiviral responses. Indeed, studies with IFN-α/β receptor knockout mice in which either subunit of the IFN-α receptor complex has been disrupted demonstrate an essential role for IFN-α signaling in the induction of antiviral resistance [51–53]. Type I IFNs (IFN-α/β) induce potent antiviral protection in a wide variety of cells. They also activate a variety of innate and adaptive immune mechanisms that eliminate viral infections. Although their primary role seems to be in mediating defense against viruses, they can also be induced by other stimuli and their expanding functional repertoire is generating a reassessment of our understanding of the type I IFNs.

In view of the exquisite susceptibility to viruses by mice with disruptions in the type I IFN system, it was surprising when a novel family of cytokines, the IFN-λs, was found to have potent antiviral activity [2, 3, 54]. IFN-λ-induced antiviral activity has been demonstrated against encephalomyocarditis virus (EMCV) and vesicular stomatitis virus (VSV) in different cell types. IFN-λs, similar to the type I IFNs, can also increase the expression of class I MHC (HLA) molecules, presumably providing more efficient targets for immune recognition of virus-infected cells. Thus, IFN-λs appear to be *bona fide* antiviral proteins, with a role in the physiological defense against viruses.

Consistent with their antiviral activity the IFN-λ genes are coexpressed together with other type I IFNs (IFN-α/β) in virus-infected cells [2, 3]. Virtually any cell type can express IFN-α/β following viral infection and, presumably, most viruses induce IFN-λ expression (Fig. 6.4). For example, we showed that four distinct viruses (Sindbis virus, Dengue virus, VSV and EMCV) induced coexpression of IFN-λs in several different cell types, including HeLa (cervical epithelial carcinoma), HT-29 (colorectal carcinoma) and HuH7 (hepatoma) [3]. More recently, it was shown that infection of human epithelial cells by respiratory syncytial virus induces coexpression of type I (IFN-α/β) and type III (IFN-λ) IFNs [55]. Other viruses, such as influenza virus and Sendai virus have also been shown to induce expression of IFN-λs in human monocyte-derived DCs (mDCs) [56, 57]. Furthermore, in one study, it was shown that IFN-α amplifies induction of IFN-λ by influenza or Sendai virus [56]. The ability of IFN-α to upregulate induction of IFN-λ appears to be due to the ability IFN-α to upregulate expression of the Toll-like receptor (TLR) and IRF-7 genes.

It was also shown that viral infection and treatment with diverse TLR agonists induces differential expression of the IFN-α/β and -λ genes in pDCs and mDCs [57]. pDCs are a highly specialized and relatively rare cell population that can produce large amounts of IFN-α/β in response to environmental challenges such as viral infection. A recent report by Coccia et al. [57] showed that influenza virus infection of pDCs or mDCs induces coexpression of all of the IFN-α subtypes and IFN-β as well as the three IFN-λ subtypes. Certain TLR agonists such as CpG DNA which signals via TLR-9 in plasmacytoid DCs also induced coexpression of IFN-α, -β and -λ. In contrast, other TLR agonists such as LPS and poly I:C which signal via TLR-4 and -3, respectively, induced expression of IFN-β and -λ, but did not elicit expression of IFN-α in monocyte-derived DCs. We have also observed differential induction of IFN-α/β and -λ by LPS and poly I:C in monocyte-derived DCs (unpublished observations). It is apparent therefore that live virus infection induces coexpression of IFN-α/β and -λ, whereas as other microbial agents or microbial components such as bacterial DNA, endotoxin and dsRNA elicit a more selective expression of these IFN types. The similar pattern of expression of type I and III IFNs is consistent with the existence of upstream regulatory elements in the IFN-λ genes similar to some of those found in the type I IFNs.

6.6
The Murine IFN-λ Antiviral System

In light of the common antiviral activities of IFN-α/β and -λs, it is somewhat puzzling that mice with a disrupted type I IFN system are still highly susceptible to viral infections. Why doesn't the IFN-λ antiviral system confer protection against viruses in such mice? One possibility is that the IFN-λ system is not fully functional in mice. However, mouse IFN-λs and -λR1 orthologs do exist, and are functionally quite similar to their human counterparts (Fig. 6.5 and data not shown).

The murine and human IFN-λ ligands are almost identical with the exception of several amino acids. The murine IFN-λ genes are clustered on chromosome 7A3. Their intron/exon structures are very similar to those of the human genes. A currently available sequence of the mouse genome suggests a similar organization of the mouse IFN-λ locus in comparison with the human IFN-λ locus. There are two genes representing mouse IFN-λ2 and -λ3 gene orthologs encoding intact proteins. Both of these have been cloned and their gene products are active on the human IFN-λ receptor complex (unpublished results). In contrast, the mouse IFN-λ1 gene has been extensively mutated, and does not encode a functional protein (unpublished data).

Fig. 6.5. Evolution of the IFN family. (A) Alignments of the amino acid sequences of human (h), mouse (m) and chicken (c) IFN-λs, as well as zebrafish (f) IFNs. The consensus sequence is shown on the bottom. Identical amino acids corresponding to the consensus sequence are shown in black outline with white lettering. Similar amino acids are shown in gray outline with white lettering. Amino acid residues are numbered starting from the first Met residue (signal peptide amino acids are included). Positions of corresponding introns are indicated by arrows. (B) Human type I IFNs were added into the alignment and the results were utilized to generate a phylogenetic tree for these cytokines. Only one IFN-α was used in alignment since 13 IFN-α species are nearly identical. Because of the low sequence identity, the tree is subject to small changes in the alignment and is therefore instructive, not definitive. The intron/exon structure of genes encoding fish IFNs is identical to those of IFN-λ genes as schematically shown in (C). In contrast, mammalian as well as bird type I IFN genes do not have introns.

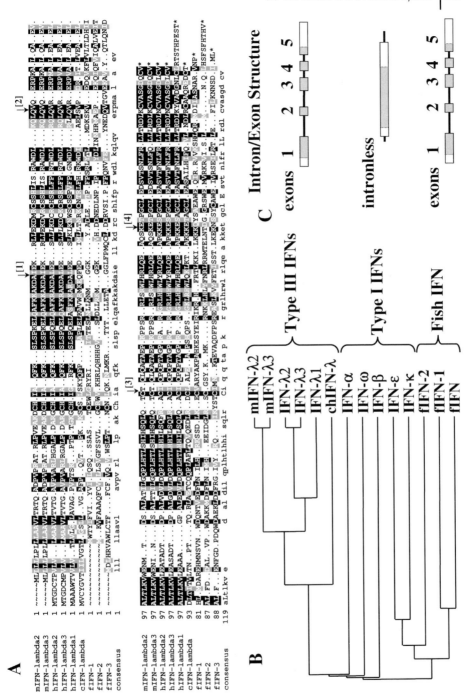

Functional differences between the murine and human IFN-λ systems could arise from subtle differences in the intracellular domains of their receptors (IFN-λR1). There is a stretch of negatively charged residues close to the end of the human receptor intracellular domain. This region in the mouse receptor is significantly altered by a short insertion and substitutions of several amino acid residues, resulting in a longer and more negatively charged region in the mouse receptor [9]. In addition, while two out of three tyrosine residues of the human receptor intracellular domain are conserved in the mouse ortholog, the mouse receptor contains three additional tyrosine residues. Thus, there is a possibility that distinct subsets of signaling molecules might be activated through mouse or human receptors. Nevertheless, both tyrosine-based motifs which are involved in STAT activation [46] are preserved in the mouse ortholog. However, the differences in other tyrosine-containing motifs between the murine and human orthologs may lead to mouse and human IFN-λs conferring overlapping, but distinct, biological activities.

Although there are only two functional IFN-λ genes encoded in the mouse genome, the murine IFN-λ system appears to be intact, and able to mediate the antiviral response both *in vitro* and *in vivo* [58]. Murine IFN-λs can protect B16 melanoma cells against the cytopathic effect of VSV infection (unpublished). Although, the expression of murine IFN-λs by recombinant vaccinia viruses (vIFN-λs) did not impair the ability of the virus to grow in culture, it strongly attenuated the virus in mouse infection models. Mice infected intranasally with vIFN-λs did not show any signs of illness or weight loss. In addition vIFN-λ2 was cleared more rapidly from infected lungs and, in contrast to control viruses, was unable to disseminate to the brain. Attenuation of vIFN-λ2 was associated with increases in both lymphocytes in bronchial alveolar lavages (BAL) and $CD4^+$ T cells in total lung lymphocyte preparations. Thus IFN-λs possess the potent antiviral and immunostimulatory activity against poxvirus infection *in vivo*.

Other possible explanations for the inability of the IFN-λ system to render mice with a disrupted type I IFN system protection against viral challenges are that either the kinetics of induction or the potency of the IFN-λ system are insufficient as a "stand-alone" antiviral system for the viruses tested in the absence of a functional type I IFN system.

6.7
Evolution of the IFN Family

Additional insight into the evolutionary relationships may come from examining related cytokines in other organisms. Analysis of the available sequenced genomes reveals that IFN-λ orthologs can be predicted in numerous mammalian species. The IFN-λ antiviral system can be traced as far back in evolution as birds. The chicken genome contains at least one IFN-λ gene (Fig. 6.5) demonstrating that the IFN-λ system has been well preserved in evolution.

Interestingly, zebrafish (*Danio rerio*) IFN possessing antiviral activity has been recently identified [59]. Two additional quite distinct IFN species can be predicted in the zebrafish genome (unpublished data and Fig. 6.5) which can be tentatively designated as IFNs, whose antiviral activity remains to be demonstrated experimentally. Although, these fish IFNs demonstrate substantial sequence homology to both mammalian type I and type III IFNs, all fish IFNs share higher percentage of amino acid identity with mammalian type I IFNs (Fig. 6.5). Intriguingly, the intron/exon structure of genes encoding fish IFNs is identical to those of IFN-λ genes (Fig. 6.5), whereas mammalian as well as bird type I IFN genes do not have introns. Thus, to date, intronless IFN genes have not been found in fish genomes. This observation suggests that early in evolution, the proto-IFN genes possessed introns. Later, IFN genes evolved into two subfamilies: IFN-λ genes retained the ancient intron/exon structure, whereas type I IFN genes lost introns. Currently, it is not clear whether fishes have only one IFN antiviral system, and, if so, when in evolution the duplication of an antiviral system into type I and III IFN systems occurred.

6.8
Therapeutic Potential

Currently, clinical targets of type I IFN therapy include: viral infections, particularly chronic hepatitis C virus (HCV) or HBV infections; various cancers, including hairy cell and myelogenous leukemias, multiple myeloma, lymphomas, renal cell carcinoma, Kaposi' sarcoma and metastatic melanoma; and multiple sclerosis (IFN-β). However, there are severe adverse events associated with type I IFN therapy including inhibition of hematopoesis, neuropsychiatric effects (depression, anxiety) and influenza-like symptoms (myalgia, fever, fatigue). The great need for better therapeutics is particularly clear in the case of chronic HCV infections.

HCV infections have reached epidemic proportions with nearly 2% of the world population currently infected. Although recombinant IFN-α is now the mainstay for therapy of chronic HCV infection, IFN-α therapy is effective in only about half of infected persons, leaving large numbers of patients with persistent infection. In fact, over time, many patients who are treated with recombinant human IFN-α products become resistant to the biological activities of this agent and continued treatment does little to control viral replication or the morbidity associated with chronic HCV infection. Moreover, many patients are unable to receive doses of IFN-α that might more effectively treat their disease because of severe dose-limiting toxicities, including neuropsychiatric disease, myelosuppression, autoimmune disorders and debilitating flu-like symptoms. The biological basis of many of these treatment-limiting side-effects appears to result from the very broad activity of IFN-α on cells of the immune system, particularly lymphocytes and neutrophils. Consequently, there is a great need for alternative biological agents for treating this widespread disease, especially agents that elicit fewer or less severe adverse reactions. Therefore, it is important to evaluate whether IFN-λs may represent better therapeutics which may have fewer side-effects while retaining beneficial antiviral potency.

6.9
Conclusions

Along with advancements in our understanding of the complexity of virus–host interactions, IFNs continue to reveal novel features of the antiviral response. Multiple innate and adaptive immune mechanisms directed to eliminate viral infections are activated by IFNs. Discovery of IFN-λs has opened a new direction in antiviral research. The relative importance of the IFN-λs in host antiviral defense remains to be fully determined. Although acting through s unique receptor complex, type III IFNs seem to mimic type I IFNs in their pattern of expression, signaling pathways and biological activities. However, different antiviral potency and differential expression of receptor subunits suggest that the type I and III IFN antiviral systems do not merely replicate each other. Perhaps at this point the IFN-λ antiviral system should be viewed as relatively weak, tissue-specific antiviral system.

References

1 Isaacs A, Lindenmann J. Virus interference. I. The interferon. *Proc Roy Soc Lond B* **1957**, *147*, 258–267.
2 Sheppard P, Kindsvogel W, Xu W, Henderson K, Schlutsmeyer S, Whitmore TE, Kuestner R, Garrigues U, Birks C, Roraback J, Ostrander C, Dong D, Shin J, Presnell S, Fox B, Haldeman B, Cooper E, Taft D, Gilbert T, Grant FJ, Tackett M, Krivan W, McKnight G, Clegg C, Foster D, Klucher KM. IL-28, IL-29 and their class II cytokine receptor IL-28R. *Nat Immunol* **2003**, *4*, 63–68.
3 Kotenko SV, Gallagher G, Baurin VV, Lewis-Antes A, Shen M, Shah NK, Langer JA, Sheikh F, Dickensheets H, Donnelly RP. IFN-lambdas mediate antiviral protection through a distinct class II cytokine receptor complex. *Nat Immunol* **2003**, *4*, 69–77.
4 Renauld JC. Class II cytokine receptors and their ligands: key antiviral and inflammatory modulators. *Nat Rev Immunol* **2003**, *3*, 667–676.
5 Pestka S, Krause CD, Sarkar D, Walter MR, Shi Y, Fisher PB. Interleukin-10 and related cytokines and receptors. *Annu Rev Immunol* **2004**, *22*, 929–979.
6 Pestka S, Krause CD, Walter MR. Interferons, interferon-like cytokines, and their receptors. *Immunol Rev* **2004**, *202*, 8–32.
7 Kotenko SV. The family of IL-10-related cytokines and their

receptors: related, but to what extent? *Cytokine Growth Factor Rev* **2002**, *13*, 223–240.

8 LANGER JA, CUTRONE EC, KOTENKO S. The Class II cytokine receptor (CRF2) family: overview and patterns of receptor–ligand interactions. *Cytokine Growth Factor Rev* **2004**, *15*, 33–48.

9 KOTENKO SV, LANGER JA. Full house: 12 receptors for 27 cytokines. *Int Immunopharmacol* **2004**, *4*, 593–608.

10 DONNELLY RP, SHEIKH F, KOTENKO SV, DICKENSHEETS H. The expanded family of class II cytokines that share the IL-10 receptor-2 (IL-10R2) chain. *J Leukoc Biol* **2004**, *76*, 314–321.

11 FICKENSCHER H, HOR S, KUPERS H, KNAPPE A, WITTMANN S, STICHT H. The interleukin-10 family of cytokines. *Trends Immunol* **2002**, *23*, 89–96.

12 DUMOUTIER L, RENAULD JC. Viral and cellular interleukin-10 (IL-10)-related cytokines: from structures to functions. *Eur Cytokine Netw* **2002**, *13*, 5–15.

13 ZDANOV A. Structural features of the interleukin-10 family of cytokines. *Curr Pharm Des* **2004**, *10*, 3873–3884.

14 WALTER MR. Structural analysis of IL-10 and Type I interferon family members and their complexes with receptor. *Adv Protein Chem* **2004**, *68*, 171–223.

15 HARDY MP, OWCZAREK CM, JERMIIN LS, EJDEBACK M, HERTZOG PJ. Characterization of the type I interferon locus and identification of novel genes. *Genomics* **2004**, *84*, 331–345.

16 LAFLEUR DW, NARDELLI B, TSAREVA T, MATHER D, FENG P, SEMENUK M, TAYLOR K, BUERGIN M, CHINCHILLA D, ROSHKE V, CHEN G, RUBEN SM, PITHA PM, COLEMAN TA, MOORE PA. Interferon-kappa, a novel type I interferon expressed in human keratinocytes. *J Biol Chem* **2001**, *276*, 39765–39771.

17 ORITANI K, MEDINA KL, TOMIYAMA Y, ISHIKAWA J, OKAJIMA Y, OGAWA M, YOKOTA T, AOYAMA K, TAKAHASHI I, KINCADE PW, MATSUZAWA Y. Limitin: an interferon-like cytokine that preferentially influences B- lymphocyte precursors. *Nat Med* **2000**, *6*, 659–666.

18 MOORE KW, DE WAAL MR, COFFMAN RL, O'GARRA A. Interleukin-10 and the interleukin-10 receptor. *Annu Rev Immunol* **2001**, *19*, 683–765.

19 AGGARWAL S, XIE MH, MARUOKA M, FOSTER J, GURNEY AL. Acinar cells of the pancreas are a target of interleukin-22. *J Interferon Cytokine Res* **2001**, *21*, 1047–1053.

20 DUMOUTIER L, VAN ROOST E, COLAU D, RENAULD JC. Human interleukin-10-related T cell-derived inducible factor: molecular cloning and functional characterization as an hepatocyte-stimulating factor. *Proc Natl Acad Sci USA* **2000**, *97*, 10144–10149.

21 WOLK K, KUNZ S, WITTE E, FRIEDRICH M, ASADULLAH K, SABAT R. IL-22 increases the innate immunity of tissues. *Immunity* **2004**, *21*, 241–254.

22 RESMINI C, SHIELDS KM, LAMBERT AJ, WONG A, PEDNAULT G, HEGEN M, FOUSER LA, PITTMAN DD. An anti-murine IL-22 monoclonal antibody decreases disease severity in a murine model of collagen induced arthritis. *Eur Cytokine Netw* **2003**, *14S*, 129.

23 Pittman DD, Goad B, Lambert AJ, Clark E, Tan XY, Spaulding V, Wang IM, Kobayashi M, Whitters M, Thibodeaux D, Leanard J, Ling V, Wu P, Annis B, Lu Z, Zollner R, Jacobs K, Fouser LA. IL-22 is a tightly-regulated IL10-like molecule that induces an acute-phase response and renal tubular basophilia. *Genes Immun* **2001**, *2*, 172.

24 Blumberg H, Conklin D, Xu WF, Grossmann A, Brender T, Carollo S, Eagan M, Foster D, Haldeman BA, Hammond A, Haugen H, Jelinek L, Kelly JD, Madden K, Maurer MF, Parrish-Novak J, Prunkard D, Sexson S, Sprecher C, Waggie K, West J, Whitmore TE, Yao L, Kuechle MK, Dale BA, Chandrasekher YA. Interleukin 20: discovery, receptor identification, and role in epidermal function. *Cell* **2001**, *104*, 9–19.

25 Xu WF, Chandrasekher Y, Haugen H, Hughes S, Dillon S, Sivakumar P, Brender T, Waggie K, Yao L, Schlutsmeyer S, Anderson M, Kindsvogel W, Chen Z, Blumberg H, Cooper KD, McCormick TS, Novak J, Clegg C, McKernan PA, Foster D. IL-20 and IL-22 in psoriasis. *Eur Cytokine Netw* **2003**, *14*, 65.

26 Dumoutier L, Lejeune D, Colau D, Renauld JC. Cloning and characterization of IL-22 binding protein, a natural antagonist of IL-10-related T cell-derived inducible factor/IL-22. *J Immunol* **2001**, *166*, 7090–7095.

27 Kotenko SV, Izotova LS, Mirochnitchenko OV, Esterova E, Dickensheets H, Donnelly RP, Pestka S. Identification, cloning, and characterization of a novel soluble receptor that binds IL-22 and neutralizes its activity. *J Immunol* **2001**, *166*, 7096–7103.

28 Xu W, Presnell SR, Parrish-Novak J, Kindsvogel W, Jaspers S, Chen Z, Dillon SR, Gao Z, Gilbert T, Madden K, Schlutsmeyer S, Yao L, Whitmore TE, Chandrasekher Y, Grant FJ, Maurer M, Jelinek L, Storey H, Brender T, Hammond A, Topouzis S, Clegg CH, Foster DC. A soluble class II cytokine receptor, IL-22RA2, is a naturally occurring IL-22 antagonist. *Proc Natl Acad Sci USA* **2001**, *98*, 9511–9516.

29 Nagalakshmi ML, Murphy E, McClanahan T, de Waal MR. Expression patterns of IL-10 ligand and receptor gene families provide leads for biological characterization. *Int Immunopharmacol* **2004**, *4*, 577–592.

30 Wolk K, Kunz S, Asadullah K, Sabat R. Cutting edge: immune cells as sources and targets of the IL-10 family members? *J Immunol* **2002**, *168*, 5397–5402.

31 Gallagher G, Dickensheets H, Eskdale J, Izotova LS, Mirochnitchenko OV, Peat JD, Vazquez N, Pestka S, Donnelly RP, Kotenko SV. Cloning, expression and initial characterization of interleukin-19 (IL-19), a novel homologue of human interleukin-10 (IL-10). *Genes Immun* **2000**, *1*, 442–450.

32 Dorman SE, Picard C, Lammas D, Heyne K, van Dissel JT, Baretto R, Rosenzweig SD, Newport M, Levin M, Roesler J, Kumararatne D, Casanova JL, Holland SM. Clinical features of dominant and recessive interferon gamma receptor 1 deficiencies. *Lancet* **2004**, *364*, 2113–2121.

33 Ikeda H, Old LJ, Schreiber RD. The roles of IFN gamma in protection against tumor development and cancer immunoediting. *Cytokine Growth Factor Rev* **2002**, *13*, 95–109.

34 Boehm U, Klamp T, Groot M, Howard JC. Cellular responses to interferon-gamma. *Annu Rev Immunol* **1997**, *15*, 749–795.

35 Fitzgerald-Bocarsly P. Natural interferon-alpha producing cells: the plasmacytoid dendritic cells. *Biotechniques* **2002**, Suppl, 16–19.

36 Dumoutier L, Lejeune D, Hor S, Fickenscher H, Renauld JC. Cloning of a new type II cytokine receptor activating signal transducer and activator of transcription (STAT)1, STAT-2 and STAT3. *Biochem J* **2003**, *370*, 391–396.

37 Kotenko SV, Krause CD, Izotova LS, Pollack BP, Wu W, Pestka S. Identification and functional characterization of a second chain of the interleukin-10 receptor complex. *EMBO J* **1997**, *16*, 5894–5903.

38 Novick D, Cohen B, Rubinstein M. The human interferon alpha/beta receptor: characterization and molecular cloning. *Cell* **1994**, *77*, 391–400.

39 Lutfalla G, Holland SJ, Cinato E, Monneron D, Reboul J, Rogers NC, Smith JM, Stark GR, Gardiner K, Kerr IM, Uzé G. Mutant U5A cells are complemented by an interferon-alpha beta receptor subunit generated by alternative processing of a new member of a cytokine receptor gene cluster. *EMBO J* **1995**, *14*, 5100–5108.

40 Domanski P, Witte M, Kellum M, Rubinstein M, Hackett R, Pitha P, Colamonici OR. Cloning and expression of a long form of the beta subunit of the interferon alpha beta receptor that is required for signaling. *J Biol Chem* **1995**, *270*, 21606–21611.

41 Sheikh F, Baurin VV, Lewis-Antes A, Shah NK, Smirnov SV, Anantha S, Dickensheets H, Dumoutier L, Renauld JC, Zdanov A, Donnelly RP, Kotenko SV. Cutting edge: IL-26 signals through a novel receptor complex composed of IL-20 receptor 1 and IL-10 receptor 2. *J Immunol* **2004**, *172*, 2006–2010.

42 Bach EA, Aguet M, Schreiber RD. The IFN gamma receptor: a paradigm for cytokine receptor signaling. *Annu Rev Immunol* **1997**, *15*, 563–591.

43 Kotenko SV, Pestka S. Jak–Stat signal transduction pathway through the eyes of cytokine class II receptor complexes. *Oncogene* **2000**, *19*, 2557–2565.

44 Pestka S, Kotenko SV, Muthukumaran G, Izotova LS, Cook JR, Garotta G. The interferon gamma (IFN-gamma) receptor: a paradigm for the multichain cytokine receptor. *Cytokine Growth Factor Rev* **1997**, *8*, 189–206.

45 Gibbs VC, Takahashi M, Aguet M, Chuntharapai A. A negative regulatory region in the intracellular domain of the human interferon-alpha receptor. *J Biol Chem* **1996**, *271*, 28710–28716.

46 Dumoutier L, Tounsi A, Michiels T, Sommereyns C, Kotenko SV, Renauld JC. Role of the interleukin (IL)-28

receptor tyrosine residues for antiviral and antiproliferative activity of IL-29/interferon-lambda 1: similarities with type I interferon signaling. *J Biol Chem* **2004**, *279*, 32269–32274.

47 Yan H, Krishnan K, Greenlund AC, Gupta S, Lim JT, Schreiber RD, Schindler CW, Krolewski JJ. Phosphorylated interferon-alpha receptor 1 subunit (IFNaR1) acts as a docking site for the latent form of the 113 kDa STAT-2 protein. *EMBO J* **1996**, *15*, 1064–1074.

48 Wagner TC, Velichko S, Vogel D, Rani MR, Leung S, Ransohoff RM, Stark GR, Perez HD, Croze E. Interferon signaling is dependent on specific tyrosines located within the intracellular domain of IFNAR2c. Expression of IFNAR2c tyrosine mutants in U5A cells. *J Biol Chem* **2002**, *277*, 1493–1499.

49 Meager A, Visvalingam K, Dilger P, Bryan D, Wadhwa M. Biological activity of interleukins-28 and -29: comparison with type I interferons. *Cytokine* **2005**, *31*, 109–118.

50 Dondi E, Pattyn E, Lutfalla G, Van OX, Uze G, Pellegrini S, Tavernier J. Down-modulation of type 1 interferon responses by receptor cross-competition for a shared Jak kinase. *J Biol Chem* **2001**, *276*, 47004–47012.

51 Hwang SY, Hertzog PJ, Holland KA, Sumarsono SH, Tymms MJ, Hamilton JA, Whitty G, Bertoncello I, Kola I. A null mutation in the gene encoding a type I interferon receptor component eliminates antiproliferative and antiviral responses to interferons alpha and beta and alters macrophage responses. *Proc Natl Acad Sci USA* **1995**, *92*, 11284–11288.

52 Muller U, Steinhoff U, Reis LF, Hemmi S, Pavlovic J, Zinkernagel RM, Aguet M. Functional role of type I and type II interferons in antiviral defense. *Science* **1994**, *264*, 1918–1921.

53 Steinhoff U, Muller U, Schertler A, Hengartner H, Aguet M, Zinkernagel RM. Antiviral protection by vesicular stomatitis virus-specific antibodies in alpha/beta interferon receptor-deficient mice. *J Virol* **1995**, *69*, 2153–2158.

54 Vilcek J. Novel interferons. *Nat Immunol* **2003**, *4*, 8–9.

55 Spann KM, Tran KC, Chi B, Rabin RL, Collins PL. Suppression of the induction of alpha, beta, and lambda interferons by the NS1 and NS2 proteins of human respiratory syncytial virus in human epithelial cells and macrophages. *J Virol* **2004**, *78*, 4363–4369.

56 Siren J, Pirhonen J, Julkunen I, Matikainen S. IFN-alpha regulates TLR-dependent gene expression of IFN-alpha, IFN-beta, IL-28, and IL-29. *J Immunol* **2005**, *174*, 1932–1937.

57 Coccia EM, Severa M, Giacomini E, Monneron D, Remoli ME, Julkunen I, Cella M, Lande R, Uze G. Viral infection and Toll-like receptor agonists induce a differential expression of type I and lambda interferons in human plasmacytoid and monocyte-derived dendritic cells. *Eur J Immunol* **2004**, *34*, 796–805.

58 Bartlett NW, Buttigieg K, Kotenko SV, Smith GL. Murine interferon lambdas (type III IFNs) exhibit potent antiviral activity *in vivo* in a poxvirus infection model. *J Gen Virol* **2005**, *86*, 1589–1596.

59 ALTMANN SM, MELLON MT, DISTEL DL, KIM CH. Molecular and functional analysis of an interferon gene from the zebrafish, *Danio rerio*. *J Virol* **2003**, *77*, 1992–2002.
60 KOTENKO SV, IZOTOVA LS, POLLACK BP, MUTHUKUMARAN G, PAUKKU K, SILVENNOINEN O, IHLE JN, PESTKA S. Other kinases can substitute for Jak2 in signal transduction by interferon-gamma. *J Biol Chem* **1996**, *271*, 17174–17182.

Section B
Biological Properties

7
Biological Actions of Type I Interferons

Melissa M. Brierley, Jyothi Kumaran and Eleanor N. Fish

7.1
Introduction

Type I interferons (IFNs), which include IFN-α, -β, -ω, -κ and -τ, are an evolutionarily conserved group of secreted cytokines that act as pleiotropic mediators of host defense and homeostasis. Binding of IFNs to specific cell surface receptors results in the activation of multiple intracellular signaling cascades, leading to the synthesis of proteins that mediate antiviral, growth-inhibitory and immunomodulatory responses.

In this chapter we will explore binding of type I IFNs with their cognate receptor complex, emphasizing key residues mediating these interactions. IFN-inducible signaling and subsequent biological outcomes elicited upon receptor engagement will also be discussed.

7.2
Sources of Type I IFN Production and Secretion

Historically, type I IFNs were classified as fibroblast IFN because they were first observed to be produced by fibroblast cells [1–4]. The advancement in type I IFN research has shown that T cells, macrophages, plasmacytoid dendritic cells (pDCs), DCs and natural killer (NK) cells can also secrete IFNs [5]. IFN-αs are also considered leukocyte IFNs because they are secreted by white blood cells [6]. A professional cell which is the primary type I IFN producer has been defined as the $CD4^+CD11c^-$ type 2 DC (pDC2) [7]. This natural IFN-producing cell (NIPC) can secrete between 200- and 1000-fold more IFN-α than any other white blood cell type.

The Interferons: Characterization and Application. Edited by Anthony Meager
Copyright © 2006 WILEY-VCH Verlag GmbH & Co. KGaA, Weinheim
ISBN: 3-527-31180-7

7.3
Type I IFN Interactions with the Receptor Complex

Type I IFNs induce signaling events using a common receptor complex comprised of two subunits denoted IFNAR-1 and -2c, which belong to the class II cytokine receptor superfamily [8]. The extracellular domain (EC) of the receptor subunits is composed of immunoglobulin-like domain units called fibronectin type III (FNIII) domains [9, 10]. The IFNAR-1 extracellular domain is composed of four FNIII domains and the IFNAR-2 EC domain is composed of two FNIII domains. Loop regions in the EC domain of the receptor subunits mediate interactions with IFNs [11].

7.3.1
Structure and Functional Regions of Type I IFNs

Based on X-ray crystallographic and nuclear magnetic resonance (NMR) data, all type I IFNs have the same overall secondary structure [12–16]. The proteins are composed of five α helices that are packed together in an up–up–down–up–down formation.

The IFN receptor binding sequences were mapped by evaluating chimeric IFN-αs in binding assays and in bioassays [17, 18]. Three active regions of the IFN molecule were defined as amino acid sequences 29–35, 78–95 and 123–140. These regions were designated IFN receptor recognition peptide (IRRP)-1, -2 and -3, respectively (Fig. 7.1). IRRP-1 and -3 are spatially proximal in the native protein, and are kept in an active biological conformation by a disulfide bond existing between Cys29 and Cys139. The IRRP-1 sequence is a loop structure between helices A and B, the IRRP-2 sequence is located in the C helix, and the IRRP-3 sequence is located in helix D and in the loop connecting helices D and E. Solvent-exposed residues in the E helix were also demonstrated to contribute significant energy to receptor binding [19–21].

7.3.2
IFN Domains Mediating Interactions with IFNAR-2

Type I IFNs bind to IFNAR-2 with K_d values in the range of approximately 2–10 nM [22–24]. IRRP-1 and -3 form the region which mediates interactions with IFNAR-2 [19, 25]. In addition to the IRRP-1 and -3 sequences, residues in the E helix also contribute significant energy to IFNAR-2 binding [19]. High-performance liquid chromatography analysis of interactions between soluble IFNAR-2 and IFN-α2 showed that a 1:1 complex is formed between these two molecules, indicating that one molecule of IFN interacts with one molecule of IFNAR-2 [24]. Therefore, by extrapolation, the IRRP-2 sequence in the C helix must interact with IFNAR-1.

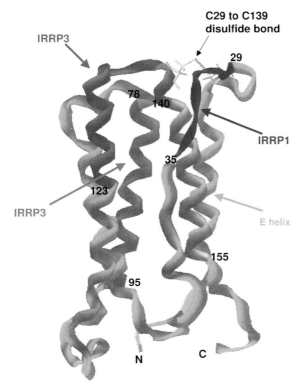

Fig. 7.1. Ribbon diagram representation of human IFN-α2 (protein data base ID 1ITF). Regions mediating binding interactions with the IFNAR subunits are highlighted. IRRP-1 (residues 29–35) is represented as a blue ribbon, IRRP-2 (residues 78–95) as a red ribbon, IRRP-3 (residues 123–140) as a purple ribbon and the E helix (residues 141–155) as an orange ribbon. The Cys29–Cys139 residues involved in the disulfide bond are represented as yellow sticks. The N- and C-termini are indicated. (This figure also appears with the color plates.)

Residues in the IFNAR-2-binding region which contribute the most binding energy were identified using human IFN-α2 and human IFNAR-2 molecules containing point mutations within putative functional regions [19]. Evaluation of these mutants using kinetic binding studies was carried out to quantify the binding energy contribution of the targeted residues (Fig. 7.2). From this analysis, 10 residues localized entirely in the IRRP-1 sequence were found to form a functional epitope on IFN-α2. Residues Leu30 and Arg33 together contribute two-thirds of the total interaction energy. Other "hotspot" residues in the ligand include Arg144, Ala145, Met148, Arg149 and Ser152 [20]. These residues are conserved amongst human IFN-αs, demonstrating the importance of the amino acid residues at these positions in mediating interactions with the receptor.

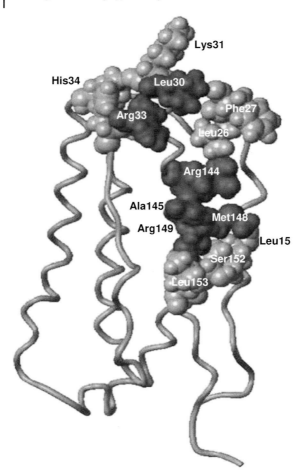

Fig. 7.2. Key residues in human IFN-α2 that contribute the most binding energy during interaction with IFNAR-2. Amino acids are represented in space-filling format. Residues that contribute more than 2 kcal mol^{-1} of energy are shaded red, and residues that contribute between 0.5 and 2 kcal mol^{-1} of energy are shaded yellow. (This figure also appears with the color plates.) Adapted from [32].

7.3.3
IFNAR-2 Domains Mediating Interactions with IFNs

IFNAR-2 site-directed mutagenesis studies and homology modeling using solved cytokine structures have identified putative residues involved in interactions with IFN-α2 and -β (Fig. 7.3). Long-range electrostatic interactions were studied by creating single-charge reverse mutants of IFNAR-2 and IFN, and determining the effect of the mutation on dissociation kinetics in binding assays [20, 32]. Based on data generated in kinetic analyses of these reverse mutants, a number of residues in the

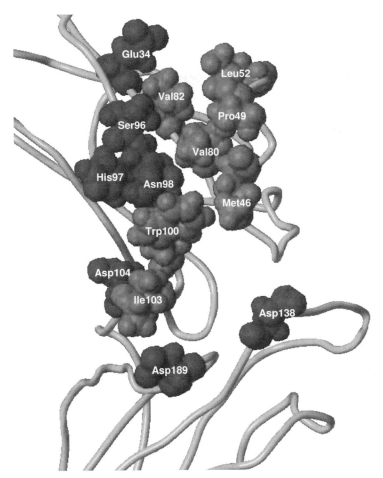

Fig. 7.3. Space-filling representation of the IFN binding surface on IFNAR-2. The elongated hydrophobic patch is comprised of residues colored in red surrounded by a ring of polar and charged residues colored in blue. (This figure also appears with the color plates.) Adapted from [21].

binding site were mutated to alanine. Dissociation kinetic measurements were used to determine the binding energy contribution of each residue. For IFN-α2, Thr44, Ile45, Met46, Ser47, Lys48, Glu77, Trp100 and Ile103 within IFNAR-2 contribute to binding, with Ile45 and Met46 contributing the most energy, and the rest contributing intermediate to marginal energy. For IFN-β, however, the binding region in IFNAR-2 shifts such that Ile45 and Trp100 contribute the most energy, Lys48 contributes intermediate energy, and Met46 contributes marginal energy. This finding is significant in that it illustrates a structural difference in receptor recognition between these IFN subtypes, which may result in the differential signaling outcome observed between IFN-αs and -β [26–30]. Blocking antibody studies with IFNAR-1 also demonstrated this difference between IFN-αs and -β. Antibodies against IFNAR-1 which block IFN-α binding and bioactivity did not affect IFN-β binding or bioactivity [31], suggesting that receptor recognition differences between the two subtypes are influencing the observed, differential signaling outcomes.

7.3.4
IFN-α Interaction with IFNAR-2

Using a panel of mutant IFN-α2 and IFNAR-2 molecules, the interaction energy between 13 residues of IFN-α2 and 11 residues of IFNAR-2 was measured [32]. From these measurements, nuclear Overhauser effect (NOE)-like constraint distances were established for five residue pairs. The largest interaction energy was measured between Arg149 on IFN-α2 and Glu77 on IFNAR-2. Other interactions include Arg144 (IFN-α2) with Met46 (IFNAR-2), Ser152 (IFN-α2) with His76 (IFNAR-2), a possible aromatic interaction between Phe27 (IFN-α2) and Tyr43 (IFNAR-2), and a potential salt bridge formation between Asp35 (IFN-α2) and Lys48 (IFNAR-2) (Fig. 7.4 and Tab. 7.1). A computer generated ligand–receptor docking model was used to demonstrate that a loop in IFNAR-2 containing "hot-

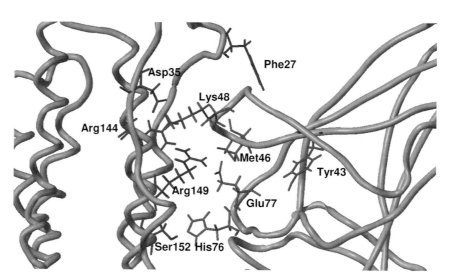

Fig. 7.4. Residues in the human IFN-α2–IFNAR-2 interface that mediate interactions between IFN and the receptor subunit. IFN-α2 is represented as a green ribbon with blue residue side-chains and IFNAR-2 is represented as a cyan ribbon with red residue side-chains. (This figure also appears with the color plates.) Adapted from [21, 32].

Tab. 7.1. Interacting residues and atoms in human IFNAR-2 and IFN-α2

IFNAR-2 residue	IFN-α2 residue	Atom ID in IFNAR-2 residue side-chain	Atom ID in IFN-α2 residue side-chain	Type of interaction
Lys48	Asp35	Nζ	Oδ1	salt bridge
His76	Ser152	Nδ1	Oγ	hydrogen bond
His76	Ser152	Nε2	Oγ	hydrogen bond
Glu77	Arg149	Oε1	Nη1	hydrogen bond
Glu77	Arg149	Oε1	Nη2	hydrogen bond

Data were derived from double-mutant cycle analysis [32]. (Adapted from [21].)

spot" residues Thr44, Ile45 and Met46 inserts into a groove formed around Ala145 in IFN-α2, which is flanked by "hotspot" residues identified in the ligand [11]. Arg33 in the ligand contributes energy to binding, although no binding partner was found on the receptor during binding studies. The docked model shows that the backbone oxygen atoms of Ser47 and Lys48 in IFNAR-2 mediate two potential hydrogen bonds with the side-chain of Arg33, explaining the lack of an Arg33 interaction partner in the receptor but its large contribution to binding energy.

7.3.5
IFN-β Interaction with IFNAR-2

The differences observed with IFN-β binding with IFNAR-2 compared with IFN-α2 binding to IFNAR-2 can be explained by residue substitution or angular orientation differences of interacting residues, based on computer modeling studies [32]. Trp100 on IFNAR-2 can potentially interact with Trp22 in IFN-β. IFN-α2 has an Ala residue at this position. This may explain why Trp100 contributes significant binding energy during IFNAR-2 interaction with IFN-β, but not with IFN-α2. Mutation of the Arg33 residue in IFN-α2 results in a 10^4-fold reduction in affinity, whereas the homologous mutation involving Arg35 in IFN-β results in only a 10-fold decrease in affinity. In the computer-generated model of IFN-β interacting with IFNAR-2, the Arg35 side-chain in IFN-β points away from IFNAR-2, most likely resulting in the observed nominal contribution by this residue to ligand affinity. The minimal contribution of Met46 in IFNAR-2 to IFN-β binding energy may be explained due to the occurrence of a valine residue at position 145 (alanine in IFN-α2). The IFNAR-2 loop deeply inserts into the groove in IFN-α2 around Ala142, so the relatively larger valine side-chain may not be able to accommodate the required interaction and so diminishes the importance of Met46 in contributing energy to IFN-β interaction with IFNAR-2.

7.3.6
Type I IFN Interactions with IFNAR-1

A solved NMR or X-ray structure of IFNAR-1 is not yet available for detailed ligand interaction studies, so mutagenesis and blocking antibody experiments have elucidated which regions of this receptor subunit are contributing to IFN binding [31, 33]. The IFNAR-2 subunit is defined as the primary binding chain, whereas IFNAR-1, which possesses little if any intrinsic affinity by itself for IFNs, plays a key role in increasing the overall affinity of IFN for its receptor complex by 10- to 20-fold [34]. The affinity of human IFNAR-1 for IFN is greater than 0.1 µM, a relatively low K_d value compared to the IFN–IFNAR-2 interaction [23, 35–38]. Antibody studies have been useful in demonstrating that epitopes found in the two membrane distal FNIII domains may be important in mediating IFNAR-1–IFN interactions [31]. Furthermore, these studies also demonstrated that different regions of IFNAR-1 contact the different IFN-α subtypes and IFN-β.

Construction of bovine–human IFNAR-1 chimeric molecules and evaluation of human IFN-α2 binding to these constructs expressed on the cell surface demonstrated that the two middle FNIII domains of IFNAR-1 are critical for moderate-affinity ligand binding, while the remaining two domains enhance binding [39, 40]. Alanine scans of bovine IFNAR-1 identified five hydrophobic residues involved in mediating binding. Residues Trp132, Phe139, Tyr160 and Trp253 are solvent exposed based on an IFNAR-1 structure generated using homology modeling [41]. The fifth residue, Tyr141, is buried beneath residues Phe139 and Tyr160, and is most likely required for proper presentation of the latter residues. These five residues are conserved in both human and murine IFNAR-1, suggesting that they are critical for function. Alanine scans of human IFNAR-1 have revealed that Tyr70 in the first FNIII domain, Trp129 in the second FNIII domain and Arg279 in the third FNIII domain also mediate binding and signaling events [33].

7.3.7
IFNAR-1 Receptor Interaction with Glycosphingolipids

Cell membrane glycosphingolipid expression, namely galabiosylceramide (Gb$_2$) and/or globotriaosylceramide (Gb$_3$), is required for proper ligand binding and transmission of IFN-induced signaling [42]. Based on primary sequence identity between IFNAR-1 and the verotoxin (VT)-1 protein produced by enterohemorrhagic *Escherichia coli*, which interacts with membrane bound glycosphingolipids through residues found in the VT-1B subunit [43], it is believed that Gb$_2$/Gb$_3$ molecules interact with residues in the most membrane distal FNIII domain of the IFNAR-1 receptor subunit. The glycosphingolipid interactions with IFNAR-1 are believed to contribute to the proper orientation of the receptor subunit so that IFN molecules are bound efficiently, allowing signaling to proceed. The glycosphingolipid–IFNAR-1 interaction may cause the extracellular domain of the IFNAR-1 subunit to fold over itself to adopt a configuration that results in proper presentation of the IFNAR-1 binding surface to the IFN molecule. Using a murine–human IFNAR-1 hybrid molecule, in which the N-terminal FNIII domain of murine IFNAR-1 was replaced with the N-terminal FNIII domain from the human protein, it was demonstrated that murine IFN-α binding and signaling through a receptor complex containing this chimeric molecule are unaffected [44], suggesting that, in the context of murine and human receptors, this domain does not contribute to the species specificity associated with type I IFNs.

7.3.8
Residues in Type I IFNs that Mediate Biological Responses

Various studies using blocking antibodies, site-directed mutagenesis, random mutagenesis and hybrid IFNs have implicated specific IFN residues in mediating antiviral and growth-inhibitory activities [45]. These residues occur in the IRRP and E helix regions of IFNs. Table 7.2 summarizes the biological activities associated with the "hotspot" residues and IRRP residues in IFNs described in this section.

There is clear evidence demonstrating that different type I IFN subtypes engage the receptor complex in a distinct manner, resulting in disparate signaling and biological outcomes [32, 46]. Blocking antibody studies and binding studies using mutated IFNs and receptor subunits show that IFN-αs and -β recognize different regions of IFNAR-1 and -2. Unlike the IFN-αs, IFN-β induces the association of both phosphorylated IFN receptor subunits [27] and is active in cells lacking TYK-2, the intracellular kinase that is non-covalently associated with IFNAR-1 [47]. Additionally, gene expression for CXCL11 (β-R1/I-TAC) [28], a natural antagonist of CC chemokine ligand (CCL5) [48], is uniquely induced by IFN-β [28]. The difference in receptor recognition by IFN-αs and -β leads to the differences in signaling outcome observed.

7.4
Type I IFN-induced Signaling Cascades

The type I IFN receptor complex transduces signals through activation of receptor-associated Janus protein tyrosine kinases (JAKs). Binding of all type I IFNs to the receptor complex induces association of the two receptor chains, and subsequent reciprocal transphosphorylation and activation of JAKs. The JAKs, TYK-2 and JAK-1, constitutively associate with the IFNAR-1 and -2c subunits of the IFN receptor, respectively [49–51].

7.4.1
The JAK–STAT (Signal Transducers and Activators of Transcription) Pathway

The JAK–STAT pathway is the primary signaling pathway for the transcriptional regulation of many IFN-stimulated genes (ISGs). Upon their activation, JAKs

Tab. 7.2. Residues in the receptor binding regions of type I IFNs that contribute to biological activity

IFN residue	Role in biological activity
Residues in all IFN-α subtypes	
Cys29, Cys139	disulfide bond critical for full IFN activity
Leu30	critical for antiviral and antiproliferative activity
Lys31	minor contribution to antiviral activity
Arg33	critical for antiviral and antiproliferative activity
His34	minor contribution to antiviral and antiproliferative activity
Asp35	minor contribution to antiviral activity
residues 78–95 (IRRP-2)	differential sensitivities of type I IFN subtypes may be attributed to this region
Asp78, Glu79	contribute to antiviral activity
Arg121, Lys122, Tyr123	critical for full activity
Phe124	contributes to antiviral activity
Gln125	critical for antiviral activity
Tyr130	contributes to antiviral activity
Leu131	critical for antiviral and antiproliferative activities
Lys134, Lys135, Tyr136	contribute to antiviral activity
Trp141	critical for antiviral activity
Ala146	critical for biological activity
Ser151	critical for biological function
Residues found in specific subtypes	
Phe27 (in all α subtypes except IFN-α1 and not in IFN-β)	necessary for full IFN activity
Asn86 and Tyr92 (in IFN-β)	association of phosphorylated receptor subunits; activity in TYK-2-deficient cells
Arg121 (in all α subtypes except IFNs-α1, -α13, -α14 and -α21)	critical for antiviral activity
Gln125 (in all α-subtypes except IFNs-α1 and -α13)	critical for antiviral activity
His138 (in IFN-β)	critical for biological function
Glu147 and Arg150 (IFN-β)	critical for biological function

Refer to [45] for a more comprehensive listing of IFN residues and their contribution to biological responses.

phosphorylate specific tyrosine residues within the intracellular domains of the receptor subunits [52]. These phosphorylated residues serve as docking sites for STAT proteins [53]. Once recruited to the receptor complex, the activated JAKs phosphorylate a single tyrosine residue within the C-terminus of the STAT protein [54]. The phosphorylated and activated STATs (Fig. 7.5) form both homodimeric and heterodimeric complexes that translocate to the nucleus, and bind specific DNA sequences within the promoter regions of ISGs to initiate transcription [55].

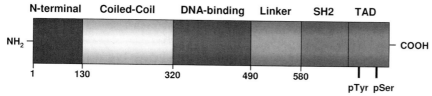

Fig. 7.5. Domain structure of STAT proteins. STAT proteins are a family of latent cytoplasmic transcription factors that serve as important mediators of cytokine, hormone and growth factor signal transduction. There are seven mammalian members of this family, STAT-1, -2, -3, -4, -5a, -5b and -6, all of which share a conserved domain-like structure. STAT proteins range in size from 748 to 851 amino acids (90–115 kDa) and consist of six different domains, each with its own defined function. The N-terminal domain (residues 1 to ~130) is involved in stabilizing STAT dimer–dimer interactions. The coiled-coil domain (~130 to ~320) is important for protein interactions. The central DNA-binding domain extends from amino acids ~320–490 and contains several residues conserved in all members of the STAT family. A linker domain exists between residues 490 and 580, and separates the DNA-binding domain from the SH2 domain. This area is comprised primarily of α-helices and appears to play a role in mediating transcription. The phosphotyrosine-binding SH2 domain is required for receptor binding and dimerization. Within this domain is a conserved tyrosine residue. Phosphorylation of this tyrosine activates the STAT molecule, allowing it to interact with the SH2 domain of another STAT. At the C-terminal end a transcriptional-activation domain (TAD) modulates the transcriptional functions of the various STAT proteins. The TAD mediates interactions of the STAT protein with a number of nuclear coactivators, facilitating chromatin modifications and transcriptional activation. (This figure also appears with the color plates.)

Upon activation, TYK-2 phosphorylates Tyr466 on IFNAR-1, generating a recruitment site for STAT-2 by means of its Src homology 2 (SH2) domain [56]. TYK-2 then phosphorylates STAT-2 on Tyr690, which acts as a recruitment site for STAT-1 [57, 58]. Subsequently STAT-1 is phosphorylated on Tyr701. The activated STAT-2:1 heterodimers dissociate from the receptor complex and translocate to the nucleus. Only the intracellular domain of the IFNAR-2c chain is required for mediating the recruitment and activation of STATs and the formation of the STAT complexes [59, 60]. Within the IFNAR-2c intracellular domain only a single tyrosine residue at either position 337 or 512 is necessary for a complete IFN response [61]. STAT-2 binds constitutively to the IFNAR-2c subunit, while STAT-1 associates with the IFNAR-2c–STAT-2 complex, suggesting that STAT-2 provides a docking site for the SH2 domain of STAT-1, linking it to the receptor [62]. The association of STAT-2 with IFNAR-2c is independent of any tyrosine phosphorylation of the receptor or STAT protein.

Another important component of the IFN receptor complex is the WD repeat-containing receptor RACK-1. This protein associates with the IFNAR-2c chain, and acts as a scaffold for the formation of a receptor complex comprised of JAK-1, TYK-2 and STAT-1 [63, 64]. A mutation in the RACK-1 binding site of IFNAR-2c abolishes both STAT-1 and -2 activation. STAT-3 is also activated upon IFN-receptor binding. For normal STAT-3 tyrosine phosphorylation, the catalytically active form of TYK-2 is essential, but receptor phosphorylation is not required [62, 65].

A major IFN-inducible complex is ISG factor 3 (ISGF-3), comprised of STAT-1 and -2, and a DNA-binding adapter protein of the IFN-regulatory factor (IRF) family, IRF-9 (p48 or ISGF-3γ) [66]. Upon nuclear import, this ISGF-3 complex binds the IFN-stimulated response element (ISRE), AGTTTN$_3$TTTC, to initiate gene transcription. Type I IFNs also induce the formation of other STAT-containing complexes: STAT-1:1, -3:3 and -5:5 homodimers as well as STAT-3:1 and -2:1 heterodimers [67–69]. These homodimers and heterodimers bind palindromic sequences, TTCN$_3$GAA, designated γ-activated sequences (GAS) located in the promoters of a different subset of ISGs. In mature B cells, IFN also induces the tyrosine phosphorylation of STAT-6, leading to the formation of a STAT-2:6 heterodimer as well as an ISGF-3-like complex comprising STAT-2 and -6, and IRF-9 [70]. IFN-α stimulation of human CD4$^+$ T cells also activates STAT-4 through recruitment to the IFNAR-complex via the C-terminus of STAT-2 [71].

In addition to tyrosine phosphorylation, several STAT members, i.e. STAT-1, -3, -4 and -5, use phosphorylation of a specific serine residue within the transcriptional activation domain to regulate transcription [72, 73]. Namely, the transcriptional activation potential of STAT-1-containing complexes, including both STAT-1 homodimers and ISGF-3, is dependent upon the phosphorylation of residue Ser727 [74, 75].

7.4.1.1 The Importance of STAT-1 and -2 in Type I IFN Signaling

Examination of the phenotypes of mice deficient in the various STAT proteins highlighted several of their specific physiological functions (Tab. 7.3) [76].

Mice lacking STAT-1 are viable and have no developmental defects [77]. However, these mice display a selective deficiency in their responses to both type I and II IFNs, making them highly susceptible to viral and other microbial infections [78]. The expression of several ISGs was severely compromised in these mice. STAT-1-deficient mice also have a greater rate of tumor development, due to de-

Tab. 7.3. Phenotypes of STAT knockout mice

Protein	Phenotype of targeted gene disruption
STAT-1	defects in type I and II IFN signaling
STAT-2	defects in type I IFN signaling
STAT-3	embryonic lethal
STAT-4	loss of IL-12 signaling
STAT-5a	defect in prolactin signaling
STAT-5b	loss of growth hormone signaling
STAT-6	loss of IL-4 responsiveness

fects in IFN responses, apoptosis and reduced expression of caspases [79]. These results demonstrate that STAT-1 has an essential and nonredundant role in IFN signaling.

STAT-2-deficient mice are developmentally normal and primarily defective in their response to IFN [80]. These mice are highly sensitive to viral infections due to their inability to activate a number of IFN-inducible antiviral genes. In addition, STAT-2-deficient mice exhibit a loss of the type I IFN autocrine/paracrine loop that regulates STAT-1 activation and subsequently mediates the activation of additional immune responses.

7.4.2
Other IFN-inducible Signaling Cascades

In addition to the JAK–STAT pathway, IFN-receptor binding results in the activation of several other intracellular signaling cascades. Together these cascades coordinate to induce the synthesis of proteins that elicit the biologic effects of IFNs (Fig. 7.6).

7.4.2.1 The CrkL Pathway

The three members of the Crk protein family, CrkL, CrkI and CrkII, are cellular homologs of the v-*crk* protooncogene [81]. Containing a single SH2 domain and two SH3 domains, both CrkL and CrkII serve as adaptors, linking tyrosine phosphorylated receptors to downstream signaling components. The SH3 domains of CrkL and CrkII bind a guanine exchange factor for Ras, Sos, and a guanine exchange factor for Rap1, C3G [82]. CrkL interacts with C3G allowing the CrkL to link the activated IFN receptor to the growth-inhibitory C3G–Rap1 pathway [83]. IFN stimulation of target cells induces tyrosine phosphorylation of CrkL and CrkII. Similarly, IFN treatment leads to the phosphorylation and activation of STAT-5, constitutively associated with IFNAR-1-bound TYK-2 [84]. Once activated, STAT-5 associates with the SH2 domain of CrkL and the resultant STAT-5–CrkL complexes translocate to the nucleus and bind GAS elements in the promoters of ISGs. Within the context of these STAT-5–CrkL DNA-binding complexes, CrkL is required for IFN-induced gene transcription via GAS elements [85]. In cells lacking CrkL, IFN-stimulated Rap1 activation is impaired, highlighting the importance of CrkL in regulating the growth-inhibitory C3G–Rap1 pathway.

7.4.2.2 The IRS Pathway

The IRS pathway operates independently of the JAK–STAT pathway and is involved in mediating IFN-α-dependent activation of phosphatidylinositol-3-kinase (PI3K) [86, 87]. High-molecular-weight IRS proteins, i.e. IRS-1 and -2, are substrates for the activated JAK kinases [88, 89]. Once phosphorylated, they provide docking sites for the SH2 domains of various signaling components, including

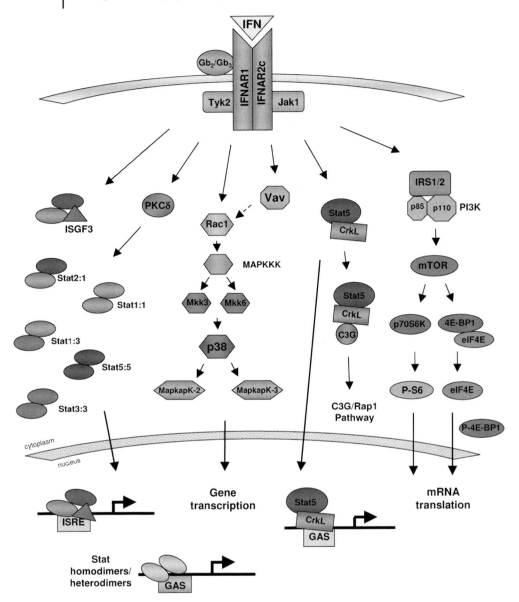

Fig. 7.6. Type I IFN-induced signaling. Engagement of the type I IFN receptor complex leads to the phosphorylation of JAK-1 and TYK-2, and activation of the JAK–STAT, p38MAPK, CrkL and IRS–PI3K pathways. (This figure also appears with the color plates.)

Grb-2, the p85 subunit of the PI3K, SHP2 and Crk, whereupon they coordinate cytokine-mediated signal transduction by linking tyrosine kinases or their substrates to downstream signaling events [90]. Specifically, IFN-α-induced JAK-1 activation results in the tyrosine phosphorylation of both IRS-1 and -2, allowing them to associate with the p85 regulatory subunit of PI3K [91, 92]. This IRS–p85 interaction leads to the activation of both the phosphatidylinositol and serine kinase activities of the p110 catalytic subunit of PI3K.

IFN stimulation of the IRS–PI3K pathway induces the activation of the downstream effector called mammalian target of rapamycin (mTOR) [93]. mTOR is a serine/threonine protein kinase which subsequently regulates the initiation of protein translation by mediating both the activation of the p70S6 kinase and the inactivation of the 4E-BP1 repressor of translation. Upon activation, the p70S6 kinase regulates the phosphorylation of serine residues on the 40S ribosomal S6 protein, which serves as a key regulator of cell–cell progression and mRNA translation. mTOR-dependent phosphorylation of 4E-BP1 allows its dissociation from eukaryotic translation initiation factor (eIF)-4E and the subsequent initiation of translation. Thus, the IFN-activated IRS–PI3K pathway provides a mechanism to regulate mRNA translation in response to type I IFNs. By promoting mRNA translation of ISGs, this pathway complements the JAK–STAT cascade in mediating the biological effects of type I IFNs.

7.4.2.3 The p38 Mitogen-activated Protein Kinases (MAPK) Pathway

MAPKs are a group of widely expressed serine/threonine kinases: the extracellular signal-regulated kinases (ERKs), the Jun N-terminal kinases (JNKs) and the p38MAPKs [94]. There are four p38MAPK isoforms (α, β, γ and δ) that are activated by a variety of stimuli, including hyperosmolarity, heat shock, radiation, other forms of cellular stress and stimulation with pro-inflammatory cytokines. Activation of p38MAPK is mediated by a series of upstream effectors, ultimately regulating the phosphorylation of a Thr–Gly–Tyr motif within the p38 protein.

Type I IFNs induce the phosphorylation and subsequent activation of p38MAPK [95, 96]. The p38MAPK pathway is important in regulating IFN-induced responses, and is essential for both ISRE- and GAS-mediated gene transcription [87, 95]. However, p38MAPK is not involved in mediating STAT complex formation, serine phosphorylation or DNA binding [87].

In addition, IFN stimulation activates upstream regulators of p38MAPK, including the small G-protein, Rac1, and both MAPK kinase (MKK)-3 and -6 [87, 97]. In cells lacking both MKK-3 and -6, IFN-dependent ISRE- and GAS-mediated transcription is impaired. Downstream effectors of p38MAPK are also activated by IFN. Specifically, IFN activates Msk-1, a kinase involved in the regulation of histone phosphorylation and chromatin remodeling [98]. IFN induces the p38MAPK-dependent activation of downstream serine kinases, MapKapK-2 and -3 [95, 98]. Moreover, p38MAPK-facilitated transcription and IFN-dependent antiviral responses are compromised in cells lacking MapKapK-2 [98]. Together these findings suggest that the role of p38MAPK in IFN-induced signaling is to regulate the activation of downstream effectors that modulate ISG transcription.

7.4.2.4 The Vav Proto-oncogene and IFN Signaling

The Vav proto-oncogene products play an important role in cellular signaling, linking cell surface receptors to various effectors functions. Best characterized as a guanine exchange factor for the Rho/Rac family of GTPases, Vav undergoes tyrosine phosphorylation in response to a number of different stimuli, including IFN [99]. Vav associates with both subunits of the type I IFN receptor, and, upon IFN stimulation, forms a complex with both TYK-2 and the Ku80 regulatory subunit of the DNA-dependent protein kinase (DNA-PK) [100, 101]. Although the role of Ku80 in IFN signaling remains unknown, Ku80 may play a role in modulating gene transcription. Moreover, IFN-induced phosphorylation activation of Vav engages the GDP/GTP exchange activity of Rac1, activating its downstream signaling. Rac1 regulates IFN-induced p38MAPK activation, suggesting that during IFN signaling Vav phosphorylation may mediate the p38MAPK signaling cascade [102, 103].

7.4.2.5 The Protein Kinase C (PKC) Family and IFN Signaling

The PKC family is a group of serine/threonine kinases that regulate cellular activities and signal transduction. This group is comprised of multiple isoforms that differ in their modular structures and regulatory functions. Classification of different PKC isotypes is based on activation requirements. The first group consists of conventional PKC (cPKC) isoforms (PKC-α, -β and -γ) that are dependent upon elevated levels of intracellular calcium and phorbol esters for their activation. The second group, called novel PKCs (nPKCs), is comprised of members (PKC-δ, -ε, -θ, -η and -μ) that require phorbol esters for their activation. The third group, named atypical PKCs (aPKCs), consists of isoforms (PKC-ζ and -λ) whose activation is independent of phorbol esters. Upon activation, the PKC proteins function as serine kinase activities, regulating the phosphorylation and activation of downstream signaling components.

Various PKC isoforms are also activated by type I IFNs. IFN activates PKC-δ to regulate the serine phosphorylation of STAT-1 [104]. Pharmacological and genetic inhibition of IFN-induced PKC-δ activation abrogates STAT-1 serine phosphorylation, restricting both GAS- and ISRE-mediated gene transcription [104]. Inhibition of IFN-activated PKC-δ also prevents the activation of p38MAPK, suggesting that, during IFN signaling, crosstalk occurs between the PKC-δ and p38MAPK pathways [104]. In T cells, IFN induces the phosphorylation and activation of PKC-θ. Inhibition of PKC-θ protein expression prevents IFN-dependent GAS-mediated gene transcription [105]. Moreover, PKC-θ is also implicated in the phosphorylation and activation of MKK-4, suggesting that IFN-dependent PKC-θ activation plays a role in the downstream engagement of MAPK signaling [105].

7.5
IFN-inducible Biological Responses

Binding of IFNs to specific cell surface receptors results in the activation of multiple intracellular signaling pathways, most notably the JAK–STAT, CrkL, IRS–PI3K and p38MAPK cascades. Together, they coordinately induce the synthesis of proteins that mediate antiviral, growth-inhibitory and immunomodulatory responses (Tab. 7.4).

7.5.1
IFN-inducible Antiviral Responses

To prevent the spread of a viral infection, infected cells activate an IFN response [106, 107] (Fig. 7.7). Specifically, virus-inducible Toll-like receptor (TLR) activation leads to the transcriptional activation of IFN-β and -α4, and subsequently the non-IFN-α4 IFN-α subtypes [108, 109]. Mice lacking IFN-β are extremely susceptible to both vaccinia and coxsackie viral infections [110, 111]. Additionally, mice with targeted deletions of different components of the IFN signaling pathway are highly susceptible to viral infections. Mice lacking IFNAR-1, or STAT-1 or -2 all exhibit increased sensitivity to infection with viruses [77, 80, 112].

Activation of the JAK–STAT pathway is important for control of viral replication. The IRS–PI3K pathway also plays a role in mediating virus-induced IFN-mediated cell death. Notably, in murine embryonic fibroblasts (MEFs), IFN-induced PI3K confers resistance to infection with encephalomyocarditis virus (EMCV) and herpes simplex virus type 1 (HSV-1), independent of the JAK–STAT pathway [113]. The p38MAPK pathway further contributes to the antiviral effects of IFN, as treatment of IFN-sensitive KT-1 cells with a pharmacological p38MAPK inhibitor decreases the IFN-stimulated antiviral response [114]. Downstream effectors of p38MAPK mediate this antiviral state, as MapKapK-2-deficient MEFs exhibit diminished IFN-induced protection against EMCV compared with MEFs expressing MapKapK-2 [98].

IFN-receptor activation leads to multiple signaling cascades, which coordinately invoke gene regulation in target cells that will create an antiviral response. IFN-inducible control of viral replication is regulated by the ISRE-mediated transcriptional activation of three key signaling effectors: double-stranded (ds) RNA-dependent protein kinase (PKR), 2′–5′-oligoadenylate synthetase (2′–5′-OAS)/endoribonuclease L (RNase L) and the myxovirus resistance protein (Mx). Studies with PKR$^{-/-}$ and RNase L$^{-/-}$ murine fibroblasts demonstrated that the IFN-induced antiviral response is severely compromised in the absence of PKR and RNase L [115].

The generation of mice triply deficient for PKR, RNase L and MxA demonstrated that the antiviral activity of type I IFNs is not confined to these three effector proteins, since these mice show residual IFN-inducible antiviral activity [116]. Other effectors implicated in IFN-induced antiviral responses include the RNA-specific

Tab. 7.4. Type I IFN-activated effectors and their contributions to IFN-inducible responses

Protein	Classification	Biological activity
Signaling intermediates		
STAT-1, -2, -3, -4, -5a, -5b and -6	transcription factor	transcriptional activation of ISGs
IRS-1/2	docking protein	mediates PI3K activation
		contributes to antiviral activity
		adapter for T cell signaling
PI3K	lipid and serine kinase	contributes to antiviral activity
		promotes survival of peripheral B cells
CrkL	adapter	transcriptional activation
		activates C3G–Rap1 cascade
		regulates cell growth
p38MAPK	serine/threonine kinase	contributes to transcriptional activation
		mediates antiviral activity and growth inhibition
ISG products		
PKR	serine/threonine kinase	antiviral activity – phosphorylates eIF2α
		regulates translation
		activates NF-κB, p38, JNK
		mediates apoptosis
2′–5′-OAS	synthetase	antiviral activity – activates RNase L
		mediates JNK activation
		mediates apoptosis
Mx proteins	GTPase	antiviral activity – block viral replication
ADAR	deaminase	antiviral activity – deaminates adenosine
		disrupts RNA editing
P56 family	IFN-induced protein with tetratricopeptide repeats	antiviral activity
		restrict translation initiation
TRAIL	ligand	antiviral activity – restricts viral infection
		mediates apoptosis
Viperin	cytoplasmic protein	antiviral activity – restricts viral replication
PLSCR-1	plasma membrane protein	antiviral activity
		amplifies ISG expression
ISG-20	3′–5′ exoribonuclease	antiviral activity
		disrupts viral RNA synthesis
p21	CDK inhibitor	regulates cell growth
		restricts phosphorylation of pocket protein
c-myc	oncogene	regulates cell growth
HIN-200 family	DNA-binding proteins	regulate cell growth
		disrupt transcriptional activation
MHC class I proteins	receptor	activate adaptive immune response
		modulate T cell responses
TLRs	receptor	mediate antimicrobial responses
IL-12, -15 and -21, and IFN-γ	cytokines	activate adaptive immune response
		modulate T cell responses
CCL5, CCL3, CXCL10	chemokines	activate adaptive immune response
		modulate T cell responses

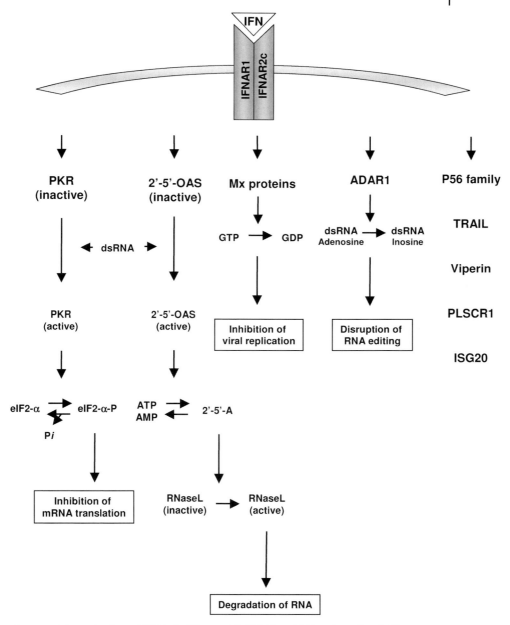

Fig. 7.7. Schematic of type I IFN-inducible antiviral responses. IFN activates the JAK–STAT pathway that induces the expression of ISGs that mediate the antiviral actions of IFN. PKR inhibits translation initiation through phosphorylation of initiation factor eIF-2α; 2′–5′-OAS and RNase L mediate RNA degradation; Mx GTPases inhibit viral assembly; ADAR-1 disrupts RNA editing. Members of the p56 family, TRAIL, Viperin, PLSCR-1 and ISG-20 also contribute to the antiviral effects of IFN. (Adapted from [107].)

adenosine deaminase (ADAR-1), the P56 family of proteins, tumor necrosis factor (TNF)-related apoptosis-inducing ligand (TRAIL), Viperin (virus-inhibitory protein, endoplasmic reticulum associated, interferon inducible), phospholipid scramblase (PLSCR-1) and ISG-20 [117–120].

7.5.1.1 PKR

PKR is an IFN-inducible serine/threonine kinase that serves as a critical mediator of IFN-inducible antiviral activity [121]. The binding of viral dsRNA intermediates to PKR induces a conformational change that effectively activates its catalytic domain [122]. Upon activation, PKR phosphorylates the α subunit of eIF-2 (eIF-2α), preventing the recycling of translation initiation factors and restricting *de novo* protein synthesis [123]. Phosphorylated eIF-2α associates with a guanidine nucleotide exchange factor, eIF-2B, blocking eIF-2-GDP to eIF-2-GTP recycling and impairing translation. In response to virus and poly(I:C), mice lacking PKR exhibit normal IFN induction, but are highly sensitive to virus infection [124]. The antiviral action of PKR is viral specific. Specifically, PKR activity is required for the IFN-mediated antiviral activities against vesicular stomatitis virus (VSV) and influenza virus, but is not essential for IFN-induced protection from EMCV or vaccinia infection [125–127].

In addition, PKR is important in dsRNA-induced signal transduction, specifically influencing the function of transcription factors STAT-1, IRF-1 and p53 [128]. PKR also stimulates Fas expression and induces apoptosis through the Fas-associated protein with death domain (FADD)/caspase-8 death signaling cascade [125, 129–131]. Thus, activated PKR mediates the expression of proapoptotic genes, leading to apoptosis of virus-infected cells and ultimately preventing viral replication.

PKR also regulates the activation of two MAPKs, p38MAPK and JNK, by proinflammatory stimuli, including lipopolysaccharide (LPS), TNF-α and interleukin (IL)-1 [132]. Indeed, PKR is an important mediator of host defense, regulating different cellular responses. PKR alone, however, is unable to mediate the full antiviral response of IFN, as mice lacking PKR have residual resistance to viral infection [124].

7.5.1.2 2′–5′-OAS/RNase L

IFNs also induce the synthesis of a group of 2′–5′-OAS enzymes that are activated by viral dsRNA intermediates [133]. Upon activation, the 2′–5′-OAS proteins catalyze the polymerization of ATP into 2′–5′-linked oligoadenylates (2′–5′-A) of different lengths. These 2′–5′-A molecules bind with high affinity to the latent RNase L, inducing its dimerization and activation. The activated RNase L cleaves single-stranded (ss) cellular and viral RNA, subsequently blocking protein synthesis. RNase L also cleaves 28S ribosomal RNA, inducing ribosome inactivation. The IFN-induced 2′–5′-OAS pathway is particularly effective in preventing picornaviral infections, including EMCV [134, 135].

RNase L activation is also necessary for activation of the JNK signaling cascade. In cells deficient in RNase L, both virally induced JNK activation and apoptosis are compromised [136]. Similarly, the apoptotic effects of IFN and 2′–5′-OAS are abrogated in cells lacking JNK-1 and -2 [136]. These findings suggest that RNase L-mediated JNK activation is an important component of the IFN-stimulated antiviral response mediating apoptosis. Additionally, the 9-2 isozyme of 2′–5′-OAS promotes apoptosis through its capacity to bind and inactivate the anti-apoptotic proteins Bcl-2 and Bcl-X_L [137]. The IFN-inducible 2′–5′-OAS pathway therefore mediates antiviral protection by directly restricting viral replication and by invoking apoptosis of virally infected cells.

7.5.1.3 Mx

Type I IFNs induce the activation of another family of antiviral mediators – Mx resistance proteins. Mx proteins belong to a superfamily of large GTPases, which include dynamins and IFN-regulated guanylate-binding proteins. Most vertebrates express one to three Mx isoforms that differ in their intracellular localization and antiviral properties [138].

In humans, two Mx isoforms, MxA and MxB, are expressed; however, only MxA possesses antiviral activity. Human MxA is a 78-kDa protein with broad antiviral effects against numerous types of viruses [139]. In mice, the nuclear Mx1 and cytosolic Mx2 proteins both confer antiviral protection. Mx1 shows specificity towards orthomyxoviruses and thogaviruses that replicate in the nucleus, whereas Mx2 targets cytosolic bunyaviruses.

GTP binding and hydrolysis of Mx proteins induces a conformational change that is critical for mediating antiviral activity [140]. The Mx proteins elicit their antiviral activity either by disrupting viral polymerase activity and blocking viral replication or by binding ribonucleoproteins and inhibiting viral RNA transcription. Mx proteins also inhibit viral assembly by preventing the transport of viral mRNA and nucleocapsid proteins within the cell. MxA specifically restricts the infection of numerous negative-stranded RNA viruses, including influenza virus, measles virus, VSV, Thogoto virus and Bunya virus [141–145]. MxA also exerts antiviral activity against positive-stranded RNA viruses, i.e. Semliki Forest virus and coxsackie virus B [146, 147]. Interestingly, for different viral targets MxA acts at different levels of the viral replication cycle. During Semliki Forest and VSV infections, MxA blocks early transcriptional events, whereas later replication stages are targeted during infection with influenza virus. Upon challenge with measles virus, MxA inhibits measles virus infection at either the viral RNA stage or the glycoprotein synthesis stage [148]. Conversely, MxA inhibits hepatitis B viral infection by blocking the synthesis of viral proteins, cytoplasmic RNAs and DNA replicative intermediates, and restricting the nuclear export of viral RNAs [149]. MxA also blocks the replication of Bunya virus by binding and sequestering viral nucleocapsid proteins within cytoplasmic complexes [140].

7.5.1.4 Other IFN-inducible Antiviral Effectors

The *Adar-1* gene is IFN-inducible and its promoter contains a functional ISRE [150]. ADAR-1 plays a role in the editing of both viral and cellular RNA by means of site-specific deamination of adenosines. More specifically, this enzyme catalyzes the covalent modification and subsequent destabilization of RNA through the conversion of adenosines to inosines [107]. Importantly, RNA editing results in changes in the peptide-coding capacity of transcripts and may control viral infectivity by preventing the translation of important virally encoded genes. Upon challenge with hepatitis δ virus (HDV), IFN-induced ADAR-1 expression resulted in a noticeable increase in RNA editing within target cells [151]. This RNA editing controls the synthesis of S and L antigens, subsequently regulating the viral life cycle [107].

The P56 family of proteins, also called IFITs (IFN-induced protein with tetratricopeptide repeats), is strongly induced in response to both viruses and IFN treatment [120]. In humans there are four related P56 proteins, HuP56 (IFIT1 or ISG56), HuP54 (IFIT2 or ISG54), HuP60 (IFIT4 or ISG60) and HuP58 (IFIT5 or ISG58). In mice, three members exist, called MuP56 (IFIT1 or ISG56), MuP54 (GARG-39) and MuP49 (GARG-49). Although their defined function remains to be determined, preliminary evidence suggests that these proteins restrict translation initiation and subsequently play a role in mediating IFN-inducible antiviral activity. Through interaction with the 48-kDa subunit of eIF-3 (eIF-3-p48), HuP56 binds a component of the translation initiation machinery, eIF-3, impairing its ability to stabilize the ternary complex containing eIF-2, GTP and Met-tRNA$_i$, thereby preventing translation [152].

IFN-β-inducible expression of TRAIL, a member of the TNF ligand superfamily, by NK cells, is ISGF3-mediated, and has been linked to anti-EMCV activity [117]. Viperin is an IFN-inducible, cytoplasmic, antiviral protein that is activated by cytomegalovirus (CMV) and the CMV envelope protein, glycoprotein B [118]. Expression of Viperin in fibroblasts prevents productive CMV infection by downregulating a number of CMV structural proteins (gB, pp28 and pp65) that are important for viral assembly and maturation. IFN-inducible PLSCR-1 is a calcium-binding plasma membrane protein that can either insert into the plasma membrane or directly bind DNA in the nucleus. Interestingly, PLSCR-1 confers IFN-inducible antiviral activity by amplifying the IFN response through induced expression of a particular set of ISGs, including ISG15, HuP54 and HuP56 [153]. ISG20 is an IFN-induced 3′–5′ exoribonuclease that acts on ssRNA and elicits antiviral activity toward RNA viruses. Notably, ISG20 confers resistance to infections with EMCV, VSV and influenza virus by disrupting mRNA synthesis and peptide production [154].

7.5.2
IFN-inducible Growth-inhibitory Responses

Accumulating evidence from *in vitro* studies implicates specific type I IFN-inducible signaling cascades as important regulators of cellular growth. Notably, the Crk and p38MAPK pathways, as well as the Vav protooncogene are key mediators of the antiproliferative responses of IFNs.

The CrkL cascade is an important mediator of the growth-inhibitory action of IFN in hematopoietic progenitors. The introduction of antisense oligonucleotides to CrkL and CrkII blocks IFN-induced growth inhibition in normal erythroid and myeloid progenitors [90]. Although the full mechanism by which Crk proteins regulate IFN-induced growth inhibition remains unknown, studies in cells derived from chronic myelogenous leukemia (CML) patients revealed that CrkL can regulate the transcriptional activity of STAT-5 [84]. Indeed, IFN treatment of these cells induces the formation of STAT-5–CrkL complexes that bind promoter sequences and activate the transcription of a subset of ISGs associated with growth inhibition, i.e. promyelocytic leukemia (PML) [84]. The *PML* gene encodes a tumor suppressor protein. Interestingly, TRAIL is a downstream transcriptional target of PML, suggesting that, in addition to its antiviral activity, TRAIL also plays an important role in IFN-induced growth inhibition and apoptosis [155].

In cells deficient in CrkL, IFN-induced activation of the GTPase, Rap1, was impaired. IFN activation of CrkL induces C3G association and the activation of Rap1, which antagonizes the Ras pathway and promotes tumor suppressor activity [156]. Thus, in IFN signaling, CrkL functions as both a nuclear adapter protein for STAT-5 and a signaling effector, linking the activated IFN receptor to the growth-inhibitory C3G–Rap1 cascade.

Studies with primary cells and cell lines derived from CML patients have revealed an essential role for the p38MAPK pathway in IFN-induced growth-inhibitory activity [157]. Pharmacological inhibition of p38MAPK activity results in partial abrogation of IFN-inducible growth inhibition of CML progenitors [114, 158]. Further studies demonstrated that IFN-induced activation of p38MAPK is required for the transcriptional activation of ISGs [87]. The implications are that the IFN-inducible Rac1–p38MAPK signaling cascade regulates the expression of genes that effect growth inhibition [114, 159].

Several IFN-activated proteins can also elicit a direct growth-inhibitory response by controlling the expression of effectors that mediate cell cycle entry, exit and progression. Notably, IFNs target essential cell cycle regulators, including the retinoblastoma gene product (Rb), cyclin A, cyclin-dependent kinase (CDK) 2, E2F, c-*myc*, as well as CDK inhibitors (CKI), specifically p15, p21 and p27 [160–162]. Each stage of the cell cycle is susceptible to IFN-inducible arrest. However, different phases are targeted in different cell types. For example, Daudi Burkitt's lymphoma cells arrest at G_0, while U-266 myeloma cells are targeted at the G_1/S transition phase [163].

In normal cells, cyclin–CDK complexes phosphorylate Rb whereupon Rb dissociates from E2F, enabling the transcription of genes involved in G_1/S phase transition. However, in cells undergoing IFN-induced G_0 arrest, Rb and p107 levels are lower. Moreover, IFN increases the levels of the CKI p21, a key inhibitor of G_1 CDK activity, which suppresses CDK-mediated phosphorylation of the pocket proteins, Rb, p107 and p130 [163–165]. These protein-bound E2F complexes act to negatively regulate the cell cycle, slowing progression into S phase [166]. IFN also downregulates cyclin E, cyclin A and CDC25, and c-*myc* expression, which correlates with the reduced DNA-binding capacity of E2F [167]. Moreover, IFN suppresses cyclin D3 expression, leading to reduced cyclin D–CDK4 and cyclin–CDK6 kinase activities.

Other effectors involved in regulating the growth-inhibitory effects of IFN are the hematopoietic IFN-inducible nuclear protein with the 200-amino-acid repeat (HIN-200) family [168]. This family of IFN-inducible proteins consists of the structurally related murine (P202, P203, P204 and P205) and human myeloid cell nuclear differentiation antigen (MNDA), absent in myeloma protein (AIM)-2 and IFN-γ-inducible protein (IFI)-16 proteins [169]. IFN treatment of Daudi cells induces MNDA and IFI16 expression [170]. Notably, IFN-mediated induction of these proteins induces a transient increase in p21 expression, activation of Rb and p130 via hypophosphorylation, as well as a decrease in c-*myc* expression [170, 171]. The P202 protein inhibits cell proliferation by associating with hypophosphorylated Rb and/or E2F, thereby preventing G_1/S phase transition. These P202–E2F complexes restrict cell growth by blocking E2F1-mediated transcriptional activation of S-phase genes [172]. Moreover, P202-bound E2F abrogates caspase-3 activation and apoptosis [173]. Another member of the HIN-200 family, IFN-inducible protein X (IFIX)-α1, restricts the growth of tumor cells in a Rb- and p53-independent manner by increasing p21 levels, and subsequently decreasing CDK2 and CDC2 kinase activity [174].

IFNs are also important mediators of apoptosis, inducing the expression of a subset of proapoptotic ISGs. IFN-stimulated apoptosis involves the activation of Fas, caspases, TRAIL, the X-linked inhibitor of apoptosis-associated factor 1 (XAF)-1, the Bcl-2 related proteins, Bak and Bax, as well as the expression of a group of regulators of IFN-induced death (RIDs), that include IP6K2 kinase [175–179]. Members of the IRF family, IRF-1, -3 and -5, also act as mediators of IFN-inducible apoptosis by transcriptionally regulating the expression of additional proapoptotic genes [180]. Although the precise mechanism of IFN-induced apoptosis is unknown, it incorporates FADD/caspase-8 signaling, ordered activation of the caspase cascade, cytochrome *c* release from mitochondria and DNA fragmentation [181].

7.5.3
IFN-inducible Immunomodulation

Type I IFNs are involved in modulating several aspects of the adaptive immune system, and in bridging the innate and adaptive immunity arms of the immune system (Fig. 7.8) [182, 183]. IFN receptor and IFN-β knockout mouse models demonstrated that the type I IFN system is important in mediating certain stages of immune system development and hematopoeisis [112, 184]. IFNAR-1-deficient mice exhibited elevated levels of myeloid lineage cells in peripheral blood and bone marrow [112]. In IFN-β-deficient mice, constitutive and induced expression of TNF-α is reduced in the spleen and by bone marrow macrophages [184]. Altered

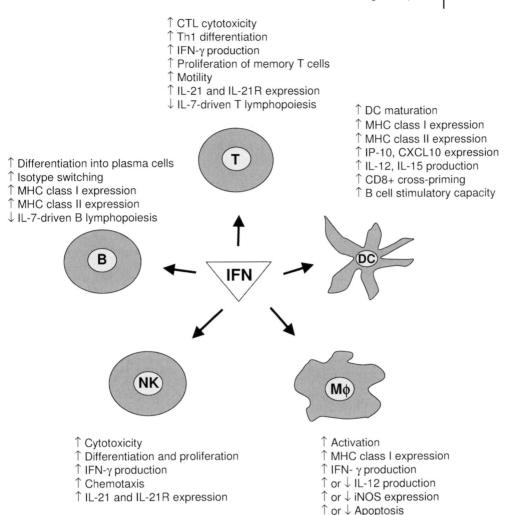

Fig. 7.8. Type I IFN-induced immunomodulatory responses. IFNs both upregulate and downregulate events in T and B lymphocytes, NK cells, macrophages (Mϕ), and DCs. (Adapted from [183].)

splenic architecture and a reduction in resident macrophages were also observed [184]. A defect in B cell maturation was also suggested due to decrease in B cell precursor numbers ($B220^{+/high}/CD43^{-}$ bone marrow-derived cells). Decreased numbers of circulating macrophages and granulocytes (IgM-, Mac-1- and Gr-1-positive cells) were also observed in IFN-β knockout mice, and this is likely due to a defect in maturation of primitive bone marrow precursors in mice.

IFNs also regulate the development and activities of numerous immune cells. Notably, IFNs modulate both T and B lymphocyte responses by promoting the proliferation of memory T cells, promoting the differentiation of type I T helper (T_h1) cells, inducing IFN-γ secretion from T cells and promoting isotype switching in B cells and differentiation into plasma cells [185–188]. Type I IFNs are potent inhibitors of IL-7-dependent growth of early B cell lineage progenitors, resulting in abrogation of B lineage differentiation at the pro-B cell stage [189–191]. Moreover, IFNs activate macrophages, activate and enhance the cytotoxicity of NK cells, and regulate the maturation and terminal differentiation of DCs [192–194].

Other functions of IFNs include their ability to upregulate MHC class I expression and subsequently promote $CD8^+$ T cell responses. IFNs also modulate the expression of components involved in antigen processing for MHC class I presentation [195]. In addition, IFNs can regulate the expression of important cytokines that modulate T cell responses, notably IL-12, -15 and -21, and IFN-γ, and the expression of chemokines that are chemoattractant for T cells, i.e. CCL-5, macrophage inflammatory protein (MIP)-1α (CCL3) and IFN-γ-inducible protein (IP)-10 (CXCL10) [196–198].

TLRs also play an important role IFN signaling [199]. Of the 10 TLR family members identified in humans (reviewed in [200, 201]), signals induced through five of them result in the production of type I IFNs. In macrophages, DCs and monocytes, engagement of TLR-3, -4, -7, -8 and -9 by dsRNA, LPS, ssRNA, imidazoquinoline-like molecules and CpG DNA, respectively, results in nuclear factor (NF)-κB, MAPK and IRF-3 activation, inducing type I IFN [202–207]. Moreover, TLR-mediated IFN induction requires the formation of a complex consisting of the Toll–IL-1 receptor domain-containing adaptor MyD88, TNF receptor-associated factor (TRAF)-6 and IRF-7 as well as TRAF-6-dependent ubiquitination [108].

7.6
Summary

Type I IFNs are cytokines that serve as potent biological response modifiers. Engagement of the receptor complex by type I IFNs induces signaling events that result in the upregulation of specific sets of ISGs, whose products mediate the characteristic biological properties associated with an IFN response. Key residues have been identified in both the IFN molecule and the receptor subunits that are critical in mediating binding interactions and biological activity. Four major pathways are activated that regulate the transcriptional activation of ISGs involved in modulating the antiviral, growth-inhibitory and immunomodulatory effects of IFN.

Microarray analysis of RNA transcripts from IFN-treated cells identified more than 300 genes that are transcriptionally activated by type I IFNs [208]. The role of many IFN-inducible signaling intermediates and effectors that regulate downstream biological outcomes has been elucidated. However, the contributions of many ISG products to the various IFN-stimulated responses remain ill defined. For example, the *6-16*, *1-8U*, *1-8D* and *ISG12* genes are highly inducible by type I IFNs [209], yet the biological functions of their protein products have yet to be determined. Gene expression studies have also identified numerous genes that are transcriptionally repressed upon IFN stimulation. These includes histone H4, insulin-like growth factor 2 and breakpoint cluster region (BCR), a serine/threonine kinase that is a GTPase-activating protein for $p21^{rac}$ [208]. Although the contribution of the downregulation of these genes in the context of IFN signaling has not been fully elucidated, their protein functions suggest that they may be involved in modulating growth-inhibitory responses.

Since their discovery in 1957, considerable progress has been made in understanding the biological actions of type I IFNs. However, the challenge remains to identify the contribution of distinct IFN-activated signaling effectors and induced ISGs to specific biological outcomes.

References

1 Isaacs A, Lindemann J. **1957**. Virus interference. I. The interferon. *Proc Roy Soc Lond B 147*, 258–67.
2 Havell EA, Hayes TG, Vilcek J. **1978**. Synthesis of two distinct interferons by human fibroblasts. *Virology 89*, 330–4.
3 Hayes TG. **1980**. Chou–Fasman analysis of the secondary structure of F and Le interferons. *Biochem Biophys Res Commun 95*, 872–9.
4 Owerbach D, Rutter WJ, Shows TB, et al. **1981**. Leukocyte and fibroblast interferon genes are located on human chromosome 9. *Proc Natl Acad Sci USA 78*, 3123–7.
5 Brassard DL, Grace MJ, Bordens RW. **2002**. Interferon-α as an immunotherapeutic protein. *J Leukocyte Biol 71*, 565–81.
6 Foster GR, Finter NB. **1998**. Are all type I human interferons equivalent? *J Viral Hepat 5*, 143–52.
7 Siegal FP, Kadowaki N, Shodell M, et al. **1999**. The nature of the principal type 1 interferon-producing cells in human blood. *Science 284*, 1835–7.
8 Kotenko SV, Pestka S. **2000**. Jak–Stat signal transduction pathway through the eyes of cytokine class II receptor complexes. *Oncogene 19*, 2557–65.
9 Bazan JF. **1990**. Structural design and molecular evolution of a cytokine receptor. *Proc Natl Acad Sci USA 87*, 6934–8.
10 Eid P, Langer JA, Bailly G, et al. **2000**. Localization of a receptor nonapeptide with a possible role in the binding of the type I interferons. *Eur Cytokine Netw 11*, 560–73.
11 Chill JH, Nivasch R, Levy R, et al. **2002**. The human interferon receptor: NMR-based modeling, mapping of the IFN-alpha 2 binding site, and observed ligand-induced tightening. *Biochemistry 41*, 3575–85.
12 Senda T, Shimazu T, Matsuda S, et al. **1992**. Three-

dimensional crystal structure of recombinant murine interferon-beta. *EMBO J 11*, 3193–201.

13 SENDA T, SAITOH S, MITSUI Y. **1995**. Refined crystal structure of recombinant murine interferon-beta at 2.15 Å resolution. *J Mol Biol 253*, 187–207.

14 RADHAKRISHNAN R, WALTER LJ, HRUZA A, et al. **1996**. Zinc mediated dimer of human interferon-alpha 2b revealed by X-ray crystallography. *Structure 4*, 1453–63.

15 KARPUSAS M, NOLTE M, BENTON CB, et al. **1997**. The crystal structure of human interferon beta at 2.2-Å resolution. *Proc Natl Acad Sci USA 94*, 11813–8.

16 RADHAKRISHNAN R, WALTER LJ, SUBRAMANIAM PS, et al. **1999**. Crystal structure of ovine interferon-tau at 2.1 Å resolution. *J Mol Biol 286*, 151–62.

17 FISH EN, BANERJEE K, STEBBING N. **1989**. The role of three domains in the biological activity of human interferon-alpha. *J Interferon Res 9*, 97–114.

18 FISH EN. **1992**. Definition of receptor binding domains in interferon-alpha. *J Interferon Res 12*, 257–66.

19 PIEHLER J, SCHREIBER G. **1999**. Mutational and structural analysis of the binding interface between type I interferons and their receptor IFNAR2. *J Mol Biol 294*, 223–37.

20 PIEHLER J, ROISMAN LC, SCHREIBER G. **2000**. New structural and functional aspects of the type I interferon–receptor interaction revealed by comprehensive mutational analysis of the binding interface. *J Biol Chem 275*, 40425–33.

21 CHILL JH, QUADT SR, LEVY R, et al. **2003**. The human type I interferon receptor: NMR structure reveals the molecular basis of ligand binding. *Structure 11*, 791–802.

22 NOVICK D, COHEN B, RUBINSTEIN M. **1994**. The human interferon alpha/beta receptor: characterization and molecular cloning. *Cell 77*, 391–400.

23 COHEN B, NOVICK D, BARAK S, et al. **1995**. Ligand-induced association of the type I interferon receptor components. *Mol Cell Biol 15*, 4208–14.

24 PIEHLER J, SCHREIBER G. **1999**. Biophysical analysis of the interaction of human ifnar2 expressed in *E. coli* with IFNalpha2. *J Mol Biol 289*, 57–67.

25 RUNKEL L, DEDIOS C, KARPUSAS M, et al. **2000**. Systematic mutational mapping of sites on human interferon-beta-1a that are important for receptor binding and functional activity. *Biochemistry 39*, 2538–51.

26 VELAZQUEZ L, FELLOUS M, STARK GR, et al. **1992**. A protein tyrosine kinase in the interferon alpha/beta signaling pathway. *Cell 70*, 313–22.

27 PLATANIAS LC, UDDIN S, DOMANSKI P, et al. **1996**. Differences in interferon alpha and beta signaling. Interferon beta selectively induces the interaction of the alpha and betaL subunits of the type I interferon receptor. *J Biol Chem 271*, 23630–3.

28 RANI MR, FOSTER GR, LEUNG S, et al. **1996**. Characterization of beta-R1, a gene that is selectively induced by interferon beta (IFN-beta) compared with IFN-alpha. *J Biol Chem 271*, 22878–84.

29 SANCEAU J, HISCOTT J, DELATTRE O, et al. **2000**. IFN-beta induces serine phosphorylation of Stat-1 in Ewing's sarcoma cells and mediates apoptosis via induction of IRF-1 and activation of caspase-7. *Oncogene 19*, 3372–83.

30 DAMDINSUREN B, NAGANO H, SAKON M, et al. **2003**. Interferon-beta is more potent than interferon-alpha in inhibition of human hepatocellular carcinoma cell growth when used alone and in combination with anticancer drugs. *Ann Surg Oncol 10*, 1184–90.

31 LU J, CHUNTHARAPAI A, BECK J, et al. **1998**. Structure–function study of the extracellular domain of the human IFN-alpha receptor (hIFNAR1) using blocking monoclonal antibodies: the role of domains 1 and 2. *J Immunol 160*, 1782–8.

32 ROISMAN LC, PIEHLER J, TROSSET JY, et al. **2001**. Structure of the interferon-receptor complex determined by distance constraints from double-mutant cycles and flexible docking. *Proc Natl Acad Sci USA 98*, 13231–6.

33 CAJEAN-FEROLDI C, NOSAL F, NARDEUX PC, et al. **2004**. Identification of residues of the IFNAR1 chain of the type I human interferon receptor critical for ligand binding and biological activity. *Biochemistry 43*, 12498–512.

34 CUTRONE EC, LANGER JA. **1997**. Contributions of cloned type I interferon receptor subunits to differential ligand binding. *FEBS Lett 404*, 197–202.

35 LIM JK, XIONG J, CARRASCO N, et al. **1994**. Intrinsic ligand binding properties of the human and bovine alpha-interferon receptors. *FEBS Lett 350*, 281–6.

36 HWANG SY, HOLLAND KA, KOLA I, et al. **1996**. Binding of interferon-alpha and -beta to a component of the human type I interferon receptor expressed in simian cells. *Int J Biochem Cell Biol 28*, 911–6.

37 NGUYEN NY, SACKETT D, HIRATA RD, et al. **1996**. Isolation of a biologically active soluble human interferon-alpha receptor–GST fusion protein expressed in *Escherichia coli*. *J Interferon Cytokine Res 16*, 835–44.

38 YOON S, HIRATA RD, NGUYEN NY, et al. **2000**. Expression and biological activity of two recombinant polypeptides related to subunit 1 of the interferon-alpha receptor. *Braz J Med Biol Res 33*, 771–8.

39 LANGER JA, YANG J, CARMILLO P, et al. **1998**. Bovine type I interferon receptor protein BoIFNAR-1 has high-affinity and broad specificity for human type I interferons. *FEBS Lett 421*, 131–5.

40 GOLDMAN LA, CUTRONE EC, DANG A, et al. **1998**. Mapping human interferon-alpha (IFN-alpha 2) binding determinants of the type I interferon receptor subunit IFNAR-1 with human/bovine IFNAR-1 chimeras. *Biochemistry 37*, 13003–10.

41 CUTRONE EC, LANGER JA. **2001**. Identification of critical residues in bovine IFNAR-1 responsible for interferon binding. *J Biol Chem 276*, 17140–8.

42 GHISLAIN J, LINGWOOD CA, FISH EN. **1994**. Evidence for glycosphingolipid modification of the type 1 IFN receptor. *J Immunol 153*, 3655–63.

43 KHINE AA, LINGWOOD CA. 2000. Functional significance of globotriaosyl ceramide in interferon-alpha$_2$/type 1 interferon receptor-mediated antiviral activity. *J Cell Physiol* 182, 97–108.

44 KUMARAN J, COLAMONICI OR, FISH EN. 2000. Structure–function study of the extracellular domain of the human type I interferon receptor (IFNAR)-1 subunit. *J Interferon Cytokine Res* 20, 479–85.

45 VISCOMI GC. 1997. Structure–activity of type I interferons. *Biotherapy* 10, 59–86.

46 DOMANSKI P, NADEAU OW, PLATANIAS LC, et al. 1998. Differential use of the betaL subunit of the type I interferon (IFN) receptor determines signaling specificity for IFNalpha2 and IFNbeta. *J Biol Chem* 273, 3144–7.

47 RUNKEL L, PFEFFER L, LEWERENZ M, et al. 1998. Differences in activity between alpha and beta type I interferons explored by mutational analysis. *J Biol Chem* 273, 8003–8.

48 PETKOVIC V, MOGHINI C, PAOLETTI S, et al. 2004. I-TAC/CXCL11 is a natural antagonist for CCR5. *J Leukoc Biol* 76, 701–8.

49 COLAMONICI O, YAN H, DOMANSKI P, et al. 1994. Direct binding to and tyrosine phosphorylation of the alpha subunit of the type I interferon receptor by p135^{tyk2} tyrosine kinase. *Mol Cell Biol* 14, 8133–42.

50 DOMANSKI P, FISH E, NADEAU OW, et al. 1997. A region of the beta subunit of the interferon alpha receptor different from box 1 interacts with Jak1 and is sufficient to activate the Jak–Stat pathway and induce an antiviral state. *J Biol Chem* 272, 26388–93.

51 MULLER M, BRISCOE J, LAXTON C, et al. 1993. The protein tyrosine kinase JAK1 complements defects in interferon-alpha/beta and -gamma signal transduction. *Nature* 366, 129–35.

52 PLATANIAS LC, UDDIN S, COLAMONICI OR. 1994. Tyrosine phosphorylation of the alpha and beta subunits of the type I interferon receptor. Interferon-beta selectively induces tyrosine phosphorylation of an alpha subunit-associated protein. *J Biol Chem* 269, 17761–4.

53 SCHINDLER C, SHUAI K, PREZIOSO VR, et al. 1992. Interferon-dependent tyrosine phosphorylation of a latent cytoplasmic transcription factor. *Science* 257, 809–13.

54 FU XY, SCHINDLER C, IMPROTA T, et al. 1992. The proteins of ISGF-3, the interferon alpha-induced transcriptional activator, define a gene family involved in signal transduction. *Proc Natl Acad Sci USA* 89, 7840–3.

55 DARNELL JE, JR. 1997. STATs and gene regulation. *Science* 277, 1630–5.

56 YAN H, KRISHNAN K, GREENLUND AC, et al. 1996. Phosphorylated interferon-alpha receptor 1 subunit (IFNaR1) acts as a docking site for the latent form of the 113 kDa STAT-2 protein. *EMBO J* 15, 1064–74.

57 QURESHI SA, LEUNG S, KERR IM, et al. 1996. Function of Stat2 protein in transcriptional activation by alpha interferon. *Mol Cell Biol* 16, 288–93.

58 LEUNG S, QURESHI SA, KERR IM, et al. 1995. Role of STAT-2

in the alpha interferon signaling pathway. *Mol Cell Biol* 15, 1312–7.

59 COLAMONICI OR, PLATANIAS LC, DOMANSKI P, et al. **1995**. Transmembrane signaling by the alpha subunit of the type I interferon receptor is essential for activation of the JAK kinases and the transcriptional factor ISGF-3. *J Biol Chem* 270, 8188–93.

60 KOTENKO SV, IZOTOVA LS, MIROCHNITCHENKO OV, et al. **1999**. The intracellular domain of interferon-alpha receptor 2c (IFN-alphaR2c) chain is responsible for Stat activation. *Proc Natl Acad Sci USA* 96, 5007–12.

61 VELICHKO S, WAGNER TC, TURKSON J, et al. **2002**. STAT-3 activation by type I interferons is dependent on specific tyrosines located in the cytoplasmic domain of interferon receptor chain 2c. Activation of multiple STATS proceeds through the redundant usage of two tyrosine residues. *J Biol Chem* 277, 35635–41.

62 NADEAU OW, DOMANSKI P, USACHEVA A, et al. **1999**. The proximal tyrosines of the cytoplasmic domain of the beta chain of the type I interferon receptor are essential for signal transducer and activator of transcription (Stat) 2 activation. Evidence that two Stat2 sites are required to reach a threshold of interferon alpha-induced Stat2 tyrosine phosphorylation that allows normal formation of interferon-stimulated gene factor 3. *J Biol Chem* 274, 4045–52.

63 USACHEVA A, SMITH R, MINSHALL R, et al. **2001**. The WD motif-containing protein receptor for activated protein kinase C (RACK1) is required for recruitment and activation of signal transducer and activator of transcription 1 through the type I interferon receptor. *J Biol Chem* 276, 22948–53.

64 USACHEVA A, TIAN X, SANDOVAL R, et al. **2003**. The WD motif-containing protein RACK-1 functions as a scaffold protein within the type I IFN receptor-signaling complex. *J Immunol* 171, 2989–94.

65 YANG CH, SHI W, BASU L, et al. **1996**. Direct association of STAT-3 with the IFNAR-1 chain of the human type I interferon receptor. *J Biol Chem* 271, 8057–61.

66 FU XY, KESSLER DS, VEALS SA, et al. **1990**. ISGF-3, the transcriptional activator induced by interferon alpha, consists of multiple interacting polypeptide chains. *Proc Natl Acad Sci USA* 87, 8555–9.

67 LI X, LEUNG S, QURESHI S, et al. **1996**. Formation of STAT-1–STAT-2 heterodimers and their role in the activation of IRF-1 gene transcription by interferon-alpha. *J Biol Chem* 271, 5790–4.

68 GHISLAIN JJ, FISH EN. **1996**. Application of genomic DNA affinity chromatography identifies multiple interferon-alpha-regulated Stat2 complexes. *J Biol Chem* 271, 12408–13.

69 BRIERLEY MM, FISH EN. **2005**. Functional relevance of the conserved DNA-binding domain of STAT-2. *J Biol Chem* 280, 13029–36.

70 GUPTA S, JIANG M, PERNIS AB. **1999**. IFN-alpha activates Stat6 and leads to the formation of Stat2:Stat6 complexes in B cells. *J Immunol* 163, 3834–41.

71 Farrar JD, Smith JD, Murphy TL, et al. **2000**. Recruitment of Stat4 to the human interferon-alpha/beta receptor requires activated Stat2. *J Biol Chem 275*, 2693–7.

72 Zhang X, Blenis J, Li HC, et al. **1995**. Requirement of serine phosphorylation for formation of STAT-promoter complexes. *Science 267*, 1990–4.

73 Horvath CM. **2000**. STAT proteins and transcriptional responses to extracellular signals. *Trends Biochem Sci 25*, 496–502.

74 Decker T, Kovarik P. **2000**. Serine phosphorylation of STATs. *Oncogene 19*, 2628–37.

75 Pilz A, Ramsauer K, Heidari H, et al. **2003**. Phosphorylation of the Stat1 transactivating domain is required for the response to type I interferons. *EMBO Rep 4*, 368–73.

76 Akira S. **1999**. Functional roles of STAT family proteins: lessons from knockout mice. *Stem Cells 17*, 138–46.

77 Meraz MA, White JM, Sheehan KC, et al. **1996**. Targeted disruption of the Stat1 gene in mice reveals unexpected physiologic specificity in the JAK–STAT signaling pathway. *Cell 84*, 431–42.

78 Durbin JE, Hackenmiller R, Simon MC, et al. **1996**. Targeted disruption of the mouse Stat1 gene results in compromised innate immunity to viral disease. *Cell 84*, 443–50.

79 Kaplan DH, Shankaran V, Dighe AS, et al. **1998**. Demonstration of an interferon gamma-dependent tumor surveillance system in immunocompetent mice. *Proc Natl Acad Sci USA 95*, 7556–61.

80 Park C, Li S, Cha E, et al. **2000**. Immune response in Stat2 knockout mice. *Immunity 13*, 795–804.

81 Feller SM. **2001**. Crk family adaptors-signalling complex formation and biological roles. *Oncogene 20*, 6348–71.

82 Gotoh T, Hattori S, Nakamura S, et al. **1995**. Identification of Rap1 as a target for the Crk SH3 domain-binding guanine nucleotide-releasing factor C3G. *Mol Cell Biol 15*, 6746–53.

83 Ahmad S, Alsayed YM, Druker BJ, et al. **1997**. The type I interferon receptor mediates tyrosine phosphorylation of the CrkL adaptor protein. *J Biol Chem 272*, 29991–4.

84 Fish EN, Uddin S, Korkmaz M, et al. **1999**. Activation of a CrkL–stat5 signaling complex by type I interferons. *J Biol Chem 274*, 571–3.

85 Lekmine F, Sassano A, Uddin S, et al. **2002**. The CrkL adapter protein is required for type I interferon-dependent gene transcription and activation of the small G-protein Rap1. *Biochem Biophys Res Commun 291*, 744–50.

86 Uddin S, Chamdin A, Platanias LC. **1995**. Interaction of the transcriptional activator Stat-2 with the type I interferon receptor. *J Biol Chem 270*, 24627–30.

87 Uddin S, Lekmine F, Sharma N, et al. **2000**. The Rac1/p38 mitogen-activated protein kinase pathway is required for interferon alpha-dependent transcriptional activation but not serine phosphorylation of Stat proteins. *J Biol Chem 275*, 27634–40.

88 Platanias LC, Uddin S, Yetter A, et al. **1996**. The type I

interferon receptor mediates tyrosine phosphorylation of insulin receptor substrate 2. *J Biol Chem* 271, 278–82.
89 BURFOOT MS, ROGERS NC, WATLING D, et al. **1997**. Janus kinase-dependent activation of insulin receptor substrate 1 in response to interleukin-4, oncostatin M, and the interferons. *J Biol Chem* 272, 24183–90.
90 PLATANIAS LC, UDDIN S, BRUNO E, et al. **1999**. CrkL and CrkII participate in the generation of the growth inhibitory effects of interferons on primary hematopoietic progenitors. *Exp Hematol* 27, 1315–21.
91 UDDIN S, FISH EN, SHER D, et al. **1997**. The IRS-pathway operates distinctively from the Stat-pathway in hematopoietic cells and transduces common and distinct signals during engagement of the insulin or interferon-alpha receptors. *Blood* 90, 2574–82.
92 UDDIN S, YENUSH L, SUN XJ, et al. **1995**. Interferon-alpha engages the insulin receptor substrate-1 to associate with the phosphatidylinositol 3′-kinase. *J Biol Chem* 270, 15938–41.
93 LEKMINE F, UDDIN S, SASSANO A, et al. **2003**. Activation of the p70 S6 kinase and phosphorylation of the 4E-BP1 repressor of mRNA translation by type I interferons. *J Biol Chem* 278, 27772–80.
94 SCHAEFFER HJ, WEBER MJ. **1999**. Mitogen-activated protein kinases: specific messages from ubiquitous messengers. *Mol Cell Biol* 19, 2435–44.
95 UDDIN S, MAJCHRZAK B, WOODSON J, et al. **1999**. Activation of the p38 mitogen-activated protein kinase by type I interferons. *J Biol Chem* 274, 30127–31.
96 GOH KC, HAQUE SJ, WILLIAMS BR. **1999**. p38MAP kinase is required for STAT-1 serine phosphorylation and transcriptional activation induced by interferons. *EMBO J* 18, 5601–8.
97 LI Y, BATRA S, SASSANO A, et al. **2005**. Activation of mitogen-activated protein kinase kinase (Mkk) 3 and Mkk6 by type I interferons. *J Biol Chem* 280, 10001–10.
98 LI Y, SASSANO A, MAJCHRZAK B, et al. **2004**. Role of p38alpha Map kinase in Type I interferon signaling. *J Biol Chem* 279, 970–9.
99 PLATANIAS LC, SWEET ME. **1994**. Interferon alpha induces rapid tyrosine phosphorylation of the *vav* proto-oncogene product in hematopoietic cells. *J Biol Chem* 269, 3143–6.
100 MICOUIN A, WIETZERBIN J, STEUNOU V, et al. **2000**. p95^{vav} associates with the type I interferon (IFN) receptor and contributes to the antiproliferative effect of IFN-alpha in megakaryocytic cell lines. *Oncogene* 19, 387–94.
101 ADAM L, BANDYOPADHYAY D, KUMAR R. **2000**. Interferon-alpha signaling promotes nucleus-to-cytoplasmic redistribution of p95Vav, and formation of a multisubunit complex involving Vav, Ku80, and Tyk2. *Biochem Biophys Res Commun* 267, 692–6.
102 CRESPO P, BUSTELO XR, AARONSON DS, et al. **1996**. Rac-1 dependent stimulation of the JNK/SAPK signaling pathway by Vav. *Oncogene* 13, 455–60.
103 CRESPO P, SCHUEBEL KE, OSTROM AA, et al. **1997**. Phosphotyrosine-dependent activation of Rac-1 GDP/GTP

exchange by the *vav* proto-oncogene product. *Nature* 385, 169–72.

104 UDDIN S, SASSANO A, DEB DK, et al. **2002**. Protein kinase C-delta (PKC-delta) is activated by type I interferons and mediates phosphorylation of Stat1 on serine 727. *J Biol Chem* 277, 14408–16.

105 SRIVASTAVA KK, BATRA S, SASSANO A, et al. **2004**. Engagement of protein kinase C-theta in interferon signaling in T-cells. *J Biol Chem* 279, 29911–20.

106 SEN GC. **2001**. Viruses and interferons. *Annu Rev Microbiol* 55, 255–81.

107 SAMUEL CE. **2001**. Antiviral actions of interferons. *Clin Microbiol Rev* 14, 778–809, table of contents.

108 KAWAI T, SATO S, ISHII KJ, et al. **2004**. Interferon-alpha induction through Toll-like receptors involves a direct interaction of IRF7 with MyD88 and TRAF6. *Nat Immunol* 5, 1061–8.

109 MALMGAARD L. **2004**. Induction and regulation of IFNs during viral infections. *J Interferon Cytokine Res* 24, 439–54.

110 DEONARAIN R, ALCAMI A, ALEXIOU M, et al. **2000**. Impaired antiviral response and alpha/beta interferon induction in mice lacking beta interferon. *J Virol* 74, 3404–9.

111 DEONARAIN R, CERULLO D, FUSE K, et al. **2004**. Protective role for interferon-β in coxsackievirus B3 infection. *Circulation* 110, 3540–3.

112 HWANG SY, HERTZOG PJ, HOLLAND KA, et al. **1995**. A null mutation in the gene encoding a type I interferon receptor component eliminates antiproliferative and antiviral responses to interferons alpha and beta and alters macrophage responses. *Proc Natl Acad Sci USA* 92, 11284–8.

113 PREJEAN C, COLAMONICI OR. **2000**. Role of the cytoplasmic domains of the type I interferon receptor subunits in signaling. *Semin Cancer Biol* 10, 83–92.

114 MAYER IA, VERMA A, GRUMBACH IM, et al. **2001**. The p38MAPK pathway mediates the growth inhibitory effects of interferon-alpha in BCR–ABL-expressing cells. *J Biol Chem* 276, 28570–7.

115 KHABAR KS, DHALLA M, SIDDIQUI Y, et al. **2000**. Effect of deficiency of the double-stranded RNA-dependent protein kinase, PKR, on antiviral resistance in the presence or absence of ribonuclease L: HSV-1 replication is particularly sensitive to deficiency of the major IFN-mediated enzymes. *J Interferon Cytokine Res* 20, 653–9.

116 ZHOU A, PARANJAPE JM, DER SD, et al. **1999**. Interferon action in triply deficient mice reveals the existence of alternative antiviral pathways. *Virology* 258, 435–40.

117 SATO K, HIDA S, TAKAYANAGI H, et al. **2001**. Antiviral response by natural killer cells through TRAIL gene induction by IFN-alpha/beta. *Eur J Immunol* 31, 3138–46.

118 CHIN KC, CRESSWELL P. **2001**. Viperin (cig5), an IFN-inducible antiviral protein directly induced by human cytomegalovirus. *Proc Natl Acad Sci USA* 98, 15125–30.

119 ZHOU Q, ZHAO J, AL-ZOGHAIBI F, et al. **2000**. Transcriptional control of the human plasma membrane phospholipid

scramblase 1 gene is mediated by interferon-alpha. *Blood* 95, 2593–9.

120 SARKAR SN, SEN GC. **2004**. Novel functions of proteins encoded by viral stress-inducible genes. *Pharmacol Ther 103*, 245–59.

121 CLEMENS MJ, ELIA A. **1997**. The double-stranded RNA-dependent protein kinase PKR: structure and function. *J Interferon Cytokine Res 17*, 503–24.

122 BARBER GN, TOMITA J, HOVANESSIAN AG, et al. **1991**. Functional expression and characterization of the interferon-induced double-stranded RNA activated P68 protein kinase from *Escherichia coli*. *Biochemistry 30*, 10356–61.

123 MEURS EF, WATANABE Y, KADEREIT S, et al. **1992**. Constitutive expression of human double-stranded RNA-activated p68 kinase in murine cells mediates phosphorylation of eukaryotic initiation factor 2 and partial resistance to encephalomyocarditis virus growth. *J Virol 66*, 5805–14.

124 YANG YL, REIS LF, PAVLOVIC J, et al. **1995**. Deficient signaling in mice devoid of double-stranded RNA-dependent protein kinase. *EMBO J 14*, 6095–106.

125 BALACHANDRAN S, ROBERTS PC, BROWN LE, et al. **2000**. Essential role for the dsRNA-dependent protein kinase PKR in innate immunity to viral infection. *Immunity 13*, 129–41.

126 STOJDL DF, ABRAHAM N, KNOWLES S, et al. **2000**. The murine double-stranded RNA-dependent protein kinase PKR is required for resistance to vesicular stomatitis virus. *J Virol 74*, 9580–5.

127 ABRAHAM N, STOJDL DF, DUNCAN PI, et al. **1999**. Characterization of transgenic mice with targeted disruption of the catalytic domain of the double-stranded RNA-dependent protein kinase, PKR. *J Biol Chem 274*, 5953–62.

128 WILLIAMS BR. **1999**. PKR; a sentinel kinase for cellular stress. *Oncogene 18*, 6112–20.

129 BALACHANDRAN S, KIM CN, YEH WC, et al. **1998**. Activation of the dsRNA-dependent protein kinase, PKR, induces apoptosis through FADD-mediated death signaling. *EMBO J 17*, 6888–902.

130 DONZE O, DOSTIE J, SONENBERG N. **1999**. Regulatable expression of the interferon-induced double-stranded RNA dependent protein kinase PKR induces apoptosis and fas receptor expression. *Virology 256*, 322–9.

131 GIL J, ALCAMI J, ESTEBAN M. **2000**. Activation of NF-kappa B by the dsRNA-dependent protein kinase, PKR involves the I kappa B kinase complex. *Oncogene 19*, 1369–78.

132 GOH KC, DEVEER MJ, WILLIAMS BR. **2000**. The protein kinase PKR is required for p38MAPK activation and the innate immune response to bacterial endotoxin. *EMBO J 19*, 4292–7.

133 REBOUILLAT D, HOVANESSIAN AG. **1999**. The human 2′,5′-oligoadenylate synthetase family: interferon-induced proteins with unique enzymatic properties. *J Interferon Cytokine Res 19*, 295–308.

134 HASSEL BA, ZHOU A, SOTOMAYOR C, et al. **1993**. A dominant negative mutant of 2–5A-dependent RNase suppresses antiproliferative and antiviral effects of interferon. *EMBO J 12*, 3297–304.

135 GHOSH A, SARKAR SN, SEN GC. **2000**. Cell growth regulatory and antiviral effects of the P69 isozyme of 2–5 (A) synthetase. *Virology 266*, 319–28.

136 LI G, XIANG Y, SABAPATHY K, et al. **2004**. An apoptotic signaling pathway in the interferon antiviral response mediated by RNase L and c-Jun NH_2-terminal kinase. *J Biol Chem 279*, 1123–31.

137 GHOSH A, SARKAR SN, ROWE TM, et al. **2001**. A specific isozyme of $2'$–$5'$ oligoadenylate synthetase is a dual function proapoptotic protein of the Bcl-2 family. *J Biol Chem 276*, 25447–55.

138 LEE SH, VIDAL SM. **2002**. Functional diversity of Mx proteins: variations on a theme of host resistance to infection. *Genome Res 12*, 527–30.

139 MACMICKING JD. **2004**. IFN-inducible GTPases and immunity to intracellular pathogens. *Trends Immunol 25*, 601–9.

140 KOCHS G, HAENER M, AEBI U, et al. **2002**. Self-assembly of human MxA GTPase into highly ordered dynamin-like oligomers. *J Biol Chem 277*, 14172–6.

141 PAVLOVIC J, ZURCHER T, HALLER O, et al. **1990**. Resistance to influenza virus and vesicular stomatitis virus conferred by expression of human MxA protein. *J Virol 64*, 3370–5.

142 PAVLOVIC J, ARZET HA, HEFTI HP, et al. **1995**. Enhanced virus resistance of transgenic mice expressing the human MxA protein. *J Virol 69*, 4506–10.

143 FRESE M, KOCHS G, MEIER-DIETER U, et al. **1995**. Human MxA protein inhibits tick-borne Thogoto virus but not Dhori virus. *J Virol 69*, 3904–9.

144 FRESE M, KOCHS G, FELDMANN H, et al. **1996**. Inhibition of bunyaviruses, phleboviruses, and hantaviruses by human MxA protein. *J Virol 70*, 915–23.

145 SCHWEMMLE M, RICHTER MF, HERRMANN C, et al. **1995**. Unexpected structural requirements for GTPase activity of the interferon-induced MxA protein. *J Biol Chem 270*, 13518–23.

146 LANDIS H, SIMON-JODICKE A, KLOTI A, et al. **1998**. Human MxA protein confers resistance to Semliki Forest virus and inhibits the amplification of a Semliki Forest virus-based replicon in the absence of viral structural proteins. *J Virol 72*, 1516–22.

147 CHIEUX V, CHEHADEH W, HARVEY J, et al. **2001**. Inhibition of Coxsackievirus B4 replication in stably transfected cells expressing human MxA protein. *Virology 283*, 84–92.

148 SCHNEIDER-SCHAULIES S, SCHNEIDER-SCHAULIES J, SCHUSTER A, et al. **1994**. Cell type-specific MxA-mediated inhibition of measles virus transcription in human brain cells. *J Virol 68*, 6910–7.

149 GORDIEN E, ROSMORDUC O, PELTEKIAN C, et al. **2001**. Inhibition of hepatitis B virus replication by the interferon-inducible MxA protein. *J Virol 75*, 2684–91.

150 GEORGE CX, SAMUEL CE. **1999**. Characterization of the 5'-flanking region of the human RNA-specific adenosine deaminase ADAR1 gene and identification of an interferon-inducible ADAR1 promoter. *Gene 229*, 203–13.

151 HARTWIG D, SCHOENEICH L, GREEVE J, et al. **2004**. Interferon-alpha stimulation of liver cells enhances hepatitis delta virus RNA editing in early infection. *J Hepatol 41*, 667–72.

152 HUI DJ, BHASKER CR, MERRICK WC, et al. **2003**. Viral stress-inducible protein p56 inhibits translation by blocking the interaction of eIF3 with the ternary complex eIF2.GTP.Met-tRNAi. *J Biol Chem 278*, 39477–82.

153 DONG B, ZHOU Q, ZHAO J, et al. **2004**. Phospholipid scramblase 1 potentiates the antiviral activity of interferon. *J Virol 78*, 8983–93.

154 ESPERT L, DEGOLS G, GONGORA C, et al. **2003**. ISG20, a new interferon-induced RNase specific for single-stranded RNA, defines an alternative antiviral pathway against RNA genomic viruses. *J Biol Chem 278*, 16151–8.

155 CROWDER C, DAHLE O, DAVIS RE, et al. **2005**. PML mediates IFN-α-induced apoptosis in myeloma by regulating TRAIL induction. *Blood 105*, 1280–7.

156 GRUMBACH IM, MAYER IA, UDDIN S, et al. **2001**. Engagement of the CrkL adaptor in interferon alpha signalling in BCR-ABL-expressing cells. *Br J Haematol 112*, 327–36.

157 VERMA A, DEB DK, SASSANO A, et al. **2002**. Cutting edge: activation of the p38 mitogen-activated protein kinase signaling pathway mediates cytokine-induced hemopoietic suppression in aplastic anemia. *J Immunol 168*, 5984–8.

158 VERMA A, DEB DK, SASSANO A, et al. **2002**. Activation of the p38 mitogen-activated protein kinase mediates the suppressive effects of type I interferons and transforming growth factor-beta on normal hematopoiesis. *J Biol Chem 277*, 7726–35.

159 VERMA A, MOHINDRU M, DEB DK, et al. **2002**. Activation of Rac1 and the p38 mitogen-activated protein kinase pathway in response to arsenic trioxide. *J Biol Chem 277*, 44988–95.

160 EINAT M, RESNITZKY D, KIMCHI A. **1985**. Close link between reduction of c-*myc* expression by interferon and G_0/G_1 arrest. *Nature 313*, 597–600.

161 IWASE S, FURUKAWA Y, KIKUCHI J, et al. **1997**. Modulation of E2F activity is linked to interferon-induced growth suppression of hematopoietic cells. *J Biol Chem 272*, 12406–14.

162 SANGFELT O, ERICKSON S, GRANDER D. **2000**. Mechanisms of interferon-induced cell cycle arrest. *Front Biosci 5*, D479–87.

163 SANGFELT O, ERICKSON S, CASTRO J, et al. **1999**. Molecular mechanisms underlying interferon-alpha-induced G_0/G_1 arrest: CKI-mediated regulation of G1 Cdk-complexes and activation of pocket proteins. *Oncogene 18*, 2798–810.

164 THOMAS NS, PIZZEY AR, TIWARI S, et al. **1998**. p130, p107, and pRb are differentially regulated in proliferating cells and during cell cycle arrest by alpha-interferon. *J Biol Chem 273*, 23659–67.

165 SZEPS M, ERICKSON S, GRUBER A, et al. **2003**. Effects of interferon-alpha on cell cycle regulatory proteins in leukemic cells. *Leuk Lymphoma 44*, 1019–25.

166 FURUKAWA Y, IWASE S, KIKUCHI J, et al. **1999**. Transcriptional repression of the E2F-1 gene by interferon-alpha is mediated through induction of E2F-4/pRB and E2F-4/p130 complexes. *Oncogene 18*, 2003–14.

167 KIMCHI A. **1992**. Cytokine triggered molecular pathways that control cell cycle arrest. *J Cell Biochem 50*, 1–9.

168 DAWSON MJ, TRAPANI JA. **1996**. HIN-200: a novel family of IFN-inducible nuclear proteins expressed in leukocytes. *J Leukoc Biol 60*, 310–6.

169 ASEFA B, KLARMANN KD, COPELAND NG, et al. **2004**. The interferon-inducible p200 family of proteins: a perspective on their roles in cell cycle regulation and differentiation. *Blood Cells Mol Dis 32*, 155–67.

170 GENG Y, CHOUBEY D. **2000**. Differential induction of the 200-family proteins in Daudi Burkitt's lymphoma cells by interferon-alpha. *J Biol Regul Homeost Agents 14*, 263–8.

171 GUTTERMAN JU, CHOUBEY D. **1999**. Retardation of cell proliferation after expression of p202 accompanies an increase in $p21^{WAF1/CIP1}$. *Cell Growth Differ 10*, 93–100.

172 CHOUBEY D, LI SJ, DATTA B, et al. **1996**. Inhibition of E2F-mediated transcription by p202. *EMBO J 15*, 5668–78.

173 YAN DH, ABRAMIAN A, LI Z, et al. **2003**. P202, an interferon-inducible protein, inhibits E2F1-mediated apoptosis in prostate cancer cells. *Biochem Biophys Res Commun 303*, 219–22.

174 DING Y, WANG L, SU LK, et al. **2004**. Antitumor activity of IFIX, a novel interferon-inducible HIN-200 gene, in breast cancer. *Oncogene 23*, 4556–66.

175 SELLERI C, SATO T, DEL VECCHIO L, et al. **1997**. Involvement of Fas-mediated apoptosis in the inhibitory effects of interferon-alpha in chronic myelogenous leukemia. *Blood 89*, 957–64.

176 CHAWLA-SARKAR M, LEAMAN DW, BORDEN EC. **2001**. Preferential induction of apoptosis by interferon (IFN)-beta compared with IFN-alpha2: correlation with TRAIL/Apo2L induction in melanoma cell lines. *Clin Cancer Res 7*, 1821–31.

177 LEAMAN DW, CHAWLA-SARKAR M, VYAS K, et al. **2002**. Identification of X-linked inhibitor of apoptosis-associated factor-1 as an interferon-stimulated gene that augments TRAIL Apo2L-induced apoptosis. *J Biol Chem 277*, 28504–11.

178 CLEMENS MJ. **2003**. Interferons and apoptosis. *J Interferon Cytokine Res 23*, 277–92.

179 MORRISON BH, BAUER JA, KALVAKOLANU DV, et al. **2001**. Inositol hexakisphosphate kinase 2 mediates growth suppressive and apoptotic effects of interferon-beta in ovarian carcinoma cells. *J Biol Chem 276*, 24965–70.

180 KUNZI MS, PITHA PM. **2003**. Interferon targeted genes in host defense. *Autoimmunity 36*, 457–61.

181 CHAWLA-SARKAR M, LINDNER DJ, LIU YF, et al. **2003**. Apoptosis and interferons: role of interferon-stimulated genes as mediators of apoptosis. *Apoptosis 8*, 237–49.

182 LEVY DE, MARIE I, PRAKASH A. **2003**. Ringing the interferon alarm: differential regulation of gene expression at the interface between innate and adaptive immunity. *Curr Opin Immunol 15*, 52–8.

183 BOGDAN C, MATTNER J, SCHLEICHER U. 2004. The role of type I interferons in non-viral infections. *Immunol Rev 202*, 33–48.

184 DEONARAIN R, VERMA A, PORTER AC, et al. 2003. Critical roles for IFN-beta in lymphoid development, myelopoiesis, and tumor development: links to tumor necrosis factor alpha. *Proc Natl Acad Sci USA 100*, 13453–8.

185 SARENEVA T, MATIKAINEN S, KURIMOTO M, et al. 1998. Influenza A virus induced IFN-a/b and IL-18 synergistically enhance IFN-g gene expression in human T cells. *J Immunol 160*, 6032–8.

186 MARRACK P, KAPPLER J, MITCHELL T. 1999. Type I interferons keep activated T cells alive. *J Exp Med 189*, 521–30.

187 TOUGH DF, SUN S, ZHANG X, et al. 1999. Stimulation of naive and memory T cells by cytokines. *Immunol Rev 170*, 39–47.

188 PARRONCHI P, DE CARLI M, MANETTI R, et al. 1992. IL-4 and IFN (alpha and gamma) exert opposite regulatory effects on the development of cytolytic potential by T_h1 or T_h2 human T cell clones. *J Immunol 149*, 2977–83.

189 WANG J, LIN Q, LANGSTON H, et al. 1995. Resident bone marrow macrophages produce type 1 interferons that can selectively inhibit interleukin-7-driven growth of B lineage cells. *Immunity 3*, 475–84.

190 SU DM, WANG J, LIN Q, et al. 1997. Interferons alpha/beta inhibit IL-7-induced proliferation of $CD4^-$ $CD8^-$ $CD3^-$ $CD44^+$ $CD25^+$ thymocytes, but do not inhibit that of $CD4^-$ $CD8^-$ $CD3^-$ $CD44^-$ $CD25^-$ thymocytes. *Immunology 90*, 543–9.

191 SHIMODA K, KAMESAKI K, NUMATA A, et al. 2002. Cutting edge: Tyk2 is required for the induction and nuclear translocation of Daxx which regulates IFN-α-induced suppression of B lymphocyte formation. *J Immunol 169*, 4707–11.

192 BIRON CA, NGUYEN KB, PIEN GC, et al. 1999. Natural killer cells in antiviral defense: function and regulation by innate cytokines. *Annu Rev Immunol 17*, 189–220.

193 HONDA K, SAKAGUCHI S, NAKAJIMA C, et al. 2003. Selective contribution of IFN-alpha/beta signaling to the maturation of dendritic cells induced by double-stranded RNA or viral infection. *Proc Natl Acad Sci USA 100*, 10872–7.

194 HONDA K, MIZUTANI T, TANIGUCHI T. 2004. Negative regulation of IFN-alpha/beta signaling by IFN regulatory factor 2 for homeostatic development of dendritic cells. *Proc Natl Acad Sci USA 101*, 2416–21.

195 GUIDOTTI LG, CHISARI FV. 2001. Noncytolytic control of viral infections by the innate and adaptive immune response. *Annu Rev Immunol 19*, 65–91.

196 BIRON CA. 2001. Interferons alpha and beta as immune regulators – a new look. *Immunity 14*, 661–4.

197 CREMER I, GHYSDAEL J, VIEILLARD V. 2002. A non-classical ISRE/ISGF-3 pathway mediates induction of RANTES gene transcription by type I IFNs. *FEBS Lett 511*, 41–5.

198 STRENGELL M, JULKUNEN I, MATIKAINEN S. 2004. IFN-alpha regulates IL-21 and IL-21R expression in human NK and T cells. *J Leukoc Biol 76*, 416–22.

199 Takeuchi O, Hemmi H, Akira S. **2004**. Interferon response induced by Toll-like receptor signaling. *J Endotoxin Res 10*, 252–6.

200 Re F, Strominger JL. **2004**. Heterogeneity of TLR-induced responses in dendritic cells: from innate to adaptive immunity. *Immunobiology 209*, 191–8.

201 Iwasaki A, Medzhitov R. **2004**. Toll-like receptor control of the adaptive immune responses. *Nat Immunol 5*, 987–95.

202 Sarkar SN, Smith HL, Rowe TM, et al. **2003**. Double-stranded RNA signaling by Toll-like receptor 3 requires specific tyrosine residues in its cytoplasmic domain. *J Biol Chem 278*, 4393–6.

203 Toshchakov V, Jones BW, Perera PY, et al. **2002**. TLR4, but not TLR2, mediates IFN-beta-induced STAT-1alpha/beta-dependent gene expression in macrophages. *Nat Immunol 3*, 392–8.

204 Ito T, Amakawa R, Kaisho T, et al. **2002**. Interferon-alpha and interleukin-12 are induced differentially by Toll-like receptor 7 ligands in human blood dendritic cell subsets. *J Exp Med 195*, 1507–12.

205 Heil F, Hemmi H, Hochrein H, et al. **2004**. Species-specific recognition of single-stranded RNA via toll-like receptor 7 and 8. *Science 303*, 1526–9.

206 Lian ZX, Kikuchi K, Yang GX, et al. **2004**. Expansion of bone marrow IFN-alpha-producing dendritic cells in New Zealand Black (NZB) mice: high level expression of TLR9 and secretion of IFN-alpha in NZB bone marrow. *J Immunol 173*, 5283–9.

207 Hochrein H, Schlatter B, O'Keeffe M, et al. **2004**. Herpes simplex virus type-1 induces IFN-alpha production via Toll-like receptor 9-dependent and -independent pathways. *Proc Natl Acad Sci USA 101*, 11416–21.

208 Der SD, Zhou A, Williams BR, et al. **1998**. Identification of genes differentially regulated by interferon alpha, beta, or gamma using oligonucleotide arrays. *Proc Natl Acad Sci USA 95*, 15623–8.

209 Martensen PM, Justesen J. **2004**. Small ISGs coming forward. *J Interferon Cytokine Res 24*, 1–19.

8
Interferons and Apoptosis – Recent Developments

Michael J. Clemens and Ian W. Jeffrey

8.1
Introduction

Recent years have seen the accumulation of a considerable amount of evidence indicating that the interferons (IFNs) not only influence the ability of viruses to replicate in their host cells, but also affect a range of functions in uninfected cells. Many of these IFN effects involve the regulation of cell proliferation and differentiation, as well as cell survival and cell death. The decision of cells to survive or to die by apoptosis (or indeed by the alternative pathway of necrosis) can have a profound influence on development, tissue growth and repair, immune function, and neurological activity.

Although the IFNs are not particularly potent inducers of cell death, in contrast to other cytokines such as tumor necrosis factor (TNF)-α or TNF-related apoptosis-inducing ligand (TRAIL), they can have important and subtle effects on the process (Fig. 8.1). Treatment of cells with IFNs alone often induces only a modest extent of apoptosis, which occurs after relatively long times (although there are exceptions; see, e.g. [1]). However, in many systems the IFNs have a permissive or sensitizing role when cells are additionally exposed to other stimuli such as stress-inducing agents or promoters of cell differentiation [2]. In this respect the IFNs fulfill an important protective function in facilitating cell death in the face of circumstances such as viral infection or tumor growth that would otherwise be harmful to the organism as a whole.

There have been several good reviews on the regulation of apoptosis by the IFNs over the last few years (e.g. [3–8]), but the subject is fast moving and there is now a need to summarize recent developments. In 2003, one of us published a review on selected aspects of the field [9] and the present chapter is an attempt to cover some of the new findings that have emerged in the ensuing period.

The Interferons: Characterization and Application. Edited by Anthony Meager
Copyright © 2006 WILEY-VCH Verlag GmbH & Co. KGaA, Weinheim
ISBN: 3-527-31180-7

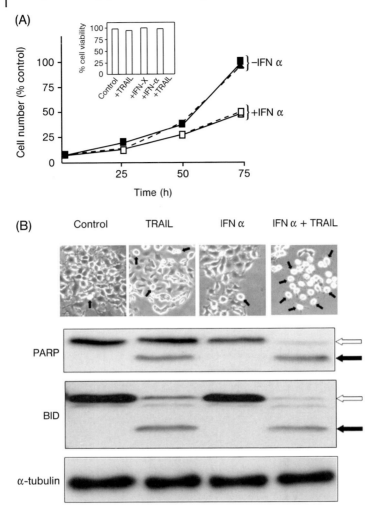

Fig. 8.1. IFN-α inhibits the proliferation of MCF-7 breast cancer cells without inducing apoptosis, but sensitizes cells to the apoptotic effect of TRAIL. (A) MCF-7 cells were incubated in the absence (closed symbols) or presence (open symbols) of human IFN-α2b (1000 U/ml) for the times indicated. TRAIL (167 ng mL^{-1}) was added for the last 5 h where shown (dashed lines). Cells were counted and the numbers are expressed as percentage of the number of untreated cells at 72 h (means of three determinations). (Inset) Cell viabilities were determined by Trypan blue exclusion at 72 h. (B) Cells treated for 72 h as described in (A) were examined by phase-contrast microscopy. Examples of apoptotic cells are indicated by arrowheads. Total cytoplasmic extracts were prepared and analyzed by SDS–gel electrophoresis, followed by immunoblotting for the caspase substrates poly(ADP ribose) polymerase (PARP) and BID. Levels of α-tubulin were also determined as a control for equal protein loading. The positions of the full-length forms of PARP and BID and their caspase-generated cleavage products are indicated by the open and closed arrows, respectively. The data are representative of the results from several independent experiments. These observations show that IFN-α inhibits cell proliferation without causing apoptosis, as

8.2
The Role of IFN-regulated Genes in the Control of Apoptosis

8.2.1
Proapoptotic Genes Induced by IFNs

All the current evidence suggests that, as with the other biological effects of IFNs, the regulation of apoptosis is mediated by changes in gene expression that occur in response to these cytokines. A recent review [8] summarizes many of the effects of IFNs on gene expression that are relevant to cell survival. Several of the genes that are up- or downregulated have been identified from microarray studies and at least 15 proapoptotic gene products have been shown to be induced by IFNs [10]. These include TRAIL [11–15], as well as the receptors for this cytokine [16], the related Fas ligand and its receptor (CD95) [17, 18], several procaspases [17, 19, 20], and proteins that regulate apoptosis such as the X-linked inhibitor of apoptosis (XIAP) inhibitor XAF-1 [21]. In addition, several proteins with important roles in signal transduction pathways that influence apoptosis are induced by IFNs. These include serine/threonine kinase (DAP) kinase [22], IFN-regulatory factors (IRFs) [8, 23] and the promyelocytic leukemia (PML) gene product [15, 24]. The promoters of several of the genes that code for regulators of apoptosis contain IFN-stimulated response elements (ISRE) or γ-activated sequence (GAS) elements [20], thus conferring IFN-sensitivity. Conversely, the expression of some antiapoptotic genes can be repressed by IFNs [23, 25], although the mechanisms involved are not so well established.

8.2.2
p53 and IFN-induced Apoptosis

The IFNs can elicit a cytostatic as well as an apoptotic effect. Indeed, the former is probably a more widespread cellular response to exposure to IFNs and it is likely that many cell types that show growth inhibition respond primarily in this way, at least in the absence of other proapoptotic factors. A recent study [26] suggests that the signaling pathways involved in the two types of response are quite distinct. Sandoval et al. report that the *INK4A* locus, which encodes the ARF proteins [27], is required for type I IFN-induced apoptosis, but not for inhibition of cell proliferation. ARF upregulates the level of p53 via its interaction with the p53 inhibitor

Fig. 8.1. (*continued*) judged by the maintenance of cell viability, normal cellular morphology and the lack of cleavage of PARP or BID at 72 h. Treatment with TRAIL for 5 h causes changes in cell morphology and partial caspase-mediated cleavage of PARP and BID, consistent with the early stages of apoptosis. IFN-α treatment enhances these effects of TRAIL. However, cell viability is only decreased at later times (not shown).

MDM2 [28]. p53 is in fact also induced at the transcriptional level by type I IFNs (although there are disagreements in the literature as to whether the mechanism is independent of ARF) and some studies indicate that p53 is required for the apoptotic response to these cytokines [24]. However, a cell line that is ARF-positive, but p53-negative, was still able to undergo type I IFN-induced cell death and dominant-negative p53 did not block this effect of IFN [26]. Moreover, type II IFN (IFN-γ) has been shown to induce apoptosis by a pathway that does not require p53 [24]. This effect may involve the CD95-mediated extrinsic pathway, rather than the p53-mediated intrinsic pathway, whereas the type I response may require both. Both type I and II IFNs markedly upregulate CD95 expression, as does p53 [29, 30]. Interestingly, in cells in which p53 has been inactivated [by the human papilloma virus (HPV) protein E6] IFN-α is unable to activate CD95 gene expression, whereas IFN-γ is still able to do so [24]. This probably explains the p53 requirement for type I IFN-induced apoptosis in some systems. Reports that signal transducer and activator of transcription (STAT)-1 and protein kinase R (PKR), which are activated and induced by IFN-α, respectively, can participate in p53 activation [31, 32] suggest that the tumor suppressor may be regulated by IFN at the post-translational level as well as at the transcriptional level. The IFN-activated pathways involving p53 and CD95 that impinge on cell death are summarized in Fig. 8.2. In contrast to apoptosis, inhibition of cell proliferation by type I IFNs in embryonic fibroblasts does not require either p53 or ARF [26], indicating that different mechanisms must be involved.

8.2.3
The Ribonuclease (RNase) L System

Prominent among other IFN-induced gene products that regulate apoptosis are the protein kinase PKR and components of the 2'–5'-oligoadenylate synthetase (2'–5'-OAS)/RNase L system. PKR will be discussed in more detail below (Section 8.3). Overexpression of RNase L or activation of the endogenous enzyme is alone sufficient to induce cell death [33–35], whereas expression of a dominant-negative form protects cells against the effects of double-stranded (ds) RNA and other proapoptotic agents in IFN-treated cells [33, 36]. In addition, apoptosis is inhibited in the tissues of RNase L-knockout mice [37]. The apoptosis that occurs in response to activation of the 2'–5'-OAS/RNase L pathway involves the Jun N-terminal kinase (JNK) stress-activated kinase, and results in mitochondrial cytochrome c release and caspase activation [34, 38]. This pathway may be a response to ribosomal damage, since RNase L cleaves rRNA at specific sites and a similar effect can be elicited by other ribotoxic agents such as ricin [39]. It is also possible that RNase L could degrade specific mRNAs encoding rapidly turning over antiapoptotic proteins, thus providing a further basis for the apoptotic response to activation of the 2'–5'-OAS/ RNase L pathway. However, to date there is no evidence for this and the levels of

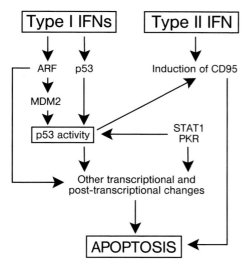

Fig. 8.2. Involvement of p53 and CD95 in pathways implicated in the induction of apoptosis by type I and II IFNs. Type I IFNs induce both ARF and p53, and ARF protein also activates p53 by binding to the p53 inhibitor MDM2. CD95 is induced by type I IFNs by a p53-dependent mechanism and by IFN-γ via a p53-independent pathway. Expression of CD95 sensitizes cells to apoptosis following exposure to the corresponding ligand for this receptor. Cell death can also be promoted by other transcriptional and post-transcriptional events mediated by p53. In addition, ARF, STAT-1 and PKR may themselves regulate apoptosis by p53-independent as well as p53-dependent mechanisms. See [24, 26] for more details.

expression of proteins of the Bcl-2 family have been reported to be unaffected by changes in RNase L activity [36].

A recent study has shown that, in prostate cancer cells, RNA interference-mediated downregulation of RNase L renders cells resistant to a normally potent combination of inducers of cell death (a topoisomerase I inhibitor plus TRAIL) [40]. In this work involvement of JNK in the apoptotic response to these inducers was again implicated. These findings suggest a required permissive role for RNase L in the links between DNA damage and/or TRAIL signaling and JNK activation, leading to apoptosis. It is significant that the *HPC1* locus, which governs hereditary predisposition to prostate cancer, has been linked to the *RNASEL* gene [41]. This suggests that resistance to apoptosis in some forms of prostate cancer may be caused by a lack of RNase L activity. However, prostate cancer cells are in general more sensitive than normal prostate epithelial cells to induction of apoptosis by 2'–5'-oligoadenylates, suggesting possible therapeutic applications of these nucleotides for the reactivation of RNase L in those tumors that express the enzyme.

8.3
The Protein Kinase PKR and the Phosphorylation of Polypeptide Chain Initiation Factor eIF2α

8.3.1
Regulation of Apoptosis by PKR

Major roles in the control of apoptosis have been demonstrated for PKR [4, 5, 9, 42–44]. This IFN-inducible protein kinase is activated by low concentrations of dsRNA (or RNA molecules with extensive secondary structure) [45]. A recent report [46] has further defined the structural requirements for RNA molecules to be good activators of PKR and has identified a minimal 16-bp dsRNA stem flanked by 10- to 15-nucleotide single-stranded regions. Following dsRNA binding the kinase dimerizes and undergoes autophosphorylation at several sites, and this results in its activation [47, 48]. There are circumstances, however, where PKR may become active in the absence of dsRNA. A model has recently been proposed in which the enzyme, at high concentrations, can dimerize spontaneously, resulting in a dsRNA-independent mechanism of autophosphorylation and activation [49]. Whether this "chain reaction" mode of PKR activation occurs *in vivo* is unclear. Nevertheless, a physiological dsRNA-independent mechanism of activation has been identified, involving the protein PACT/RAX, which binds directly to PKR [50, 51]. PACT may be important in regulating PKR-dependent apoptosis in cells treated with agents that cause increased production of ceramide [52].

There is extensive evidence that the activation of PKR can both induce apoptosis and enhance the process when it is initiated by other agents [4, 9, 43, 44, 53–62]. However, cells that are deficient for the kinase [63] or that contain a dominant-negative form of it show resistance to the proapoptotic effects of several agents [9, 44, 53, 54, 56, 58, 64]. Interestingly, a recent report [65] has shown that two proteins also involved in the cellular response to DNA damage bind to an element in the promoter of the PKR gene. However, these proteins are constitutively associated with the promoter and it is not clear whether PKR can be induced by agents and conditions that cause DNA strand breaks. Such stresses are potentially proapoptotic, but the possibility of a role for PKR in the response to DNA damage needs further investigation.

8.3.2
The Role of eIF2α Phosphorylation

It remains to be clarified as to whether the phosphorylation of the α subunit of protein synthesis initiation factor eIF2 is sufficient to mediate the proapoptotic effects of PKR. This protein is a well-characterized substrate of PKR and its phosphorylation results in inhibition of the recycling of eIF2 between successive rounds of polypeptide chain initiation [42]. The kinase has a number of other possible substrates and signaling targets [59, 66–70], and it is also conceivable

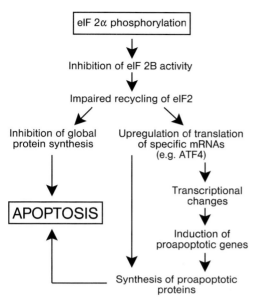

Fig. 8.3. Molecular events linking eIF2α phosphorylation with the induction of apoptosis. Phosphorylation of the α subunit of protein synthesis initiation factor eIF2 impairs the recycling of this factor between successive rounds of translation as a result of inhibition of the guanine nucleotide exchange activity of the factor eIF2B [42]. This results in a downregulation of the overall rate of protein synthesis, but can also lead to the translational upregulation of the expression of specific proteins such as the transcription factor ATF4 [78]. Both the inhibition of global protein synthesis and the possible upregulation of proapoptotic proteins can result in an enhanced rate of cell death. Increased synthesis of apoptosis-inducing proteins may be either mediated directly, by changes in the rates of translation of such proteins from preexisting mRNAs, or indirectly, as a result of altered patterns of gene expression at the transcriptional level.

that dominant-negative PKR mutants might interfere with the functions of other dsRNA-regulated proteins or proapoptotic pathways. However, a role for eIF2 seems likely, given that expression of either an inhibitor of eIF2α phosphorylation or of the nonphosphorylatable S51A mutant of the factor can protect cells from apoptosis [58, 59, 71]. Moreover, a phosphomimetic (S51D) mutant of eIF2α has the opposite effect and is able to promote cell death [58].

Exactly how the phosphorylation of eIF2α can enhance apoptosis remains to be established. A number of possible mechanisms, which are not mutually exclusive, are shown in Fig. 8.3. One possibility is that downregulation of translation selectively lowers the levels of key antiapoptotic proteins that have short half-lives (e.g. Bcl-2 [53]). Such a mechanism has recently been implicated in the eIF2α phosphorylation-induced loss of the inhibitor of nuclear factor (NF)-κB, IκB, leading to the activation of NF-κB [72, 73] (note, however, that NF-κB often exerts an antiapoptotic effect on cells rather than promoting cell death [74]). It is also possible that inhibition of translation following the phosphorylation of eIF2α could paradoxically result in the enhanced production of proapoptotic proteins [e.g. CD95, FADD (Fas-associated death domain protein), Bax, Bad and caspase-8 or -9] or their inducers (e.g. the transcription factor ATF4). Some of these proteins are poorly expressed in cells lacking PKR activity [53, 61, 75, 76]. There are precedents for the selective upregulation of expression of some proteins at the translational level following the phosphorylation of eIF2α (for the mechanisms involved, see [77, 78]). One such protein is indeed ATF4, which in turn causes enhanced expression of the proapoptotic CHOP protein at the transcriptional level. A recent study of the induction of apoptosis by proteasome inhibitors indicates a critical role for eIF2α phosphorylation and CHOP expression in the activation of caspases and subsequent cell death. However, in this case the protein kinase responsible was shown to be GCN2 rather than PKR [79]. Such phenomena as these could explain why the phosphorylation of eIF2α (whether by PKR or other kinases that can target the initiation factor) is necessary for the activation of caspases-8 and -9 and the induction of apoptosis following exposure of cells to a variety of death inducers [76, 80].

The ability of activated PKR and phosphorylated eIF2α to exert the proapoptotic effects described above may well explain the tumor suppressor roles proposed for these proteins. There are still only a few examples where mutations in PKR have been associated with malignant phenotypes and these are predominantly from various types of leukemia [81–83]. However, experimental systems have suggested the importance of eIF2α phosphorylation for protection of cells against malignancy. A nonphosphorylatable form of the protein is able to transform NIH 3T3 fibroblasts [84] and can cooperate with the telomerase component hTERT and the SV-40 large T antigen in transforming primary human kidney cells [85]. Nevertheless, expression of the mutant initiation factor alone was not sufficient to inhibit apoptosis of 3T3L1 cells treated with various inducers of cell death; nor was it able to bring about transformation of the primary kidney cells [85].

Further evidence for the importance of PKR activity and eIF2α phosphorylation in controlling apoptosis and cell transformation comes from studies on tumor viruses. The herpesvirus HHV8 (Kaposi's sarcoma-associated herpes virus) encodes the LANA2 protein (also called vIRF3) which has been shown to inhibit PKR-mediated (but not RNase L-mediated) apoptosis [86]. The HPV protein E6 (which is also an inhibitor of p53) rescues cells from PKR-induced inhibition of protein synthesis and induction of cell death by facilitating the dephosphorylation of eIF2α by the GADD34/protein phosphatase 1 complex [87]. This effect may contribute to the established oncogenic properties of the HPV E6 protein.

The relative inability of transformed cells to regulate protein synthesis and apoptosis through the eIF2α phosphorylation pathway can, however, be exploited to allow cells to be killed by an oncolytic virus such as vesicular stomatitis virus (VSV). The replication of this virus is normally very sensitive to inhibition of translation by phosphorylated eIF2α [88], but where this regulation is defective (either because phosphorylation does not occur or because protein synthesis can take place in spite of high levels of phosphorylation [89]), VSV will replicate readily and kill the cells [90]. This property has good potential for exploitation as a selective therapy for cancer.

8.4
IFNs and the Apoptotic Effects of TRAIL

It has been known for some time that IFNs can enhance the apoptotic effects of TRAIL, sensitizing cells to the actions of this cytokine or even conferring the ability to respond to TRAIL on cells previously resistant to it. TRAIL has continued to attract major attention as a potential therapeutic agent, not only because it can induce apoptosis selectively in tumor cells [91, 92], but also because it can play a role in the protection of cells against virus infection [93].

8.4.1
TRAIL Induction

Many reports indicate that both type I and II IFNs induce the production and secretion of TRAIL, as well as being able to synergize with these cytokines to enhance cell death (reviewed in [9]). The ability of IFNs to induce TRAIL at the transcriptional level, probably via activation of the PML gene [15] and the transcription factors STAT-1 [12] and/or IRF-1 [7, 94, 95], may account for many of the proapoptotic effects of IFNs on transformed or infected cells. Moreover, data now suggest that expression of TRAIL may also be involved in the IFN-mediated activation and/or apoptosis of several cell types in the immune system (e.g. natural killer cells, cytotoxic T cells and dendritic cells, as well as neutrophils, monocytes and leukemic cells derived from them [14, 16, 96]).

8.4.2
TRAIL Activity

As well as being able to induce TRAIL, the IFNs sensitize cells to the proapoptotic (and other) effects of this cytokine (Fig. 8.1). This is of great interest and potential importance since it suggests a way in which otherwise TRAIL-resistant tumors or other pathogenic cells may be induced into apoptosis. For example, some resistant human hepatoma cell lines have been shown to become responsive to TRAIL following pretreatment with IFN-α [97]. There are probably several mechanisms by which such sensitization occurs. In the case of type I IFNs, upregulation of the TRAIL receptor-2 (DR5) has been observed [97]. TRAIL-mediated caspase activation is also enhanced in cells previously exposed to IFN. Conversely, there is downregulation of proteins that may inhibit apoptosis, such as survivin [97]. The sensitizing effect of type I IFN for apoptosis of hepatoma cells also extends to CD95-induced cell death – a phenomenon that may be related to the beneficial effects of IFN combined with ribavirin against hepatitis C virus infection [98].

IFN-β can also exert "anti-antiapoptotic" activity via the induction of the *XAF1* gene in cells that are otherwise resistant to TRAIL [21]. *XAF1* encodes a protein that neutralizes the antiapoptotic effect of the inhibitor of apoptosis protein XIAP.

In some cases cells that are resistant to the proapoptotic effects of IFN-β have a block in the synthesis of the XAF1 protein at a post-transcriptional level. IFN-α can also inhibit the activation by TRAIL of NF-κB. This transcription factor has a generally antiapoptotic effect. These mechanisms may also operate in cases where cells undergo apoptosis following treatment with IFN alone, with endogenously induced TRAIL still providing the proximal death-inducing stimulus [12, 15, 99].

Interestingly, IFN-α treatment also sensitizes cells to early effects of TRAIL that are not merely a consequence of ongoing apoptosis, such as the downregulation of overall protein synthesis (Fig. 8.4). The latter effect precedes the loss of cell viability. The inhibition of translation by TRAIL, both in the presence and absence of IFN-α, is a caspase-dependent process [76]. This suggests that the same mechanisms of sensitization by IFN pretreatment, probably involving enhanced signaling and/or activation of caspase-8 at the level of TRAIL receptor(s), are responsible for both the early downregulation of protein synthesis and the later induction of apoptosis.

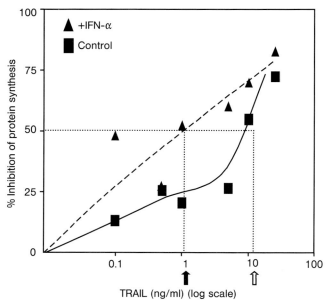

Fig. 8.4. IFN-α sensitizes cells to inhibition of protein synthesis by TRAIL. HeLa cells were cultured for 72 h in the presence or absence of human IFN-α2b (1000 U mL^{-1}) and then further treated with increasing concentrations of TRAIL for the last 5 h as indicated. Overall protein synthesis was then measured by the incorporation of [^{35}S]methionine into acid-insoluble material for the last 40 min. The data were calculated as c.p.m. incorporated µg protein^{-1} and are expressed as percentage inhibition of protein synthesis, relative to untreated cells. The values are means of three determinations. The concentrations of TRAIL producing 50% inhibition, in the absence or presence of prior IFN treatment, are shown by the open and closed arrows, respectively. These results indicate that IFNα pretreatment reduces the concentration of TRAIL required to inhibit overall protein synthesis by 50% by a factor of 10.

IFN-γ too can sensitize cells, and in some cases reverse the resistance of tumor cells, to TRAIL-mediated apoptosis [92, 100, 101]. In the case of neuroblastoma cells reversal of resistance has been attributed to transcriptional activation of the procaspase-8 gene, which contains an IFN-γ-responsive GAS element [20]. Expression of the TRAIL gene is also enhanced. However TRAIL resistance persists in some cases, due to the lack of expression of TRAIL receptors. Interestingly, the latter may be inducible with chemotherapeutic agents such as adriamycin or etoposide [20], providing a molecular basis for a potential three way synergism between TRAIL, IFN-γ and such drugs in the treatment of neuroblastomas. In the case of thyroid anaplastic carcinoma, IFN-γ has been shown to upregulate expression of the proapoptotic Bak gene, an effect which appears to be both necessary and sufficient to bring about sensitivity to TRAIL [100]. Conversely, IFN-γ inhibits the upregulation by TRAIL of the inhibitor of apoptosis protein, IAP-2 [102]. In cells where this occurs such an effect may constitute a basis for synergism between IFN-γ and TRAIL in the promotion of cell death [25, 102]. The increased apoptosis seen in response to IFN-γ plus TRAIL is characterized by elevated activity of caspases-8 and -9, with degradation of Bid and the translocation of Bax to mitochondria.

TRAIL is also induced by IFN-γ in neutrophils [16] and Ewing's sarcoma cells [103], and in the latter system there is evidence for interactions between IFN-γ and -α and between IFN-γ and TNF-α in enhancing apoptosis [13]. Cells isolated from xenografted Ewing's tumors that were resistant to TRAIL (probably due to receptor downregulation) could be restored to sensitivity by IFN-γ treatment, with concomitantly increased expression of TRAIL receptors and caspase-8 [104]. This allowed successful treatment of these tumors by the combination of cytokines in an *in vivo* model, whereas administration of TRAIL or a TRAIL receptor agonist alone was relatively ineffective.

The death-inducing effects of the combination of IFN-γ and retinoic acid against both PML and breast cancer cells also involve TRAIL, which is endogenously produced as a result of *de novo* transcription. The mechanism of TRAIL gene expression requires the tumor suppressor protein IRF-1 [95]. One recent report has shown that IRF-1 is induced by IFN-γ in a system where there was IFN-mediated enhancement of the effect of TRAIL without any changes in the levels of TRAIL receptors or a number of key regulators of apoptosis [101]. Ectopic expression of IRF-1 was able to mimic the effect of IFN in increasing the cellular response to TRAIL and IRF-1 was shown to be necessary for the latter effect. Consistent with these findings, the ability of IFN-γ to sensitize Ewing's sarcoma cells to the apoptotic effects of IFN-α or TNF-α has been attributed to its activity in sustaining the expression of IRF-1 [13].

There is also an interesting correlation between the sensitization of cells by IFNs to an apoptotic inducer (albeit in this case a CD95 agonist rather than TRAIL) and their position in the cell cycle. Jedema et al. [105] report that both type I and II IFNs recruit human myeloid leukemia cells from G_0 into G_1 (without stimulating proliferation) and that cells in the latter phase are more sensitive than those in G_0 or G_2/M phase to CD95 stimulation. This phenomenon could provide a link between the cytostatic effect of IFNs (at least where this leads to accumulation of cells in G_1) and increased sensitivity of cells to apoptosis.

8.5
Signal Transduction Pathways for IFN-mediated Effects on Apoptosis

The initial signaling mechanisms that link IFN receptors to the regulation of apoptosis appear to be very similar to the pathways by which all IFN-sensitive genes are controlled [i.e. via Janus kinase (JAK)- and STAT-dependent transcriptional regulation]. However, it is possible that some IFN-induced events involved in the regulation of apoptosis (e.g. via activation of STAT-3) do not require new transcription [106]. It is therefore of interest to analyze what signal transduction pathways are regulated as a result of STAT activation, either directly or downstream of changes in gene expression, that lead to an effect on apoptosis.

A number of studies have addressed the mechanism of activation of NF-κB by IFNs and here there is evidence that both phosphatidylinositol-3-kinase (PI3K) and protein kinase B (PKB, also known as Akt) are required. The latter enzyme activates the phosphorylation-dependent degradation of the inhibitor of NF-κB, IκBα [107]. It is of note that NF-κB activation results in a predominantly antiapoptotic cellular response and this mechanism may serve to limit the extent of IFN-induced apoptosis. Indeed such a response could account for the very limited and delayed extent of cell death that occurs when cells are exposed to IFNs in the absence of any other proapoptotic stimulus. PI3K and PKB are well known antiapoptotic factors in most eukaryotic cells and exert their effects by several mechanisms in addition to NF-κB activation [108]. However, IFN-α has also been reported to inhibit NF-κB activation in hepatitis B virus-infected cells, thus negating an antiapoptotic activity of this virus [109]. Moreover, the activation of PI3K by IFN-α has been reported to be *required* for IFN-induced apoptosis rather than blocking the process [110]. This death-inducing response was suggested to involve the protein kinase mTOR (mammalian target of rapamycin) downstream of PI3K, since inhibition of mTOR impaired the effect.

Both IFN-α and IFN-γ also cause increased phosphorylation of the mTOR substrates p70S6 kinase and 4E-BP1, via pathways involving the insulin receptor substrate (IRS) proteins and PI3K [110–112]. This is somewhat paradoxical since stimulation of mTOR activity and the regulation of its target substrates by enhanced phosphorylation are normally associated with cell survival. Conversely, 4E-BP1 has been shown to be promote apoptosis when in its hypophosphorylated form. Under these conditions 4E-BP1 is able to sequester and inhibit its target protein, the mRNA cap-binding initiation factor eIF4E, which has antiapoptotic activity when overexpressed [113]. The level of expression of 4E-BP1 is a determinant of cellular sensitivity to the mTOR inhibitor rapamycin – an agent which can exhibit antiproliferative and proapoptotic effects [114]. Thus it is difficult to reconcile the activation of mTOR or the enhancement of 4E-BP1 phosphorylation by IFNs with the proapoptotic effects of these cytokines. However, it is possible that mTOR can also activate some proapoptotic targets such as p53 [115]. It is also conceivable that alterations in the pattern of translation following IFN-induced phosphorylation of ribosomal protein S6 and/or the release of eIF4E from inhibition by 4E-BP1 might upregulate the expression of proapoptotic proteins at the translational level.

Regulation of the mitogen-activated protein kinase (MAPK) pathways is also a feature of IFN action that may be relevant to the control of apoptosis. A number of studies have shown that both type I and II IFNs activate the p38 stress-activated MAPK cascade [11, 106, 116], and p38 is believed to be responsible for the phosphorylation of Ser727 on STAT-1 that is induced by IFN action. Pharmacological inhibition of p38 not only blocks this phosphorylation event, but also decreases the induction of TRAIL by IFN-γ [11]. A recent study has shown that IFN-α treatment activates the MAPK kinases MKK-3 and -6, and one or both of these enzymes can in turn activate p38 MAPK [117]. Interestingly, this work has also suggested that this pathway is necessary for transcriptional regulation by type I IFNs, by a mechanism that is independent of the STAT proteins.

The proapoptotic effects of p38 stress-activated MAPK regulation by IFNs may be counteracted by activation of the extracellular-regulated kinase (ERK) branch of the MAPK family [106]. The latter is under the control of mitogenic growth factors and cytokines such as epidermal growth factor (EGF), which stimulate ERK activation via Ras signaling. Thus it may be possible in some instances to increase the apoptotic effect of IFNs by inhibiting the Ras–ERK pathway (e.g. by expression of dominant-negative mutants or by use of inhibitors of the ERK kinase MEK-1 [118]).

8.6
The Antiapoptotic Effects of IFNs

Given that IFNs are good at sensitizing cells to apoptosis in many different systems, it is curious that these cytokines are able to protect cells from death in certain situations. Examples of this are seen in the enhancement by both IFN-α and -γ of the survival of B-chronic lymphocytic leukemia cells, with upregulation of the antiapoptotic protein Bcl-2 [119]. IFN-γ also prevents apoptosis in Epstein–Barr virus-infected NK cell leukemia cells [120]. IFN-α has been reported to exhibit a JAK-2-dependent antiapoptotic effect on primary hepatic stellate cells [121], but in this case the effect was antagonized by IFN-γ [122]. The latter report suggested that the balance between upregulation and downregulation of the chaperone protein heat shock protein (HSP) 70 by IFN-α and -γ, respectively, was the critical factor in determining the cellular outcome. HSP70 may exert its effects on apoptosis by inhibiting caspase activation, possibly by regulating JNK kinase [123].

There may also be roles for protein kinase C and the Mcl-1 gene in the inhibition by IFN-α of CD95-mediated or interleukin-6 deprivation-induced apoptosis in myeloma cells [124, 125]. Furthermore, both type I and II IFNs can upregulate the cyclin-dependent protein kinase inhibitor (CKI) p21 [126] and the level of expression of this protein can determine whether cells exhibit a cell cycle block rather than entering into apoptosis [127, 128]. In the case of type I IFNs, as described earlier, the ability to activate NF-κB may also be important in prolonging cell survival. The PI3K pathway may be involved in type I IFN-induced survival of primary B cells and activated T cells or neutrophils [129, 130], although in the latter case PKB, a downstream target of PI3K, is not stimulated.

As well as the mechanisms involved, the physiological significance of the anti-apoptotic effects of the IFNs needs to be established since the ability to protect cells from death does not fit readily with the concept that IFNs have antitumor activity. However, in the immune system such effects do make physiological sense since increased production of IFNs in response to viral infections may help to prolong the survival of cells that are required to mount cell-mediated immune responses. As an example, T cells that have been activated by antigen remain viable as a result of the actions of type I IFNs [131–133]. Note, however, that the ability to maintain the viability of leukemia or lymphoma cells [119, 120] may predicate against the clinical use of IFNs for malignancies of this type.

8.7
Conclusions

In this chapter, we have described a variety of ways in which the IFNs can regulate apoptosis in mammalian cells. The challenge now, beyond increasing our understanding of the mechanisms involved, is to apply this knowledge for the improvement of the clinical efficacy of IFN therapy in cancer and serious viral infections. Due to the relatively low toxicity of TRAIL, the combination of this cytokine with type I or II IFNs holds great promise for the selective targeting of pathogenic cells. It may well be that other combined treatments involving IFNs will also prove effective, especially where malignant or infected cells are more susceptible than normal cells to the agents involved. We can therefore look forward to a new era in which the IFNs are not so much "magic bullets" themselves as the weapons that allow the latter to be fired with more deadly accuracy.

Acknowledgments

The work in our laboratory is funded by the Leukemia Research Fund, the Association for International Cancer Research and the Cancer Prevention Research Trust.

References

1 T. Panaretakis, K. Pokrovskaja, M. C. Shoshan, D. Grander, *Oncogene* **2003**, *22*, 4543–4556.
2 Y. Honma, Y. Ishii, Y. Yamamoto-Yamaguchi, T. Sassa, K. Asahi, *Cancer Res* **2003**, *63*, 3659–3666.
3 D. Grandér, O. Sangfelt, S. Erickson, *Eur J Haematol* **1997**, *59*, 129–135.
4 S. L. Tan, M. G. Katze, *J Interferon Cytokine Res* **1999**, *19*, 543–554.
5 B. R. G. Williams, *Oncogene* **1999**, *18*, 6112–6120.
6 M. J. De Veer, M. Holko, M. Frevel, E. Walker, S. Der, J. M. Paranjape, R. H. Silverman, B. R. G. Williams, *J Leukoc Biol* **2001**, *69*, 912–920.
7 G. Romeo, G. Fiorucci, M. V. Chiantore, Z. A. Percario, S. Vannucchi, E. Affabris, *J Interferon Cytokine Res* **2002**, *22*, 39–47.
8 M. Chawla-Sarkar, D. J. Lindner, Y. F. Liu, B. R. Williams, G. C. Sen, R. H. Silverman, E. C. Borden, *Apoptosis* **2003**, *8*, 237–249.
9 M. J. Clemens, *J Interferon Cytokine Res* **2003**, *23*, 277–292.
10 S. D. Der, A. M. Zhou, B. R. G. Williams, R. H. Silverman, *Proc Natl Acad Sci USA* **1998**, *95*, 15623–15628.
11 J. Lee, J. S. Shin, J. Y. Park, D. Kwon, S. J. Choi, S. J. Kim, I. H. Choi, *J Neurosci Res* **2003**, *74*, 884–890.
12 E. A. Choi, H. Lei, D. J. Maron, J. M. Wilson, J. Barsoum, D. L. Fraker, W. S. El Deiry, F. R. Spitz, *Cancer Res* **2003**, *63*, 5299–5307.
13 A. Abadie, F. Besançon, J. Wietzerbin, *Oncogene* **2004**, *23*, 4911–4920.
14 C. Tecchio, V. Huber, P. Scapini, F. Calzetti, D. Margotto, G. Todeschini, L. Pilla, G. Martinelli, G. Pizzolo, L. Rivoltini, M. A. Cassatella, *Blood* **2004**, *103*, 3837–3844.
15 C. Crowder, O. Dahle, R. E. Davis, O. S. Gabrielsen, S. Rudikoff, *Blood* **2005**, *105*, 1280–1287.
16 H. Kamohara, W. Matsuyama, O. Shimozato, K. Abe, C. Galligan, S. Hashimoto, K. Matsushima, T. Yoshimura, *Immunol* **2004**, *111*, 186–194.
17 C. Choi, E. Jeong, E. N. Benveniste, *J Neurooncol* **2004**, *67*, 167–176.
18 K. A. Kirou, R. K. Vakkalanka, M. J. Butler, M. K. Crow, *Clin Immunol* **2000**, *95*, 218–226.
19 S. Fulda, K. M. Debatin, *Oncogene* **2002**, *21*, 2295–2308.
20 X. Z. Yang, M. S. Merchant, M. E. Romero, M. Tsokos, L. H. Wexler, U. Kontny, C. L. Mackall, C. J. Thiele, *Cancer Res* **2003**, *63*, 1122–1129.
21 D. W. Leaman, M. Chawla-Sarkar, K. Vyas, M. Reheman, K. Tamai, S. Toji, E. C. Borden, *J Biol Chem* **2002**, *277*, 28504–28511.
22 T. Raveh, A. Kimchi, *Exp Cell Res* **2001**, *264*, 185–192.
23 J. Sancéau, J. Hiscott, O. Delattre, J. Wietzerbin, *Oncogene* **2000**, *19*, 3372–3383.

24 C. PORTA, R. HADJ-SLIMANE, M. NEJMEDDINE, M. PAMPIN, M. G. TOVEY, L. ESPERT, S. ALVAREZ, M. K. CHELBI-ALIX, *Oncogene* **2005**, *24*, 605–615.

25 M. KAMACHI, A. KAWAKAMI, S. YAMASAKI, A. HIDA, T. NAKASHIMA, H. NAKAMURA, H. IDA, M. FURUYAMA, K. NAKASHIMA, K. SHIBATOMI, T. MIYASHITA, K. MIGITA, K. EGUCHI, *J Lab Clin Med* **2002**, *139*, 13–19.

26 R. SANDOVAL, J. XUE, M. PILKINTON, D. SALVI, H. KIYOKAWA, O. R. COLAMONICI, *J Biol Chem* **2004**, *279*, 32275–32280.

27 A. SATYANARAYANA, K. L. RUDOLPH, *J Clin Invest* **2004**, *114*, 1237–1240.

28 T. IWAKUMA, G. LOZANO, *Mol Cancer Res* **2003**, *1*, 993–1000.

29 M. BENNETT, K. MACDONALD, S. W. CHAN, J. P. LUZIO, R. SIMARI, P. WEISSBERG, *Science* **1998**, *282*, 290–293.

30 M. MULLER, S. WILDER, D. BANNASCH, D. ISRAELI, K. LEHLBACH, M. LI-WEBER, S. L. FRIEDMAN, P. R. GALLE, W. STREMMEL, M. OREN, P. H. KRAMMER, *J Exp Med* **1998**, *188*, 2033–2045.

31 P. A. TOWNSEND, T. M. SCARABELLI, S. M. DAVIDSON, R. A. KNIGHT, D. S. LATCHMAN, A. STEPHANOU, *J Biol Chem* **2004**, *279*, 5811–5820.

32 A. R. CUDDIHY, A. H. T. WONG, N. W. N. TAM, S. Y. LI, A. E. KOROMILAS, *Oncogene* **1999**, *18*, 2690–2702.

33 J. C. CASTELLI, B. A. HASSEL, K. A. WOOD, X. L. LI, K. AMEMIYA, M. C. DALAKAS, P. F. TORRENCE, R. J. YOULE, *J Exp Med* **1997**, *186*, 967–972.

34 L. RUSCH, A. M. ZHOU, R. H. SILVERMAN, *J Interferon Cytokine Res* **2000**, *20*, 1091–1100.

35 M. DÍAZ-GUERRA, C. RIVAS, M. ESTEBAN, *Virology* **1997**, *236*, 354–363.

36 J. C. CASTELLI, B. A. HASSEL, A. MARAN, J. PARANJAPE, J. A. HEWITT, X. L. LI, Y. T. HSU, R. H. SILVERMAN, R. J. YOULE, *Cell Death Differ* **1998**, *5*, 313–320.

37 A. M. ZHOU, J. PARANJAPE, T. L. BROWN, H. Q. NIE, S. NAIK, B. H. DONG, A. S. CHANG, B. TRAPP, R. FAIRCHILD, C. COLMENARES, R. H. SILVERMAN, *EMBO J* **1997**, *16*, 6355–6363.

38 G. Q. LI, Y. XIANG, K. SABAPATHY, R. H. SILVERMAN, *J Biol Chem* **2004**, *279*, 1123–1131.

39 M. S. IORDANOV, D. PRIBNOW, J. L. MAGUN, T. H. DINH, J. A. PEARSON, S. L. CHEN, B. E. MAGUN, *Mol Cell Biol* **1997**, *17*, 3373–3381.

40 K. MALATHI, J. M. PARANJAPE, R. GANAPATHI, R. H. SILVERMAN, *Cancer Res* **2004**, *64*, 9144–9151.

41 J. CARPTEN, N. NUPPONEN, S. ISAACS, R. SOOD, C. ROBBINS, J. XU, M. FARUQUE, T. MOSES, C. EWING, E. GILLANDERS, P. HU, P. BUJNOVSZKY, I. MAKALOWSKA, A. BAFFOE-BONNIE, D. FAITH, J. SMITH, D. STEPHAN, K. WILEY, M. BROWNSTEIN, D. GILDEA, B. KELLY, R. JENKINS, G. HOSTETTER, M. MATIKAINEN, J. SCHLEUTKER, K. KLINGER, T. CONNORS, Y. XIANG, Z. WANG, A. DE MARZO, N. PAPADOPOULOS, O. P. KALLIONIEMI, R. BURK, D. MEYERS, H. GRONBERG, P. MELTZER, R. SILVERMAN, J. BAILEY-WILSON, P. WALSH, W. ISAACS, J. TRENT, *Nat Genet* **2002**, *30*, 181–184.

42 M. J. CLEMENS, in: J. W. B. HERSHEY, M. B. MATHEWS, N.

Sonenberg (Eds.), *Translational Control*, Cold Spring Harbor Laboratory Press, Cold Spring Harbor, NY, **1996**, pp. 139–172.

43 R. Jagus, B. Joshi, G. N. Barber, *Int J Biochem Cell Biol* **1999**, *31*, 123–138.

44 M. C. Yeung, J. Liu, A. S. Lau, *Proc Natl Acad Sci USA* **1996**, *93*, 12451–12455.

45 S. Nanduri, B. W. Carpick, Y. W. Yang, B. R. G. Williams, J. Qin, *EMBO J* **1998**, *17*, 5458–5465.

46 X. Zheng, P. C. Bevilacqua, *RNA* **2004**, *10*, 1934–1945.

47 D. R. Taylor, S. B. Lee, P. R. Romano, D. R. Marshak, A. G. Hinnebusch, M. Esteban, M. B. Mathews, *Mol Cell Biol* **1996**, *16*, 6295–6302.

48 P. R. Romano, M. T. Garcia-Barrio, X. L. Zhang, Q. Z. Wang, D. R. Taylor, F. Zhang, C. Herring, M. B. Mathews, J. Qin, A. G. Hinnebusch, *Mol Cell Biol* **1998**, *18*, 2282–2297.

49 P. A. Lemaire, J. Lary, J. L. Cole, *J Mol Biol* **2005**, *345*, 81–90.

50 R. C. Patel, G. C. Sen, *EMBO J* **1998**, *17*, 4379–4390.

51 T. Ito, M. L. Yang, W. S. May, *J Biol Chem* **1999**, *274*, 15427–15432.

52 P. P. Ruvolo, F. Q. Gao, W. L. Blalock, X. M. Deng, W. S. May, *J Biol Chem* **2001**, *276*, 11754–11758.

53 S. Balachandran, C. N. Kim, W. C. Yeh, T. W. Mak, K. Bhalla, G. N. Barber, *EMBO J* **1998**, *17*, 6888–6902.

54 T. Takizawa, C. Tatematsu, Y. Nakanishi, *J Biochem (Tokyo)* **1999**, *125*, 391–398.

55 O. Donzé, J. Dostie, N. Sonenberg, *Virology* **1999**, *256*, 322–329.

56 S. D. Der, Y. L. Yang, C. Weissmann, B. R. G. Williams, *Proc Natl Acad Sci USA* **1997**, *94*, 3279–3283.

57 N. M. Tang, M. J. Korth, M. Gale, Jr., M. Wambach, S. D. Der, S. K. Bandyopadhyay, B. R. G. Williams, M. G. Katze, *Mol Cell Biol* **1999**, *19*, 4757–4765.

58 S. P. Srivastava, K. U. Kumar, R. J. Kaufman, *J Biol Chem* **1998**, *273*, 2416–2423.

59 J. Gil, J. Alcamí, M. Esteban, *Mol Cell Biol* **1999**, *19*, 4653–4663.

60 S. B. Lee, D. Rodríguez, J. R. Rodríguez, M. Esteban, *Virology* **1997**, *231*, 81–88.

61 J. Gil, M. Esteban, *Oncogene* **2000**, *19*, 3665–3674.

62 T. Takizawa, K. Ohashi, Y. Nakanishi, *J Virol* **1996**, *70*, 8128–8132.

63 A. Kumar, Y. L. Yang, V. Flati, S. Der, S. Kadereit, A. Deb, J. Haque, L. Reis, C. Weissmann, B. R. G. Williams, *EMBO J* **1997**, *16*, 406–416.

64 M. C. Yeung, A. S. Lau, *J Biol Chem* **1998**, *273*, 25198–25202.

65 S. Das, S. V. Ward, D. Markle, C. E. Samuel, *J Biol Chem* **2004**, *279*, 7313–7321.

66 Z. Xu, B. R. G. Williams, *Mol Cell Biol* **2000**, *20*, 5285–5299.

67 L. M. Parker, I. Fierro-Monti, M. B. Mathews, *J Biol Chem* **2001**, *276*, 32522–32530.

68 R. C. Patel, D. J. Vestal, Z. Xu, S. Bandyopadhyay, W. D. Guo, S. M. Erme, B. R. G. Williams, G. C. Sen, *J Biol Chem* **1999**, *274*, 20432–20437.

69 S. R. Brand, R. Kobayashi, M. B. Mathews, *J Biol Chem* **1997**, *272*, 8388–8395.
70 A. R. Cuddihy, S. Y. Li, N. W. N. Tam, A. H. T. Wong, Y. Taya, N. Abraham, J. C. Bell, A. E. Koromilas, *Mol Cell Biol* **1999**, *19*, 2475–2484.
71 B. Datta, R. Datta, *Exp Cell Res* **1999**, *246*, 376–383.
72 H. Y. Jiang, S. A. Wek, B. C. McGrath, D. Scheuner, R. J. Kaufman, D. R. Cavener, R. C. Wek, *Mol Cell Biol* **2003**, *23*, 5651–5663.
73 J. Deng, P. D. Lu, Y. Zhang, D. Scheuner, R. J. Kaufman, N. Sonenberg, H. P. Harding, D. Ron, *Mol Cell Biol* **2004**, *24*, 10161–10168.
74 M. S. Hayden, S. Ghosh, *Genes Dev* **2004**, *18*, 2195–2224.
75 S. Balachandran, P. C. Roberts, T. Kipperman, K. N. Bhalla, R. W. Compans, D. R. Archer, G. N. Barber, *J Virol* **2000**, *74*, 1513–1523.
76 I. W. Jeffrey, M. Bushell, V. J. Tilleray, S. Morley, M. J. Clemens, *Cancer Res* **2002**, *62*, 2272–2280.
77 A. G. Hinnebusch, *Trends Biochem Sci* **1994**, *19*, 409–414.
78 P. D. Lu, H. P. Harding, D. Ron, *J Cell Biol* **2004**, *167*, 27–33.
79 H. Y. Jiang, R. C. Wek, *J Biol Chem* **2005**.
80 J. Gil, M. A. Garcia, M. Esteban, *FEBS Lett* **2002**, *529*, 249–255.
81 N. Abraham, M. L. Jaramillo, P. I. Duncan, N. Méthot, P. L. Icely, D. F. Stojdl, G. N. Barber, J. C. Bell, *Exp Cell Res* **1998**, *244*, 394–404.
82 S. I. Hii, L. Hardy, T. Crough, E. J. Payne, K. Grimmett, D. Gill, N. A. J. McMillan, *Int J Cancer* **2004**, *109*, 329–335.
83 J. M. Murad, L. G. Tone, L. R. de Souza, F. L. De Lucca, *Blood Cells Mol Dis* **2005**, *34*, 1–5.
84 O. Donzé, R. Jagus, A. E. Koromilas, J. W. B. Hershey, N. Sonenberg, *EMBO J* **1995**, *14*, 3828–3834.
85 D. J. Perkins, G. N. Barber, *Mol Cell Biol* **2004**, *24*, 2025–2040.
86 M. Esteban, M. A. García, E. Domingo-Gil, J. Arroyo, C. Nombela, C. Rivas, *J Gen Virol* **2003**, *84*, 1463–1470.
87 S. Kazemi, S. Papadopoulou, S. Y. Li, Q. Z. Su, S. O. Wang, A. Yoshimura, G. Matlashewski, T. E. Dever, A. E. Koromilas, *Mol Cell Biol* **2004**, *24*, 3415–3429.
88 D. F. Stojdl, N. Abraham, S. Knowles, R. Marius, A. Brasey, B. D. Lichty, E. G. Brown, N. Sonenberg, J. C. Bell, *J Virol* **2000**, *74*, 9580–9585.
89 S. Balachandran, G. N. Barber, *Cancer Cell* **2004**, *5*, 51–65.
90 G. N. Barber, *Viral Immunol* **2004**, *17*, 516–527.
91 A. Almasan, A. Ashkenazi, *Cytokine Growth Factor Rev* **2003**, *14*, 337–348.
92 C. Ruiz de Almodovar, A. Lopez-Rivas, C. Ruiz-Ruiz, *Vitam Horm* **2004**, *67*, 291–318.
93 J. Strater, P. Moller, *Vitam Horm* **2004**, *67*, 257–274.
94 A. Kröger, M. Köster, K. Schroeder, H. Hauser, P. P. Mueller, *J Interferon Cytokine Res* **2002**, *22*, 5–14.
95 N. Clarke, A. M. Jimenez-Lara, E. Voltz, H. Gronemeyer, *EMBO J* **2004**, *23*, 3051–3060.

96 T. Yamamoto, H. Nagano, H. Sakon, H. Wada, H. Eguchi, M. Kondo, B. Damdinsuren, H. Ota, M. Nakamura, H. Wada, S. Marubashi, A. Miyamoto, K. Dono, K. Umeshita, S. Nakamori, H. Yagita, M. Monden, *Clin Cancer Res* **2004**, *10*, 7884–7895.

97 M. Shigeno, K. Nako, T. Ichikawa, K. Suzuki, A. Kawakami, S. Abiru, S. Miyazoe, Y. Nakagawa, H. Ishikawa, K. Hamasaki, K. Nakata, N. Ishii, K. Eguchi, *Oncogene* **2003**, *22*, 1653–1662.

98 S. F. Schlosser, M. Schuler, C. P. Berg, K. Lauber, K. Schulze-Osthoff, F. W. Schmahl, S. Wesselborg, *Antimicrob Agents Chemother* **2003**, *47*, 1912–1921.

99 A. Papageorgiou, L. Lashinger, R. Millikan, H. B. Grossman, W. Benedict, C. P. Dinney, D. J. McConkey, *Cancer Res* **2004**, *64*, 8973–8979.

100 S. H. Wang, E. Mezosi, J. M. Wolf, Z. Y. Cao, S. Utsugi, P. G. Gauger, G. M. Doherty, J. R. Baker, Jr., *Oncogene* **2004**, *23*, 928–935.

101 S. Y. Park, J. W. Seol, Y. J. Lee, J. H. Cho, H. S. Kang, I. S. Kim, S. H. Park, T. H. Kim, J. H. Yim, M. Kim, T. R. Billiar, D. W. Seol, *Eur J Biochem* **2004**, *271*, 4222–4228.

102 S. Y. Park, T. R. Billiar, D. W. Seol, *Biochem Biophys Res Commun* **2002**, *291*, 233–236.

103 A. Abadie, J. Wietzerbin, *Ann N Y Acad Sci* **2003**, *1010*, 117–120.

104 M. S. Merchant, X. Yang, F. Melchionda, M. Romero, R. Klein, C. J. Thiele, M. Tsokos, H. U. Kontny, C. L. Mackall, *Cancer Res* **2004**, *64*, 8349–8356.

105 I. Jedema, R. M. Barge, R. Willemze, J. H. Falkenburg, *Leukemia* **2003**, *17*, 576–584.

106 M. Caraglia, M. Marra, G. Pelaia, R. Maselli, M. Caputi, S. A. Marsico, A. Abbruzzese, *J Cell Physiol* **2005**, *202*, 323–335.

107 C. H. Yang, A. Murti, S. R. Pfeffer, J. G. Kim, D. B. Donner, L. M. Pfeffer, *J Biol Chem* **2001**, *276*, 13756–13761.

108 P. Sen, S. Mukherjee, D. Ray, S. Raha, *Mol Cell Biochem* **2003**, *253*, 241–246.

109 K. Ohata, T. Ichikawa, K. Nakao, M. Shigeno, D. Nishimura, H. Ishikawa, K. Hamasaki, K. Eguchi, *FEBS Lett* **2003**, *553*, 304–308.

110 L. Thyrell, L. Hjortsberg, V. Arulampalam, T. Panaretakis, S. Uhles, M. Dagnell, B. Zhivotovsky, I. Leibiger, D. Grandér, K. Pokrovskaja, *J Biol Chem* **2004**, *279*, 24152–24162.

111 F. Lekmine, S. Uddin, A. Sassano, S. Parmar, S. M. Brachmann, B. Majchrzak, N. Sonenberg, N. Hay, E. N. Fish, L. C. Platanias, *J Biol Chem* **2003**, *278*, 27772–27780.

112 F. Lekmine, A. Sassano, S. Uddin, J. Smith, B. Majchrzak, S. M. Brachmann, N. Hay, E. N. Fish, L. C. Platanias, *Exp Cell Res* **2004**, *295*, 173–182.

113 V. A. Polunovsky, A. C. Gingras, N. Sonenberg, M. Peterson, A. Tan, J. B. Rubins, J. C. Manivel, P. B. Bitterman, *J Biol Chem* **2000**, *275*, 24776–24780.

114 M. B. Dilling, G. S. Germain, L. Dudkin, A. L. Jayaraman,

X. W. Zhang, F. C. Harwood, P. J. Houghton, *J Biol Chem* **2002**, *277*, 13907–13917.
115 M. Castedo, K. F. Ferri, J. Blanco, T. Roumier, N. Larochette, J. Barretina, A. Amendola, R. Nardacci, D. Métivier, J. A. Este, M. Piacentini, G. Kroemer, *J Exp Med* **2001**, *194*, 1097–1110.
116 S. Parmar, L. C. Platanias, *Curr Opin Oncol* **2003**, *15*, 431–439.
117 Y. Li, S. Batra, A. Sassano, B. Majchrzak, D. E. Levy, M. Gaestel, E. N. Fish, R. J. Davis, L. C. Platanias, *J Biol Chem* **2005**, *280*, 10001–10010.
118 M. Caraglia, P. Tagliaferri, M. Marra, G. Giuberti, A. Budillon, E. D. Gennaro, S. Pepe, G. Vitale, S. Improta, P. Tassone, S. Venuta, A. R. Bianco, A. Abbruzzese, *Cell Death Differ* **2003**, *10*, 218–229.
119 A. P. Jewell, *Leuk Lymphoma* **1996**, *21*, 43–47.
120 S. Mizuno, K. Akashi, K. Ohshima, H. Iwasaki, T. Miyamoto, N. Uchida, T. Shibuya, M. Harada, M. Kikuchi, Y. Niho, *Blood* **1999**, *93*, 3494–3504.
121 B. Saile, C. Eisenbach, H. El Armouche, K. Neubauer, G. Ramadori, *Eur J Cell Biol* **2003**, *82*, 31–41.
122 B. Saile, C. Eisenbach, J. Dudas, H. El Armouche, G. Ramadori, *Eur J Cell Biol* **2004**, *83*, 469–476.
123 J. S. Lee, J. J. Lee, J. S. Seo, *J Biol Chem* **2005**, *280*, 6634–6641.
124 A. Egle, A. Villunger, M. Kos, G. Böck, J. Gruber, B. Auer, R. Greil, *Eur J Immunol* **1996**, *26*, 3119–3126.
125 M. Jourdan, J. De Vos, N. Mechti, B. Klein, *Cell Death Differ* **2000**, *7*, 1244–1252.
126 P. S. Subramaniam, P. E. Cruz, A. C. Hobeika, H. M. Johnson, *Oncogene* **1998**, *16*, 1885–1890.
127 V. Giandomenico, G. Vaccari, G. Fiorucci, Z. Percario, S. Vannucchi, P. Matarrese, W. Malorni, G. Romeo, E. Affabris, *Eur Cytokine Netw* **1998**, *9*, 619–631.
128 Y. K. Zhang, N. Fujita, T. Tsuruo, *Oncogene* **1999**, *18*, 1131–1138.
129 K. Ruuth, L. Carlsson, B. Hallberg, E. Lundgren, *Biochem Biophys Res Commun* **2001**, *284*, 583–586.
130 D. Scheel-Toellner, K. Q. Wang, N. V. Henriquez, P. R. Webb, R. Craddock, D. Pilling, A. N. Akbar, M. Salmon, J. M. Lord, *Eur J Immunol* **2002**, *32*, 486–493.
131 D. Pilling, A. N. Akbar, J. Girdlestone, C. H. Orteu, N. J. Borthwick, N. Amft, D. Scheel-Toellner, C. D. Buckley, M. Salmon, *Eur J Immunol* **1999**, *29*, 1041–1050.
132 P. Marrack, J. Kappler, T. Mitchell, *J Exp Med* **1999**, *189*, 521–529.
133 Y. Refaeli, L. Van Parijs, S. I. Alexander, A. K. Abbas, *J Exp Med* **2002**, *196*, 999–1005.

9
Viral Defense Mechanisms against Interferon

Santanu Bose and Amiya K. Banerjee

9.1
Introduction

Innate immunity represents the first line of defense by the cells against invading pathogens including viruses before an orchestrated adaptive immune response involving immune cell priming and antibody production is launched against such pathogens [1–6]. Thus, the complex evolutionary conserved defense responses by mammals rely on the communication between the innate and adaptive arms of the immune system to efficiently combat the manifestation of pathogenesis and systemic spread of the viruses within the infected organism. In light of this significance, a great deal of study has focused on the host molecules that are the key mediators and regulators during innate immune response against a variety of pathogens including viruses.

In this chapter, we will discuss studies on the innate immune defense elicited by the host following infection with RNA and DNA viruses, as well as the virus's defense mechanisms to subvert the cell's innate immune response. The primary focus of this chapter will be on negative sense single-stranded (ss) RNA viruses and double-stranded (ds) DNA viruses, since the majority of studies related to the defense mechanisms of cells and viruses have been carried out using these classes of viruses. We will also summarize studies with several dsRNA virus and RNA/DNA reverse-transcribing viruses, particularly hepatitis C virus (HCV), human immunodeficiency virus (HIV)-1 and hepatitis B virus (HBV).

9.2
Innate Immune Antiviral Defense Mechanisms of Host Cells

One of the principal features of innate immunity is the activation of nuclear factor (NF)-κB family of transcriptional factors [7–9]. In uninfected cells, NF-κB family of proteins exists as heterodimers or homodimers and sequestered in the cytoplasm by virtue of their association with members of the IκB family of proteins. A number of stimulatory signals, including virus infection trigger a signaling cascade (Fig. 9.1) involving activation of the protein kinase, the inhibitor of IκB kinase (IKK) and subsequent phosphorylation of IκB, culminates in the degradation of IκB proteins via the ubiquitin–26S proteosome pathway. This event unmasks the nuclear localization signal (NLS) of NF-κB, leading to its translocation to the nucleus and binding to the promoters of target genes. The signaling cascade leading to NF-κB activation has been reviewed in detail elsewhere [7–9].

The activation and modulation of NF-κB pathway following viral infection is known to be mediated primarily by three mechanisms (Fig. 9.1): (i) activation of NF-κB-inducing membrane receptors [Toll-like receptors (TLR)] [10–12], (ii) intracellular expression of virus protein(s) [13, 14] and (ii) dsRNA (an intermediate in RNA virus replication)-mediated activation of dsRNA-dependent protein kinase (PKR) [15, 16]. TLRs (TLR-1 to -10) are evolutionary conserved membrane anchored receptors that were originally identified in *Drosophila* as an array of antifungal mediators. Subsequent studies revealed that these molecules are also utilized by both Gram-positive and -negative bacteria for NF-κB activation in mammalian cells [17, 18]. Surprisingly, recent studies have reported that several viruses utilize TLR to activate NF-κB [10–12] and TLR-3 is specifically activated by dsRNA [19]. An array of signaling proteins originating from the cell surface following activation of TLRs induces NF-κB [9]. Activation of the TLR (TLR-4 or -2) complex comprising of CD14 and MD2 on the cell surface following binding of viral envelope protein(s) [10–12] results in recruitment of cytosolic adaptor proteins MyD88 (myeloid differentiation primary response gene 88) and IRAK [interleukin (IL)-1 receptor-associated kinase] in the membrane [20]. Activation of IRAK results in phosphorylation of tumor necrosis factor (TNF) receptor-associated factor (TRAF)-6, which relays signals through the TAK1–TAB1–TAB2 complex to IKK. Apart from the cell surface molecules (TLRs), intracellular protein, i.e. PKR, activates NF-κB following virus infection [15, 16]. PKR is a serine/threonine kinase that is activated following binding to dsRNA, a byproduct of viral replication [21]. Activation of PKR, involving its dimerization and autophosphorylation, leads to its phys-

Fig. 9.1. A schematic diagram depicting the innate immune antiviral response elicited by NF-κB and IRF-3. Virus-mediated NF-κB activation by TLR(s), dsRNA and/or virus protein(s) (via the MyD88-dependent or -independent/IKK/IκB pathway) leads to the production of IFN-β and/or TNF-α/IL-1β, following binding of NF-κB to the IFN-β and TNF-α/IL-1β promoters, respectively. Likewise, activation of IRF-3 by the viruses [via dsRNA and/or viral protein(s)] following induction of the IKK-related kinase complex (IKKε and TBK-1) results in the production of IFN-α/β by virtue of IRF-3-mediated transactivation of IFN-α/β genes. The antiviral cytokine, IFN-α/β, then primes the uninfected cells (paracrine action) following binding to the IFN receptors (IFNAR-1 and -2), resulting in the activation of the JAK–STAT pathway and expression of antiviral proteins like PKR, 2′–5′-OAS and Mx. Similar to IFN, TNF-α and IL-1β prime uninfected cells [following binding to the TNF-α receptor (TNFR) and IL-1 receptor (IL-1R)] to activate NF-κB signaling and establish an antiviral state independent of IFNs. Apart from the JAK–STAT pathway, in infected cells HPIV-3-mediated activation of NF-κB also confers an antiviral state independent of IFNs.

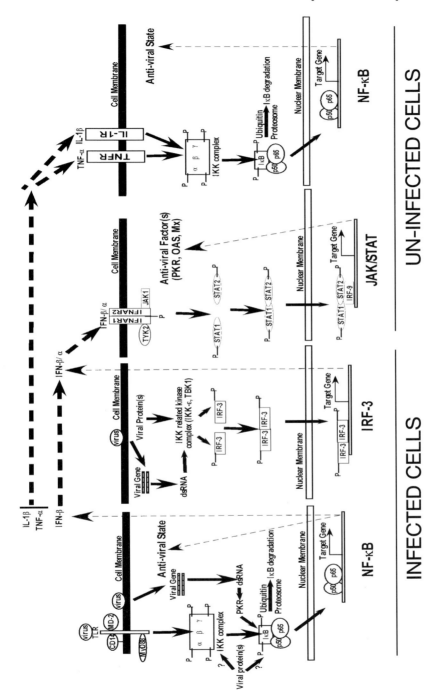

ical association and activation of IKK. However, it has been demonstrated that the catalytic domain of PKR is required for IKK activation [21]. Although dsRNA is able to activate NF-κB by virtue of its binding to PKR, recent studies have demonstrated a PKR-independent activation of NF-κB by dsRNA [22]. This idea was recently supported by demonstration that mammalian TLR-3 activates NF-κB following its interaction with dsRNA [19]. A description of the proteins involved in the TLR signaling cascade is shown in Fig. 9.1 (reviewed in [17, 18]).

The genes activated by NF-κB play an important role in both adaptive [chemokines, adhesion molecules and major histocompatibility complex I (MHCI)] and innate [cytokines like interferon (IFN)-β] immune responses [23]. IFN-β, an antiviral cytokine produced upon NF-κB activation, is a critical innate immune molecule restricting virus replication in uninfected cells by a paracrine mechanism via activation of the well-conserved Janus kinase (JAK)–signal transducers and activators of transcription (STAT) (see below) pathway (Fig. 9.1) [1]. NF-κB also regulates the expression of additional antiviral factor(s) such as tumor necrosis factor (TNF)-α and IL-1β, which was shown by us and others to posses potent antiviral properties against several cytoplasmic RNA viruses [24, 25]. Recently, we demonstrated that NF-κB activation by such viruses or by TNF-α and IL-1β directly confers an innate antiviral state within the cell independent of IFN [25] (Fig. 9.1) (see below).

Apart from NF-κB, virus infection directly produces IFN-α/β following activation of IFN regulatory factor (IRF)-3 [26] (Fig. 9.1). IRF-3 is expressed constitutively in a variety of cells and the mRNA levels are not altered following virus infection. However, the activation of IRF-3 occurs following virus induced phosphorylation by a IKK-related serine/threonine kinase complex [IKKε and TANK-binding kinase (TBK)-1] which phosphorylates within the C-terminus of IRF-3 on Ser385, 386, 396, 398, 402 and 405, and Thr404 [26, 27]. The phosphorylation facilitates its homodimerization and translocation to the nucleus. Moreover, phosphorylation of IRF-3 results in its association with histone acetyltransferase nuclear proteins cAMP response element-binding (CREB)-binding protein (CBP) and p300. This association causes IRF-3, which normally shuttles into and out of the nucleus to become predominantly nuclear [26]. In the nucleus, it binds to the IFN-stimulated response element (ISRE) of the IFN-stimulated gene (ISG)-15 promoter to stimulate transcription of IFN-α/β genes (Fig. 9.1). The secreted IFN then mediate its antiviral action via a paracrine loop (Fig. 9.1). Unlike NF-κB, the mechanism utilized by viruses to activate IRF-3 is not clear. However, recent studies have suggested the involvement of dsRNA and the virally encoded nucleocapsid protein during this process [28, 29]. Apart from IRF-3, which constitutes an important-virus induced factor involved in IFN production, several other members of IRF family are also involved in this process. These members, including IRF-1, -7 and -5, were shown to be selectively activated by various viruses for efficient IFN production and optimal host defense [30–34].

IFNs are antiviral and antiproliferative cytokines that mediate their effect through a well-conserved signaling pathway [35–37]. In humans and mice, type I and II IFNs include IFN-α/β (IFN-I) and IFN-γ (IFN-II), respectively. Although

IFN-I is produced by all cells under appropriate conditions, IFN-II are produced by limited number of cells including the immune cells [activated natural killer, T helper 1 and dendritic cells (DCs)]. Since IFN-I constitutes an important host innate response against viruses, in contrast to the IFN-II-mediated adaptive response, accordingly our chapter focuses on the IFN-I pathway only. The major molecular players involved in transducing the IFN-I antiviral signal are depicted in Fig. 9.1 (reviewed in [1]). IFN-α/β induces its antiviral activity by binding to IFN receptor on the cell surface, which leads to activation of receptor-associated JAKs, specifically JAK-1 and TYK-2 [35–37]. Consequently, the STATs are activated by phosphorylation, leading to formation of ISG factor (ISGF)-3 complex comprising of STAT-1 and -2, and p48. Following the engagement of IFN receptors (IFNAR-1 and -2) by IFN-I, STAT-1 and -2 are phosphorylated at Tyr701 and 692, respectively. Once phosphorylated, the STAT-1 and -2 heterodimer translocates to the nucleus and associates with p48 (IRF-9). This complex (STAT-1 and -2, and IRF-9), termed ISGF-3, associates with the ISREs to activate ISGs. The three well-established antiviral proteins involved in IFN-mediated inhibition of virus infection are (i) the $2'$–$5'$-oligoadenylate synthetase ($2'$–$5'$-OAS)/ribonuclease (RNase) L pathway [38], which degrades viral RNAs following dsRNA activation, (ii) PKR, which inhibits mRNA translation by phosphorylating translation initiation factor eIF-2α [39], and (iii) the Mx proteins (Mx1 in mice and MxA in humans) possessing GTPase activity, which restricts virus infection at several stages including primary transcription, transcription and intracellular trafficking of viral proteins/genome [40]. Apart from the three well-established cytoplasmic IFN-induced antiviral proteins (PKR, $2'$–$5'$-OAS and Mx), our recent studies have demonstrated that IFN also induces soluble secreted antiviral proteins that inhibit virus entry (Bose and Banerjee, unpublished).

9.3
Evasion of IFN-mediated Antiviral Response

Since IFN-I plays an important role in the innate antiviral response by the cell, this chapter will discuss studies dealing with the strategies employed by different viruses to counteract IFN-mediated antiviral action for their replicative advantage. During the last decade a great deal of studies have been performed to understand such mechanisms employed by viruses to evade the host-induced defense apparatus. It was apparent from these studies that almost all families of mammalian viruses are armed with weapons to protect themselves against the IFN system. The viruses have evolved to inhibit IFN-dependent antiviral response at various stages, including (i) inhibition in IFN synthesis, (ii) blocking IFN signaling, (iii) inhibiting the function of IFN-induced antiviral proteins and (iv) producing IFN receptor decoy molecules to prevent induction of IFN signaling. The specific mechanism utilized by different RNA and DNA viruses to defeat IFN system is discussed below.

9.3.1
Nonsegmented Negative-strand RNA Viruses

Nonsegmented negative strand RNA viruses are enveloped single stranded RNA containing viruses that replicates in the cytoplasm of infected cells. This class of viruses belongs to an important family of human and animal pathogens and is divided into three families, paramyxoviruses [41], filoviruses [42] and rhabdoviruses [43]. The human pathogens cause high morbidity and mortality among adults, infants, children and immuno-ompromised adults. In this chapter we will focus on five members of the Paramyxoviridae family [41] including, Sendai virus of mice (SV), human parainfluenza virus types 3 (HPIV-3), human parainfluenza virus types 2 (HPIV-2), simian virus 5 (SV5), mumps virus (MUV), avian Newcastle disease virus (NDV), measles virus (MV), human (RSV) and bovine respiratory syncytial virus (bRSV), and Nipah and Hendra viruses. We will also discuss two viruses, Ebola virus and rabies virus, belonging to the filovirus [42] and rhabdovirus [43] families, respectively. Several paramyxoviruses encode an assortment of genes that are involved in modulating the host defense mechanism for its own replicative advantages. These include V (SV5, MV, MUV, Nipah and Hendra viruses), C (SV, HPIV, MV), NS1 and NS2 (RSV and bRSV) and W (Nipah virus) proteins. Apart from paramyxoviruses, Ebola (filovirus) and rabies (rhabdovirus) viruses encode VP35 and P (phosphoprotein) proteins, respectively, to counteract the innate defense mechanism. The mechanism(s) utilized by these virally encoded antagonist proteins to negatively modulate the IFN-mediated innate immune response are discussed below.

9.3.1.1 Paramyxoviruses

SV SV is known to activate NF-κB following infection [44, 45]; however, the mechanism and signaling components involved in the activation process have not yet been elucidated. IRF-3 activation by SV has been observed in infected cells and it is suggested that dsRNA generated during the virus replication may activate IRF-3 [46–48]. SV phosphorylates multiple serine and threonine residues located at the C-terminal of IRF-3 to activate it, leading to its translocation to the nucleus and transcription of target genes. Phosphorylation of IRF-3 also results in the virus-mediated degradation of IRF-3 [47]. The inter-relationship between the activation and degradation of IRF-3 by SV at the molecular and functional level is not yet clear. A study has also suggested that activated IRF-3 may mediate SV-induced apoptosis following activation of caspases-8, -9 and -3 [46]. It is important to mention that recently TLR-7 (via MyD88) was shown to be utilized for induction of IFN gene by several nonsegmented negative-strand viruses and ssRNAs in DC cells [49, 50]. Moreover TLR-8 was identified as a ssRNA-interacting protein [49, 50]. The critical requirements of these TLRs during NF-κB activation in nonimmune cells by different ssRNA viruses are still to be validated.

Although SV induces NF-κB and IRF-3 that result in the production of IFNs, the

Tab. 9.1. Accessory and nonstructural proteins of the nonsegmented negative-strand RNA viruses

Virus	RNA editing			Overlapping ORF				Individual ORF	
	+0G	+1 (or +4)G	+2G	C'	C	Y1	Y2		
SV	P	V	W	C'	C	Y1	Y2		
SV5	V	I	P	–	–	–	–		
HPIV-3	P	V	D	–	C	–	–		
HPIV-2	V	I	P	–	–	–	–		
MUV	V	I	P	–	–	–	–		
NDV	P	V	I	–	–	–	–		
MV	P	V	W	–	C	–	–		
RSV								NS1	NS2

W and I ORFs are terminated by a stop codon shortly following the editing site.

virus has evolved a strategy to overcome the antiviral action of IFNs [51]. The circumvention of the antiviral action of IFNs is achieved following expression of an array of proteins that block the antiviral signaling pathway [52] (Tabs. 9.1 and 9.2). Recent advances in reverse genetics by which recombinant negative-strand RNA viruses are produced from infectious cDNA [53] have paved the way for the identification of the major virally encoded proteins that mediate the anti-IFN action in infected cells. For SV these IFN antagonist accessory proteins are encoded by either overlapping reading frames (C', C, Y1 and Y2 proteins, collectively called C proteins) or in-frame open reading frames (ORFs) (X proteins) present within the P (phosphoprotein) gene or RNA editing (pseudotemplated addition of nucleotides) (V and W proteins) of the P gene [41] (Tab. 9.1). Thus, the P protein (part of the viral transcriptase complex) of nonsegmented negative-strand RNA viruses not only serves as an important viral factor required for transcription/replication, but several proteins expressed from the P gene are critical modulators of IFN-mediated innate antiviral response.

While recombinant SV lacking the V protein suppressed the IFN-stimulated gene products similar to the wild-type SV, the recombinant virus lacking C (C, C', Y1 and Y2) protein lost its ability to suppress IFN signaling pathway [54]. Initially it was demonstrated that in infected cells, C proteins physically interact with STAT-1 [55] – a critical protein in the IFN antiviral signaling pathway. Further studies utilizing wild-type and recombinant SV lacking the C proteins revealed that C proteins inhibit Tyr701 phosphorylation of STAT-1 [56–58], an event critical for launching the IFN-mediated antiviral response. The same studies also demonstrated that expression of individual C proteins in *trans* inhibited IFN signaling by preventing Tyr701 phosphorylation of STAT-1 [52]. However, only the larger C proteins (C and C′) induced STAT-1 instability as a result of degradation via ubiquitination [56]. Thus, it seems that although both longer (C and C′) and shorter (Y1 and Y2) proteins are capable of inhibiting IFN signaling probably by preventing STAT-1 phosphorylation, the longer C proteins have evolved an additional mechanism whereby they degrade STAT-1. It is speculated that the involvement of SV C protein in STAT-1 degradation is cell-type dependent, since in the majority of cell lines inhibition of STAT-1 tyrosine phosphorylation constitutes the primary mechanism utilized by SV C protein to antagonize IFN action [59]. In that context, a recent study [60] demonstrated that inhibition of tyrosine phosphorylation of STAT-1 and its degradation by SV C protein represents two separate events – the former being IFN signaling dependent, while the latter is independent of IFN-mediated induction of the JAK–STAT pathway. Apart from inhibition of STAT-1 tyrosine phosphorylation and its degradation, it was also suggested that SV may counteract IFN action by preventing Ser727 phosphorylation of STAT-1α [61], a downstream mediator for IFN signaling, and partially inhibiting the tyrosine phosphorylation of IFNR-2 associated tyrosine kinase, TYK-2 [62]. These alternative mechanisms may reflect the virus's ability to utilize them based on cell type and/or during various conditions of growth. Apart from the ability of C and V proteins to inactivate/degrade STAT molecules, SV C and V proteins also block IFN-β synthesis/production by inhibiting phosphorylation/activation of IRF-3 [63].

Although the paramyxoviruses have evolved a strategy to counteract the antiviral response of IFN (Tab. 9.2), it is important to keep in mind that pretreatment of cells with IFN prior to virus infection drastically restricts virus infection [64, 65]. These results suggested that the virus fails to antagonize the antiviral state when the IFN signaling pathway is induced prior to virus replication. The IFN-dependent cellular factors rendering the antiviral state are yet another interesting area of research which is beyond the scope of this review. At least three IFN-I induced antiviral proteins including PKR, 2′–5′-OAS and Mx are known to inhibit the infection of a broad spectrum of viruses [64, 65]. However, to date, no definitive IFN-responsive factors have been demonstrated to confer antiviral status against SV.

SV5 In contrast to SV, SV5 does not encode C proteins (Tab. 9.1). However, SV5, like SV, blocks IFN-I signaling and inhibits ISGF-3 complex formation [61]. The inhibition of IFN signaling by SV5 was shown to be mediated by the SV5 V protein [66], which physically interacts with the STAT proteins *in vitro* and *in vivo* [67] (Tab. 9.2). This interaction leads to degradation of STAT-1, but not STAT-2, via the 26S proteosomal pathway [66–68]. It is worth mentioning that species-specific STAT-2 protein is also required to efficiently block IFN signaling [69]. This phenomenon was confirmed by the observation that SV5, which induces STAT-1 degradation and inhibits IFN signaling in human cells, fails to do so in murine cells [51]. However, expression of human STAT-2 in murine cells in the presence of SV5 V protein reverted the phenotype leading to efficient murine STAT-1 degradation and inhibi-

Tab. 9.2. Strategies of negative-sense ssRNA viruses to inhibit IFN-I signaling

Virus	IFN blocker	Mechanism
HPIV-3	C protein	IFN signaling (STAT-1 phosphorylation inhibition)
HPIV-2	V protein	IFN signaling (STAT-2 degradation)
SV	C protein	IFN signaling (STAT-1 degradation and phosphorylation inhibition, and TYK-2 and STAT-1α phosphorylation inhibition) IFN synthesis (IRF-3 inhibition)
	V protein	IFN synthesis (IRF-3 inhibition)
MV	V protein	IFN synthesis (unknown) IFN signaling (STAT-1 and -2 phosphorylation and nuclear translocation inhibition)
MUV	V protein	IFN signaling (STAT-1 degradation and phosphorylation inhibition, and STAT-2 phosphorylation inhibition)
SV5	V protein	IFN signaling (STAT-1 degradation)
NDV	V protein	IFN signaling (STAT-1 degradation)
RSV	NS1 and NS2 proteins unknown	IFN synthesis (IRF-3 inhibition) IFN signaling (STAT-2 degradation)
bRSV	NS1 and NS2 proteins	IFN synthesis (IRF-3 inhibition)
Nipah	V protein	IFN signaling (STAT-1 and -2 phosphorylation and nuclear translocation inhibition) IFN expression (unknown)
	W protein	IFN signaling (STAT-1 nuclear sequestration) IFN synthesis (IRF-3 inhibition)
	P protein	IFN signaling (STAT-1 nuclear translocation inhibition)
Hendra	V protein	IFN signaling (STAT-1 and -2 nuclear translocation inhibition)
Ebola	VP35 protein	IFN synthesis (IRF-3 inhibition)
Rabies	P protein	IFN synthesis (IRF-3 inhibition)
Influenza A	NS1 protein	IFN synthesis (IRF-3 and -7, and NF-κB inhibition) IFN signaling (PKR inhibition)
	P58IPK (cellular protein)	IFN signaling (PKR inhibition)
Rift Valley fever	NSs protein	IFN synthesis (transcriptional suppression of IFN promoter)
Bunyamwera	NSs protein	IFN synthesis (RNA polymerase II phosphorylation inhibition)

tion of IFN signaling. These results suggest that STAT-2 could act as a species-specific host range determinant for SV5 and human, but not murine, confers growth advantage for SV5. Moreover, STAT-2 protein is critically required to create a degradation-permissive environment for efficient degradation of STAT-1 by SV5 [67].

The direct involvement of V protein in inducing STAT-1 degradation was tested by using recombinant SV5 lacking V protein [70, 71]. In contrast to the wild-type virus, virus lacking V failed to degrade STAT-1 in infected cells [70]. Interestingly, the virus lacking V had a better replicative advantage compared to the wild-type virus, since intracellular viral mRNA and proteins as well as the virus titer were significantly augmented following infection with V lacking virus. In addition, the SV5 lacking V induced apoptosis in various cell lines tested, suggesting that apart from antagonizing the IFN action, the SV5 V protein may also play a role in viral gene expression and virus induced apoptosis [70]. At this time it is not known whether SV5 activates NF-κB and/or IRF-3. This line of study will certainly be interesting in light of the mechanism involved in the production of IFN by SV5.

HPIV-3 and -2 Recent studies from our laboratory have demonstrated rapid activation (30 min post-infection) of NF-κB following HPIV-3 infection of human lung epithelial A549 cells prior to the initiation of its replicative cycle [25]. Surprisingly, we found that inactivation of HPIV-3-induced NF-κB activation by either pyrollidine dithiocarbamate (PDTC, a general NF-κB inhibitor), expression of dominant-negative IκB super-repressor or infection of IKKγ null cells led to a dramatic increase in virus replication and cytopathogenicity independent of IFN-I production. Moreover, expression of dominant-negative MyD88, an adaptor protein for TLRs, also stimulated HPIV-3 replication, suggesting that activation of NF-κB probably occurs via the TLR/MyD88/IKKγ/IκB signaling cascade [25]. It is not known which of the HPIV-3 antigen(s) are involved in NF-κB activation via specific TLR(s) (TLR-1 to -10). Further studies demonstrated a direct antiviral role of NF-κB during innate host defense against viruses by establishing an intracellular antiviral state independent of the paracrine action of the IFN-induced JAK–STAT antiviral pathway. HPIV-3 is known to produce IFN-I following infection *in vitro* or *in vivo* [72–74]. The mechanism underlying the production of these antiviral cytokines is still not clear. The activation of IRF-3 by parainfluenza viruses has yet to be reported. Thus, it seems that both NF-κB and IFN-I play a central role in innate antiviral response; however, the mechanisms of attaining the antiviral state are distinct in both cases.

Although both HPIV-3 and -2 are capable of producing IFN-I, similar to the other members of the paramyxovirus family (SV and SV5), both have evolved to counteract the antiviral action of IFN [61] (Tabs. 9.1 and 9.2). However, in contrast to SV, HPIV-2 only encodes V, but not C, proteins, and HPIV-3 encodes only the longer form of C protein and a novel protein D (encoded when the P ORF is fused to the D ORF by transcriptional editing), found only in bovine and human PIV-3 (Tab. 9.1). Interestingly, although HPIV-3 V protein has not been detected biochemically, it may be synthesized, since disruption of a putative V ORF affects HPIV-3 replication [75]. HPIV-3 infection appeared to block IFN signaling by reducing the levels of Ser727 phosphorylation of STAT-1α, a downstream mediator of IFN signaling, whereas there was no effect on the Tyr701 phosphorylation of STAT-1 – a hallmark of IFN receptor activation [61]. In contrast, HPIV-2 did not alter STAT-1 serine and tyrosine phosphorylation; it specifically degraded STAT-2 and the presence of STAT-1 in the host cells was obligatory for degradation of STAT-2 [61, 69, 76]. The inhibition of IFN signaling by HPIV-2 was also shown to be mediated by the HPIV-2 V protein [68, 77], which physically interacts with the STAT proteins *in vitro* and *in vivo* [67], and such interaction leads to STAT-2 degradation by a proteosome-dependent pathway. The importance of HPIV-2 V protein in blocking IFN-I signaling and pathogenesis of HPIV-2 were also confirmed by utilizing recombinant HPIV-2 lacking V protein [78]. However, in contrast to human cells, STAT-2 was detected in various animal cells expressing HPIV-2 V protein [79]. Thus, STAT-2 degradation by HPIV-2 V protein is species specific and, thus, such a mechanism may determine tissue/cellular tropism of HPIV-2.

Recently, the role of HPIV-3 C protein in IFN-I-antagonizing function was determined by our laboratory [80]. Studies with HPIV-3 C protein overexpressing stable cells revealed that C protein inhibits both IFN-α- and -γ-mediated antiviral responses. The molecular mechanism of IFN-antagonizing activity of C protein constitutes inhibition in the phosphorylation of STAT-1 [80]. In addition to antagonizing IFN activity, C protein is also critical for HPIV-3 replication/transcription and pathogenesis. Recombinant HPIV-3 lacking C protein was attenuated significantly *in vitro* and *in vivo* (rodents and primates) [75]. In contrast, the growth of recombinant HPIV-3 lacking V and D proteins either individually or in groups was not altered both *in vivo* and *in vitro*, but double-mutant (lacking both V and D) viruses were highly attenuated *in vivo* [75]. Interestingly, we recently identified HPIV-3 C as a viral transcriptional suppressor, since it inhibited HPIV-3 transcription [81]. Moreover, the association of HPIV-3 C protein with the viral polymerase L protein was also observed recently [82] (Banerjee, unpublished observation). Further studies to elucidate the functional significance of these observations during the viral life cycle will be an interesting area of research.

MUV Compared to the other viruses in the paramyxovirus family, studies addressing the innate immune response and its counteraction by virus accessory proteins are limited with MUV. MUV, similar to other members of the paramyxovirus family, activates NF-κB in glial cells, but not in neuronal cells, via IκB degradation [83]. Apart from this report, the precise mechanism involved in NF-κB activation by MUV is still unclear.

The activation of IRF-3 by MUV has not been reported yet. However, several studies have reported that MUV is capable of suppressing the IFN-I signaling pathway [84, 85]. Upon IFN-I treatment, persistently MUV-infected cells failed to activate IFN-I-induced signaling components including STAT-1α and -2, and p48, resulting in poor induction of IFN-I inducible genes (PKR, MxA and 2′–5′-OAS)

[84]. Further studies confirmed that the V protein of MUV is responsible for antagonizing IFN-I signaling [86] (Tabs. 9.1 and 9.2). Expression of the wild-type or C-terminal cysteine-rich domain of MUV V protein alone in cells in *trans* suppressed IFN-I-mediated induction of STAT-1α and -2, and p48 [86]. It was further suggested that at least the suppression of STAT-1α by MUV following IFN-I treatment could be due to the post-transcriptional inhibition of STAT-1α protein [85]. Moreover, the V protein of MUV was demonstrated to physically associate with STAT proteins [87]. Recently, a study has revealed a possible mechanism that may be utilized by MUV V protein to antagonize IFN action [88]. This study has demonstrated that MUV V protein physically associates with IRAK-1, a cellular adaptor protein that links STAT-1 with the activated IFN receptor (bound to IFN). However, IRAK-1 possess a higher affinity for MUV V protein compared to STAT-1 and as a result, in MUV-infected cells, IRAK-1 is sequestered by MUV V protein. Such sequestration leads to disruption of the STAT-1–IRAK-1–IFN receptor complex formation that is required for initiation of the JAK–STAT signaling pathway. Another mechanism involving STAT-1 inactivation by V protein involves degradation of the STAT-1 molecule by proteosomal pathway [89, 90]. It was demonstrated that V protein directly associates with STAT-1 and the cysteine-rich domain of V is essential for STAT-1 degradation [89]. Yet another study reported that MUV V protein also blocks tyrosine phosphorylation of STAT-1 and -2 proteins [91]. Thus MUV V protein has evolved to inactivate JAK–STAT signaling by deregulating STAT-1 expression/function by various mechanisms. These studies, in accordance with the other paramyxoviruses, have thus established the IFN-I antagonist role of one of the accessory proteins, V protein of MUV, in blocking IFN-I signaling (Tabs. 9.1 and 9.2).

NDV NDV, an avian virus, induces NF-κB by a mechanism(s) yet to be elucidated [92]. Interestingly, NDV induced IFN-I production following infection by a mechanism dependent on the intracellular expression of the envelope protein HN, but not F [93, 94]. The production of IFN-I was dependent on IRF-3 activation by NDV and its translocation to the nucleus from the cytoplasm [95]. Recently, it was demonstrated that NF-κB- and IRF-3-mediated IFN-I production by NDV was dependent on the interaction of a dsRNA-binding protein, PACT (protein activator for PKR) [29], suggesting that induction of NF-κB and IRF-3 by NDV requires virus replication to generate dsRNA and activation of PKR.

It was demonstrated that NDV is capable of antagonizing the IFN-I-mediated antiviral signaling pathway [96]. NDV encodes three accessory proteins, V, I and W [97] (Tab. 9.1), and a recent study [96] demonstrated that the NDV V protein posses anti-IFN-I antagonist function. Although the mechanism by which V protein exerts its antagonist function is not clear, it is speculated that it could be due to degradation of STAT-1 [98]. Moreover, it was shown that recombinant NDV lacking V protein was severely restricted in replication compared to the wild-type virus [99], suggesting that the V protein is essential for virus replication. These studies have suggested that V protein may suppress the antiviral action of IFN-I produced by NDV-infected cells, thus serving as an important virulence factor.

MV MV is known to induce NF-κB very rapidly in lung epithelial cells, similar to HPIV-3 [100]. It was shown that in human monocytes, wild-type MV activated TLR-2 which in turn induced NF-κB [12]. The activation of TLR-2 is a property of MV envelope H (hemagglutinin) protein, since expression of H alone in *trans* activated TLR-2. It is important to note that although both wild-type and vaccine strains activate NF-κB, only the wild-type is capable of activating TLR [12]. Future studies are needed to unravel the mechanism(s) involved in NF-κB activation by the vaccine strain and the molecular basis for the difference in strain-specific activation of TLR.

Similar to the difference observed in the mode of NF-κB activation by the wild-type versus the vaccine MV strain, the wild-type MV produced much less IFN-I compared to the vaccine strain [101]. The production of IFN-I by MV, like SV, was due to the activation of IRF-3 [102]. The nucleocapsid (N) protein of MV was shown to physically interact with IRF-3 and expression of N protein alone in the absence of additional MV proteins activated IRF-3 [28]. These results along with the NF-κB activation studies by MV demonstrate the ability of host cells to exploit MV proteins (N or H) to initiate IRF-3 and NF-κB-mediated induction of innate immune response.

The major IFN antagonizing MV accessory (C and V proteins) protein is V protein [103, 104] (Tab. 9.1). Although MV encodes V and C proteins, V protein was shown to inhibit STAT-1 and -2 phosphorylation [105]. In addition, direct binding of V protein with STAT molecules results in a defect in nuclear translocation of STAT-1 and -2 [106]. In contrast to V protein, C protein lacked its ability to block IFN signaling [105]. However, a study has indicated that C protein inhibits IFN-α/β promoter activity [107]. Whether, C protein exerts its antagonism function by preventing IFN-β expression is yet to be confirmed. V and C proteins also play important roles as a virulence factors and in *in vivo* pathogenesis of MV. Recombinant MV, lacking either C or V proteins, had similar growth kinetics as the wild-type virus in cultured cells [104]. However, infection of mice harboring human thymus/liver implants with MV lacking either V or C proteins showed not only lower titers compared to the wild-type [108, 109], but the systemic spread of MV was also restricted [110, 111]. As a result, the mice infected with MV lacking V or C had significantly milder clinical symptoms and lower mortality compared to the wild-type MV [110]. These studies have established an important role for V and C as virulence factors during MV pathogenesis.

RSV and bRSV Similar to SV, MV and HPIV-3, RSV induces NF-κB in human lung epithelial cells – the cells that are the natural host for RSV during infection [112–115]. In these cells, unlike HPIV-3, RSV induced NF-κB late in infection and viral replication is a prerequisite for NF-κB induction [112, 114]. The induction of NF-κB by RSV involved phosphorylation and degradation of the NF-κB inhibitory subunit, IκB [116, 117]. Interestingly, RSV seems to utilize a novel degradation pathway independent of 26S proteosome [118]. Several studies have reported that NF-κB induction by RSV may also require cellular protein kinase C (PKC) [13] and activation of phosphatidylinositol-3-kinase (PI3K) pathway [119]. Apart from these downstream mediators, recently, RSV and specifically RSV envelope protein F (fusion protein) was shown to induce NF-κB via TLR-4 along with its binding partners MyD88 and CD14 [10, 11]. In vivo, infection of TLR-4 null mice resulted in an impaired adaptive immune response against RSV resulting in delayed virus clearance, suggesting that RSV F protein interacts with the membrane-bound TLR-4 to activate NF-κB [10, 11]. However, TLR-4 was shown to be utilized by RSV in immune cells including macrophages. Whether TLR-4 is also utilized by RSV in human lung epithelial cells, the target cells for initial infection, has not yet been studied. These studies along with the dissection of the molecular mediator(s) and mechanism(s) involved during NF-κB activation by RSV will shed important insights on this important innate immune response pathway.

As discussed above, HPIV-3 rapidly induced NF-κB upon infection [25], which is in contrast to the replication-dependent activation of NF-κB by RSV [114]. Moreover, while inhibition of NF-κB induction by HPIV-3 resulted in a profound increase in virus replication (see above), similar treatment had no effect on RSV infection. These results suggested that not only NF-κB activation leads to the establishment of an antiviral state in infected cells (for HPIV-3), but the viruses (HPIV-3 and RSV) may have evolved to modulate the antiviral activity of NF-κB for replicative advantage by manipulating the time frame of NF-κB induction following infection. Based on this scenario, pretreatment of human lung epithelial cells with potent NF-κB-inducing cytokines, TNF-α and IL-1β, resulted in inhibition of RSV replication in a NF-κB-dependent manner [25]. Moreover, similar to HPIV-3 (see above), the inhibition of RSV replication following NF-κB activation was due to the establishment of an antiviral state intracellularly which was independent of the paracrine action of IFN-I. These studies have established, in addition to the IFN-induced JAK–STAT antiviral pathway, a TNF-α- IL-1β- and/or HPIV-3-induced NF-κB pathway as a novel antiviral pathway operative in infected cells as part of the innate immune response [25]. Thus, it seems that activation of NF-κB, in addition to producing IFN-I as part of the antiviral response, directly establishes an intracellular antiviral state in infected cells as observed with HPIV-3 and in uninfected cells following paracrine action of TNF-α/IL-1β produced from RSV-infected cells [25, 120, 121].

In contrast to TNF-α and IL-1β, in human lung epithelial cells, RSV is resistant to the antiviral effect of IFN-I, since pretreatment of cells with IFN-I, unlike HPIV-3 or VSV, failed to inhibit RSV replication significantly [122], suggesting that RSV infection may counteract the IFN-I-mediated activation of JAK–STAT antiviral pathway by utilizing mechanisms similar to SV, SV5, and HPIV-3 and -2. Recently, the two RSV nonstructural proteins, NS1 and NS2, that mediate antagonism against IFN-I were characterized (Tabs. 9.1 and 9.2). These proteins are encoded by two individual ORFs and are the major viral proteins in infected cells as a result of their genes being positioned at the extreme 3′ end of the RSV genome [123, 124]. The IFN-I antagonist functions of NS1 and NS2 were first demonstrated by utilizing recombinant bRSV lacking these genes, and genetically modified rabies virus expressing the bRSV NS1 and NS2 proteins [125]. The recombinant bRSVs harboring either single (NS1 or NS2) or double (NS1 and NS2) deletions were severely restricted in IFN nonproducing Vero cells in the presence of IFN-I compared to the wild-type bRSV [125]. Likewise, IFN-sensitive rabies virus expressing bRSV NS1 and/or NS2 was highly resistant to the pretreatment of cells with IFN-I compared to the wild-type virus [125]. The antagonism function of bRSV NS1 and NS2 proteins was shown to be as a consequence of impaired phosphorylation/activation of IRF-3, which leads to suppression in IFN-β production [126]. Similar IFN-I-antagonizing functions of human RSV NS1 and NS2 proteins were reported based on the protection of a heterologous IFN-I-sensitive rhabdovirus expressing human RSV NS1 and NS2 proteins from the antiviral action of IFN-I [127]. Similar to bRSV NS proteins, human RSV NS1 and NS2 proteins also suppressed activation and nuclear translocation of IRF-3 [128]. The *in vivo* relevance of NS proteins in counteracting IFN defense mechanism was borne out by a recent study showing that suppressing NS1 protein levels in RSV-infected mice by small interfering RNA (siRNA) technology results in a significant decrease in viral titer and inflammation [129]. Apart from the ability of NS proteins to block IFN-β induction, infection of respiratory epithelial cells with human RSV results in proteosome-mediated degradation of STAT-2 protein [130]. The RSV antigen and mechanism responsible for inhibiting JAK–STAT signaling cascade has not been identified yet.

Nipah and Hendra Viruses Nipah virus encodes two accessory proteins, V and W, to counteract IFN action. The V protein of Nipah virus possesses a conserved cysteine-rich domain that is preserved among other paramyxoviruses. The molecular mechanism by which V protein inhibits IFN signaling constitutes binding with STAT-1 and -2 proteins [131]. The sequestration of STAT proteins in a tripartite complex (STAT-1 and -2, and V) results in inhibition of nuclear translocation of STAT proteins [131] and efficient phosphorylation [132]. In contrast to cytoplasmic localization of V protein, Nipah virus W protein is present in the nucleus. It was demonstrated that W protein sequesters STAT-1 in the nucleus to prevent transcription of IFN-dependent genes [133]. Interestingly, the P protein of Nipah virus was also identified as an IFN antagonist, which along with the V protein retains STAT-1 in the cytoplasm, thus preventing its nuclear targeting [133]. Apart from blocking JAK–STAT activation, V and W proteins also prevent activation of the IFN-β promoter [134]. Although the mechanism by which V protein exerts its inhibitory activity on IFN-β production is not clear, W protein expression results in impaired phosphorylation of IRF-3 [135]. Similar to Nipah virus V protein, Hendra virus V protein also sequesters STAT-1 and -2 by forming a high-molecular-weight complex in the cytoplasm [136], which results in lack of nuclear transport of STATs and activation of STAT dependent genes.

9.3.1.2 Filovirus

Ebola Virus Ebola virus encodes VP35 protein that prevents activation of IRF-3 and subsequent production of IFN-β [137]. Recently, it was demonstrated that the C-terminal basic amino acid motif of VP35 is essential for inhibiting phosphorylation of IRF-3 [138].

9.3.1.3 Rhabdovirus

Rabies Virus Rabies virus has evolved a nonconventional way of counteracting IFN response. In contrast to paramyxoviruses, which encode several IFN antagonist accessory proteins from the P gene, rabies virus utilizes full-length P protein (a subunit of viral transcriptase complex) to antagonize IFN signaling [139]. Specifically, P protein of rabies virus prevents IFN-β production from infected cells, by preventing phosphorylation/activation of IRF-3 [139]. The P protein inhibits the IRF-3 upstream kinase pathway to prevent IRF-3 phosphorylation in infected cells. Thus, P protein of rabies virus not only serves as an important factor for virus transcription/replication [43], but it also posses IFN-antagonizing function.

9.3.2
Segmented Negative-strand RNA Viruses

Segmented negative-strand RNA viruses are single-stranded enveloped viruses that replicate in the nucleus of infected cells. These classes of viruses are divided into three families: orthomyxoviruses [140], bunyaviruses [141] and arenaviruses. Influenza viruses belonging to the orthomyxo family have caused wide-spread mortality among humans. Influenza virus A isolated from avian and mammalian species are a major cause of human pandemics that result in high mortality. Recently, in 1997, highly pathogenic influenza virus A was transmitted from chicken to humans in Hong Kong, resulting in many casualties [140]. The threat of another influenza virus pandemic lingers due to the ability of these viruses to re-assort envelope genes from various strains by antigenic drift, and lack of effective antiviral therapy and vaccines to counteract these viruses. Viruses belonging to the bunyavirus family are also highly pathogenic human viruses that cause hemorrhagic fevers [141]. The segmented negative-strand RNA viruses, similar to their nonsegmented counterparts, have developed an efficient mechanism to evade the antiviral response of IFN. These viruses, although lacking the wide assortments of IFN antagonist proteins expressed by nonsegmented RNA viruses, efficiently utilize their nonstructural proteins to block IFN production (Tab. 9.2) as detailed below.

9.3.2.1 Orthomyxovirus

Influenza Virus Influenza viruses are highly sensitive to the antiviral effect of exogenously added IFN [142]. Interestingly, similar to our recent studies, TNF-α also acted as a potent antiviral cytokine against influenza A virus infection [143]. However, influenza viruses have evolved a strategy to evade the IFN response by preventing IFN-α/β production from infected cells. Influenza A virus expresses its only nonstructural protein NS1 to block the IFN-α/β response. NS1 protein possessing dsRNA-binding ability was shown to inhibit synthesis/production of IFN-α/β [144–147]. By using NS1-deleted influenza A viruses and following expression of NS1 protein in *trans*, it was demonstrated that this protein acts as an IFN antagonist by inhibiting induction of NF-κB [146], and IRF-7 [148] and -3 [147]. Apart from preventing IFN production, the NS1 protein also acts as a PKR inhibitor [149] by sequestering dsRNA, the molecule essential for PKR activation. Interestingly, the dsRNA-binding domain of NS1 was also critical for inhibiting IFN synthesis/IRF-3 activation [150]. The importance of NS1 during virus pathogenesis and evasion of innate immunity was apparent from studies showing that highly virulent H5N1 influenza A virus strains are resistant to the antiviral effect of IFN and TNF-α due to a point mutation in the NS1 gene [151, 152]. Although the precise mechanism by which NS1 inhibits antiviral function of IFN is not clear, the cumulative studies have concluded that NS1 protein plays a critical role in evading the IFN-mediated innate immune antiviral response by (i) blocking IFN production, (ii) inhibiting the antiviral function of IFN and (ii) inhibiting PKR activation. It is important to mention that although inhibition of IRF-3 activation by NS1 in virus-infected cells was documented by several studies, numerous reports have shown that IRF-3 is activated in influenza A virus-infected cells [153, 154]. Currently the difference in IRF-3 activation status reported by various groups is not clear and, therefore, further studies are required to explain the contrasting results. Apart from NS1 protein, influenza A viruses also activate a host protein to block IFN response. A cellular protein, p58IPK, that is activated by the virus directly binds to PKR to inhibit its activation [155, 156]. In contrast to the NS1 protein of influenza A virus, the NS1 protein counterpart of influenza B virus was shown to interact with and inhibit the conjugation of an IFN-stimulated gene, ISG15 [157]. Such mechanism may have a role in the induction of IFN-mediated antiviral response.

9.3.2.2 Bunyaviruses

Rift Valley Fever Virus Rift Valley fever virus expresses a nonstructural protein, NSs, which prevents synthesis of IFN-α/β [158]. The exact mechanism by which Rift Valley fever virus achieves inhibition in IFN-α/β synthesis was recently reported. It was observed that while NSs protein does not inhibit IRF-3 activation/nuclear translocation and NF-κB induction, it prevented transcription from a constitutive promoter [159]. These studies suggested that NSs protein does not specifically inhibit IFN-β-specific transcription factors, but it blocks IFN gene expression by blocking transcription.

Bunyamwera Virus Bunyamwera virus antagonizes the IFN response by an unconventional mechanism. Bunyamwera virus nonstructural protein NSs prevents expression of the IFN-β gene [160] by inhibiting phosphorylation of RNA polymerase II [161]. The general block in transcriptional activity due to deactivation of RNA polymerase II leads to prevention in IFN-β synthesis in infected cells [161].

9.3.3
Positive-sense ssRNA Viruses

This class of viruses constitutes several highly pathogenic viruses, including polio, foot and mouth disease virus (FMDV), West Nile virus (WNV), dengue viruse and HCV. The replication/transcription of the genome of these viruses occurs in the cytoplasm and therefore, several strategies are employed to overcome an IFN-dependent antiviral response (Tab. 9.3). In this section we will discuss several viruses belonging to three families of positive sense RNA viruses: picornaviruses, flaviviruses and coronaviruses.

9.3.3.1 Picornaviruses

Poliovirus Poliovirus belonging to the family of picornaviruses utilizes its own replication strategy to inhibit IFN secretion. Poliovirus infection is associated with a shut-off of the host translational machinery and a resulting block in the cellular secretory pathway, the latter being mediated by the viral protein 3A. As a consequence of these events, IFN-α/β is not secreted from poliovirus-infected cells [162]. In addition, a recent study revealed a unique mechanism by which poliovirus inhibits NF-κB activation [163]. In infected cells, poliovirus-encoded protease 3C specifically cleaves the p65/RelA subunit of NF-κB to render NF-κB complex formation inactive.

FMDV Similar to poliovirus, FMDV encodes the L protease, which cleaves cellular proteins required for cellular mRNA translation. As a result of cellular translational inhibition, IFN-α/β is not synthesized in FMDV-infected cells [164].

9.3.3.2 Flaviviruses

HCV HCV, a flavivirus, is the leading cause of chronic liver disease [165]. Although currently IFN-α treatment serves as a therapy for HCV infection, the virus

Tab. 9.3. Strategies of positive-sense, dsRNA viruses, and RNA and DNA reverse-transcribing viruses to inhibit IFN-I signaling

Virus	IFN blocker	Mechanism
Positive sense single-stranded RNA viruses		
Poliovirus	3C protease	IFN synthesis (NF-κB inhibition)
	3A protease[a]	IFN synthesis/secretion
FMDV	L protein[a]	IFN synthesis/secretion
HCV	NS3 and NS3/4A proteins	IFN synthesis (IRF-3 phosphorylation inhibition)
	NS5A and E2 proteins	IFN signaling (PKR inhibition)
	NS3/4A and core/E2/E1 proteins	IFN signaling (STAT-1 degradation)
WNV	nonstructural proteins	IFN signaling (STAT-1 and -2, JAK-1, and TYK-2 phosphorylation inhibition)
	unknown	IFN synthesis (delayed IRF-3 phosphorylation)
SARS-CoV	unknown	IFN synthesis (IRF-3 inhibition)
Double-stranded RNA virus		
Reovirus	σ3 protein	IFN signaling (PKR inhibition)
RNA and DNA reverse transcribing viruses		
HIV-1	TAT protein and TAR RNA	IFN signaling (PKR inhibition)
	unknown	IFN signaling (2′–5′-OAS/RNase L inhibition)
HBV	core antigen and terminal protein	IFN synthesis
	unknown	IFN signaling (Mx inhibition)

[a] Poliovirus and FMDV utilizes their natural replication strategy involving shutting down of host mRNA translation to inhibit IFN synthesis and secretion.

has developed a mechanism to antagonize IFN signaling and inhibit activation of IFN-induced antiviral protein PKR. HCV is known to activate NF-κB in infected cells via the virus-encoded nonstructural proteins NS5A and NS3 and core protein [166, 167]. Although HCV induces NF-κB, the virus has evolved to inhibit activation of IRF-3, which is required for optimal IFN-α/β synthesis. NS3 protein inactivates IRF-3 by virtue of binding to TBK-1, the kinase essential for phosphorylation of IRF-3 [168]. Association of NS3 with TBK-1 prevents binding of TBK-1 with IRF-3 and subsequent activation. In addition to NS3, HCV serine protease NS3/4A also plays an important role in deregulating IFN-β production by blocking IRF-3 phosphorylation [169, 170]. Recent studies [170] have shown that the mode of inhibitory activity of NS3/4A constitutes proteolytic degradation of Toll–IL-1 receptor domain-containing adaptor inducing IFN-β (TRIF) protein – an adaptor protein essential for linking TLR-3 to kinases responsible for IRF-3 phosphorylation/activation. Thus, HCV has designed a mechanism to inhibit IRF-3 directly (via NS3) and indirectly by blocking upstream pathways (via NS3/4A) required for activation of IRF-3 kinases.

In addition to counteracting IFN-β production, resistance of HCV against IFN-α is due to its ability to inhibit induction of the JAK–STAT pathway. Studies with cells expressing the HCV genome have revealed that virus-encoded proteins inhibit the JAK–STAT pathway downstream of STAT tyrosine phosphorylation [171]. Apart from blocking the IFN-induced JAK–STAT pathway, two viral proteins also directly manipulate PKR protein to counteract its antiviral activity. HCV nonstructural protein NS5A interacts with PKR to inhibit its activity by preventing dimerization of PKR [172]. In addition, E2 glycoprotein of HCV competes with eIF-2α for binding to PKR and thus prevents PKR from inhibiting translation of viral proteins [173]. Further studies to establish the involvement of NS5A protein during IFN antagonism function are required, since recent studies with the HCV replicon system have shown that NS5A protein is not essential for the antiviral activity of IFN-α against HCV [174]. Apart from inhibiting IFN-induced antiviral proteins, NS3/4A and core/E2/E1 proteins of HCV directly inhibit IFN signaling by reducing the expression of STAT-1 protein following its degradation by the proteosome-dependent pathway [175]. At this time the exact mechanism involved in STAT-1 degradation is not clear; however, HCV core protein was shown to bind to STAT-1. Whether direct binding of HCV proteins with STAT-1 results in its degradation is yet to be examined.

WNV Another important member of the flavivirus family is WNV – a highly pathogenic human virus causing deadly neurological diseases upon infection [176]. Although WNV induces NF-κB in infected cells [177], WNV have also evolved to counteract IFN-mediated antiviral action by delaying the activation of IRF-3 and producing JAK–STAT pathway antagonist proteins in infected cells. A recent report [178] has shown that in WNV-infected cells IRF-3 activation is delayed, which may serve as a mechanism to postpone expression of IFN-dependent antiviral proteins. Moreover, two strains (New York strain 99 and Kunjin strain) of WNV block IFN-α/β signaling by inhibiting STAT-1 and STAT-2 phosphorylation and nuclear translocation [179]. The lack of STAT phosphorylation could be as a result of impaired phosphorylation and activation of Janus kinases JAK-1 and TYK-2 in WNV-infected cells, as suggested recently [180]. The WNV proteins responsible for inhibiting JAK–STAT activation constitute WNV-encoded nonstructural proteins, NS2A, NS2B, NS3, NS4A and NS4B [179].

9.3.3.3 Coronavirus

Severe Acute Respiratory Syndrome (SARS) Coronavirus (SARS-CoV) Recently, SARS-CoV was identified as the etiological agent that causes severe respiratory

9.3 Evasion of IFN-mediated Antiviral Response | 247

disease in humans [181]. Although SARS-CoV is highly sensitive to exogenously added IFN [182], the virus has developed a strategy to escape the IFN response by preventing secretion/synthesis of IFN-β from infected cells. It was shown that IFN-β mRNA levels and IFN-β promoter activity were undetectable in SARS-CoV-infected cells [183]. Although the precise mechanism employed by the virus to suppress IFN-β synthesis is not clear, it seems that IRF-3 activation is compromised in virus-infected cells [183]. Further studies are required to dissect the precise mechanism involved in suppression of IFN-β transactivation and the viral factors responsible for such action.

9.3.4
dsRNA Viruses

Reovirus   Reovirus is an important virus responsible for the development of viral myocarditis [184]. It has been shown that the cardiac protective response following reovirus infection is determined by the ability of reovirus to induce IFN-β and IFN-β-dependent gene expression [185]. However, such a response was cell-type dependent. The cell-type-dependent mechanism of the IFN-β induction profile by reovirus is not clear, although activation of NF-κB [186] and IRF-3 [187] was observed in reovirus-infected cells. So far it has not been documented if reovirus completely shuts off IFN-β production and/or inhibits IFN-mediated antiviral signaling. However, one study has shown that PKR function is impaired in reovirus-infected cells. It was suggested that reovirus-encoded Í3 protein binds to dsRNA and such binding abolishes the ability of dsRNA to serve as the substrate for PKR activation [188] (Tab. 9.3).

9.3.5
RNA and DNA Reverse-transcribing Viruses

9.3.5.1   Retrovirus

HIV-1   HIV-1 is an important human retrovirus responsible for the world-wide AIDS epidemic [189]. HIV-1 is highly sensitive to IFN treatment, since activation of the JAK–STAT pathway restricts virus replication [190]. Moreover, HIV-1 induces NF-κB, and IRF-1 and -2 for utilization of these proteins as transcriptional activators of the viral genome [191, 192]. However, HIV-1 has designed a strategy to counteract the IFN-mediated defense function by inhibiting the function of two IFN-induced antiviral proteins, i.e. PKR and 2′–5′-OAS (2′–5′-OAS/RNase L pathway) (Tab. 9.3). TAT protein encoded by HIV-1 degrades PKR by an unknown mechanism [193]. Moreover, TAT protein also acts as a decoy for eIF-2α, the natural substrate of PKR, and thus inhibits PKR function [194]. Similarly, HIV-1 infection induces an RNase L inhibitor protein that downregulates RNase L activity [195]. The precise mechanism by which this protein operates to inactivate RNase L is yet to be resolved. Apart from proteins, HIV-1 TAR RNA, which has partial double-stranded structure, binds to PKR to inhibit its activation [196].

9.3.5.2 Hepadnavirus

HBV HBV, belonging to the hepdanavirus, is highly sensitive to IFN treatment [197]. Although IFN exerts its potent antiviral activity on HBV, the virus has developed a strategy to evade such a response by inhibiting IFN-β synthesis in infected cells (Tab. 9.3). The inhibition in IFN synthesis is rendered by the HBV ORF-C product (core antigen) and the viral terminal protein [198, 199]. The precise mechanism by which these factors operate to block IFN production is not clear. Interestingly, HBV also counteracts IFN-induced antiviral protein activity by downregulating MxA expression [200]. It was demonstrated that the HBV-encoded precore/core proteins directly bind to the MxA promoter to modulate its expression.

9.3.6
dsDNA Viruses

9.3.6.1 Adenovirus

Adenoviruses [201] have evolved several mechanisms to evade the IFN-α/β antiviral system (Tab. 9.4). These viruses encode proteins that inhibit activation of IRF-3 and NF-κB in infected cells, resulting in inefficient expression of the IFN-β gene. Adenovirus E3 protein prevents induction of NF-κB by preventing its nuclear translocation and activation of IKK – the kinase responsible for phosphorylating IκB [202]. Similarly, IRF-3-mediated IFN-β gene expression is blocked in infected cells by a novel mechanism. CBP/p300 coactivator binds to phosphorylated IRF-3 in the nucleus to activate IRF-3-stimulated gene expression [26, 27]. However, adenovirus E1A protein competes with IRF-3 for binding to CBP/p300 and, thus, inhibits transcription of IFN-β gene [203]. The important role of E1A protein in antagonizing IFN function is apparent from the ability of E1A protein to directly counteract IFN signaling [204]. E1A acts as a potent inhibitor of the JAK–STAT pathway by preventing ISGF-3 complex formation by virtue of downregulating the expression of STAT-1α and IRF-9 (p48) proteins [205]. In addition, E1A also interacts directly with STAT-1 to inhibit its activity [206]. A second viral factor that counteracts the antiviral activity of IFN constitutes virus-encoded VAI viral RNA which inhibits PKR activation [207]. VAI RNA does not encode any viral protein and is transcribed by the cellular RNA polymerase III. This RNA in infected cells acts like dsRNA by directly binding to PKR. However, in contrast to dsRNA, PKR bound to VAI RNA fails to activate itself, resulting in its inhibition.

9.3.6.2 Poxvirus

Poxviruses [208] are DNA viruses that replicate in the cytoplasm of infected cells. Vaccinia virus, a vaccine strain of poxvirus utilized for the eradication of smallpox,

Tab. 9.4. Strategies of dsDNA viruses to inhibit IFN-I signaling

Virus	IFN blocker	Mechanism
Adenovirus	E3 protein	IFN synthesis (NF-κB inhibition)
	E1A protein	IFN synthesis (IRF-3 inhibition), IFN signaling (down-regulation of IRF-9 and STAT-1α)
	VAI RNAs	IFN signaling (PKR inhibition)
Vaccinia	B18R protein	IFN signaling (IFN receptor decoy)
	K1L protein	IFN synthesis (NF-κB inhibition)
	E3L protein	IFN signaling (PKR and 2'–5'-OAS/RNase L inhibition), IFN synthesis (IRF-3 and -7 inhibition)
	K3L protein	IFN signaling (PKR inhibition)
	A46R, A52R and N1L proteins	IFN synthesis (NF-κB and IRF-3 inhibition; TLR inhibitor)
HSV	ICP0 protein	IFN synthesis (IRF-3 inhibition)
	ICP34.5 and US11 proteins	IFN signaling (PKR inhibition)
	unknown	IFN signaling (2'–5'-OAS/RNase L inhibition)
CMV	IE86 protein	IFN synthesis
	ppUL83 protein	IFN synthesis (IRF-3 inhibition)
	unknown	IFN signaling (downregulation of IRF-9, JAK-1)
HHV-8/KSHV	vIRF 1–4 proteins	IFN synthesis (IRF-1/3/7 inhibition)
	vIRF-2 protein	IFN signaling (PKR inhibition)
EBV	EBNA-2 protein	IFN signaling
	EBER RNAs	IFN signaling (PKR inhibition)
HPV	E6 and E7 proteins	IFN synthesis (NF-κB inhibition)
	E6 protein	IFN synthesis (IRF-3 inhibition), IFN signaling (STAT-1 and -2, and and TYK-2 phosphorylation inhibition)
	E7 protein	IFN signaling (IRF-9/p48 nuclear translocation inhibition)

serves as a model virus to study poxvirus biology. Vaccina virus has evolved to evade IFN response by several mechanisms (Tab. 9.4). The virus encodes K1L gene product that prevents activation of NF-κB by virtue of inhibiting IκBα degradation [209]. In addition to blocking IFN-β synthesis (by blocking NF-κB activation), the virus has designed a "clever" way to prevent initiation of IFN signaling. Vaccinia virus encodes a decoy protein B18R to evade IFN signaling. B18R is a soluble secreted homolog of the IFN receptor [210, 211]. Secretion of B18R from virus-infected cells results in the extracellular sequestration of IFN-α/β, since B18R is a soluble form of the IFN receptor. Such competition with the cell surface

IFN receptor for IFN binding results in failure of IFN to associate with the "physiological receptor" on the plasma membrane. As a consequence, IFN is unable to induce the JAK–STAT antiviral signaling cascade. Production of extracellular decoy molecules is indeed a "smart" approach, since secreted B18R will not only prevent IFN signaling in infected cells, but such an effect will also be extended to noninfected neighboring cells.

Apart from B18R, vaccinia viruses also encode two proteins, E3L and K3L, that antagonize the function/activation of IFN-induced antiviral proteins. E3L, a dsRNA-binding protein, is involved in inhibiting the activation of PKR and 2'-5'-OAS [212, 213]. Since dsRNA is an important cofactor for PKR and 2'-5'-OAS activation, sequestration of dsRNA by E3L results in failure of activation of PKR and 2'-5'-OAS in infected cells. Alternative mechanisms employed by E3L to inhibit IFN signaling include: (i) inactivation of PKR following direct interaction of E3L with PKR [214] and (ii) inhibition of IRF-3/7 pathways to suppress IFN-α/β synthesis [148]. In contrast to E3L, the IFN antagonistic activity of K3L is only directed against PKR [215]. K3L serves as a substrate analog for PKR and thus it competes with eIF-2α for PKR binding. As a result of the competition from K3L, PKR is unable to phosphorylate eIF-2α for inhibiting translation. In addition, vaccinia virus encodes two proteins, A46R and A52R, that inhibit NF-κB and IRF-3 induction via TLR activation [216, 217]. Recently, another TLR antagonist was identified in vaccinia virus-infected cells. It was shown that N1L protein of vaccinia virus suppresses TLR signaling essential for NF-κB and IRF-3 activation [218].

It is interesting to note that E3L and the NS1 protein of influenza A virus share a high degree of similarity in antagonizing the IFN response. These similarities include, (i) binding to dsRNA, (ii) preventing dsRNA-mediated activation of PKR and (iii) inhibiting synthesis of IFN-α/β. The similarity in the mechanism of IFN antagonism function of vaccinia and influenza viruses also includes the ability to inactivate PKR function by these two viruses. Both viruses express proteins, p58IPK (cellular protein induced in influenza virus-infected cells) and K3L (vaccinia virus-encoded protein) proteins, which directly interact with PKR to prevent its activation. Therefore, vaccinia and influenza viruses have evolved two mechanisms to inhibit PKR activation: (i) E3L and NS1 proteins that bind dsRNA, and (ii) K3L and p58IPK proteins that directly interact with PKR. It is important to mention that apart from vaccinia virus, other poxviruses, including, myxoma, monkeypox, cowpox, swinepox and ectromelia viruses, also encode an assortment of IFN antagonist molecules. Readers are referred to two review articles elaborating the viral immune evasion molecules of different poxviruses [219, 220].

9.3.6.3 Herpesvirus

Like poxviruses, herpesviruses are DNA viruses; however, in contrast to poxviruses, they replicate in the nucleus of infected cells [221]. Herpesviruses utilize various strategies to evade the IFN response (Tab. 9.4). These include counteracting IFN signaling and inhibiting the function of IFN-induced antiviral proteins. This section covers herpes simplex virus (HSV), human cytomegalovirus (CMV), human herpesvirus 8/Kaposi's Sarcoma-associated herpesvirus (KSHV) and Epstein–Barr virus (EBV).

HSV HSV activates NF-κB in infected cells following expression of ICP0 and BICP0 proteins [222]. In addition to viral proteins, HSV genomic DNA interacts with TLR-9 [223]. Such interaction may also lead to activation of NF-κB in infected cells. In contrast to NF-κB activation, HSV prevents nuclear translocation of IRF-3. The viral encoded immediate-early protein ICP0 is responsible for inhibiting nuclear accumulation of IRF-3 [224]. In addition to preventing efficient IFN-β synthesis, HSV also encodes a protein named ICP34.5 to inhibit the PKR antiviral system [225]. However, in contrast to vaccinia virus E3L and K3L proteins, which prevent phosphorylation of eIF-2α by PKR, ICP34.5 protein recruits a cellular phosphatase to dephosphorylate eIF-2α [225]. As a result of dephosphorylation, PKR fails to inhibit translation of cellular and viral mRNAs. The importance of ICP34.5 in virus pathogenesis is borne out by the observation that recombinant HSV lacking ICP34.5 protein was attenuated [226]. However, the attenuated phenotype could be overcome by extragenic mutations in the HSV genome resulting in expression of the viral protein US11 early during infection [226]. US11 protein is expressed late during viral infection and the reversion of the attenuated phenotype in early US11-expressing viruses is due to inhibition of PKR by US11, which directly interacts with PKR to inhibit its activity [226]. Based on these interesting observations it was speculated that US11 represents a factor that was utilized by HSV to counteract the IFN-mediated defense mechanism early during evolution; however, with time, such a function was replaced by ICP34.5. Apart from PKR, HSV also inhibits the IFN-induced 2′–5′-OAS/RNase L pathway [227]. Although the exact mechanism of 2′–5′-OAS/RNase L pathway inhibition is not clear, it is suggested that 2′–5′-oligoadenylate derivatives are produced in infected cells, which bind to RNase L to block its enzymatic activity.

CMV CMV, belonging to the herpesvirus family, also counteracts IFN signaling. Although CMV activates NF-κB via the TLR-2/CD14 pathway [228], the virus has developed a strategy to suppress IFN-β production (Tab. 9.4). It was observed that CMV immediate-early-2 gene product IE86 blocks induction of IFN-β in infected cells [229]. Moreover, CMV structural protein pp65 (ppUL83) inhibits activation of IRF-3 by preventing its nuclear accumulation and incomplete phosphorylation [230]. Even though CMV prevents IFN-β production, interestingly the virus induces an IFN response early during infection by virtue of interaction of the glycoprotein gB with the cell surface [231]. Although the virus activates IFN signaling, it also at the same time has evolved to block the downstream signaling cascade required to launch an antiviral function. CMV achieves its goal of interfering with IFN signaling by downregulating the expression of JAK-1 and IRF-9 (p48) in infected cells [232, 233]. Since JAK-1 and IRF-9 are critical factors required for activation of the ISGF-3 complex, its impaired expression in infected cells inhibits IFN-

α/β signaling. Based on the dual mechanism employed by CMV during infection, it is speculated that while activation of IFN signaling is advantageous for viral survival, the virus has designed a ploy to shut-down the antiviral arm of the IFN signaling.

KSHV NF-κB activation by KSHV FLICE-inhibitory protein (vFLIP) is critical for induction of cellular transformation [234]. However, KSHV has developed a mechanism to defend itself against IFN action by encoding four different sets of IRF homologs (vIRF-1, -2, -3 and -4). It is not clear how these homologs participate in inactivation of the IFN response. However, several of these homologs may function as dominant-negative IRFs (IRF-1, -3 and -7) [235, 236]. Therefore, expression of these homologs will result in inefficient synthesis and production of IFN-α/β, since IRFs are critical regulators of IFN expression. Apart from exerting a dominant-negative function, vIRF-1 is known to interact and inhibit p300, a histone acetyltransferase, whose activity is essential for chromatin remodeling during transcription [237]. Thus, transcriptional block induced by vIRF-1 may also result in nonexpression of IFN-dependent antiviral genes. Moreover, direct inactivation of antiviral proteins is achieved by vIRF-2, which interacts with PKR to inhibit its activation [238] (Tab. 9.4).

EBV Apart from KSHV, another γ-herpesvirus, EBV, activates NF-κB in infected cells via latent membrane protein (LMP)-1 [239, 240]. However, EBV also encodes EBNA-2 protein which prevents transcriptional activation of IFN-dependent genes by an unknown mechanism [241]. In addition, EBV encodes a small RNA, EBER-1, which interacts with PKR to inhibit its activation [242] (Tab. 9.4).

9.3.6.4 Papillomavirus

Human papillomaviruses (HPVs) [243] have evolved to evade IFN-mediated innate antiviral response primarily by two mechanisms: (i) inhibiting IFN-β production from infected cells and (ii) counteracting the IFN-induced JAK–STAT signaling (Tab. 9.4). HPV restricts IFN-β production by preventing activation of NF-κB and IRF-3 in infected cells [244, 245]. Virus-encoded E6 and E7 oncoproteins are potent inhibitors of NF-κB activation [244]. E7 protein associates with IKK to impair its kinase activity, resulting in improper phosphorylation and degradation of IκB. In contrast to E7, E6 protein inhibits NF-κB-dependent transcriptional activity in the nucleus. In addition to inhibiting expression of NF-κB-stimulated genes, E6 protein also binds directly with IRF-3 [245] to inhibit its transcriptional activation function, resulting in inhibition in IFN-β synthesis. Apart from blocking IFN-β production, E6 and E7 proteins also inhibit IFN-induced JAK–STAT signaling. E7 protein directly binds to IRF-9/p48 (the DNA-binding partner of ISGF-3) to prevent its nuclear translocation [246]. In contrast to E7 protein, which targets p48/IRF-9, E6 protein expression leads to decreased tyrosine phosphorylation of STAT-1 and -2, and TYK-2 [247]. The impaired phosphorylation status of STATs is due to the ability of E6 protein to directly interact with TYK-2 to prevent its activation/phosphorylation following IFN-α/β treatment [247].

9.4 Concluding Remarks

The innate immune response against viruses represents an important host defense mechanism. The importance of this mechanism is borne out by the fact that it not only restricts virus replication during the initial stages of infection, but the innate response leads to the establishment of the adaptive immune response against viruses as well. Many RNA and DNA viruses initiate the innate immune response following activation of NF-κB and/or IRF-3 by several mechanisms, as discussed earlier in the chapter (Fig. 9.1). These two critical transcription factors are known to play an important role in establishing the innate immune response following production of antiviral cytokine IFN-I from infected cells. The IFN-I produced from infected cells by virtue of the paracrine action exerts an antiviral state in non-infected cells by inducing the JAK–STAT pathway (Fig. 9.1). Interestingly, apart from the IFN-induced JAK–STAT pathway, activation of NF-κB by either HPIV-3 or induced by TNF-α/IL-1β directly confers an intracellular antiviral state independent of IFN-I. Although the host innate defense is armed with innate antiviral factors, the viruses for their replicative advantage have evolved to counteract their actions by either modulating the antiviral signaling (JAK–STAT) cascade at various stages (Tabs. 9.1–9.5) or manipulating the time frame of induction of antiviral molecule(s) (NF-κB) following infection [25]. The IFN antiviral system is a critical innate defense mechanism that is operative since early evolution to defend eukaryotic organisms from viral infection. The IFN-mediated antiviral response is elicited by three major mechanisms: (i) ability of the host to sense virus infection resulting in IFN secretion, (ii) association of IFN with its receptor leading to the activation of the JAK–STAT pathway and expression of IFN-induced antiviral factors, and (iii) antiviral activity of IFN stimulated genes that results in virus replication inhibition at various stages of the viral life cycle. However, most viruses replicate efficiently in their hosts as a consequence of encoding IFN antagonist factors that block various steps of the IFN system. These steps include: (i) inhibition in IFN synthesis, (ii) blocking IFN signaling, (iii) inhibiting the function of IFN-induced antiviral proteins, and (iv) producing IFN receptor decoy molecules to prevent induction of IFN signaling. Different viruses that counteract various steps of the IFN system are detailed in Tab. 9.5. IFN antagonists of different viruses that modulate the IFN system for their replicative advantage are described in Tabs. 9.1–9.4. Thus, this continued battle between the host, striving to restrict virus replication by initiating an innate response, and the virus's struggle to antagonize the host's effort, eventually decides who will ultimately win. Future studies will no doubt shed new light on this important host–virus interaction and will advance our understanding

Tab. 9.5. Various stages of the IFN antiviral system blocked by different viruses

Mechanism of action	Viruses
Inhibition in IFN synthesis/production	HBV
	HPV
	HHV-8/KSHV
	influenza
	MV
Inhibition in IFN signaling	adenovirus
	HBV
	HPV
	EBV
	CMV
	HHV-8/KSHV
	SV
	SV5
	RSV
	MUV
	HCV
	Ebola
	HPIV-3
	HPIV-2
	influenza
Inhibition in IFN-induced antiviral protein functions	adenovirus
	HSV
	EBV
	HCV
	HIV-1
	vaccinia
	poliovirus
	influenza
	reovirus
IFN receptor decoys	vaccinia

of the body's immune defense mechanisms, which have a great potential for the development of novel targeted antiviral therapies.

References

1 DECKER, T., STOCKINGER, S., KARAGHIOSOFF, M., MULLER, M., KOVARIK, P. 2002. IFNs and STATs in innate immunity of microorganisms. *J Clin Invest* 109, 1271–1277.

2 CAAMANO, J., HUNTER, C. A. **2002**. NF-kappaB family of transcription factors: central regulators of innate and adaptive immune functions. *Clin Microbiol Rev* 15, 414–429.

3 BOSE, S., BANERJEE, A. K. **2003**. Innate immune response against non-segmented negative strand RNA viruses. *J Interferon Cytokine Res* 23, 401–412.

4 SEN, G. C. **2001**. Viruses and interferons. *Annu Rev Microbiol* 55, 255–281.

5 HAMERMAN, J. A., OGASAWARA, K., LANIER, L. L. **2005**. NK cells in innate immunity. *Curr Opin Immunol* 17, 29–35.

6 PASARE, C., MEDZHITOV, R. **2005**. Toll-like receptors: linking innate and adaptive immunity. *Adv Exp Med Biol* 560, 11–18.

7 BALDWIN, A. S., JR. **2001**. Series introduction: the transcription factor NF-kappaB and human disease. *J Clin Invest* 107, 3–6.

8 GHOSH, S., KARIN, M. **2002**. Missing pieces in the NF-κB puzzle. *Cell* 109, S81–S96.

9 SILVERMAN, N., MANIATIS, T. **2001**. NF-κB signaling pathways in mammalian and insect innate immunity. *Genes Dev* 15, 2321–2342.

10 HAYNES, L. M., MOORE, D. D., KURT-JONES, E. A., FINBERG, R. W., ANDERSON, L. J., TRIPP, R. A. **2001**. Involvement of toll-like receptor 4 in innate immunity to respiratory syncytial virus. *J Virol* 75, 10730–10737.

11 KURT-JONES, E. A., POPOVA, L., KWINN, L., HAYNES, L. M., JONES, L. P., TRIPP, R. A., WALSH, E. E., FREEMAN, M. W., GOLENBOCK, D. T., ANDERSON, L. J., FINBERG, R. W. **2000**. Pattern recognition receptors TLR4 and CD14 mediate response to respiratory syncytial virus. *Nat Immunol* 1, 398–401.

12 BIEBACK, K., LEIN, E., KLAGGE, I. M., AVOTA, E, SCHNEIDER-SCHAYKUESM, J., DUPREX, W. P., WAGNER, H., KIRSCHNING, C. J., TER MEULEN, V., SCHNEIDER-SCHAULIES, S. **2002**. Hemagglutinin protein of wild-type measles virus activates toll-like receptor 2 signaling. *J Virol* 76, 8729–8736.

13 BITKO, V., BARIK, S. **1998**. Persistent activation of RelA by respiratory syncytial virus involves protein kinase C, underphosphorylated IkappaBbeta, and sequestration of protein phosphatase 2A by the viral phosphoprotein. *J Virol* 72, 5610–5618.

14 PAHL, H. L., BAEUERLE, P. A. **1995**. Expression of influenza virus hemagglutinin activates transcription factor NF-kappa B. *J Virol* 69, 1480–1484.

15 DEMARCHI, F., GUTIERREZ, M. I., GIACCA, M. **1999**. Human immunodeficiency virus type 1 tat protein activates transcription factor NF-kappaB through the cellular interferon-inducible, double-stranded RNA-dependent protein kinase, PKR. *J Virol* 73, 7080–7086.

16 ZAMANIAN-DARYOUSH, M., MOGENSEN, T. H., DIDONATO, J. A., WILLIAMS, B. R. **2000**. NF-kappaB activation by double-stranded-RNA-activated protein kinase (PKR) is mediated through NF-kappaB-inducing kinase and IkappaB kinase. *Mol Cell Biol* 20, 1278–1290.

17 O'NEILL, L. **2000**. The toll/interleukin-1 receptor domain: a

molecular switch for inflammation and host defense. *Biochem Soc Trans* 28, 557–563.

18 ADEREM, A., ULEVITCH, R. J. **2000**. Toll-like receptors in the induction of the innate immune response. *Nature* 406, 782–787.

19 ALEXOPOULOU, L., HOLT, A. C., MEDZHITOV, R., FLAVELL, R. A. **2001**. Recognition of double-stranded RNA and activation of NF-kappaB by Toll-like receptor 3. *Nature* 413, 732–738.

20 MUZIO, M., NI, J., FENG, P., DIXIT, V. M. **1997**. IRAK (Pelle) family member IRAK-2 and MyD88 as proximal mediators of IL-1 signaling. *Science* 278, 1612–1615.

21 GIL, J., RULLAS, J., GARCIA, M. A., ALCAMI, J., ESTEBAN, M. **2001**. The catalytic activity of dsRNA-dependent protein kinase, PKR, is required for NF-κB activation. *Oncogene* 20, 385–394.

22 IORDANOV, M. S., WONG, J., BELL, J. C., MAGUN, B. E. **2001**. Activation of NF-kappa B by double strand RNA (dsRNA) in the absence of protein kinase R and Rnase L demonstrates the existence of two separate dsRNA-triggered antiviral programs. *Mol Cell Biol* 21, 61–72.

23 MOGENSEN, T. H., PALUDAN, S. R. **2001**. Molecular pathways in virus-induced cytokine production. *Microb Mol Biol Rev* 65, 131–150.

24 NEUZIL, K. M., TANG, Y. W., GRAHAM, B. S. **1996**. Protective role of TNF-alpha in respiratory syncytial virus infection *in vitro* and *in vivo*. *Am J Med Sci* 311, 201–204.

25 BOSE, S., KAR, N., MAITRA, R., DIDONATO, J. A., BANERJEE, A. K. **2003**. Temporal activation of NF-kappa B regulates an interferon independent innate anti-viral response against cytoplasmic RNA viruses. *Proc Natl Acad Sci USA* 100, 10890–10895.

26 HISCOTT, J., GRANDVAUX, N., SHARMA, S., TENOEVER, B. R., SERVANT, M. J., LIN, R. **2003**. Convergence of the NF-kappaB and interferon signaling pathways in the regulation of antiviral defense and apoptosis. *Ann NY Acad Sci* 1010, 237–248.

27 SHARMA, S., TENOEVER, B. R., GRANDVAUX, N., ZHOU, G. P., LIN, R., HISCOTT, J. **2003**. Triggering the interferon antiviral response through an IKK-related pathway. *Science* 300, 1148–1151.

28 tenOEVER, B. R., SERVANT, M. J., GRANDVAUX, N., LIN, R., HISCOTT, J. **2002**. Recognition of the measles virus nucleocapsid as a mechanism of IRF-3 activation. *J Virol* 76, 3659–3669.

29 IWAMURA, T., YONEYAMA, M., KOIZUMI, N., OKABE, Y., NAMIKI, H., SAMUEL, C. E., FUJITA, T. **2001**. PACT, a double-stranded RNA binding protein acts as a positive regulator for type I interferon gene induced by Newcastle disease virus. *Biochem Biophys Res Commun* 282, 515–523.

30 KIM, T. K., MANIATIS, T. **1997**. The mechanism of transcriptional synergy of an *in vitro* assembled interferon-beta enhanceosome. *Mol Cell* 1, 119–129.

31 MIYAMOTO, M., FUJITA, T., KIMURA, Y., MARUYAMA, M., HARADA, H., SUDO, Y., MIYATA, T., TANIGUCHI, T. **1988**. Regulated expression of a gene encoding a nuclear factor, IRF-

1, that specifically binds to IFN-beta gene regulatory elements. *Cell* 54, 903–913.

32 TAKEUCHI, R., TSUTSUMI, H., OSAKI, M., HASEYAMA, K., MIZUE, N., CHIBA, S. **1998**. Respiratory syncytial virus infection of human alveolar epithelial cells enhances interferon regulatory factor 1 and interleukun-1 beta-converting enzyme gene expression but does not cause apoptosis. *J Virol* 72, 4498–4502.

33 MARIE, I., DURBIN, J. E., LEVY, D. E. **1998**. Differential viral induction of distinct interferon-alpha genes by positive feedback through interferon regulatory factor-7. *EMBO J* 17, 6660–6669.

34 BARNES, B. J., MOORE, P. A., PITHA, P. M. **2001**. Virus-specific activation of a novel interferon regulatory factor, IRF-5, results in the induction of distinct interferon alpha genes. *J Biol Chem* 276, 23382–23390.

35 DARNELL, J. E., KERR, I. M., STARK, G. R. **1994**. JAK–STAT pathways and transcriptional activation in response to IFNs and other extracellular signaling proteins. *Science* 264, 1415–1421.

36 STARK, G. R., KERR, I. M., WILLIAMS, B. R. G., SILVERMAN, R. H., SCHREIBER, R. D. **1998**. How cells respond to interferons. *Annu Rev Biochem* 67, 227–264.

37 BIRON, C. A., SEN, G. C. **2001**. Interferons and other cytokines. In: D. M. KNIPE, P. M. HOWLEY (Eds.), *Fields Virology*. Philadelphia, PA: Lippincott Williams & Wilkins, p. 321.

38 SILVERMAN, R. H. **1997**. 2–5A-dependent RNase L: a regulated endoribonuclease in the interferon system. In: G. D'ALESSIO, J. F. RIORDAN (Eds.), *Ribonucleases: Structure and Function*. New York: Academic Press, pp. 515–551.

39 CLEMENS, M. J., ELIA, A. **1997**. The double-stranded RNA-dependent protein kinase PKR: structure and function. *J Interferon Cytokine Res* 17, 503–524.

40 HALLER, O., KOCHS, G. **2002**. Interferon-induced Mx proteins: dynamin-like GTPases with antiviral activity. *Traffic* 10, 710–717.

41 LAMB, RA., KOLAKOFSKY, D. **2001**. Paramyxoviridae: the viruses and their replication. In: D. M. KNIPE, P. M. HOWLEY (Eds.), *Fields Virology*. Philadelphia, PA: Lippincott Williams & Wilkins, p. 1305.

42 SANCHEZ, A., KHAN, A. S., ZAKI, S. R., NABEL, G. J. **2001**. Filoviridae: Marburg and Ebola viruses. In: D. M. KNIPE, P. M. HOWLEY (Eds.), *Fields Virology*. Philadelphia, PA: Lippincott Williams & Wilkins, p. 1279.

43 ROSE, J. K., WHITT, M. A. **2001**. Rhabdoviridae: the viruses and their replication. In: D. M. KNIPE, P. M. HOWLEY (Eds.), *Fields Virology*. Philadelphia, PA: Lippincott Williams & Wilkins, p. 1221.

44 GAROUFALIS, E., KWAN, I., LIN, R., MUSTAFA, A., PEPIN, N., ROULSTON, A., LACOSTE, J., HISCOTT, J. **1994**. Viral induction of the human beta interferon promoter: modulation of transcription by NF-kappa B/rel proteins and interferon regulatory factors. *J Virol* 68, 4707–4715.

45 MATIKAINEN, S., PIRHONEN, J., MIETTINEN, M., LEHTONEN, A., GOVENIUS-VINTOLA, C., SARENEVA, T., JULKUNEN, I. **2000**. Influenza A and Sendai viruses induced differential chemokine gene expression and transcription factor activation in human macrophages. *Virology 10*, 138–147.

46 HEYLBROECK, C., BALACHANDRAN, S., SERVANT, M. J., DELUCA, C., BARBER, G. N., LIN, R., HISCOTT, J. **2000**. The IRF-3 transcription factor mediates Sendai virus-induced apoptosis. *J Virol 74*, 3781–3792.

47 LIN, R., HEYLBROECK, C., PITHA, P. M., HISCOTT, J. **1998**. Virus-dependent phosphorylation of the IRF-3 transcription factor regulates nuclear translocation, transactivation potential, and proteasome-mediated degradation. *Mol Cell Biol 18*, 2986–2996.

48 LIN, R., MAMANE, Y., HISCOTT, J. **1999**. Structural and functional analysis of interferon regulatory factor 3: localization of the transactivation and autoinhibitory domains. *Mol Cell Biol 19*, 2465–2474.

49 LUND, J. M., ALEXOPOULOU, L., SATO, A., KAROW, M., ADAMS, N. C., GALE, N. W., IWASAKI, A., FLAVELL, R. A. **2004**. Recognition of single-stranded RNA viruses by Toll-like receptor 7. *Proc Natl Acad Sci USA 101*, 5598–5603.

50 HEIL, F., HEMMI, H., HOCHREIN, H., AMPENBERGER, F., KIRSCHNING, C., AKIRA, S., LIPFORD, G., WAGNER, H., BAUER, S. **2004**. Species-specific recognition of single-stranded RNA via toll-like receptor 7 and 8. *Science 303*, 1526–1529.

51 DIDCOCK, L., YOUNG, D. F., GOODBOURN, S., RANDALL, R. E. **1999**. Sendai virus and simian virus 5 block activation of interferon responsive genes: importance for virus pathogenesis. *J Virol 73*, 3125–3133.

52 KATO, A., OHNISHI, Y., KOHASE, M., SAITO, S., TASHIRO, M., NAGAI, Y. **2001**. Y2, the smallest of the Sendai virus C proteins, is fully capable of both counteracting the antiviral action of interferons and inhibiting viral RNA synthesis. *J Virol 75*, 3802–3810.

53 NEUMANN, G., WHITT, M. A., KAWAOKA, Y. **2002**. A decade after the generation of a negative-sense RNA virus from cloned cDNA – what have we learned? *J Gen Virol 83*, 2635–2662.

54 GOTOH, B., TAKEUCHI, K., KOMATSU, T., YOKOO, J., KIMURA, Y., KUROTANI, A., KATO, A., NAGAI, Y. **1999**. Knockout of the Sendai virus C gene eliminates the viral ability to prevent the interferon-alpha/beta-mediated responses. *FEBS Lett 459*, 205–210.

55 TAKEUCHI, K., KOMATSU, T., YOKOO, J., KATO, A., SHIODA, T., NAGAI, Y., GOTOH, B. **2001**. Sendai virus C protein physically associates with Stat1. *Genes Cells 6*, 545–557.

56 GARCIN, D., CURRAN, J., ITOH, M., KOLAKOFSKY, D. **2001**. Longer and shorter forms of Sendai virus C proteins play different roles in modulating the cellular antiviral response. *J Virol 75*, 6800–6807.

57 GARCIN, D., MARQ, J. B., STRAHLE, L., LE MERCIER, P., KOLAKOFSKY, D. **2002**. All four Sendai virus C proteins bind Stat1: but only the larger forms also induce its monoubiquitination and degradation. *Virology 295*, 256–265.

58 KOMATSU, T., TAKEUCHI, K., YOKOO, J., GOTOH, B. 2002. Sendai virus C protein impairs both phosphorylation and dephosphorylation processes of Stat1. *FEBS Lett 511*, 139–144.

59 GOTOH, B., KOMATSU, T., TAKEUCHI, K., YOKOO, J. 2002. Paramyxovirus strategies for evading the interferon response. *Rev Med Virol 12*, 337–357.

60 GARCIN, D., MARQ, J. B., GOODBOURN, S., KOLAKOFSKY, D. 2003. The amino-terminal extensions of the longer Sendai virus C proteins modulate pY701–STAT1 and bulk STAT1 levels independently of interferon signaling. *J Virol 77*, 2321–2329.

61 YOUNG, D. F., DIDCOCK, L., GOODBOURN, S., RANDALL, R. E. 2000. Paramyxoviridae use distinct virus-specific mechanisms to circumvent the interferon response. *Virology 269*, 383–390.

62 KOMATSU, T., TAKEUCHI, K., YOKOO, J., TANAKA, Y., GOTOH, B. 2000. Sendai virus blocks alpha interferon signaling to signal transducers and activators of transcription. *J Virol 74*, 2477–2480.

63 KOMATSU, T., TAKEUCHI, K., YOKOO, J., GOTOH, B. 2004. C and V proteins of Sendai virus target signaling pathways leading to IRF-3 activation for the negative regulation of interferon-beta production. *Virology 325*, 137–148.

64 SAMUEL, C. E. 2001. Antiviral actions of interferons. *Clin Microbiol Rev 14*, 778–809.

65 STAEHELI, P. 1990. Interferon-induced proteins and the antiviral state. *Adv Virus Res 38*, 147–200.

66 DIDCOCK, L., YOUNG, D. F., GOODBOURN, S., RANDALL, R. E. 1999. The V protein of simian virus 5 inhibits interferon signaling by targeting STAT1 for proteosome mediated degradation. *J Virol 73*, 9928–9933.

67 PARISIEN, J. P., LAU, J. F., RODRIGUEZ, J. J., ULANE, C. M., HORVATH, C. M. 2002. Selective STAT protein degradation induced by paramyxoviruses requires both STAT1 and STAT2 but is independent of alpha/beta interferon signal transduction. *J Virol 76*, 4190–4198.

68 ANDREJEVA, J., YOUNG, D. F., GOODBOURN, S., RANDALL, R. E. 2002. Degradation of STAT1 and STAT2 by the V proteins of simian virus 5 and human parainfluenza virus type 2, respectively: consequences for virus replication in the presence of alpha/beta and gamma interferons. *J Virol 76*, 2159–2167.

69 PARISIEN, J. P., LAU, J. F., HORVATH, C. M. 2002. STAT2 acts as a host range determinant for species-specific paramyxovirus interferon antagonism and simian virus 5 replication. *J Virol 76*, 6435–6441.

70 WANSLEY, E. K., PARKS, G. D. 2002. Naturally occurring substitutions in the p/v gene convert the noncytopathic paramyxovirus simian virus 5 into a virus that induces alpha/beta interferon synthesis and cell death. *J Virol 76*, 10109–10121.

71 HE, B., PATERSON, R. G., STOCK, N., DURBIN, J. E., DURBIN, R. K., GOODBOURN, S., RANDALL, R. E., LAMB, R. A. 2002. Recovery of paramyxovirus simian virus 5 with a V protein lacking the conserved cysteine rich domain: the

multifunctional V protein blocks both interferon beta induction and interferon signaling. *Virology 303*, 15–32.

72 GAO, J., DE, B. P., BANERJEE, A. K. 1999. Human parainfluenza virus type 3 up-regulates major histocompatibility complex class I and II expression on respiratory epithelial cells: involvement of a STAT1- and CIITA-independent pathway. *J Virol 73*, 1411–1418.

73 GAO, J., CHOUDHARY, S., BANERJEE, A. K., DE, B. P. 2000. Human parainfluenza virus type 3 upregulates ICAM-1 (CD54) expression in a cytokine-independent manner. *Gene Expr 9*, 115–121.

74 HARMON, A. T., HARMON, M. W., GLEZEN, W. P. 1982. Evidence of interferon production in the hamster lung after primary or secondary exposure to parainfluenza virus type 3. *Am Rev Respir Dis 125*, 706–711.

75 DURBIN, A. P., MCAULIFFE, J. M., COLLINS, P. L., MURPHY, B. R. 1999. Mutations in the C, D, and V open reading frames of human parainfluenza virus type 3 attenuate replication in rodents and primates. *Virology 261*, 319–330.

76 PARISIEN, J. P., LAU, J. F., RODRIGUEZ, J. J., SULLIVAN, B. M., MOSCONA, A., PARKS, G. D., LAMB, R. A., HORVATH, C. M. 2001. The V protein of human parainfluenza virus 2 antagonizes type I interferon responses by destabilizing signal transducer and activator of transcription 2. *Virology 283*, 230–239.

77 NISHIO, M., TSURUDOME, M., ITO, M., KAWANO, M., KOMADA, H., ITO, Y. 2001. High resistance of human parainfluenza type 2 virus protein-expressing cells to the antiviral and anti-cell proliferative activities of alpha/beta interferons: cysteine-rich V-specific domain is required for high resistance to the interferons. *J Virol 75*, 9165–9176.

78 KAWANO, M., KAITO, M., KOZUKA, Y., KOMADA, H., NODA, N., NANBA, K., TSURUDOME, M., ITO, M., NISHIO, M., ITO, Y. 2001. Recovery of infectious human parainfluenza type 2 from cDNA clones and properties of the defective virus without V-specific cysteine-rich domain. *Virology 284*, 99–112.

79 PRECIOUS, B., YOUNG, D. F., ANDREJEVA, L., GOODBOURN, S., RANDALL, R. E. 2005. In vitro and *in vivo* specificity of ubiquitination and degradation of STAT1 and STAT2 by the V proteins of the paramyxoviruses simian virus 5 and human parainfluenza virus type 2. *J Gen Virol 86*, 151–8.

80 MALUR, A. G., CHATTOPADHYAY, S., MAITRA, R. K., BANERJEE, A. K. 2005. Inhibition of STAT1 phosphorylation by human parainfluenza virus type 3 C protein. *J Virol 79*, 7877–7882.

81 MALUR, A. G., HOFFMAN, M. A., BANERJEE, A. K. 2004. The human parainfluenza virus type 3 (HPIV 3) C protein inhibits viral transcription. *Virus Res 99*, 199–204.

82 SMALLWOOD, S., MOYER, S. A. 2004. The L polymerase protein of parainfluenza virus 3 forms an oligomer and can interact with the heterologous Sendai virus L, P and C proteins. *Virology 318*, 439–450.

83 DHIB-JALBUT, S., XIA, J., RANGAVIGGULA, H., FANG, Y. Y., LEE, T. 1999. Failure of measles virus to activate nuclear factor-kappa B in neuronal cells: implications on the immune

response to viral infections in the central nervous system. *J Immunol* 162, 4024–4029.

84 FUJII, N., YOKOSAWA, N., SHIRAKAWA, S. **1999**. Suppression of interferon response gene expression in cells persistently infected with mumps virus, and restoration from its suppression by treatment with ribavirin. *Virus Res* 65, 175–185.

85 YOKOSAWA, N., KUBOTA, T., FUJII, N. **1998**. Poor induction of interferon-induced 2′,5′-oligoadenylate synthetase (2–5 AS) in cells persistently infected with mumps virus is caused by decrease of STAT-1 alpha. *Arch Virol* 143, 1985–1992.

86 KUBOTA, T., YOKOSAWA, N., YOKOTA, S., FUJII, N. **2001**. C terminal CYS-RICH region of mumps virus structural V protein correlates with block of interferon alpha and gamma signal transduction pathway through decrease of STAT1-alpha. *Biochem Biophys Res Commun* 283, 255–259.

87 NISHIO, M., GARCIN, D., SIMONET, V., KOLAKOFSKY, D. **2002**. The carboxyl segment of the mumps virus V protein associates with Stat proteins *in vitro* via a tryptophan-rich motif. *Virology* 300, 92–99.

88 KUBOTA, T., YOKOSAWA, N., YOKOTA, S., FUJII, N. **2002**. Association of mumps virus V protein with RACK1 results in dissociation of STAT-1 from the alpha interferon receptor complex. *J Virol* 76, 12676–12682.

89 YOKOSAWA, N., YOKOTA, S., KUBOTA, T., FUJII, N. **2002**. C-terminal region of STAT-1alpha is not necessary for its ubiquitination and degradation caused by Mumps virus V protein. *J Virol* 76, 12683–12690.

90 YOKOTA, S., YOKOSAWA, N., KUBOTA, T., OKABAYASHI, T., ARATA, S., FUJII, N. **2003**. Suppression of thermotolerance in mumps virus-infected cells is caused by lack of HSP27 induction contributed by STAT-1. *J Biol Chem* 278, 41654–41660.

91 KUBOTA, T., YOKOSAWA, N., YOKOTA, S., FUJII, N., TASHIRO, M., KATO, A. **2005**. Mumps virus V protein antagonizes interferon without the complete degradation of STAT1. *J Virol* 79, 4451–4457.

92 UMANSKY, V., SHATROV, V. A., LEHMANN, V., SCHIRRMACHER, V. **1996**. Induction of NO synthesis in macrophages by Newcastle disease virus is associated with activation of nuclear factor-kappa B. *Int Immunol* 8, 491–498.

93 ZENG, J., FOURNIER, P., SCHIRRMACHER, V. **2002**. Induction of interferon-alpha and tumor necrosis factor-related apoptosis-inducing ligand in human blood mononuclear cells by hemagglutinin–neuraminidase but not F protein of Newcastle disease virus. *Virology* 297, 19–30.

94 JESTIN, V., CHERBONNEL, M. **1991**. Interferon-induction in mouse spleen cells by the Newcastle disease virus (NDV) HN protein. *Ann Rech Vet* 22, 365–372.

95 JUANG, Y. T., AU, W. C., LOWTHER, W., HISCOTT, J., PITHA, P. M. **1999**. Lipopolysaccharide inhibits virus-mediated induction of interferon genes by disruption of nuclear transport of interferon regulatory factors 3 and 7. *J Biol Chem* 274, 18060–18066.

96 PARK, M. S., SHAW, M. L., MUNOZ-JORDAN, J., CROS, J. F., NAKAYA, T., BOUVIER, N., PALESE, P., GARCIA-SASTRE, A., BASLER, C. F. 2003. Newcastle disease virus (NDV) based assay demonstrates interferon antagonist activity for the NDV V protein and the Nipah virus V, W, and C proteins. *J Virol 77*, 1501–1511.

97 STEWARD, M., SAMSON, A. C., ERRINGTON, W., EMMERSON, P. T. 1995. The Newcastle disease virus V protein binds zinc. *Arch Virol 140*, 1321–1328.

98 HUANG, Z., KRISNAMURTHY, S., PANDA, A., SAMAL, S. K. 2003. Newcastle disease virus V protein is associated with viral pathogenesis and functions as an alpha interferon antagonist. *J Virol 77*, 8676–8685.

99 MEBATSION, T., VERSTEGEN, S., DE VANN, L. T., ROMER-OBERDORFER, A., SCHRIER, C. C. 2001. A recombinant Newcastle disease virus with low level V protein expression is immunogenic and lacks pathogenicity for chicken embryos. *J Virol 75*, 420–428.

100 HELIN, E., VAINIONPAA, R., HYYPIA, T., JULKUNEN, I., MATIKAINEN, S. 2001. Measles virus activates NF-kappa B and STAT transcription factors and production of IFN-alpha/beta and IL-6 in the human lung epithelial cell line A549. *Virology 290*, 1–10.

101 NANICHE, D., YEH, A., ETO, D., MANCHESTER, M., FRIEDMAN, R. M., OLDSTONE, M. B. 2000. Evasion of host defenses by measles virus: wild-type measles virus infection interferes with induction of alpha/beta interferon production. *J Virol 74*, 7478–7484.

102 SERVANT, M. J., TENOEVER, B., LEPAGE, C., CONTI, L., GESSANI, S., JULKUNEN, I., LIN, R., HISCOTT, J. 2001. Identification of distinct signaling pathways leading to the phosphorylation of interferon regulatory factor 3. *J Biol Chem 276*, 355–363.

103 LISTON, P., BRIEDIS, D. J. 1994. Measles virus V protein binds zinc. *Virology 198*, 399–404.

104 SCHNEIDER, H., KAELIN, K., BILLETER, M. A. 1997. Recombinant measles viruses defective for RNA editing and V protein synthesis are viable in cultured cells. *Virology 227*, 314–322.

105 TEKEUCHI, K., KADOTA, S. I., TAKEDA, M., MIYAJIMA, N., NAGATA, K. 2003. Measles virus V protein blocks interferon (IFN)-alpha/beta but not IFN-gamma signaling by inhibiting STAT1 and STAT2 phosphorylation. *FEBS Lett 5452–5453*, 177–182.

106 PALOSAARI, H., PARISIEN, J. P., RODRIGUEZ, J. J., ULANE, C. M., HORVATH, C. M. 2003. STAT protein interference and suppression of cytokine signal transduction by Measles virus V protein. *J Virol 77*, 7635–7644.

107 SHAFFER, J. A., BELLINI, W. J., ROTA, P. A. 2003. The C response of Measles virus inhibits the type 1 interferon response. *Virology 315*, 389–397.

108 VALSAMAKIS, A., SCHNEIDER, H., AUWAERTER, P. G., KANESHIMA, H., BILLETER, M. A., GRIFFIN, D. E. 1998. Recombinant measles viruses with mutations in the C, V, or F gene have altered growth phenotypes *in vivo*. *J Virol 72*, 7754–7761.

109 Tober, C., Seufert, M., Schneider, H., Billeter, M. A., Johnston, I. C., Niewiesk, S., Ter Meulen, V., Schneider-Schaulies, S. **1998**. Expression of measles virus V protein is associated with pathogenicity and control of viral RNA synthesis. *J Virol* 72, 8124–8132.

110 Patterson, J. B., Thomas, D., Lewicki, H., Billeter, M. A., Oldstone, M. B. **2000**. V and C proteins of measles virus function as virulence factors *in vivo*. *Virology* 267, 80–89.

111 Mrkic, B., Odermatt, B., Klein, M. A., Billeter, M. A., Pavlovic, J., Cattaneo, R. **2000**. Lymphatic dissemination and comparative pathology of recombinant measles viruses in genetically modified mice. *J Virol* 74, 1364–1372.

112 Bitko, V., Velazquez, A., Yang, L., Yang, Y. C., Barik, S. **1997**. Transcriptional induction of multiple cytokines by human respiratory syncytial virus requires activation of NF-kappa B and is inhibited by sodium salicylate and aspirin. *Virology* 232, 369–378.

113 Mastronarde, J. G., He, B., Monick, M. M., Mukaida, N., Matsushima, K., Hunninghake, G. W. **1996**. Induction of interleukin (IL)-8 gene expression by respiratory syncytial virus involves activation of nuclear factor (NF)-kappa B and NF-IL-6. *J Infect Dis* 174, 262–267.

114 Fiedler, M. A., Wernke-Dollries, K., Stark, J. M. **1996**. Inhibition of viral replication reverses respiratory syncytial virus-induced NF-kappaB activation and interleukin-8 gene expression in A549 cells. *J Virol* 70, 9079–9082.

115 Tian, B., Zhang, Y., Luxon, B. A., Garofalo, R. P., Casola, A., Sinha, M., Brasier, A. R. **2002**. Identification of NF-kappaB-dependent gene networks in respiratory syncytial virus-infected cells. *J Virol* 76, 6800–6814.

116 Fiedler, M. A., Wernke-Dollries, K. **1999**. Incomplete regulation of NF-kappaB by IkappaBalpha during respiratory syncytial virus infection in A549 cells. *J Virol* 73, 4502–4507.

117 Thomas, L. H., Friedland, J. S., Sharland, M., Becker, S. **1998**. Respiratory syncytial virus-induced RANTES production from human bronchial epithelial cells is dependent on nuclear factor-kappa B nuclear binding and is inhibited by adenovirus-mediated expression of inhibitor of kappa B alpha. *J Immunol* 161, 1007–1016.

118 Jamaluddin, M., Casola, A., Garofalo, R. P., Han, Y., Elliott, T., Ogra, P. L., Brasier, A. R. **1998**. The major component of IkappaBalpha proteolysis occurs independently of the proteasome pathway in respiratory syncytial virus-infected pulmonary epithelial cells. *J Virol* 72, 4849–4857.

119 Thomas, K. W., Monick, M. M., Staber, J. M., Yarovinsky, T., Carter, A. B., Hunninghake, G. W. **2002**. Respiratory syncytial virus inhibits apoptosis and induces NF-kappa B activity through a phosphatidylinositol 3-kinase dependent pathway. *J Biol Chem* 277, 492–501.

120 Tsutsumi, H., Takeuchi, R., Ohsaki, M., Seki, K., Chiba, S. **1999**. Respiratory syncytial virus infection of human respiratory epithelial cells enhances inducible nitric oxide synthase gene expression. *J Leukoc Biol* 66, 99–104.

121 Patel, J. A., Jiang, Z., Nakajima, N., Kunimoto, M. **1998**.

Autocrine regulation of interleukin-8 by interleukin-1alpha in respiratory syncytial virus-infected pulmonary epithelial cells in vitro. *Immunology* 95, 501–506.

122 ATREYA, P. L., KULKARNI, S. **1999**. Respiratory syncytial virus strain A2 is resistant to the antiviral effects of type I interferons and human MxA. *Virology* 261, 227–241.

123 COLLINS, P. L., CHANOCK, R. M., MURPHY, B. R. **2001**. Respiratory syncytial virus. In: D. M. KNIPE, P. M. HOWLEY (Eds.), *Fields Virology*. Philadelphia, PA: Lippincott Williams & Wilkins, p. 1443.

124 JOHNSON, P. R., COLLINS, P. L. **1989**. The 1B (NS2), 1C (NS1) and N proteins of human respiratory syncytial virus (RSV) of antigenic subgroups A and B: sequence conservation and divergence within RSV genomic RNA. *J Gen Virol* 70, 1539–1547.

125 SCHLENDER, J., BOSSERT, B., BUCHHOLZ, U., CONZELMANN, K. K. **2000**. Bovine respiratory syncytial virus nonstructural proteins NS1 and NS2 cooperatively antagonize alpha/beta interferon-induced antiviral response. *J Virol* 74, 8234–8242.

126 BOSSERT, B., MAROZIN, S., CONZELMANN, K. K. **2003**. Non structural proteins NS1and NS2 of bovine respiratory syncytial virus block activation of interferon regulatory factor 3. *J Virol* 77, 8661–8668.

127 BOSSERT, B., COZELMANN, K. K. **2002**. Respiratory syncytial virus (RSV) nonstructural (NS) proteins as host range determinants: a chimeric bovine RSV with NS genes from human RSV is attenuated in interferon-competent bovine cells. *J Virol* 76, 4287–4293.

128 SPANN, K. M., TRAN, K. C., COLLINS, P. L. **2005**. Effects of non structural proteins NS1 and NS2 of human respiratory syncytial virus on interferon regulatory factor 3, NF-kappa B, and proinflammatory cytokines. *J Virol* 79, 5353–5362.

129 ZHANG, W., YANG, H., KONG, X., MOHAPATRA, S., SAN JUAN-VERGARA, H., HELLERMANN, G., BEHERA, S., SINGAM, R., LOCKEY, R. F., MOHAPATRA, S. S. **2005**. Inhibition of respiratory syncytial virus infection with intranasal siRNA nanoparticles targeting the viral NS1 gene. *Nat Med* 11, 56–62.

130 RAMASWAMY, M., SHI, L., MONICK, M. M., HUNNINGHAKE, G. W., LOOK, D. C. **2004**. Specific inhibition of type 1 interferon signal transduction by respiratory syncytial virus. *Am J Respir Cell Biol* 30, 893–900.

131 RODRIGUEZ, J. J., CRUZ, C. D., HORVATH, C. M. **2004**. Identification of the nuclear export signal and STAT-binding domains of the Nipah virus V protein reveals mechanisms underlying interferon evasion. *J Virol* 78, 5358–5367.

132 RODRIGUEZ, J. J., PARISIEN, J. P., HORVATH, C. M. **2002**. Nipah virus V protein evades alpha and gamma interferons by preventing STAT1 and STAT2 activation and nuclear accumulation. *J Virol* 76, 11476–11483.

133 SHAW, M. L., GARCIA-SASTRE, A., PALESE, P., BASLER, C. F. **2004**. Nipah virus V and W proteins have a common STAT1-binding domain yet inhibit STAT1 activation from the cytoplasmic and nuclear compartments, respectively. *J Virol* 78, 5633–5641.

134 PARK, M. S., SHAW, M. L., MUNOZ-JORDAN, J., CROS, J. F., NAKAYA, T., BOUVIER, N., PALESE, P., GARCIA-SASTRE, A., BASLER, C. F. 2003. Newcastle disease virus (NDV)-based assay demonstrates interferon-antagonist for the NDV V protein and the Nipah virus V, W, and C proteins. *J Virol* 77, 1501–1511.

135 SHAW, M. L., CARDENAS, W. B., ZAMARIN, D., PALESE, P., BASLER, C. F. 2005. Nuclear localization of the Nipah W protein allows for inhibition of both-toll-like receptor 3-triggered signaling pathways. *J Virol* 79, 6078–6088.

136 RODRIGUEZZ, J. J., WANG, L. F., HORVATH, C. M. 2003. Hendra virus V protein inhibits interferon signaling by preventing STAT1 and STAT2 nuclear accumulation. *J Virol* 77, 11842–11845.

137 BASLER, C. F., MIKULASOVA, A., MARTINEZ-SABRIDO, L., PARAGAS, J., MUHLBERGER, E., KLENK, H. D., PALESE, P., GARCIA-SASTRE, A. 2003. The Ebola virus VP35 protein inhibits activation of interferon regulatory factor 3. *J Virol* 77, 7945–7956.

138 HARTMAN, A. L., TOWNER, J. S., NICHOL, S. T. 2004. A C-terminal basic amino acid motif of Zaire ebolavirus VP35 is essential for type 1 interferon antagonism and displays high identity with the RNA-binding domain of another interferon antagonist, the NS1 protein of influenza A virus. *Virology* 328, 177–184.

139 BRZOZKA, K., FINKE, S., CONZELAMANN, K. K. 2005. Identification of the rabies virus alpha/beta interferon antagonist: phosphoprotein P interferes with phosphorylation of interferon regulatory factor 3. *J Virol* 79, 7673–7681.

140 LAMB, R. A., KRUG, R. M. 2001. Orthomyxoviridae: the viruses and their replication. In: D. M. KNIPE, P. M. HOWLEY (Eds.), *Fields Virology*. Philadelphia, PA: Lippincott Williams & Wilkins, p. 1487.

141 SCHMALJOHN, C. S., HOOPER, J. W. 2001. Bunyaviridae: the viruses and their replication. In: D. M. KNIPE, P. M. HOWLEY (Eds.), *Fields Virology*. Philadelphia, PA: Lippincott Williams & Wilkins, p. 1581.

142 STRUBE, M., BODO, G., JUNGWIRTH, C. 1985. Sensitivity of ortho- and paramyxovirus replication to human interferon alpha. *Mol Biol Rep* 10, 237–243.

143 SEO, S. H., WEBSTER, R. G. 2002. Tumor necrosis factor alpha exerts powerful anti-influenza virus effects in lung epithelial cells. *J Virol* 76, 1071–1076.

144 DONELAN, N. R., DAUBER, B., WANG, X., BASLER, C. F., WOLFF, T., GARCIA-SASTRE, A. 2004. The N-and C-terminal domains of the NSI protein of influenza B virus can independently inhibit IRF-3 and beta interferon promoter activation. *J Virol* 78, 11574–11582.

145 GARCIA-SASTRE, A., EGOROV, A., MATASSOV, D., BRANDT, S., LEVY, D. E., DURBIN, J. E., PALESE, P., MUSTER, T. 1998. Influenza A virus lacking the NS1 gene replicates in interferon deficient systems. *Virology* 252, 324–330.

146 WANG, X., LI, M., ZHENG, H., MUSTER, T., PALESE, P., BEG, A. A., GARSIA-SASTRTE, A. 2000. Influenza A virus NS1 protein

prevents activation of NF-kappa B and induction of alpha/beta interferon. *J Virol 74*, 11566–11573.

147 Talon, J., Horvath, C. M., Polley, R., Basler, C. F., Muster, T., Palese, P., Garcia Sastre, A. **2000**. Activation of interferon regulatory factor 3 is inhibited by the influenza A virus NS1 protein. *J Virol 74*, 7989–7996.

148 Smith, E. J., Marie, I., Prakash, A., Garcia-Sastre, A., Levy, D. E. **2001**. IRF3 and IRF7 phosphorylation in virus-infected cells does not require double-stranded RNA-dependent protein kinase R or IκB kinase but is blocked by vaccinia virus E3L protein. *J Biol Chem 276*, 8951–8957.

149 Bergmann, M., Garcia-Sastre, A., Carnero, E., Pehamberger, H., Wolff, K., Palese, P., Muster, T. **2000**. Influenza virus NS1 protein counteracts PKR-mediated inhibition of replication. *J Virol 74*, 6203–6206.

150 Donelan, N. R., Basler, C. F., Garcia-Sastre, A. **2003**. A recombinant influenza A virus expressing an RNA-binding-defective NS1 protein induces high levels of beta interferon and is attenuated in mice. *J Virol 77*, 13257–13266.

151 Seo, S. H., Hoffmann, E., Webster, R. G. **2002**. Lethal H5N1 influenza viruses escape host escape anti-viral cytokine responses. *Nat Med 8*, 950–954.

152 Seo, S. H., Hoffmann, E., Webster, R. G. **2004**. The NS1 gene of H5N1 influenza viruses circumvents the host anti-viral cytokine responses. *Virus Res 103*, 107–113.

153 Kim, M.-J., Latham, A. G., Krug, R. M. **2002**. Human influenza viruses activate an interferon-independent transcription of cellular antiviral genes: the outcome with influenza A virus is unique. *Proc Natl Acad Sci USA 99*, 10096–10101.

154 Noah, D. L., Twu, K. Y., Krug, R. M. **2003**. Cellular antiviral responses against influenza A virus are countered at the post-transcriptional level by the viral NS1A protein via its binding to a cellular protein required for the 3′ end processing of cellular pre-mRNAs. *Virology 307*, 386–395.

155 Lee, T. G., Tomita, J., Hovanessian, A. G., Katze, M. G. **1990**. Purification and partial characterization of a cellular inhibitor of the interferon-induced protein kinase of M_r 68,000 from influenza virus-infected cells. *Proc Natl Acad Sci USA 87*, 6208–6212.

156 Gale, M., Tan, S. L., Wambach, M., Katze, M. G. **1996**. Interaction of the interferon-induced PKR protein kinase with inhibitory proteins P58IPK and vaccinia virus K3L is mediated by unique domains: implications for kinase regulation. *Mol Cell Biol 16*, 4172–4181.

157 Yuan, W., Krug, R. M. **2001**. Influenza B virus NS1 protein inhibits conjugation of the interferon (IFN)-induced ubiquitin-like ISG15 protein. *EMBO J 20*, 362–371.

158 Bouloy, M., Janzen, C., Vialat, P., Khun, H., Pavlovic, J., Huerre, M., Haller, O. **2001**. Genetic evidence for an interferon-antagonistic function of Rift Valley Fever virus nonstructural protein NSs. *J Virol 75*, 1371–1377.

159 Billecocq, A., Spiegel, M., Kohl, A., Weber, F., Bouloy, M., Haller, O. **2004**. NSs protein of Rift Valley Fever virus blocks

interferon production by inhibiting host gene transcription. *J Virol* 78, 9798–9806.

160 WEBER, F., BRIDGEN, A., FAZAKERLEY, J. K., STREITENFELD, H., KESSLER, N., RANDALL, R. E., ELLIOTT, R. M. 2002. Bunyamwera bunyavirus nonstructural protein NSs counteracts the induction of alpha/beta interferon. *J Virol* 76, 7949–7955

161 THOMAS, D., BLAKQORI, G., WAGNER, V., BANHOLZER, M., KESSLER, N., ELLIOT, R. M., HALLER, O., WEBER, F. 2004. Inhibition of RNA polymerase II phosphorylation by a viral interferon antagonist. *J Biol Chem* 279, 31471–31477.

162 DODD, D. A., GIDDING, T. H., JR., KIRKEGAARD, K. 2001. Poliovirus 3A protein limits interleukin-6 (IL-6), IL-8, and beta interferon secretion during viral infection. *J Virol* 75, 8158–8165.

163 NEZNANOV, N., CHUMAKOV, K. M., NEZNANOV, L., ALSAMAN, A., BANERJEE, A. K., GUDKOV, A. V. 2005. Proteolytic cleavage of the p65-RelA subunit of NF-kappa B during poliovirus infection. *J Biol Chem* 280, 24153–24158.

164 CHINSANGARAM, J., PICCONE, M. E., GRUBMAN, M. J. 1999. Ability of foot and mouth disease virus to form plaques in cell culture is associated with suppression of alpha/beta interferon. *J Virol* 73, 9891–9898.

165 DOLGANIUE, A., KODYS, K., KOPASZ, A., MARSHALL, C., DO, T., ROMIES, L., JR., MANDREKAR, P., ZAPP, M., SZABO, G. 2003. Hepatitis C virus core and nonstructural protein 3 proteins induce pro- and anti-inflammatory cytokines and inhibit dendritic cell differentiation. *J Immunol* 170, 5615–5624.

166 GONG, G., WARIS, G., TANVEER, R., SIDDIQUI, A. 2001. Human hepatitis C virus NS5A protein alters intracellular calcium levels, induces oxidative stress, and activates STAT-3 and NF-kappa B. *Proc Natl Acad Sci USA* 98, 9599–9604.

167 OTSUKA, M., KATO, N., MORIYAMA, M., TANIGUCHI, H., WANG, Y., DHAREL, N., KAWABE, T., OMATA, M. 2005. Interaction between the HCV NS3 protein and the host TBK-1 protein leads to inhibition of cellular antiviral responses. *Hepatology* 41, 1004–1012.

168 FOY, E., LI, K., WANG, C., SUMPTER, R., JR., IKEDA, M., LEMON, S. M., GALE, M., JR. 2003. Regulation of interferon regulatory factor-3 by the hepatitis virus serine protease. *Science* 300, 1145–1148.

169 BREIMAN, A., GRANVAUX, N., LIN, R., OTTONE, C., AKIRA, S., YONEYAMA, M., FUJITA, T., HISCOTT, J., MEURS, E. F. 2005. Inhibition of RIG-1 dependent signaling to the interferon pathway during hepatitis C virus expression and restoration of signaling by IKKepsilon. *J Virol* 79, 3969–3978.

170 LI, K., FOY, E., FERREON, J. C., NAKAMURA, M., FERREON, A. C., IKEDA, M., RAY, S. C., GALE, M., JR., LEMON, S. M. 2005. Immune evasion by hepatitis C virus NS3/4Aprotease-mediated cleavage of the Toll-like receptor 3 adaptor protein TRIF. *Proc Natl Acad Sci USA* 102, 2992–2997.

171 HEIM, M. H., MORADPOUR, D., BLUM, H. E. 1999. Expression of hepatitis C virus proteins inhibits signal transduction through the Jak–STAT pathway. *J Virol* 73, 8469–8475.

172 GALE, M. J., BLAKELY, C. M., KWIECISZEWSKI, B., TAN, S. L., DOSSETT, M., TANG, N. M., KORTH, M. J., POLYAK, S. J., GRETCH, D. R., KATZE, M. G. **1998**. Control of PKR protein kinase by hepatitis C virus nonstructural 5A protein: molecular mechanisms of kinase regulation. *Mol Cell Biol 18*, 5208–5218.

173 TAYLOR, D. R., SHI, S. T., ROMANO, P. R., BARBER, G. N., LAI, M. M. **1999**. Inhibition of the interferon-inducible protein kinase PKR by HCV E2 protein. *Science 285*, 107–110.

174 AUS DEM SIEPEN, M., LOHMANN, V., WIESE, M., ROSS, S., ROGGENDORF, M., VIAZOV, S. **2005**. Nonstructural protein 5A does not contribute to the resistance of hepatitis C virus replication to interferon alpha in cell culture. *Virology 336*, 131–136.

175 LIN, W., CHOE, W. H., HIASA, Y., KAMEGAYA, Y., BLACKARD, J. T., SCHMIDT, E. V., CHUNG, R. T. **2005**. Hepatitis C virus expression suppresses interferon signaling by degrading STAT1. *Gastroenterology 128*, 1034–1041.

176 GOULD, L. H., FIKRIG, E. **2004**. West Nile virus: a growing concern? *J Clin Invest 113*, 1102–1107.

177 CHENG, Y., KING, N. J., KESSON, A. M. **2004**. Major histocompatibility complex class 1 (MHC-1) induction by West Nile virus: involvement of 2 signaling pathways in MHC-I up-regulation. *J Infect Dis 189*, 658–668.

178 FREDERICKSEN, B. L., SMITH, M., KATZE, M. G., SHI, P. Y., GALE, M., JR. **2004**. The host response to West Virus infection limits viral spread through the activation of the interferon regulatory factor 3 pathway. *J Virol 78*, 7737–7747.

179 LIU, W. J., WANG, X. J., MOKHONOV, V. V., SHI, P. Y., RANDALL, R., KHROMYKH, A. A. **2005**. Inhibition of interferon signaling by the New York 99 strain and Kunjin subtype of West Nile virus involves blockage of STAT1 and STAT2 activation by nonstructural proteins. *J Virol 79*, 1934–1942.

180 GUO, J. T., HAYASHI, J., SEEGER, C. **2005**. West Nile virus inhibits the signal transduction pathway of alpha interferon. *J Virol 79*, 1343–1350.

181 HOLMES, K. V. **2003**. SARS coronavirus: a new challenge for prevention and therapy. *J Clin Invest 111*, 1605–1609.

182 ZHENG, B., HE, M. L., WONG, K. L., LUM, C. T., POON, L. L., PENG, Y., GUAN, Y., LIN, M. C., KUNG, H. F. **2004**. Potent inhibition of SARS-associated coronavirus (SCOV) infection and replication by type I interferons (IFN-alpha/beta) but not by type II interferon (IFN-gamma). *J Interferon Cytokine Res 24*, 388–390.

183 SPIEGEL, M., PICHLMAIR, A., MARTINEZ-SOBRIDO, L., CROS, J., GARCIA-SASTRE, A., HALLER, O., WEBER, F. **2005**. Inhibition of Beta interferon induction by severe acute respiratory syndrome coronavirus suggests a two-step model for activation of interferon regulatory factor 3. *J Virol 79*, 2079–2086.

184 NIBERT, M. L., SCHIFF, L. A. **2001**. Reoviruses and their replication. In: D. M. KNIPE, P. M. HOWLEY (Eds.), *Fields Virology*. Philadelphia, PA: Lippincott Williams & Wilkins, p. 1679.

185 STEWART, M. J., SMOAK, K., BLUM, M. A., SHERRY, B. **2005**. Basal and reovirus – induced beta interferon (IFN-beta) and

186 CLARKE, P., DEBIASI, R. L., MEINTZER, S. M., ROBINSON, B. A., TYLER, K. L. **2005**. Inhibition of NF-kappa B and cFLIP expression contribute to viral-induced apoptosis. *Apoptosis* 10, 513–24.

187 NOAH, D. L., BLUM, M. A., SHERRY, B. **1999**. Interferon regulatory factor 3 is required for viral induction of beta interferon in primary cardiac myocyte cultures. *J Virol* 73, 10208–10213.

188 IMANI, F., JACOBS, B. L. **1988**. Inhibitory activity for the interferon-induced protein kinase is associated with the reovirus serotype 1 sigma 3 protein. *Proc Natl Acad Sci USA* 85, 7887–7891.

189 FREED, E. O., MARTIN, M. A. **2001**. HIVs and their replication. In: D. M. KNIPE, P. M. HOWLEY (Eds.), *Fields Virology*. Philadelphia, PA: Lippincott Williams & Wilkins, p. 1971.

190 KORTH, M. J., TAYLOR, M. D., KATZE, M. G. **1998**. Interferon inhibits the replication of HIV-1, SIV, and SHIV chimeric viruses by distinct mechanisms. *Virology* 247, 265–273.

191 LIU, J., PERKINS, N. D., SCHMID, R. M., NABEL, G. J. **1992**. Specific NF-kappa B subunits act in concert with Tat to stimulate human immunodeficiency virus type 1 transcription. *J Virol* 66, 3883–3887.

192 MARSILI, G., BORSETTI, A., SAGARBANTI, M., REMOLI, A. L., RIDOLFI, B., STELLACCI, E., ENSOLL, B., BATTISTINI, A. **2003**. On the role of interferon regulatory factors in HIV-1 replication. *Ann NY Acad Sci* 1010, 29–42.

193 ROY, S., KATZE, M. G., PARKIN, N. T., EDERY, I., HOVANESSIAN, A. G., SONENBERG, N. **1990**. Control of the interferon-induced 68-kilodalton protein by the HIV-1tat gene product. *Science* 247, 1216–1219.

194 BRAND, S. R., KOBAYASHI, R., MATHEWS, M. B. **1997**. The Tat protein of human immunodeficiency virus type 1 is a substrate and inhibitor of the interferon-induced, virally activated protein kinase, PKR. *J Biol Chem* 272, 8388–8395.

195 MARTINAND, C., MONTAVON, C., SALEHZADA, T., SILHOL, M., LEBLEU, B., BISBAL, C. **1999**. RNase L inhibitor is induced during human immunodeficiency virus type 1 infection and down regulates the 2–5A/RNase L pathway in human T cells. *J Virol* 73, 290–296.

196 GUNNERY, S., RICE, A. P., ROBERTSON, H. D., MATHEWS, M. B. **1990**. Tat-responsive region RNA of human immunodeficiency virus 1 can prevent activation of the double stranded-RNA-activated protein kinase. *Proc Natl Acad Sci USA* 87, 8687–8691.

197 DAVIS, M. G., JANSEN, R. W. **1994**. Inhibition of hepatitis B virus in tissue culture by alpha interferon. *Antimicrob Agents Chemother* 38, 2921–2924.

198 WHITTEN, T. M., QUETS, A. T., SCHLOEMER, R. H. **1991**. Identification of the hepatitis B virus factor that inhibits expression of the beta interferon gene. *J Virol* 65, 4699–4704.

199 FOSTER, G. R., ACKROLL, A. M., GOLDIN, R. D., KERR, I. M., THOMAS, H. C., STARK, G. R. **1991**. Expression of the terminal

protein region of hepatitis B virus inhibits cellular responses to interferons alpha and gamma and double-stranded RNA. *Proc Natl Acad Sci USA* 88, 2888–2892.

200 FERNANDEZ, M., QUIROGA, J. A., CARRENO, V. **2003**. Hepatitis B virus down regulates the human interferon-inducible MxA promoter through direct interaction of precore/core proteins. *J Gen Virol* 84, 2073–2082.

201 SHENK, T. E. **2001**. Adenoviridae: the viruses and their replication. In: D. M. KNIPE, P. M. HOWLEY (Eds.), *Fields Virology*. Philadelphia, PA: Lippincott Williams & Wilkins, p. 2265.

202 FRIEDMAN, J. M., HORWITZ, M. S. **2002**. Inhibition of tumor necrosis factor alpha-induced NF-kappa B activation by the adenovirus E3-10.4/14.5K complex. *J Virol* 76, 5515–5521.

203 JUANG, Y. T., LOWTHER, W., KELLUM, M., AU, WC., LIN, R., HISCOTT, J., PITHA, P. M. **1998**. Primary activation of interferon A and interferon B gene transcription by interferon regulatory factor 3. *Proc Natl Acad Sci USA* 95, 9837–9842.

204 LEONARD, G. T., SEN, G. C. **1997**. Restoration of interferon responses of adenovirus E1A-expressing HT1080 cell lines by overexpression of p48 protein. *J Virol* 71, 5095–5101.

205 LEONARD, G. T., SEN, G. C. **1996**. Effects of adenovirus E1A protein on interferon – signaling. *Virology* 224, 25–33.

206 LOOK, D. C., ROSWIT, W. T., FRICK, A. G., GRIS-ALEVY, Y., DICKHAUS, D. M., WALTER, M. J., HOLTZMAN, M. J. **1998**. Direct suppression of Stat1 function during adenoviral infection. *Immunity* 9, 871–880.

207 MATHEWS, M. B., SHENK, T. **1991**. Adenovirus virus-associated RNA and translation control. *J Virol* 65, 5657–5662.

208 MOSS, B. **2001**. Poxviridae: the viruses and their replication. In: D. M. KNIPE, P. M. HOWLEY (Eds.), *Fields Virology*. Philadelphia, PA: Lippincott Williams & Wilkins, p. 2849.

209 SHISLER, J. L., JIN, X. L. **2004**. The vaccinia virus K1L gene product inhibits host NF-kappa B activation by preventing IkappaBalpha degradation. *J Virol* 78, 3553–3560.

210 SYMONS, J. A., ALCAMI, A., SMITH, G. L. **1995**. Vaccinia virus encodes a soluble type I interferon receptor of novel structure and broad species specificity. *Cell* 81, 551–560.

211 COLAMONICI, O. R., DOMANSKI, P., SWETZER, S. M., LARNER, A., BULLER, R. M. **1995**. Vaccinia virus B18R gene encodes a type I interferon-binding protein that blocks interferon alpha transmembrane signaling. *J Biol Chem* 270, 15974–15978.

212 CHANG, H. W., WATSON, J. C., JACOBS, B. L. **1992**. The E3L gene of vaccinia virus encodes an inhibitor of the interferon-induced, double-stranded RNA-dependent protein kinase. *Proc Natl Acad Sci USA* 89, 4825–4829.

213 RIVAS, C., GIL, J., MELKOVA, Z., ESTEBAN, M., DIAZ-GUERRA, M. **1998**. Vaccinia virus E3L protein is an inhibitor of the interferon (i.f.n.)-induced 2–5A synthetase enzyme. *Virology* 243, 406–414.

214 SHARP, T. V., MOONAM, F., ROMASHKO, A., JOSHI, B., BARBER, G. N., JAGUS, R. **1998**. The vaccinia virus E3L gene product interacts with both the regulatory and the substrate binding

regions of PKR: implications for PKR autoregulation. *Virology* 250, 302–315.

215 DAVIES, M. V., ELROY-STEIN, O., JAGUS, R., MOSS, B., KAUKMAN, R. J. **1992**. The vaccinia virus K3L gene product potentiates translation by inhibiting double-stranded-RNA-activated protein kinase and phosphorylation of the alpha subunit of eukaryotic initiation factor 2. *J Virol* 66, 1943–1950.

216 BOWIE, A., KISS-TOTH, E., SYMONS, J. A., SMITH, G. L., DOWER, S. K., O'NEILL, L. A. **2000**. A46R and A52R from vaccinia virus are antagonists of host IL-1 and toll-like receptor signaling. *Proc Natl Acad Sci USA* 97, 10162–10167.

217 STACK, J., HAGA, I. R., SCHRODER, M., BARTLETT, N. W., MALONEY, G., READING, P. C., FITZGERALD, K. A., SMITH, G. L., BOWIE, A. G. **2005**. Vaccinia virus protein A46R targets multiple Toll-like-interleukin-1 receptor adaptors and contributes to virulence. *J Exp Med* 201, 1007–1018.

218 DIPERNA, G., STACK, J., BOWIE, A. G., BOYD, A., KOTWAL, G., ZHANG, Z., ARVIKAR, S., LATZ, E., FITZGERALD, K. A., MARSHALL, W. L. **2004**. Poxvirus protein N1L targets the I-kappaB kinase complex, inhibits signaling to NF-kappaB by the tumor necrosis factor superfamily of receptors, and inhibits NF-kappaB and IRF3 signaling by toll-like receptors. *J Biol Chem* 279, 36570–36578.

219 SEET, B. T., JOHNSTON, J. B., BRUNETTI, C. R., BARRETT, J. W., EVERETT, H., CAMERON, C., SYPULA, J., NAZARIAN, S. H., LUCAS, A., MCFADDEN, G. **2003**. Poxviruses and immune evasion. *Annu Rev Immunol* 21, 377–423.

220 JOHNSTON, J. B., MCFADDEN, G. **2003**. Poxvirus immunomodulatory strategies: current perspectives. *J Virol* 77, 6093–6100.

221 ROIZMAN, B., PELLETT, P. E. **2001**. The family herpesviridae: a brief introduction. In: D. M. KNIPE, P. M. HOWLEY (Eds.), *Fields Virology*. Philadelphia, PA: Lippincott Williams & Wilkins, p. 2381.

222 DIAO, L., ZHANG, B., FAN, J., GAO, X., SUN, S., YANG, K., XIN, D., JIN, N., GENG, Y., WANG, C. **2005**. Herpes virus proteins ICP0 and BICP0 can activate NF-kappa B by catalyzing Ikappa Balpha ubiquitination. *Cell Signal* 17, 217–229.

223 LUND, J., SATO, A., AKIRA, S., MEDZHITOV, R., IWASKI, A. **2003**. Toll-like receptor 9-mediated recognition of herpes simplex virus-2 by plasmacytoid dendritic cells. *J Exp Med* 198, 513–520.

224 MELROE, G. T., DELUCA, N. A., KNIPE, D. M. **2004**. Herpes simplex virus 1 has multiple mechanisms for blocking virus-induced interferon production. *J Virol* 78, 8411–8420.

225 JING, X., CERVENY, M., YANG, K., HE, B. **2004**. Replication of herpes simplex virus 1 depends on the gamma 134.5 functions that facilitate virus response to interferon and egress in the different stages of productive infection. *J Virol* 78, 7653–7666.

226 CASSADY, K. A., GROSS, M., ROIZMAN, B. **1998**. The herpes simplex virus US11 protein effectively compensates for the $\gamma 1(34.5)$ gene if present before activation of protein kinase R by precluding its phosphorylation and that of the α subunit of eukaryotic translation initiation factor 2. *J Virol* 72, 8620–8626.

227 CAYLEY, P. J., DAVIES, J. A., MCCULLAGH, K. G., KERR, I. M. **1984**. Activation of the ppp(A2′p)nA system in interferon-treated, herpes simplex virus-infected cells and evidence for novel inhibitors of the ppp(A2′p)nA-dependent RNase. *Eur J Biochem* 143, 165–174.

228 COMPTON, T., KURT-JONES, E. A., BOEHME, K. W., BELKO, J., LATZ, E., GOLENBOCK, D. T., FINBERG, R. W. **2003**. Human cytomegalovirus activates inflammatory cytokine responses via CD14 and Toll-like receptor 2. *J Virol* 77, 4588–4596.

229 TAYOR, R. T., BRESNAHAN, W. A. **2005**. Human cytomegalovirus immediate-early 2 gene expression blocks virus-induced beta interferon production. *J Virol* 79, 3873–3877.

230 ABATE, D. A., WATANBE, S., MACARSKI, E. S. **2004**. Major human cytomegalovirus structural proteins pp65 (ppUL83) prevents interferon response factor 3 activation in the interferon response. *J Virol* 78, 10995–11006.

231 SIMMEN, K. A., SINGH, J., LUUKKONEN, B. G., LOPPER, M., BITTNER, A., MILLER, N. E., JACKSON, M. R., CROMPTON, T., FRUH, K. **2001**. Global modulation of cellular transcription by human cytomegalovirus is initiated by viral glycoprotein B. *Proc Natl Acad Sci USA* 98, 7140–7145.

232 MILLER, D. M., ZHANG, Y., RAHILL, B. M., WALDMAN, W. J., SEDMAK, D. D. **1999**. Human cytomegalovirus inhibits IFN-α-stimulated antiviral and immunoregulatory responses by blocking multiple levels of IFN-α signal transduction. *J Immunol* 162, 6107–6113.

233 MILLER, D. M., RAHILL, B. M., BOSS, J. M., LAIRMORE, M. D., DIRBIN, J. E., WALDMAN, J. W., SEDMAK, D. D. **1998**. Human cytomegalovirus inhibits major histocompatibility complex class II expression by disruption of the Jak/Stat pathway. *J Exp Med* 187, 675–683.

234 SUN, Q., ZACHARIAH, S., CHAUDHURY, P. M. **2003**. The human herpes virus 8-encoded viral FLICE-inhibitory protein induces cellular transformation via NF-kappaB activation. *J Biol Chem* 278, 52437–52445.

235 ZIMRING, J. C., GOODBOURN, S., OFFERMANN, M. K. **1998**. Human herpesvirus 8 encodes an interferon regulatory factor (IRF) homolog that represses IRF-1-mediated transcription. *J Virol* 72, 701–707.

236 LUBYOVA, B., PITHA, P. M. **2000**. Characterization of a novel human herpesvirus 8-encoded protein, vIRF-3, that shows homology to viral and cellular interferon regulatory factors. *J Virol* 74, 8194–8201.

237 BURYSEK, L., YEOW, W. S., LUBYOVA, B., KELLUM, M., SCHAFER, S. L., HUANG, Y. Q., PITHA, P. M. **1999**. Functional analysis of human herpesvirus 8-encoded viral interferon regulatory factor 1 and its association with cellular interferon regulatory factors and p300. *J Virol* 73, 7334–7342.

238 BURYSEK, L., PITHA, P. M. **2001**. Latently expressed human herpesvirus 8-encoded interferon regulatory factor 2 inhibits double-stranded RNA-activated protein kinase. *J Virol* 75, 2345–2352.

239 IZUMI, K. M., KIEFF, E. D. **1997**. The Epstein–Barr virus oncogene product latent membrane protein 1 engages the tumor necrosis factor receptor-associated death domain protein to mediate B lymphocyte growth transformation and activate NF-kappaB. *Proc Natl Acad Sci USA* 94, 12592–12597.

240 HUEN, D. S., HENDERSON, S. A., CROOM-CARTER, D., ROWE, M. **1995**. The Epstein–Barr virus latent membrane protein-1 (LMP1) mediates activation of NF-kappa B and cell surface phenotype via two effector regions in its carboxy-terminal cytoplasmic domain. *Oncogene* 10, 549–560.

241 KANDA, K., DECKER, T., AMAN, P., WAHLSTROM, M., KALLIN, B. **1992**. The EBNA2-related resistance towards alpha interferon (IFN-alpha) in Burkitt's lymphoma cells effects induction of IFN-induced genes but not the activation of transcription factor ISGF-3. *Mol Cell Biol* 12, 4930–4936.

242 CLARKE, P. A., SCHWEMMLE, M., SCHICKINGER, J., HILSE, K., CLEMENS, M. J. **1991**. Binding of Epstein–Barr virus small RNA EBER-1 to the double-stranded RNA-activated protein kinase DAI. *Nucleic Acids Res* 19, 243–248.

243 HOWLEY, P. M., LOWY, D. R. **2001**. Papillomaviruses and their replication. In: D. M. KNIPE, P. M. HOWLEY (Eds.), *Fields Virology*. Philadelphia, PA: Lippincott Williams & Wilkins, p. 2197.

244 SPITKOVASKY, D., HEHNER, S. P., HOFMANN, T. G., MOLLER, A., SCHMITZ, M. L. **2002**. The human papillomavirus oncoprotein E7 attenuates NF-kappa B activation by targeting the Ikappa Bkinase complex. *J Biol Chem* 277, 25576–25582.

245 RONCO, L. V., KARPOVA, A. Y., VIDAL, M., HOWLEY, P. M. **1998**. Human papillomavirus 16 E6 oncoprotein binds to interferon regulatory factor-3 and inhibits its transcriptional activity. *Genes Dev* 12, 2061–2072.

246 BARNARD, P., MCMILLAN, N. A. **1999**. The human papillomavirus E7 oncoprotein abrogates signaling mediated by interferon-alpha. *Virology* 259, 305–313.

247 LI, S., LABRECQUE, S., GAUZZI, M. C., CUDDIHY, A. R., WONG, A. R., PELLEGRINI, S., MATLASHEWSKI, G. J., KOROMILAS, A. E. **1999**. The human papilloma virus (HPV)-18 E6 oncoprotein physically associates with Tyk2 and impairs Jak–STAT activation by interferon-alpha. *Oncogene* 18, 5727–5737.

**Section C
Clinical Applications**

10
Overview of Clinical Applications of Type I Interferons

Frank Müller

10.1
Introduction

Isaacs and Lindenmann, who studied what had been a subject for research since 1935 [1], i.e. the phenomenon of viral interference, reported in 1957 a factor that conferred the property of viral interference and coined the name interferon (IFN) to describe this new substance [2].

In the following decades many attempts were undertaken in order to purify IFN to homogeneity. As a result, IFN was found to be heterogeneous, comprising a family of distinct, but structurally related, IFN molecules [3]. Today these are classified as type I (IFN-α, β, ω and -τ), all of which bind to the same receptor consisting of IFNAR-1 and -2 [4]. A single antigenically distinct type II (IFN-γ) IFN has also been characterized.

As a response to stimulation by a various pathogens, type I IFNs are produced and secreted by virtually all eukaryotic cells. They can be mass produced by biotechnological processes, and both natural and recombinant IFNs have received regulatory approval in many countries for the treatment of diseases, including chronic hepatitis (CH), malignant melanoma, non-Hodgkin's lymphoma (NHL), chronic myelogenous leukemia (CML), multiple sclerosis (MS) and others [5] (see Section 10.5).

10.2
Biological Effects

IFNs belong to the network of cytokines that are involved in the control of cellular functions and replication, and that become actively engaged in host defense during infection. Type I IFNs induce a wide range of biological effects, as listed below, and are used in the clinic as antiviral and antitumor agents.

The Interferons: Characterization and Application. Edited by Anthony Meager
Copyright © 2006 WILEY-VCH Verlag GmbH & Co. KGaA, Weinheim
ISBN: 3-527-31180-7

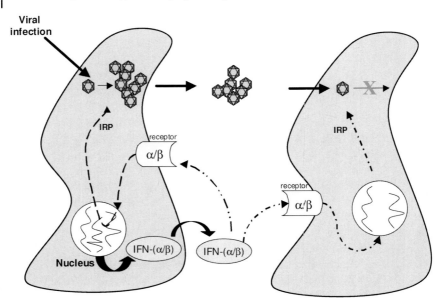

Fig. 10.1. Induction and mechanism of IFN; IRP = IFN-regulated proteins. (Modified from [62].)

- Inhibition of viral replication
- Inhibition of cell proliferation
- Regulation of cell differentiation
- Activation/propagation of natural killer cells and cytotoxic cell activity resulting in antitumor effects, for example

IFNs are also used to treat certain autoimmune diseases, especially MS, and show interesting results in rheumatoid arthritis [6, 79].

The IFNs themselves do not inhibit viral propagation, but they induce an antiviral status by triggering the expression of IFN-regulated proteins in the surrounding noninfected cells (Fig. 10.1).

10.3
Type I IFN Products Currently Available or Under Development

The techniques of modern biotechnology have made it possible to isolate and to generate highly purified IFN-α and -β. Since the first approval of two recombinant IFN-α (2a and 2b) preparations in 1986 by the FDA, many other IFN products have been developed, and substantial research has been done on their clinical efficiency in nonmalignant and malignant diseases.

10.3 Type I IFN Products Currently Available or Under Development

Table 10.1 gives an overview of currently available IFNs or those in various clinical phases, including their field of application.

Several different expression systems are used for the commercial production of Type 1 IFNs. The first approved products on the market were Roferon A (IFN-α2a; Hoffmann-La Roche) and Intron A (IFN-α2b; Schering-Plough), both of them produced in *Escherichia coli*.

Infergen is an aqueous formulation of recombinant, nonglycosylated, N-terminal methionine, "consensus" IFN-α also expressed in *E. coli*. The amino acid sequence of Infergen was designed by selecting at each sequence position the most commonly occurring amino acid at that position among the various subtypes of natural human IFN-α. Four additional changes were made to facilitate molecular construction [10].

Today, IFNs are also available which have their origin of production in yeasts [IFN-α2a, *Hansenula polymorpha* (Rhein Minapharm) and IFN-α2b, *Pichia pastoris* (Shanta Biotechnics)]. These expression systems show economic advantages, process-related simplifications, and, in contrast to *E. coli*, they have the capability to secrete proteins and to perform modifications and processing steps linked to the secretory pathway – most importantly, the formation of disulfide bonds and glycosylation [13].

Alferon N (Hemispherx Biopharma) is a nonrecombinant IFN product derived from human leukocytes and contains a mixture of natural IFN-α subtypes [12].

Turning to IFN-β, IFN-β1b, the so-called Betaf(s)eron (Schering), is again an *E. coli* product, whereas IFN-β1a [Rebif (Ares-Serono) and Avonex (Biogen-Idec)] is derived from mammalian cell culture using transfected [Chinese hamster ovary (CHO)] cells. IFN-β1a is glycosylated, whereas IFN-β1b is not; additionally, the latter has cysteine (position 17) replaced by serine. Without this modification, the undesired, but thermodynamically favored, disulfide bond between Cys17 and Cys41 can be formed during molecular refolding in the downstream processing of *E. coli*-derived IFN-β-1b. This leads to an incorrect conformation that shows an "antigenic character" and is less active than the natural IFN-β with its disulfide bond between Cys31 and Cys141. IFN-β1b has a lower specific activity than that of IFN-β1a *in vitro*.

Soluferon (Fig. 10.2) is also an IFN-β, but due to molecular modeling this IFN-β variant has the advantage of a higher solubility than the natural molecule because of its significantly lower hydrophobic character. As a result of these modifications, an improvement of the bioavailability of the IFN-β "mutein" and also an increase of pharmacological stability is expected [107, 108].

Because "pegylation" appears particularly useful for therapeutic proteins that are required for a sustained period of time, as the IFNs are, in the early 1990s a first attempt was made to develop a pegylated IFN-α [16, 17]. It took several years of improvement, but in 2000 the first pegylated IFN (PegIntron; pegylated IFN-α2b) was approved, followed by a second one, called Pegasys (pegylated IFN-α2a). PegIntron is a 12-kDa linear pegylated IFN-α2b and Pegasys a 40-kDa branched pegylated IFN-α2a [17, 18].

Tab. 10.1. Overview of currently available IFN products [6–10, 119, 120, 123, 124]

Trade name/product	Organism	Company	Approval/clinical phase	Indication/application
Pegasys (pegylated IFN-α2a)	E. coli	Hoffmann-La Roche	2002 (EU)/2002 (US)	CHC
			2005 (EU)	CHB
PegIntron (pegylated rh-IFN-α2b)	E. coli	Schering-Plough/Enzon	2000 (EU)/2001 (US)	hepatitis C
			phase III	CHB
			phase III	CML
			phase III	malignant melanoma
			phase I	solid tumors
Viraferon PEG (pegylated rh-IFN-α2b)	E. coli	Schering-Plough	2000 (EU)	CHB/C
Roferon A (IFN-α2a)	E. coli	Hoffmann-La Roche	1986 (US)	HCL
			1988	KS
			1995	CML
			1996	hepatitis C
Intron A (IFN-α2b)	E. coli	Schering-Plough	1986 (US)	HCL
			1988	genital warts
			1988	KS
			1991	hepatitis C
			1992	hepatitis B
			1995	malignant melanoma
				follicular lymphoma
			1997	CML
Viraferon (IFN-α2b)	E. coli	Schering-Plough	2000 (EU)	CHB/C

10.3 Type I IFN Products Currently Available or Under Development

Product	Source	Company	Approval	Indication
Alfatronol (IFN-α2b)	E. coli	Schering-Plough	1999 (EU/US)	hepatitis B, hepatitis C, HCL, CML, multiple myeloma, follicular lymphoma, carcinoid tumors, malignant melanoma
Infergen (IFN-alfacon-1)	E. coli	Amgen/Yamanouchi Europe/Intermune	1997 (US)/1999 (EU)	CHC
PEG-alfacon-1 (pegylated IFN-alfacon-1)	E. coli	Intermune	phase I	hepatitis C
Alferon N (IFN-αn3)	human leukocytes	Hemispherx Biopharma	1989	genital warts (condyloma acuminata)
Alferon LDO (IFN-αn3)	human leukocytes	Hemispherx Biopharma	phase I/II	HIV infection
Alferon (IFN-αn3)	human leukocytes	Hemispherx Biopharma	phase II	West Nile virus encephalitis
Virtron (IFN-α2b)	E. coli	Schering-Plough	(1999 EU/US)	Treatment of CHB and CHCC

Tab. 10.1 *(continued)*

Trade name/product	Organism	Company	Approval/clinical phase	Indication/application
Reiferon (IFN-α2a)	*Hansenula polymorpha*	Minapharm/Rhein Biotech	2004 (Egypt)	CHC
				CHB
				HCL
				KS
				CML
				follicular lymphoma
				cutaneous T cell lymphoma
				malignant melanoma
				RCC
Shanferon (IFN-α2b)	*Pichia pastoris*	Shanta Biotechnics	2002 (India)	CML
				CHC
				CHB
Recombinant IFN-α2b	*Lemna* (algae)	Biolex	First trials started in Q1/2005	hepatitis C
Veldona (natural-oral IFN-α)	Ball-1 cells (lymphoblastoid cell line)	Amarillo Bioscience	phase III	Sjögren's syndrome
			phase II	fibromyalgia syndrome
			phase II	pulmonary fibrosis
			phase II	oral warts (HIV-positive)
			phase II/III	Behçets disease
Albuferon-α (Albumin-IFN-α2a)	CHO (?)	Human Genome Science	phase II b	CHC

r-IFN-α2b XL	E. coli	Flamel Technologies	phase I/II	hepatitis B hepatitis C
Rebif (IFN-β-1a)	CHO	Ares-Serono	1998 (EU) 2002 (US)	RRMS
r-IFN-β	CHO	Ares-Serono	phase III	CHC in Asian patients
Avonex (IFN-β-1a)	CHO	Biogen IDEC	1997 (EU) 1996 (US)	RRMS
			phase II	chronic inflammatory demyelinating polyneuropathy (CIDP)
Betaferon (IFN-β-1b)	E. coli	Schering AG	1995 (EU)	RRMS
Betaseron (IFN-β-1b)	E. coli	Berlex Laboratories/Chiron	1993 (US)	RRMS
Albuferon-β (Albumin-IFN-β-1a)	CHO(?)	Human Genome Science	preclinical	no data available
Improved/modified interferon β	no data available	Maxygen/Biovitrum	preclinical	MS
Soluferon (modified IFN-β)	CHO	Vakzine Management/ProBioGen	development	RRMS

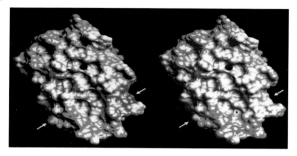

Fig. 10.2. A comparison of the surface area of the wild-type IFN-β and the modified IFN-β (Soluferon). www.igb.fraunhofer.de. (This figure also appears with the color plates.)

Fusion proteins, such as Albuferon-α and -β, in which of human serum albumin (HSA) is conjugated to IFN, also show improved pharmacokinetic and pharmacodynamic profiles due to prolonged serum half-life [119, 120]. Albuferon-α is in clinical phase II and Albuferon-β in the preclinical phase.

10.4
Pharmacokinetics

10.4.1
IFN-α

IFN-α is normally administered by intramuscular (i.m.) or subcutaneous (s.c.) injection. Maximal plasma concentration of the IFN-α occurs after 1–8 h [4]. For the i.m. route of administration, the peak plasma level was observed earlier (after 4 h) than for the s.c. route (after 7.5–8 h).

IFN-α is totally filtered out in the kidney glomeruli and undergoes proteolytic degradation during renal tubular reabsorption, causing a negligible amount of IFN-α to remain in systemic circulation. Liver metabolism and subsequent biliary excretion are minor pathways in the elimination process [106].

A major drawback to the therapies with IFNs (α and β), however, is their short serum half-life and rapid clearance from the circulation. The reported terminal elimination half-life of IFN-α ranges from 4 to 16 h [17, 20]. After 24 h, IFN-α is undetectable in the bloodstream.

Compared with unmodified IFN-α, pegylated IFNs display a significantly (approximately 10-fold) prolonged plasma half-life [18]. Polyethylene glycol (PEG) shows little toxicity and is eliminated from the body intact either by the kidneys (for PEGs < 30 kDa) or in the feces (for PEGs > 20 kDa). PEG lacks immunogenicity and antibodies to PEG are generated in rabbits only if PEG is combined with highly immunogenic proteins. There are no reports of the generation of antibodies to PEG under routine clinical administration of pegylated proteins [19].

The amount of IFN-α used in clinical treatments is disease dependent. Several oncological indications require daily dosing, while viral indications may require only thrice-weekly dosing. The IFN treatment for these disease categories may range from periods of several months to a year or longer [17]. Pegylated IFNs generally follow a once weekly administration schedule.

Nonpegylated IFNs and pegylated IFNs have comparable safety and pharmacodynamic profiles [18]. However, the potential for inducing neutralizing antibodies (NAbs) may differ. For example for Intron A this potential was found to be lower (fewer than 1% patients treated developed Nabs) than for Roferon A (about 25%) before the formulation of the latter was changed; subsequently "new formulation" Roferon A has shown much reduced immunogenicity [10].

10.4.2
IFN-β

IFN-β is metabolized in the liver. Peak serum concentrations were achieved in 3–15 h in healthy volunteers following a 60 μg i.m. dose. It remains controversial whether the administration route has an impact on the bioavailability or not. For IFN-β1a it was shown that it has a 3-fold increase in bioavailability with i.m. injection as compared with s.c. route [106]. The elimination half-life varies from 1 to 10 h [4, 106]. Obert and Pöhlau arrived at the conclusion that both administration routes lead to comparable bioavailabilties [111]. Sometimes IFN-β is also administered by intravenous (i.v.) or intratumoral (i.t.) routes.

Biologic response marker levels (e.g. neopterin, tryptophan and β_2-microglobulin) increase within 12 h of injection and remain elevated for at least 4 days; they usually peak within 48 h of administration. Injection site reactions show that IFN-β1a is better tolerated than IFN-β1b [109, 110].

The incidence of Nabs to IFN-β is found to be within the range of 2–24% for IFN-β1a preparations and 31–58% for IFN-β1b [111–113] (see Chapter 13).

10.5
Clinical Applications of Type I Interferons

The first clinical studies were undertaken in the late 1970s with crude preparations derived from virus-induced leukocytes that were less than 1% IFN by weight and consisted of a mixture of various type I IFNs. Kari Cantell in Helsinki was nevertheless able to produce appreciable amounts of that crude IFN. This "low-level-pure" IFN was the first and, until that time, the only IFN available for human clinical trials [136]. However, these preparations indicated the therapeutic potential of IFNs [3]. Studies found that IFN-α could induce regression of tumors in significant numbers of patients.

IFN-α was first cloned by Weissman et al. at the end of 1979 and IFN-β was cloned for the first time by Taniguchi et al. in September 1979 [136]. From 1978 onwards substantial progress was made in the purification of IFNs: Pestka et al. found a strategy to purify human leukocyte IFN (now known as IFN-α) to homogeneity [114] by using reverse-phase high-performance liquid chromatography technology.

Today there are many fields of application for IFNs and there are thousands of people worldwide – suffering from certain types of cancer, viral diseases or autoimmune diseases – who benefit from the plentiful availability of IFNs. The most prominent areas for medical applications and the role of IFNs for each of the diseases is described in the sections.

10.5.1
Chronic hepatitis B (CHB)

Viral hepatitis infections are one of the major public health concerns worldwide and CHB is one of the chief causes of adult mortality. Millions around the globe are infected with the hepatitis B virus (HBV; Fig. 10.3), but most people who become infected get rid of the virus within 6 months. A short infection is known as an "acute" case of hepatitis B. About 10% of people infected with HBV develop a chronic, life-long infection; approximately 350–400 million people are affected.

A proportion of these people (Fig. 10.4) develop chronic liver disease and are

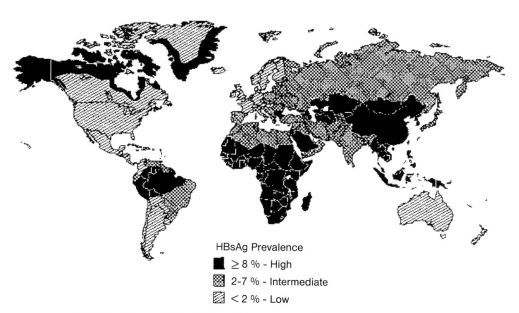

Fig. 10.3. Geographic distribution of chronic HBV infection [61].

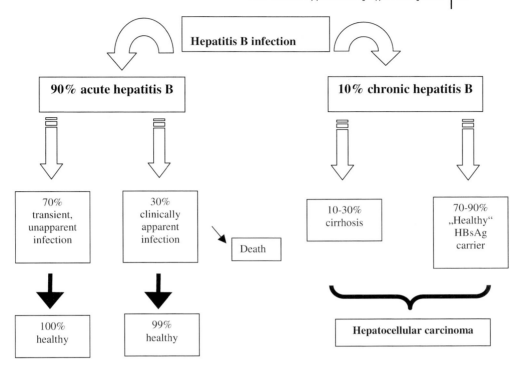

Fig. 10.4. Course of hepatitis B infection. (Modified from [6]).

therefore at risk of progression to cirrhosis or hepatocellular carcinoma (HCC). It has been estimated that about 0.5% of all people infected with HBV develop HBV-associated HCC 20–40 years after infection [60].

The HBV belongs to the family of hepadnaviruses and is a double-stranded, enveloped DNA virus, which replicates in the liver and causes hepatic dysfunction [61, 62]. The virus particle is 42 nm in diameter and is composed of a 27-nm nucleocapsid core (HbcAg) surrounded by an outer lipoprotein coat containing the surface antigen (HBsAg). It has been found that it is not the virus itself that damages the hepatocytes, but the host response to the infection, e.g. apoptotic signals mediated by the CD95 receptor–ligand system [63].

IFN-α has been shown to be an effective agent to suppress the replication of HBV. The application of IFN-α tends to decrease the levels of serum HBV DNA and HBeAg in the blood as several clinical trials have clearly shown. Between 15 and 40% of treated patients respond with a long-term remission [64–67].

There are several different categories of patients and the efficacy of IFN-α treatment in CHB depends on what category the patient belongs to. A very detailed and precise overview on this topic is given by Lok and McMahon [70]. The following categories have to be taken into consideration:

- HBeAg-positive patients
- HBeAg-negative patients
- Nonresponders
- Positive clinical cirrhosis patients

The dose regimen by s.c. injection of IFN-α in therapy for treatment of CHB is normally:

- Duration of IFN-α administration: 4–6 months
- 5 MU m^{-2} daily or 10 MU m^{-2} thrice weekly (for children 6 MU m^{-2} thrice weekly)

A response to therapy is monitored by a decrease of HBV-DNA (undetectable levels in unamplified assays; below 10^5 copies mL^{-1}) and the loss of HBeAg in patients who were initially HBeAg-positive. Furthermore, the decrease of serum alanine aminotransferase (ALT) levels as a biochemical response is also monitored.

In February 2005, pegylated IFN-α2a (Pegasys) was approved for the treatment of CHB. Pegasys is also well known as a highly effective treatment of CHC (see Section 10.5.2). The approval was granted for two categories of the disease – HBeAg-positive and HBeAg-negative CHB.

In comparison to the conventional, unpegylated IFN-α, the virological responses were significantly higher with pegylated IFN-α2a (Pegasys). At the end of follow-up, HBeAg was cleared in 37, 35 and 29% of patients receiving Pegasys at doses of 90, 180 and 270 µg, respectively, compared with only 25% of patients on conventional IFN-α2a therapy [71]. Furthermore it was found that the remission rate was significantly higher than observed in a monotherapy with Lamivudine (Epivir-HBV or 3TC). Lamivudine is the (–)-enantiomer of 2′,3′-dideoxy-3′-thiacytidine (3TC). Incorporation of the active triphosphate (3TC-TP) into growing DNA chains results in premature chain termination and thereby inhibits HBV-DNA synthesis [70]. Forty-three percent of Pegasys-treated patients showed a reduction of HBV-DNA to less than 2×10^4 copies mL^{-1}, whereas only 29% of patients who were administered Lamivudine achieved this reduction [68].

10.5.2
Chronic hepatitis C (CHC)

Infections with the hepatitis C virus (HCV) are common worldwide. The WHO estimates that about 3% of the world's population are HCV carriers. Progression to CHC occurs in up to 85% of infected people; approximately 20% of these develop liver cirrhosis and 2–5% contract HCC [93].

The HCV was found in 1989 by Michael Houghten et al. and is an enveloped RNA virus with a diameter of 60–80 nm that belongs to the Flaviviridae family. Eleven major HCV genotypes and several subtypes are known today. These genotypes differ by 30–50% in their nucleotide sequences. Furthermore, these viruses

Tab. 10.2. Location of the various HCV genotypes (adapted from [91])

Genotype	Location
1a, 1b, 1c	Western Europe, Scandinavia, Eastern Europe, USA, Japan, Hong Kong
2a, 2b, 2c	Western Europe, Scandinavia, USA, Japan, Hong Kong
3a, 3b	Western Europe, Scandinavia, USA
4a	Middle East, Egypt
5a	South Africa
6a	Hong Kong

have a high tendency to mutate. Table 10.2 indicates the geographic location of the different genotypes [62, 91, 92].

The genetic heterogeneity of HCV plus the significant antigenic variation among virus quasispecies within individual patients probably explains the difficulties of finding a unified effective therapy and accounts for the problems in the development of prophylactic vaccines. Nevertheless, propitious effects of IFN-α in hepatitis C treatment were first reported by Hoofnagle et al. in 1986, but only 15–20% of patients with CHC achieved a sustained virological response when IFN-α was used in a monotherapeutic approach [94]. In 1998, a combination therapy with IFN-α and ribavirin was introduced. Ribavirin is a synthetic guanosine analog that was initially developed for the treatment of HIV. Given alone ribavirin has no effect on HCV-RNA, but it was found to reduce the serum ALT levels when given 600–1200 mg daily for a period of 12–36 months [96]. This observation led to clinical studies of the combination of IFN-α and ribavirin, which have shown that a sustained response rate of 38–43% is achieved in patients treated for 12 months [95].

An additional improvement in the treatment of hepatitis C has been was achieved with the development of pegylated IFNs. The advantages of pegylated proteins are well known and established in many cases [98–100]. The responses to pegylated IFNs in a monotherapy are in the same range of those reported in therapy with IFN-α combined with ribavirin [101, 102].

The current "gold standard" for treatment of CHC is a combination therapy of pegylated IFN-α and ribavirin. In contrast to unmodified IFNs, the pegylated IFNs have a significantly prolonged, approximately 10-fold greater, plasma half-life [18]. As a consequence the frequency of "drug" administration is reduced to a once-weekly regime from the thrice-weekly regime used for the unmodified IFNs. Their diminished/slower clearance increases the efficacy for the combination therapy with ribavirin. PegIntron-ribavirin is known as Rebetol (Schering-Plough) and Pegasys-ribavirin is known as Copegus (Hoffmann-La Roche).

Both pegylated IFNs are given weekly by s.c. injection; examples of the dose levels are given in Tab. 10.3. On average, sustained virological responses in 50–60% of patients have been found in either Rebetol or Copegus therapies. However, HCV genotype 1 is less responsive (cure rates 42–52%) and requires 12 months for

Tab. 10.3. Regimen for combination therapy of pegylated IFN with Ribavirin

IFN	Ribavirin
Pegylated IFN-α2a: 180 µg	Genotype 1, 1000 mg day^{-1} if <75 kg Genotype 1, 1200 mg day^{-1} if >75 kg Genotype 2/3, 800 mg day^{-1}
Pegylated IFN-α2b: 1.5 µg kg^{-1}	Genotype 1, 1000 mg day^{-1} if <75 kg Genotype 1, 1200 mg day^{-1} if >75 kg Genotype 2/3, 800 mg day^{-1}

therapy compared to the responsiveness of the genotypes 2 and 3 (cure rates 78–84%; therapy duration 6 months) [93, 103, 104].

10.5.3
Chronic hepatitis D (CHD)

CHD is caused by the hepatitis δ virus (HDV). Rizzetto et al. discovered HDV in 1977 [80]. It is a small, defective, single-stranded RNA virus of 36–43 nm that only replicates efficiently in the presence of HBsAg. CHD leads to cirrhosis in up to 70% of patients and is a severe form of liver disease [81–84]. Worldwide, approximately 10–15 million people are infected by HDV [85]. Diagnosis is based on finding both anti-HDV and HBsAg in serum of patients with chronic liver disease. This diagnosis is important because the dose and regimen of IFN-α differs from the therapy used in CHB [81].

Pilot studies as well as randomized controlled trials indicated that in CHD treatment requires high doses of IFN-α (5 MU day^{-1} or 9–10 MU thrice weekly) for a prolonged period of time. High dosages were effective in reducing disease symptoms in 25–50% of patients. For instance, one CHD patient treated in this manner and monitored for 12 years was found to have lost HDV RNA, HBV DNA and HBsAg at the end of treatment. In one randomized controlled study, IFN-α was given at 9 MU thrice weekly for 12 months and 36% of patients showed remission in disease. A dose of 3 MU thrice weekly was shown to be less effective [81, 84, 86–91].

10.5.4
Hairy Cell Leukemia (HCL)

HCL (or leukemic reticuloendotheliosis) is a rare malignant lymphoproliferative disorder. It is an uncommon form of leukemia that is a slowly progressive cancer of lymphocytes. Malignant cancer cells are found in both the peripheral blood and bone marrow. HCL is characterized by a reduction in all types of blood cells. HCL cells have fine projections from their surface that create an impression of the cells appearing "hairy" [22].

People affected with this form of leukemia have an increased risk of acquiring infections because of the lowered resistance to infection. The cause of this disease is still unknown. The risk of men being affected is 5 times higher than for women and the median age of onset is 55. Other risk factors are unknown.

In 1984, it was first reported that patients with HCL showed a response to IFN-α therapy. In 1986, both IFN-α2a and -α2b were approved by the FDA for treating HCL. IFN-α therapy is associated with high overall response rates; normalization of peripheral blood (including hemoglobin levels, white blood cells, neutrophils, monocytes and platelets) is usually manifested in more than 80% of treated patients. A study using Roferon A, for example, showed that out of 75 treated patients, 61% achieved a complete or partial response, 28% had a minor remission and 11% remained stable. None had worsening of disease. Responding patients also showed a marked decrease in infection episodes and improvement in performance status [23].

In the standard treatment of HCL, IFN-α is normally administered over a 2-year period at 3 MU/day. Relapse of almost all patients occurs upon cessation of therapy. Today, IFN-α has been superceded by more effective therapies for HCL, e.g. by 2-deoxycoformycin or Cladribine [4].

10.5.5
Renal Cell Carcinoma (RCC)

RCC, a form of kidney cancer that involves cancerous changes in the cells of the renal tubule, is the most common type of kidney cancer in adults. RCC affects about 3:10 000 people [54]. What causes the cells to become cancerous is not known, but family history, smoking and inheritance increase the risk for RCC. The first symptom is usually blood in the urine. The cancer metastasizes easily, most often to the lungs and other organs, and about one-third of patients have metastasis at the time of diagnosis. Rare spontaneous remissions have been observed and documented [55].

In clinical trials, Ritchie and Pyrhonen [56, 57], for example, demonstrated a 10–15% objective response rate together with a modest increase in survival times in patients treated with IFN-α. Since its antitumor activity in RCC was low and there was no dramatic effect on long-term survival of patients, IFN-α has also been tested in many combination therapies, using both chemical drugs and biological agents, such as interleukin-2 [58]. As yet, none of these studies has resulted in durable clinical benefits. Since October 2003, another phase III clinical trial is underway in order to analyze the combination of IFN-α plus an anti-vascular endothelial growth factor antibody (anti-VEGF) [59]. No results have been reported yet.

10.5.6
Basal Cell Carcinoma (BCC)

Basal cells are normal skin cells. They may develop cancerous changes, causing a lump or bump that is painless. A new skin growth that ulcerates, bleeds easily or does not heal well may indicate development of basal cell skin cancer. BCC is a slow-growing skin tumor involving cancerous changes in basal skin cells. BCC is the most common type of skin cancer in humans, but has a high cure rate with conventional cancer therapies. However, neglect can allow the cancer to enlarge, causing possible disability or, in rare cases, death. This cancer usually remains local and almost never metastasizes to distant parts of the body, but it may continue to grow, and invade and destroy nearby tissues and structures, including the nerves, bones and brain.

In several clinical trials intralesional injection of IFN-α has been found to be highly effective in inducing BCC regression. Using intralesional IFN-α over a 3-week period, the overall success rate in most of the trials was between 70 and 100%.

10.5.7
Malignant Melanoma

Melanomas are malignant neoplasms arising from melanin-producing cells of the basal layer of the skin epidermis. It can also involve the pigmented portion of the eye. Melanoma is an aggressive type of cancer that can spread very rapidly. The exact cause of melanoma is unknown although risk factors have been intensively studied. Melanomas tend to occur at sites of intermittent, intensive sun exposure rather than at body sites receiving the greatest UV irradiation. Major risk factors for melanoma include changing nevus, increased age, large number of moles, history of multiple atypical moles, etc.

In treatment of malignant melanoma, IFN-α is normally used as an adjuvant therapy [24] after surgical removal of the tumor, with a margin of normal skin also removed. Surgical removal of nearby lymph nodes may accompany removal of the tumor. Radiation therapy or chemotherapy may also be an additional treatment to the IFN-α therapy. The benefit from IFN-α therapy remains controversial because concerns over toxicity have limited its application. The IFN-α treatment regimen currently most in favor is a prolonged regimen, which demonstrates a significant relapse-free benefit in some cases, but shows strong adverse effects [116–118].

10.5.8
Kaposi Sarcoma (KS)

KS is named after Moritz Kaposi, who first described it in 1872. A sarcoma is a type of cancer that develops in connective tissues such as cartilage, bone, fat, muscle, blood vessels or fibrous tissues [29]. There are several types of KS. They each differ in their patterns of symptoms and organs likely to be affected, how aggressively the cancer grows and spreads, risk factors, and other personal characteristics of patients. The treatment used and the likelihood of survival depend on the type of KS, as well as other factors.

The different types of KS are:
- Classic KS
- African (endemic) KS
- Transplant related (acquired) KS
- AIDS related (endemic) KS

KS is caused by a virus called the KS-associated herpesvirus (KSHV) [human herpesvirus 8 (HHV-8)]. HHV-8 is related to other herpes viruses, such as the viruses that cause cold sores and genital herpes, as well as Epstein–Barr virus (EBV; the "mono" virus) and cytomegalovirus (CMV). However, these other herpes viruses do not cause KS [30, 31].

A standard therapy for the treatment of Kaposi sarcoma does not exist. Apart from surgery, chemotherapy and radiation therapy, immunotherapy with IFN-α is a very effective therapy, either as monotherapy or in combination, e.g. with AZT (zidovudine) or paclitaxel. With IFNs, IFN-α2a or -α2b, it is possible to achieve a remission rate for the intermittent KS as well as for the AIDS-related KS of 45–70%. The probability of responding to IFN therapy is correlated with CD4 cells. Patients with more than 400 CD4 cells μL^{-1} show a response higher than 45%, whereas only 7% of those patients with less than 200 cells μL^{-1} showed a response [32].

A great deal of research is under way to find more effective treatments for KS, above all to reduce toxicity of treatments. The administration of pegylated IFN-α2b (PegIntron), for example, gave promising data in a study that was performed by Thomas-Gerber in 2002 [33]. Alternative therapeutic approaches using the so-called angiogenesis inhibitors, drugs that block the development of blood vessels – the main component of KS lesions – within tumors, may also provide increased clinical benefits.

10.5.9
Multiple Myeloma

Multiple myeloma (also known as myeloma or plasma cell myeloma) is a progressive hematologic (blood) disease. It is a cancer of the plasma cells – an important part of the immune system that produces immunoglobulins to help fight infections and diseases.

Multiple myeloma is characterized by excessive numbers of abnormal plasma cells in the bone marrow and overproduction of intact monoclonal immunoglobulin (IgG, IgA, IgD or IgE). Hypercalcemia, anemia, renal damage, increased susceptibility to bacterial infection and impaired production of normal immunoglobulin are common clinical manifestations of multiple myeloma. It is often also characterized by diffuse osteoporosis, usually in the pelvis, spine, ribs and skull [121].

The role of IFN in the treatment of multiple myeloma has been investigated for several decades. The benefits of therapy with IFN remain controversial because of discordant results from different clinical trials. Mandelli et al. published data from a comparison of IFN-treated patients versus non-treated and found a 22% reduced relapse rate after 33 months of follow-up [137, 138]. A moderate improvement in relapse-free survival was shown in a recent meta-analysis of 24 randomized clinical trials carried out with 4000 patients. In the same trials only a minor improvement in survival was demonstrated.

Currently, a high-dose melphalan regimen is still the standard therapy, because of its potential for inducing complete remission rates in the 40% to 50% range [139].

10.5.10
Chronic myelogenous leukemia (CML)

CML is a disease characterized by three well-defined stages:

- Initial (chronic) phase: variable duration ranging from a few months to a few years
- Accelerated phase: variable duration
- Terminal (blast) phase: median duration of 2–4 months

The blast crisis is a highly malignant, secondary acute leukemia defined by the presence of 30% or more blasts in the peripheral blood or bone marrow. The blasts affect the normal elements of the marrow, leading to bleeding or hemorrhage, infection and thrombosis in the central nervous system, which result in death of the patient.

CML is a clonal disorder that is usually easily diagnosed because the leukemic cells of more than 95% of the patients have a distinctive cytogenetic abnormality – the Philadelphia chromosome (Ph1). The Ph1 results from a reciprocal translocation between the long arms of chromosomes 9 and 22, and is demonstrable in all hematopoietic precursors. This in turn results in a fused *BCR–ABL* gene and in the production of an abnormal tyrosine kinase protein that causes the disordered myelopoiesis found in CML.

IFN is the only substance that has consistently been shown to prolong survival and decrease tumor burden, as compared to chemotherapy [21, 35, 36]. In 1994, a group of 322 patients was treated either with IFN-α2a or by conventional chemotherapy. The IFN-α2a monotherapy was associated with a major cytogenetic response of about 30% in comparison to 5% in the chemotherapy arm [37]. Several other clinical trials confirmed these results.

Combination therapy is now considered as a rational approach, since research studies have demonstrated synergistic growth inhibition of leukemic myeloid progenitors with the coadministration of IFN-α and ara-C (cytarabine) [40, 41]. In 1999, Kantarjian et al. [40] found that 92% patients treated with IFN-α at 5 MU m^{-2} and daily 10 mg ara-C showed a complete hematologic response, and 74% demonstrated a cytogenetic response. These results were significantly higher than in control groups (IFN-α monotherapy or IFN-α with intermittent ara-C). However, the introduction of Imatinib, an ABL tyrosine kinase inhibitor, has revolutionized drug treatment of CML and has in the last few years largely supplanted IFN-α therapy [141].

10.5.11
Non-Hodgkinis lymphoma (NHL)

The ailment is named after Thomas Hodgkin, who described it as a new disease as long ago as 1832. Two types of lymphomas have to be distinguished: (i) the Hodgkin lymphoma and (ii) all other lymphomas, which are collectively called NHLs [49]. NHLs are malignancies of lymphoid tissue (lymph nodes, spleen and other organs of the immune system). The NHL is classified into four different stages ranked in ascending order by the severity code [50].

A variety of factors, including infectious, physical and chemical agents, have been associated with an increased risk of developing NHL. Infectious agents such as viral infections (EBV, HIV and the human T cell leukemia virus), and bacterial infections (such as *Heliobacter pylori*) have been associated with the development of NHLs.

IFN can induce shrinkage of tumors in some NHL types. For example, IFNs have been found to be active as agents against follicular lymphoma or T cell lymphoma [51–53]. Nevertheless clinical experts remain uncertain about the benefits of IFN and there are questions about its suitability in combination with chemotherapy. IFNs generally are ineffective as single agents against intermediate- or high-grade lymphomas [122].

10.5.12
Laryngeal Papillomatosis

Laryngeal papillomatosis is a disease resulting in the growth of benign tumors inside the larynx (voice box), vocal cords or the air passages leading from the nose into the lungs (respiratory tract). It is a rare disease caused by the human papillomavirus (HPV). Although scientists are uncertain how people are infected with HPV, more than 60 types of HPVs have been identified. Tumors caused by HPVs, called papillomas, are often associated with two specific types of the virus – HPV-6 and -11. They may vary in size and grow very quickly. Eventually, these tumors may block the airway passage and cause difficulty in breathing [45].

Treatment is usually based on the cryodestruction of the lesions with a carbon dioxide laser or on mircosurgery. Once they have been resected, the tumors have a tendency to return. In addition to the above-mentioned treatments, IFN-α was successfully used by i.v. application in combination with cidofovir, whereas a monotherapy with IFN-α seems to be inefficient [46, 47]. Controversial results were reported from a multicenter phase IV study of 169 patients in September 2004 [48]. After surgical removal of the lesions, IFN-α2b was administered by the i.m. route. Seventy-three percent of the patients concluded the treatment without reappearance of new lesions, while the rest showed a significant reduction in the number and size of lesions.

10.5.13
Mycosis Fungoides (MF)

In 1806, the term "mycosis fungoides" was first used by Alibert, a French dermatologist, when he described a severe disorder in which large necrotic tumors resembling mushrooms presented on a patient's skin [43].

MF is the most common type of the relatively rare lymphomas which are known under the "umbrella" name of cutaneous T cell lymphoma. It represents a specific subset of cutaneous non-Hodgkin's lymphoma [42]. MF is usually an indolent (slow-growing) cancer, but in rare cases may become aggressive. This cancer is seen only rarely by most physicians, dermatologists or oncologists and is often hard to diagnose in its early stages.

Recombinant IFN-α has been recognized as an active substance in the treatment of MF in patients who did not respond to standard therapies. These trials demonstrated significant response rates of 45–65% in extensive pretreated patients [42].

Today IFN-α is often administered in a combination to a chemotherapy or a so-called "psoralen and UV A (PUVA)" therapy [44].

10.5.14
Condyloma Acuminata

Condyloma acuminata (genital warts) is the official name for a type of wart that appears on the genital area of men and women. It is caused by highly contagious HPV, which is generally sexually transmitted between adults. However, some people apparently can have the virus, not show any signs of condyloma and still transmit the virus to a partner, who may then develop condyloma. The incubation period typically ranges from 3 weeks to 8 months.

Alferon N, a human leukocyte-derived, multi-subspecies IFN-α is especially indicated for the intralesional treatment of refractory and recurrent external condyloma acuminata (approval in 1986). In a double-blind study with 156 patients, Alferon N produced complete or partial resolution of warts in 80% of treated patients compared with 44% of the patients receiving placebo. Patients were injected intra-

Tab. 10.4. Degree of resolution as measured by total wart volume per patient [n (%)]

Degree of resolution	Alferon N (N = 81)	Placebo (N = 75)
Complete resolution	44 (54%)[a]	15 (20%)
Partial resolution (>50%)	21 (26%)[a]	18 (24%)
Minor resolution (<50%)	12 (15%)	10 (13%)
Progression/no change	4 (5%)	32 (43%)

[a] Complete and partial resolution data show a statistically significant ($P < 0.001$) difference from placebo [34].

lesionally with a mean of 225 000 IU of Alferon N injection per wart 2 times a week for up to 8 weeks. See Tab. 10.4.

Recombinant IFN-α2b is also administered by intralesional injection and has also been approved for the treatment of genital warts [34].

10.5.15
Multiple Sclerosis (MS)

As one of the most common inflammatory diseases of the central nervous system, MS (or encephalomyelitis disseminata) affects approximately 2.5 million individuals worldwide. MS is approximately twice as common in women as in men, and the symptoms typically begin in young adulthood and are rarely present outside the age range of 20–50 years [73, 74].

A single cause for MS is not yet identified [72] and the disease seems to originate from the interaction of several environmental influences. Many studies hint that MS is an autoimmune disease whereby the body's defenses attack and destroy the myelin sheath of neuronal axons at *multiple* centers of inflammation. This results in a so-called "myelin stripping process", and leads to unprotected axons and to a significant dysfunction of the nervous system. The consequence is a progressive loss of myelin. Only a few parts of the myelin can be rebuilt and other parts of the nerve tissue are scared over (sclerosis) [73].

It is also conjectured that genetic factors and viral or bacterial infections are associated with MS and play a prominent role in its initiation and progression [75, 76]. Several scientific investigations are for example focused on the influence of *Chlamydia pneumoniae*, HHV-6 and the EBV [77]. Results presented in a recent publication [78] suggest the impact of EBV infection to the pathogenesis of MS.

MS is classified into four categories as in the Tab. 10.5.

Cytokines have been extensively studied in MS and several different IFN-β1 (a/b) preparations are established as first-line therapeutics for RRMS, but not for other categories. IFN-β has been demonstrated to be efficacious in RRMS by decreasing the relapse rate between exacerbations of the disease by approximately 30%; a re-

Tab. 10.5. Different categories in MS [77, 132]

Category	Description
Primary-progressive MS (PPMS)	PPMS is characterized by a nearly continuous neurologic deterioration from the onset of symptoms (around 15% of patients)
Progressive-relapsing MS (PRMS)	involves a progressive course from onset with occasional relapses later in disease
Relapsing-remitting MS (RRMS)	80–85% of patients have a RRMS characterized by episodes with or without full or partial recovery and clinically stable periods between these episodes
Secondary-progressive MS (SPMS)	around 50% of patients with RRMS develop many years later a slow, insidiously progressive, neurological deterioration over many years with or without clinical attacks superimposed – this is termed SPMS

duction in the number of magnetic resonance imaging (MRI) lesions has been found as well.

The recommended dosage for IFN-β-1b (Betaferon, Betaseron) is 250 µg (8 MIU) s.c. every second day. The administration for the IFN-β1a differs for the two products (Avonex, Rebif) on the market. Avonex is given in 30 µg (8 MIU) portions once a week by i.m. injection, whereas Rebif is administered s.c. 3 times a week using 22 or 44 µg (6–12 MIU) IFN-β1a.

The application for SPMS is not yet approved, but has been evaluated in several distinct studies [133, 134].

The exact mechanism of how IFN-β acts as a therapeutic agent in MS is not fully understood at present, but it has been found that IFN-β suppresses T cell proliferation, leads to a reduction of T cell migration into the central nervous system and alters the T cell cytokine secretion repertoire from relatively pro-inflammatory T helper 1 (T$_h$1) response to relatively anti-inflammatory T$_h$2 response [77, 135].

10.6
IFN Toxicity

Apart from all the positive influences IFN shows in clinical usage, there is unfortunately always an association with significant toxicological side-effects, which are largely dose dependent. Very high doses are profoundly toxic and have defined the maximum tolerated dose at around 30–50 MIU.

Toxicological side effects can be divided into the following eight groups:

- Constitutional (fatigue, headache, lightheadeness, chills, dehydration, loss of weight, anorexia, fever, myalgias)
- Neuropsychiatric (confusion, dizziness, lethargy, numbness, tingling, depression)
- Hematologic (thrombocytopenia, anemia, neutropenia, leukopenia)
- Hepatic
- Gastrointestinal (diarrhea, emesis, nausea, abdominal pain, constipation)
- Cardiovascular (hypotension, hypertension, sinus tachycardia)
- Metabolic (hypocalcemia, hyperglycemia, hypoglycemia, hypophosphatemia)
- Other (allergic reaction, pruritis, wound infection, dry skin, rash)

A further classification of toxicological phenomena can be done by division into chronic and acute appearances. Fatigue, anorexia and neuropsychiatric symptoms rank among the chronic manifestations [4]. Fever, chills, headaches, myalgias and malaise count as examples of the acute phenomena.

10.7
Type I IFNs in the Future

Although over the past decades a great deal of successful development work in the production of IFNs by biotechnological processes, as well as some success in their clinical application, has been achieved, significant hurdles to bring additional applications forward, to optimize dosages in existing therapies without serious toxicological side-effects and to develop the next generation IFNs, with an improved pharmacokinetic and pharmacodynamic profile, are still to be overcome.

One of those potential new fields of application, with improved IFN formulation and administration, could be the treatment of Sjögren's syndrome, an autoimmune disease in which the body's immune system mistakenly attacks its own moisture-producing glands. Sjögren's is one of the most prevalent autoimmune disorders, striking as many as 4 000 000 Americans, for example. Nine out of 10 patients are women. The average age of onset is the late 40s, although Sjögren's can occur in all age groups in both women and men [123, 105].

About 50% of the time Sjögren's syndrome occurs as a single pathology and 50% of the time it exists together with another connective tissue disease. The four most common diagnoses that co-exist with Sjögren's syndrome are rheumatoid arthritis, systemic lupus erythematosus, systemic sclerosis (scleroderma) and polymyositis/dermatomyositis. Sometimes researchers refer to the first type as "primary Sjögren's" and the second as "secondary Sjögren's." All instances of Sjögren's syndrome are systemic, affecting the entire body [123, 105].

The hallmark symptoms are dry eyes and dry mouth. Sjögren's may also cause dryness of other organs, affecting the kidneys, gastrointestinal tract, blood vessels, lung, liver, pancreas and the central nervous system. Many patients experience debilitating fatigue and joint pain. Symptoms can plateau, worsen or go into remission. While some sufferers experience mild symptoms, others suffer debilitating symptoms that greatly impair their quality of life [123, 105].

Amarillo Biocsciences has developed a low-dose oral IFN-α2a [123] which is now in different clinical trials. Apart from this application for Sjögren's syndrome, there are trials for therapeutic applications in fibromyalgia syndrome, pulmonary Fibrosis, oral warts (HIV-positive) and Behçets disease on the way.

Flamel Technologies is developing another next-generation formulation of native IFN-α2b with better performance (i.e. longer time of action, higher bioavailability and efficacy, reduced side-effects, and improved compliance) compared with first-generation IFNs. They use a poly-amino acid nanoparticle system (Medusa) for their product formulation. This modified IFN is currently in phase I/II clinical trials [124].

Albuferon-α and -β are novel, long-acting forms of IFN-α2a and -β from Human Genome Sciences (HGS). They are both fused to HSA, thus combining the therapeutic activity of IFN-α/β with the long half-life of HSA. In April 2005, HGS reported the results of phase II clinical trials for Albuferon-α in patients with CHC. These results showed that Albuferon is well tolerated, remains in the blood substantially longer than is reported for either recombinant- or pegylated IFN-α and exhibits robust antiviral activity [125]. Albuferon-α was detectable for up to 4 weeks and exhibited a median half-life of 148 h.

A phase III study with recombinant IFN-β in 250 patients in Asia with CHC was initiated by Serono in 2002. Results from the study completed in 2001 suggested that patients of Asian origin with this indication will benefit [126, 128–131]. In 2005, Serono also started with the development of an IFN-β that can be administered by inhalation [126].

Tak et al. [79] have considered the potential of IFN-β treatment in rheumatoid arthritis. Promising data [127] from the first clinical trials have to be confirmed by further trials; clinically meaningful benefits of IFN-β therapy remain to be shown in rheumatoid arthritis.

On the basis of the ongoing progress in the manufacturing of "improved" IFNs and many encouraging results for the therapeutic application of type 1 IFNs in a broad range of diseases, it is expected that the next years and decades should result in further enhancement of *in vivo* activity and the perfecting of the clinical usage of IFNs.

New elements involved in IFN-mediated signaling may also contribute the development of novel IFN-based or -related therapies in the future [140].

References

1 Cantell K. *The Story of Interferon (The Ups and Downs in the Life of a Scientist)*. World Scientific Publishing, Singapore, **1998**.

2 ISAACS A, LINDENMANN J. Virus interference. I. The interferon. *Proc Roy Soc Lond B* **1957**, *147*, 258–267.

3 PESTKA S. The human interferon-alpha species and hybrid proteins. *Semin Oncol* **1997**, *24 (Suppl 9)*, S9-4–S9-17.

4 JONASCH E, HALUSKA FG. Interferon in oncological practice: review of interferon biology, clinical applications, and toxicities. *The Oncologist* **2001**, *6*, 34–55.

5 PFEFFER LM, DINARELLO CA, HERBERMANN RB, et al. Biological properties of recombinant αIFN – 40th anniversary of the discovery of Interferons. *Cancer Res* **1998**, *58*, 2489–2499.

6 GELLISSEN G. Hansenula polymorpha – *Biology and Application*. Wiley-VCH, Weinheim, **2002**.

7 GELLISSEN G. *Production of Recombinant Proteins*. Wiley-VCH, Weinheim, **2005**.

8 www.phrma.org/newmedicines/resources/2004-10-25.145.pdf; accessed 27 June 2005.

9 www.gdch.de/taetigkeiten/nch/down/biopharmazeutika.pdf; accessed 27 June 2005.

10 RADER RA. *Biopharma: Biopharmaceutical Products in the US Market*. 3rd edn. BioPlan, Rockville, MD, Sept. 2004, pp. 157–190.

11 *The Cytokine Handbook*, 3rd edn. Academic Press, New York, Editor: Agnus Thomson **1998**.

12 ALFERON N; www.hemispherx.net/content/products/product_brochure.htm; accessed 27 June 2005.

13 MÜLLER F II, TIEKE A, WASCHK D, et al. Production of IFN-α-2a in *Hansenula polymorpha*. *Process Biochem* **2001**, *38*, 15–25.

14 Infergen; www.intermune.com/pdf/infergen_pi.pdf; accessed 27 June 2005.

15 www.igb.fraunhofer.de; accessed 27 June 2005.

16 WANG YS, YOUNGSTER ST, GRACE M, et al. Structural and biological characterisation of pegylated recombinant interferon alpha-2b and its therapeutic implications. *Adv Drug Del Rev* **2002**, *54*, 547–570.

17 BAILON P, PALLERONI A, CAROL A, et al. Rational design of a potent, long-lasting form of interferon: a 40 kDa branched polyethylene glycol-conjugated interferon α-2a for the treatment of hepatitis C. *Bioconjugate Chem* **2001**, *12*, 195–201.

18 GLUE P, FANG JWS, ROUZIER-PANIS R, et al. Pegylated interferon-α2b: pharmacokinetics, pharmacodynamics, and preliminary efficacy data. *Clin Pharmacol Ther* **2000**, *68*, 556–567.

19 MILTON HARRIS J, CHESS RB. Effect of pegylation on pharmaceuticals. *Nat Drug Disc Rev* **2003**, *2*, 214–221.

20 WILLS RJ. Clinical pharmacokinetics of interferons. *Clin Pharmacokinet* **1990**, *19*, 390–399.

21 www.eudra.org/humandocs/PDFs/EPAR/Introna/267299FI2.pdf; accessed 27 June 2005.

22 www.introna.com/introna/index.jsp; accessed 27 June 2005.

23 www.rocheusa.com/products/roferon/pi.pdf; accessed 27 June 2005.

24 VOLKENANDT M, HALLEK M, SCHMID-WENDTNER MH, et al. Adjuvante medikamentöse Therapie des malignen Melanoms. In: *Maligne Melanome*. Tumorzentrum, München, **2000**.

25 Kirkwood JM, Strawdermann MH, Ernsthoff MS, et al. Interferon α-2b adjuvant therapy of high risk resected cutaneous melanoma. The Eastern Cooperative Oncology Group Trial EST 1684. *J Clin Oncol* **1996**, *14*, 7–17.

26 Kirkwood JM, Ibrahim J, Sondak V, et al. Prelimanary analysis of the E1690/S9111/C9190 intergroups postoperative adjuvant trial of high and low dose IFN α2b in high risk primary or lymphnode metastatic melanoma. *Proc Am Ass Clin Oncol* **1999**, *18*, 2072.

27 Pehamberger H, Soyer HP, Steiner A, et al. Adjuvant interferon α-2a treatment in resected primary Stage II cutaneous melanoma. *J Clin Oncol* **1998**, *16*, 1425–1429.

28 Grob JJ, Dreno B, de la Salmonière P, et al. Randomized trial of IFN-α2a as adjuvant therapy in resected primary melanoma thicker than 1.5 mm without clinically detectable node metastases. *Lancet* **1998**, *351*, 1905–1910.

29 www.nlm.nih.gov/medlineplus/kaposissarcoma.html; accessed 27 June 2005.

30 Andreoni M, Sarmat L, Nicastri E, et al. Primary human Herpesvirus-8 infection in immunocompetent children. *J Am Med Ass* **2002**, *287*, 1295–1300.

31 Aoki Y, Tosato G. HIV-1 Tat enhances KSHV infectivity. *Blood* **2004**, *104*, 810–814.

32 hiv.net/2010/buch/ks.htm; accessed 27 June 2005.

33 Thoma-Greber E, Sander CA, Messer G, et al. Pegyliertes IFN-α-2b: Neue Therapie bei klassischem KS. Klinik Münchener Fortbildungswoche für Dermatologen, München, **2002**.

34 www.hemispherx.net/content/products/product_brochure.htm; accessed 27 June 2005.

35 Druker BJ, Sawyers CL, Capdeville R, et al. Chronic myelogenous leukemia. *Hematology* **2001**, 87–112.

36 Silver RT, Woolf SH, Hehlmann R, et al. An evidence based analysis of the effect of busulfan, hydroxyurea, interferon and allogeneic bone marrow transplantation in treating the chronic phase of CML. *Blood* **1999**, *94*, 1517–1536.

37 The Italian Cooperative Study Group on CML. "Interferon-alpha-2a as compared with convertion chemotheraphy for the treatment of chronic myeloid leukemia". *N Engl J Med* **1994**, *330*, 820–825.

38 Ohnishi K, Ohno R, Tomonaga M, et al. A randomized trial comparing interferon-alpha with busulfan for newly diagnosed chronic myelogenous leukemia in chronic phase. *Blood* **1995**, *86*, 906–916.

39 Broustet A, Reiffers J, Marit G, et al. Hydroxyurea versus interferon-alfa-2b in chronic myelogenous leukemia: prelimi-nary results of an open French multicentre randomized study. *Eur J Cancer* **1991**, *27 (Suppl 4)*, 18–21.

40 Kantarjian HM, Keating MJ, Estey EH, et al. Treatment of Philadelphia chromosome-positive early chronic myelogenous leukemia with daily doses of interferon-alpha and low-dose cytarabine. *J Clin Oncol* **1999**, *17*, 284–292.

41 Guilhot F, Chastang C, Michallet M, et al. Interferon alfa-2b combined with cytarabine versus interferon alone in

chronic myelogenous leukemia. French Chronic Myeloid Leukemia Study Group. *N Engl J Med* **1997**, *337*, 223–229.

42 KUZEL TM, ROENIKG HH, SAMUELSON E, JR., et al. Effectiveness of interferon-alfa-2a combined with phototherapy for Mycosis Fungoides and the Sézary Syndrome. *J Clin Oncol* **1995**, *13*, 257–263.

43 www.emedicine.com; accessed 27 June 2005.

44 www.cancer.gov/cancertopics/pdq/treatment/mycosisfungoides; accessed 27 June 2005.

45 www.nidcd.nih.gov/health/voice/laryngeal.asp; accessed 27 June 2005.

46 ARMBRUSTER C, KREUZER A, VORBACH H, et al. Successful treatment of severe respiratory papillomatosis with intravenous cidofovir and interferon α-2b. *Eur Respir J* **2001**, *17*, 830–831.

47 SNOECK R, WELLENS W, DESLOOVERE C. Treatment of severe Laryngeal Papillomatosis with intralesional injections of cidofovir. *J Med Virol* **1998**, *54*, 219–225.

48 NORDARSE-CUNÍ H, IZNAGA-MARÍN N, VIERA-ALVAREZ D, et al. Interferon alpha-2b as adjuvant treatment of recurrent respiratory papillomatosis in Cuba: National programme (1994–1999 report). *J Laryngol Otol* **2004**, *118*, 681–687.

49 www.cancer.org/docroot/CRI/content/CRI_2_4_1X_What_Is_Non_ Hodgkins_Lymphoma_32.asp; accessed 27 June 2005.

50 www.nci.nih.gov/cancertopics/pdq/treatment/adult-non-hodgkins/patient/allpages; accessed 27 June 2005.

51 ROHATINER AZS, GREGORY WM. PETERSON B. Meta-analysis to evaluate the role of interferon in Follicular Lymphoma. *J Clin Oncol* **2005**, *23*, 2215–2223.

52 LOUIE AC, GALLAGHER JC, SIKORA K, et al. Follow up observations on the effect of human leukocyte interferon in non-Hodgkin's lymphoma. *Blood* **1981**, *58*, 712–718.

53 LEAVITT J, RATANATHATHORN V, OZER H, et al. Alfa-2b Interferon in the treatment of Hodgkin disease and non-Hodgkin's lymphoma. *Semin Oncol* **1987**, *14 (Suppl 2)*, 18–23.

54 www.nlm.nih.gov/medlineplus/ency/article/000516.htm; accessed 27 June 2005.

55 OIVER RTD, NETHERSELL ABW, BOTTOMLEY JM. Unexplained spontaneous regression and alpha-interferon as treatment for metastatic renal carcinoma. *Br J Urol* **1989**, *63*, 128–131.

56 RITCHIE AWS. Interferon-alpha and survival in metastatic renal carcinoma: early results of a randomized controlled trial. Medical Research Council Renal Cancer Collaborators. *Lancet* **1999**, *353*, 14–17.

57 PYRHONEN S, SALMINEN E, RUUTU M, et al. Prospective randomized trial of interferon alfa-2a plus vinblastine versus vinblastine alone in patients with advanced renal cell cancer. *J Clin Oncol* **1999**, *17*, 2859–67.

58 NEGRIER S, ESCUDIER B, LASSET C, et al. Recombinant interleukin-2, recombinant interferon alfa-2a, or both in metastatic renal-cell carcinoma. Groupe Francais d'Ìmmunotherapie. *N Eng J Med* **1998**, *338*, 1272–1278.

59 RINI BI, HALABI S, TAYLOR J, et al. Cancer and Leukemia Group B 90206: a randomized phase III trial of interferon-α

plus anti-vascular endothelial growth factor antibody (Bevacizumab) in metastatic renal cell carcinoma. *Clin Cancer Res* **2004**, *10*, 2584–2586.

60 BUENDIA MA, PATERLINI P, TIOLLAIS P, et al. Hepatocellular carcinoma: molecular aspects. In: ZUCKERMAN AJ, THOMAS CT (Eds.), *Viral Hepatitis*. Churchill Livingstone, London, **1998**, pp. 179–200.

61 MAHONEY FJ. Update on diagnosis, management, and prevention of hepatitis B virus infection. *Clin Microbiol Rev* **1999**, *12*, 351–366.

62 BRANDIS H, KÖHLER W, EGGERS HJ, PULVERER G. *Lehrbuch der Medizinischen Mikrobiologie 7*. Gustav Fischer, Jena, **1994**.

63 GALLE PR, HOFMANN J, WALCZAK H, et al. Involvement of the CD95 (APO-1/Fas) receptor and ligand in liver damage. *J Exp Med* **1995**, *182*, 1223–1230.

64 LOK ASF, WU P-C, LAI C-L, et al. A controlled trial of interferon with or without prednisone priming for chronic hepatitis B. *Gastroenterology* **1992**, *102*, 2091–2097.

65 DI BISCEGLIE AM, BERGASA N, FONG T-L, et al. A randomized controlled trial of recombinant alpha interferon therapy for chronic hepatitis B. *Am J Gastroenterol* **1993**, *88*, 1887–1892.

66 WONG DK, CHEUNG AM, O'ROURKE K, et al. Effect of alpha-interferon treatment in patients with hepatitis B e antigen-positive chronic hepatitis B. *Ann Intern Med* **1993**, *119*, 312–323.

67 BROOK MG, KARAYIANNIS P, THOMAS HC. Which patients with chronic hepatitis B virus infection will respond to alpha-interferon therapy? *Hepatology* **1989**, *10*, 761–763.

68 MARCELIN P, et al. Peginterferon Alfa-2a alone, lamivudine alone, and the two in combination in patients with HBeAg-negative chronic hepatitis B. *N Engl J Med* **2004**, *351*, 32–43.

69 HADZIYYANNIS SJ. Hepatitis B e antigen negative chronic hepatitis B: from clinical recognition to pathogenesis and treatment. *Viral Hepatitis Rev* **1995**, *1*, 7–15.

70 LOK ASF, MCMAHON BJ. Chronic Hepatitis B. *Hepatolgy* **2001**, *34*, 1225–1241.

71 COOKSLEY WEG, PIRATVISUTH T, LEE SD, et al. Peginterferon α-2a (40 kDa): an advance in the treatment of hepatitis B e antigen-positive chronic hepatitis B. *J Viral Hepatitis* **2003**, *10*, 298–305.

72 Multiple Sclerosis International Federation; www.msif.org/en/ms_the_disease/causes_of_ms.html; accessed 27 June 2005.

73 Deutsche Multiple Sklerose Gesellschaft; www.dmsg.de/index.php?kategorie=wasistms; accessed 27 June 2005.

74 National MS Society; www.nationalmssociety.org/Sourcebook-Epidemiology.asp; accessed 27 June 2005.

75 OKSENBERG JR, BARANZI SE, BARCELLOS LF, et al. Multiple sclerosis: genomic rewards. *J Neuroimmunol* **2001**, *113*, 171–184.

76 HUNTER SF, HAFLER DA. Ubiquitous pathogens – links between infection and autoimmunity in MS? *Neurology* **2000**, *55*, 164–165.

77 KEEGAN BM, NOSEWORTHY JH. Multiple sclerosis. *Ann Rev Med* **2002**, *53*, 285–302.
78 CEPOK S, ZHOU D, SRIVASTAVA R, et al. Identification of Epstein–Barr virus proteins as putative targets of the immune response in multiple sclerosis. *J Clin Invest* **2005**, *115*, 1352–1360.
79 VAN HOLTEN J, PLATER-ZYBERK C, TAK PP. Interferon-β for treatment of rheumatoid arthritis? *Arthritis Res* **2002**, *4*, 346–352.
80 RIZZETTO M, CANESE MG, ARIC S, et al. Immunofluorescence detection of new antigen-antibody system (delta/anti-delta) associated to hepatitis B virus in liver and serum of HBsAg carriers. *Gut* **1977**, *18*, 997–1003.
81 HOOFNAGLE JH. Therapy of viral hepatitis. *Digestion* **1998**, *59*, 563–578.
82 LAI MMC. The molecular biology of hepatitis delta virus. *Annu Rev Biochem* **1995**, *64*, 259–286.
83 WHO; www.who.int/csr/disease/hepatitis/HepatitisD_whocdscsrncs2001_1.pdf; accessed 27 June 2005.
84 ÖRMECI N. Short- and long-term effects of treatment of chronic hepatitis B and delta virus by IFN. *Fundam Clin Pharmacol* **2003**, *17*, 651–658.
85 HADZIYANNIS SJ. Review: hepatitis delta. *J Gastroenterol Hepatol* **1997**, *12*, 289–298.
86 LAU DTY, KLEINER DE, PARK Y, et al. Resolution of chronic delta hepatitis after 12 years of interferon alfa therapy. *Gastroenterology* **1999**, *117*, 1229–1233.
87 THOMAS HC, FARCI P, SHEIN R, et al. Inhibition of hepatitis delta virus replication by lymphoblastoid human alpha interferon. *Prog Clin Biol Res* **1987**, *234*, 277–290.
88 HOOFNAGLE JH, MULLEN K, PETERS M, et al. Treatment of chronic delta hepatitis with recombinant human alpha interferon. *Prog Clin Biol Res* **1987**, *234*, 291–298.
89 FARCI P, MANDAS A, COIANA A, et al. Treatment of chronic hepatitis D with interferon alfa-2a. *N Engl J Med* **1994**, *330*, 88–94.
90 DI MARCO V, GIACCHINO R, TIMITILLI A, et al. Long-term interferon-α treatment of children with chronic hepatitis delta: a multicenter study. *J Viral Hepat* **1996**, *3*, 123–128.
91 WALSH K, ALEXANDER GJM. Update on chronic viral hepatitis. *Postgrad Med J* **2001**, *77*, 498–505.
92 WHO; www.who.int/csr/disease/hepatitis/Hepc.pdf; accessed 27 June 2005.
93 LO RE III V, KOSTMAN JR. Management of chronic hepatitis C. *Postgrad Med J* **2005**, *81*, 376–382.
94 HOOFNAGLE J, MULLEN K, JONES D. Treatment of chronic non-A, non-B hepatitis with recombinant human alpha-Interferon. *N Engl J Med* **1986**, *315*, 1575–1578.
95 POYNARD T, MARCELLIN P, LEE SS, et al. Randomized trial of interferon alfa 2b plus ribavirin versus interferon alfa 2b plus placebo for 48 weeks for treatment of chronic infection with hepatitis C virus. *Lancet* **1998**, *352*, 1426–1431.

96 hepatitis-c.de/ifn.htm; accessed 27 June 2005.
97 www.adis.com/files/hepC1–2-00.pdf; accessed 27 June 2005.
98 KATRE NV. Immunogenicity of recombinant IL-2 modified by covalent attachment of polyethylene glycol. *J Immunol* **1990**, *144*, 209–213.
99 AVGERINOS GC, TURNER BG, GORELICK KJ, et al. Production and clinical development of a *Hansenula polymorpha*-derived PEGylated hirudin. *Semin Thromb Hemost* **2001**, *27*, 357–72.
100 WONG SF, CHAN HO. Effects of a formulary change from granulocyte colony-stimulating factor to granulocyte-macrophage colony-stimulating factor on outcomes in patients treated with myelosuppressive chemotherapy. *Pharmacotherapy* **2005**, *25*, 372–378.
101 REDDY KR, WRIGHT TL, POCKROS PJ, et al. Efficacy and safety of pegylated (40-kd) Interferon α-2a compared with Interferon α-2a in noncirrhotic patients with chronic hepatitis C. *Hepatology* **2001**, *33*, 433–438.
102 LINDSAY KL, TREPO Ch, HEINTGES T, et al. A randomized, double-blind trial comparing pegylated Interferon alfa-2b to Interferon alfa-2b as initial treatment for chronic hepatitis C. *Hepatology* **2001**, *34*, 395–403.
103 HADZIYANNIS SJ, SETTE H JR., MORGAN TR, et al. Peginterferon-α2a and ribavirin combination therapy in chronic hepatitis C. *Ann Intern Med* **2004**, *140*, 346–355.
104 FRIED MW, SHIFFMAN ML, REDDY KR, et al. Peginterferon alfa-2a plus ribavirin for chronic hepatitis C virus infection. *N Engl J Med* **2002**, *347*, 975–982.
105 www.sjogrens.org; accessed 27 June 2005.
106 www.uspharmacist.com/oldformat.asp?url=newlook/files/Feat/interferons.htm&pub_id=8&article_ id=731; accessed 27 June 2005.
107 www.igb.fraunhofer.de/WWW/Presse/bilder/download.bis2000/IGB_IFN-beta.jpg
108 www.igb.fraunhofer.de/WWW/Presse/Jahr/2003/download.en.doc/US06572853.pdf; accessed 27 June 2005.
109 [No authors listed]. Interferon beta-1A: new preparation. A short-term impact on the course of multiple sclerosis. *Prescrire Int* **1998**, *37*, 142–143.
110 www.avonex.com; accessed 27 June 2005.
111 OBERT HJ, PÖHLAU D. *Beta-Interferon – Schwerpunkt Multiple Sklerose*. Springer, Berlin, **1999**.
112 ABDUL-AHAD AK, GALAZKA AR, REVEL M, et al. Incidence of antibodies to interferon-beta in patients treated with recombinant human interferon-beta 1a from mammalian cells. *Cytokines Cell Mol Ther* **1997**, *3*, 27–32.
113 BERTOLOTTO A, MALUCCHI S, SALA A, et al. Differential effects of three interferon betas on neutralizing antibodies in patients with multiple sclerosis: a follow up study in an independent laboratory. *J Neurol Neurosurg Psychiatry* **2002**, *73*, 148–153.
114 www.isicr.org/newsletter/isicr4.2.pdf; accessed 28 June 2005.
115 TANIGUCHI T, FUJII-KURIYAMA Y, MURAMATSU M. Molecular cloning of human interferon cDNA. *Proc Nat Acad Sci USA* **1980**, *77*, 4003–4006.
116 KIRKWOOD J, STRAWDERMAN MH, ERNSTOFF MS, et al.

Interferon alfa-2b adjuvant therapy of high-risk resected cutaneous melanoma: The Eastern Cooperative Oncology Group Trial EST 1864. *J Clin Oncol* **1996**, *14*, 7–17.

117 PEHAMBERGER H, SOYER HP, STEINER A, et al. Adjuvant interferon alfa-2a treatment in resected primary Stage II cutaneous melanoma. *J Clin Oncol* **1998**, *16*, 1425–1429.

118 KIRKWOOD J, IBRAHIM J, SONDAK V, et al. Hugh- and low-dose interferon alfa-2b in high-risk melanoma: First analysis of Intergroup Trial E1690/S9111/C9190. *J Clin Oncol* **2000**, *18*, 2444–2458.

119 SUNG C, NARDELLI B, LAFLEUR DW, et al. An IFN-β–albumin fusion protein that displays improved pharmaco- kinetic and pharmacodynamic properties in nonhuman primates. *J Interferon Cytokine Res* **2003**, *23*, 25–369.

120 Press releases Human Genome Sciences; www.hgsi.com; accessed 30 June 2005.

121 www.multiplemyeloma.org/about_myeloma/index.html; accessed 30 June 2005.

122 www.oncologychannel.com/nonhodgkins/generaltreatment.shtml bio; accessed 30 June 2005.

123 www.amarbio.com; accessed 1 July 2005.

124 www.flamel.com/techAndProd/medusa.shtml pipeline; available from Internet 2005 July 1.

125 www.hgsi.com/products/albuferon.html; accessed 1 July 2005.

126 www.serono.com; accessed 1 July 2005.

127 TAK PP, HART BA, KRAAN MC, et al. The effects of interferon-beta treatment on arthritis. *Rheumatology* **1999**, *38*, 362–369.

128 KONISHI I, HORIIKE N, MICHITAKA K, et al. Renal transplant recipient with chronic hepatitis C who obtained sustained viral response after interferon-β therapy – case report. *Intern Med* **2004**, *43*, 931–934.

129 FESTI D, SANDRI L, MAZZELLA E, et al. Safety of Interferon-β treatment for chronic HCV hepatitis. *World J Gastroenterol* **2004**, *10*, 12–16.

130 PELLICANO R, PALMAS F, CARITI G, et al. Re-treatment with interferon-beta of patients with chronic hepatitis C virus infection. *Eur J Gastroenterol Hepatol* **2002**, *14*, 1377–1382.

131 RIZZETTO M, ALBERTI A, CRAXI A, et al. Open, multicenter, randomized, controlled trial to compare safety and efficacy of r-hIFN beta 1a alone or in combination with ribavirin in HCV naive patients. *Dig Liver Dis* **2003**, *35 (Suppl)*, A14.

132 Biotechnology for multiple sclerosis – beta-interferon; www.bioimpact.org; accessed 2 July 2005.

133 The North American Study Group on Beta-Interferon 1b in Secondary progressive MS. Interferon-beta 1b in secondary progressive MS. *Neurology* **2004**, *63*, 1788–1795.

134 Secondary Progressive Efficacy Clinical Trial of Recombinant Interferon-beta-1a in MS (SPECTRIMS) Study Group. Randomized controlled trial of Interferon-beta 1a in secondary progressive MS. *Neurology* **2001**, *56*, 1496–1504.

135 YONG VW, CHABOT S, STUVE O, et al. Interferon beta in the treatment of multiple sclerosis: mechanism of action. *Neurology* **1998**, *51*, 682–689.

136 BUCKEL P (Ed.). *Recombinant Protein Drugs – Milestones in Drug Therapy.* Birkhäuser, Basel, **2001**.
137 MANDELLI F, AVVISATI G, AMADORI S, et al. Maintenance treatment with recombinant interferon alfa 2b in patients with multiple myeloma responding to conventional induction chemotherapy. *N Engl J Med* **1990**, *322*, 1430–1434.
138 FERLIN-BEZOMBES M, JOURDAN M, LIAUTARD J, et al. IFN-α is a survival factor for human myeloma cells and reduces dexamethasone-induced apoptosis. *J Immunol* **1998**, *161*, 2692–2699.
139 BARLOGIE B, SHAUGHNESSY J, TRICOT G, et al. Treatment of multiple myeloma. *Blood* **2004**, *103*, 20–32.
140 PLATANIAS LC. Mechanisms of type-1- and type-II-interferon mediated signaling. *Nat Rev Immunol* **2005**, *5*, 375–386.
141 DEININGER M, BUCHUNGER E, DRUKER BJ. The development of imatinib as a therapeutic agent for chronic myeloid leukemia. *Blood* **2005**, *105*, 2640–2653.

11
Clinical Applications of Interferon-γ

Christine W. Czarniecki and Gerald Sonnenfeld

11.1
Introduction

A review of any aspect of interferon (IFN) research invariably includes a reference to the work of Alick Isaacs and Jean Lindenmann. In 1957, they described their discovery of a substance that protected cells from virus infection and termed that substance "interferon" [1]. That report served as a major milestone laying the foundation for a field of research that, when partnered with the subsequent advances in molecular biology and biotechnology, led to the clinical availability of novel therapies for a wide range of human diseases beyond viral disease.

Today, we know that IFN is not one "substance", but a group or family of proteins that are: (i) defined by the ability to inhibit the replication of both RNA and DNA viruses, and (ii) induced by and secreted from cells that have been exposed to a variety of stimuli. Currently there are two "types" of IFNs: types I and II. There are seven "classes" of type I IFNs (IFN-α, -β, -ε, -κ, -ω, -δ and -τ), but only one type II IFN (IFN-γ). IFN-like cytokines have also been described [limitin, and interleukin (IL)-28A, -28B and -29]. For a recent review of IFNs, IFN-like cytokines and their receptors, see Pestka et al. [2].

Many of these classes of IFNs have been produced in sufficient quantities as purified products to allow evaluation in clinical trials. The therapeutic uses evaluated were based on the natural roles of these molecules – host defense and modulation of the immune system. Today, almost 50 years from the initial 1957 discovery, a number of IFNs, including modified IFNs ("second-generation" IFNs), have been approved by health authorities worldwide for use as clinical therapeutics. The ever-growing list of IFNs currently available as approved therapies in the US is provided in Tab. 11.1. Many other clinical indications have been evaluated, and the results of these phase 1, 2, and 3 clinical trials have met with various levels of success.

In 1986, Jaffe and Sherwin summarized the early clinical trials of IFN-γ conducted to evaluate the therapeutic potential of this cytokine [3]. These early studies focused on safety and pharmacokinetic parameters, and showed evidence of anti-

The Interferons: Characterization and Application. Edited by Anthony Meager
Copyright © 2006 WILEY-VCH Verlag GmbH & Co. KGaA, Weinheim
ISBN: 3-527-31180-7

Tab. 11.1. Human IFNs commercially available in the US and their FDA-approved indications

IFN type	IFN class	Trade name/ (proper name)	Description	Indication for use in the US	Manufacturer
Type I	IFN-α	Roferon A (Interferon alfa-2a)	recombinant human IFN-α2a	– chronic hepatitis C – hairy cell leukemia – chronic – myelogenous – leukemia	Hoffman La-Roche
		Intron A (Interferon alfa-2b)	recombinant human IFN-α2b	– hairy cell leukemia – AIDS-related Kaposi's sarcoma – malignant melanoma – follicular lymphoma in conjunction with chemotherapy – chronic hepatitis C – condylomata acuminata	Schering
		Infergen (Interferon alfacon-1)	recombinant consensus IFN-α	chronic hepatitis C	InterMune
		PEGASYS (Peginterferon alfa-2a)	pegylated IFN-α2a	alone and in combination with ribavirin for chronic hepatitis C	Hoffman La-Roche
		PEG-Intron (Peginterferon alfa-2b)	pegylated IFN-α2b	alone and in combination with ribavirin for chronic hepatitis C	Schering
		Alferon N (Interferon alfa-n3)	natural human leukocyte derived IFN	refractory or recurring external condylomata acuminata	Hemispherx Biopharma
	IFN-β	AVONEX (Interferon beta-1a)	recombinant human IFN-β1a	relapsing forms of multiple sclerosis	Biogen
		Betaseron (Interferon beta-1b)	recombinant human IFN-β1b	relapsing forms of multiple sclerosis	Chiron and Berlex
		Rebif (Interferon beta-1a)	recombinant human IFN-β1a	relapsing forms of multiple sclerosis	Serono
Type II	IFN-γ	ACTIMMUNE (Interferon gamma-1b)	recombinant human IFN-γ1b	delaying time to disease progression in patients with CGD or severe, malignant osteopetrosis	InterMune

cancer and anti-infective activities. Since that summary report, many additional clinical trials with IFN-γ have been conducted in a wide range of clinical indications.

In this chapter, we will review clinical trials that have been conducted with IFN-γ, focusing on its current clinical use and will review adverse effects and potential risks. As described below, the well-established anti-inflammatory, antimicrobial and antifibrotic activities attributed to IFN-γ formed the basis for clinical studies designed to evaluate its therapeutic potential in a wide range of clinical indications.

11.2
IFN-γ – The Molecule

Human IFN-γ is a glycosylated protein, 143 amino acids in length, that has minimal sequence homology with IFN-α classes or IFN-β. IFN-γ is produced endogenously by activated T ($CD4^+$ and $CD8^+$) lymphocytes and natural killer (NK) cells in response to specific mitogenic stimuli. While IFN-γ shares many activities with the type I IFNs, it also exhibits many other activities, which has led to the conclusion that its natural role is the modulation of the immune system [4, 5].

IFN-γ1b is a form of human IFN-γ produced by recombinant DNA technology. IFN-γ1b is manufactured by fermentation of genetically engineered *Escherichia coli* containing the DNA that encodes the human protein and purification by conventional column chromatography. IFN-γ1b and IFN-γ produced by human peripheral blood leukocytes (PBL-γ) have been shown to have the same primary structure, without any insertions or deletions [6, 7]. However, these two forms of IFN-γ differ in several ways: (i) PBL-γ is glycosylated and IFN-γ1b is not, (ii) IFN-γ1b is 140 amino acids in length compared to the 143 amino acid length of PBL-γ, and (iii) PBL-γ has blocked N-termini in the form of pyroglutamate residues and IFN-γ1b has methionine at the N-terminus [6, 7]. The majority of the clinical trials summarized in this chapter were conducted with this recombinant form of IFN-γ.

11.3
FDA-approved Indications: Established Benefit and Risks

11.3.1
Chronic Granulomatous Disease (CGD)

CGD is the first clinical indication for which IFN γ1b received US FDA approval/licensure [the FDA-approved product is termed Actimmune® (Interferon gamma 1b)]. This approval was granted in 1991. CGD is a rare, congenital disorder that causes patients, mainly children, to be vulnerable to severe, recurrent bacterial and fungal infections. This disease results in frequent and prolonged hospitalizations, and commonly leads to death. CGD is an inherited disorder of leukocyte function caused by defects in reduced nicotinomide adenine dinucleotide phosphate oxidase – the enzyme complex responsible for generation of superoxide [8]. The lack of superoxide and hydrogen peroxide production in phagocytes predisposes patients to bacterial and fungal infections.

Based on earlier data identifying IFN-γ as an agent able to enhance respiratory burst in phagocytic cells of cancer patients [9] and patients with lepromatous leprosy [10], clinical studies were conducted to evaluate the potential of IFN-γ to reverse the immunological defect in CGD patients.

Based on the results of a randomized, double-blind, placebo-controlled study which showed a 67% reduction in there relative risk of infection in CGD patients receiving IFN-γ1b compared to placebo [11], IFN-γ1b was approved by FDA for "reducing the frequency and the severity of serious infections associated with CGD". The recommended dosage is 50 µg m^{-2} (1 MIU m^{-2}) for patients whose body surface area is greater than 0.5 m^2 and 1.5 mg kg^{-1} per dose for patients whose body surface area is equal to or less than 0.5 m^2. Injections are to be administered subcutaneously (s.c.), 3 times a week.

In the above pivotal study, 128 eligible patients with different patterns of inheritance were enrolled and IFN-γ1b was administered s.c., 3 times a week. Patients ranged in age from 1 to 44 years with the mean age being 14.6 years. The primary endpoint was "time to serious infection" with serious infection defined as a clinical event requiring hospitalization and the use of parenteral antibiotics. Analysis of results demonstrated a 67% reduction in relative risk of serious infection in patients receiving IFN-γ1b ($n = 63$) compared to placebo ($n = 65$) across all groups. In addition, the length of hospitalization for the treatment of all clinical events provided evidence for a beneficial effect of IFN-γ1b treatment with placebo-treated patients requiring 3 times as many inpatient hospitalization days compared to IFN-γ1b treated patients.

After the above study was completed, a phase 4 open-label study designed to evaluate the long-term efficacy and toxicity of IFN-γ1b was initiated. Enrollment in this open-label study extended from 1992 to 2001 (observation of patients for up to 9 years). Seventy-six patients were evaluated in this study [12]. As in the initial study [11], "serious infection" was defined as an infection that required hospitalization or treatment with parenteral antibiotics or antifungals. Assessments included measurements designed to evaluate the effects of IFN-γ1b treatment on growth and development.

The results of this long-term follow-up study confirmed the safety and efficacy profile established in the earlier study. The persistent reduction in the frequency of serious infections and mortality was demonstrated. In addition, the investigators observed no evidence for delay in endocrinologic development, and concluded that prolonged IFN-γ1b therapy permitted normal growth and development, which is an important factor in any therapy indicated for chronic use in children.

Long-term data from patients treated with IFN-γ indicate that the natural course of CGD has changed dramatically with the availability of this cytokine as well as the trimethoprin-sulfamethoxazole, itraconazole [12]. However, the mechanism by which IFN-γ improves host defense against infections in these patients remains to be elucidated. Although reports from pilot trials of IFN-γ in CGD patients showed evidence of treatment-related enhancement of phagocyte function, including increased levels of superoxide levels and enhanced inhibition of *Staphylococcus aureus* [13, 14], data from the prospective human trial showed no reversal in the defect in superoxide production in IFN-γ-treated CGD patients [11].

11.3.2
Osteopetrosis

The second FDA-approved clinical indication for IFN-γ1b was for osteopetrosis – another life-threatening pediatric disease.

Congenital osteopetrosis is a rare osteosclerotic bone disease characterized by a defect in osteoclastic function (leading to decreased bone resorption resulting in blindness or deafness) and reduced generation of superoxide by leukocytes [15]. Patients with this disease are predisposed to pathologic fractures due to accumulation of sclerotic bone compromising marrow space and cranial-nerve foramina. Mortality caused by compromised resistance to infection, neurological deficits and bone marrow failure is common within the first decade of life. Bone marrow transplantation is curative therapy, while ameliorative treatments focused on: (i) improving hematologic function (prednisone), (ii) stimulation of osteoclast activity (high-dose calcitriol and parathormone) and (iii) improving immune function.

Based on results from the IFN-γ1b/CGD trials demonstrating reduced infections in treated patients with the rare superoxide generation defect (summarized above), and data from nonclinical studies of IFN-γ in mice with osteopetrosis and microphthalmos [16], Key et al. conducted clinical trials to evaluate IFN-γ1b treatment in patients with osteopetrosis [17, 18].

In 2000, based on data from a controlled, randomized study in which patients with severe malignant osteopetrosis were treated with IFN-γ1b (dose, route ectionand schedule as that established for CGD patients) plus calcitriol versus calcitriol alone, US FDA approval was obtained and this disease indication was added to the approved labeling for IFN-γ1b.

In that study, 16 patients (ranging from 1 month to 8 years of age) with severe, malignant osteopetrosis were randomized to receive either IFN-γ1b plus calcitriol ($n = 11$) or calcitriol alone ($n = 5$). Disease progression was defined as: (i) death, (ii) significant reduction in hemoglobin or platelet counts, (iii) serious bacterial infection requiring antibiotics or (iv) a 50-dB decrease in hearing or progressive optic atrophy. The results of this study demonstrated a significant delay in median time to disease progression in the IFN-γ1b plus calcitriol arm compared to the control arm. In addition, an analysis which combined data from a second study showed that patients receiving IFN-γ1b treatment (with or with calcitriol) for at least 6 months demonstrated reduced trabecular bone volume compared to baseline, providing evidence of increased bone resorption [18].

11.3.3
Adverse Reactions

Complete information on the risks and benefits of IFN-γ can be found in the package insert for the commercial product. The most common adverse experiences occurring with IFN-γ therapy are "flu-like" symptoms such as fever, headache, chills, myalgia or fatigue. Other most common events are rash, diarrhea, nausea and vomiting. Acetaminophen has been used successfully to prevent or partially alleviate fever and headache. Due to these flu-like symptoms, individuals receiving IFN-γ are cautioned regarding pre-existing cardiac conditions, including ischemia, congestive heart failure or arrhythmia. Reversible neutropenia and thrombocytopenia that may be severe have been observed with IFN-γ treatment. Up to 25-fold elevations of aspartate aminotransferase (AST) and/or alanine aminotransferase (ALT) have also been observed during therapy.

11.4
Infectious Diseases

Using experimental models of infectious disease, IFN-γ has been shown to exert effective antimicrobial activity against all classes of nonviral pathogens, including intracellular and extracellular parasites, fungi, and bacteria [19–23]. The innate immune system, consisting of neutrophils, monocytes and tissue-based macrophages, plays an important role in the host defense against these pathogens, and IFN-γ is a potent activator of these cells, forming a basis for evaluation of IFN-γ as a therapeutic option in the management of infectious diseases.

11.4.1
Mycobacterial Infection

IFN-γ has been studied in infectious diseases caused by several types of Mycobacteria. *In vitro* and *in vivo* studies indicate multiple pathways by which this cytokine inhibits infection with this class of microorganism. IFN-γ treatment enhances phagocytic oxidative metabolism and also leads to: (i) release of antimicrobial proteins and other cytokines; (ii) degradation of tryptophan; antigen processing and presentation; and prostaglandin synthesis [21].

11.4.1.1 Leprosy
Leprosy afflicts over 12 million people world-wide. This disease is caused by *Mycobacterium leprae* (the Hansen bacillus) and the initial infection can progress into either the tuberculoid (a contained infection) or the lepromatous form of the disease. The lepromatous form is characterized by no immune response to *M. leprae* anti-

gens, no granuloma formation and destructive tissue infiltration [21]. Lepromatous leprosy is characterized by deficient cell-mediated immune responses, high mycobacterial burden, widespread cutaneous and neurologic disease, and worse clinical outcome. The lesions in the tuberculoid form are characterized by abundance of IFN-γ and IL-2 mRNA; and lesions from the lepromatous form are characterized by abundance of IL-4, IL-5 and IL-10 mRNAs. This T helper (T_h1/T_h2) pattern and the lack of IFN-γ led to clinical trials of IFN-γ in the disseminated form of leprosy.

Results from these trials are mixed. Intradermal administration of IFN-γ led to accelerated bacillary clearance from the skin, associated with local accumulation of mononuclear cells and the killing of infected macrophages [24, 25]. However, this encouraging finding was also accompanied by the observation that treatment resulted in development of erythema nodosum leprosum (ENL) – a toxic syndrome that is characterized by excess tumor necrosis factor (TNF)-α production [26]. The inclusion of thalidomide in the treatment regimen of IFN-γ and chemotherapy resulted in a lower frequency of ENL, but also resulted in elimination of the IFN-γ-induced bacillary clearance [27].

11.4.1.2 *Mycobacterium avium* Infection

Disseminated infection with *M. avium* can occur in patients with cancer or those receiving immunosuppressive therapy and localized infection can occur in patients with underlying structural lung disease (e.g. smokers) [23]. IFN-γ has been used to treat small numbers of patients with AIDS and non-AIDS associated *M. avium* complex infection, in open-label, pilot studies [21, 28, 29]. These results showed that treatment was well tolerated and led to reduced levels of microorganisms in patients only when IFN-γ was used in combination with antimycobacterial agents. *In vitro* studies demonstrated that IFN-γ can increase the intracellular concentration of macrolide antibiotics [30], suggesting a possible mechanism for the observed enhanced antimicrobial activities with combination treatment.

11.4.1.3 **Tuberculosis (TB)**

Multidrug-resistant TB (MDR-TB) is defined as disease caused by strains of *Mycobacterium tuberculosis* that are resistant to at least isoniazid and rifampin. The emergence of these drug-resistant strains has resulted in high morbidity and mortality, and has created a major global threat. Infection is transmitted by inhalation of bacilli from infected airway secretions. Therapy is often a combination of second-line orally administered drugs (with significant systemic toxicity) and resectional surgery.

The basis for evaluation of IFN-γ as therapy in this indication stems from its role in the activation of macrophages to inhibit the growth of and kill mycobacteria, and the observed inhibitory effects on *M. avium* discussed in Section 11.4.1.2.

With the goal of activating alveolar macrophages, IFN-γ was administered by aerosolization to patients infected with MDR-TB. In an open-label trial, aerosol

IFN-γ1b (500 µg, 3 times per week for 4 weeks) was given to five patients with smears and cultures positive for pulmonary MDR-TB [31]. The aerosol route of administration was chosen with the goal of achieving high local levels of IFN-γ1b in the epithelial lining fluid and lower respiratory tract, with low systemic levels and therefore potentially lower systemic side-effects. Results suggested a decrease in mycobacterial burden as evidenced by sputum acid-fast bacillus smears becoming negative in all patients and a nonsignificant increase in time to positive culture from 17 to 24 days. Additionally, 2 months after treatment cessation, the size of cavitary lesions was reduced. However, the sputum smears of all patients reverted to positive within 1–5 months after discontinuation of treatment. The authors concluded that longer courses of IFN-γ1b treatment may be necessary for sustained effect and the low level of observed toxicity supported further investigation.

A subsequent study evaluated the effects of 6 months of aerosolized IFN-γ therapy in patients with refractory MDR-TB [32]. In an open-label, nonrandomized observational study, six patients received 2 MIU IFN-γ (100 µg), 3 times a week for 6 months. After IFN-γ inhalation treatment, sputum cultures remained positive throughout the study period. Sputum cultures for two patients were transiently negative at month 4, but reverted to positive at 6 months.

These small studies led to a double-blind placebo-controlled trial in which antimycobacterial therapy was administered with or without aerosolized IFN-γ1b to patients with MDR-TB in South Africa. Unfortunately, this study was terminated prematurely due to lack of efficacy and the unethical aspects of continued exposure to an experimental treatment in the absence of benefit [33]. This lack of efficacy led to the question of whether the diseased state of the lungs of these patients prevented adequate deposition of IFN-γ1b.

Condos et al. [33] addressed the deposition question in a study in which 14 patients with pulmonary TB received 12 treatments with IFN-γ1b aerosol (500 µg) plus antimycobacterial therapy. This study showed that while deposition was inefficient, significant distribution was achieved throughout the lower respiratory tract in all patients (with mild to moderate disease), but one patient who had advanced parenchymal and airway disease. The authors hypothesized that the lack of efficacy observed in the South Africa clinical trial may have been due to inclusion of patients with destroyed lungs in whom aerosol deposition was insufficient. The results of this study have led the authors to conduct a randomized clinical trial of antimycobacterial therapy alone or with aerosolized IFN-γ1b (for 4 months) in TB patients with mild to moderate cavitary disease. Results from this study have not yet been reported.

Suarez-Mendez et al. [34] reported results of an open-label pilot study in which eight patients received daily intramuscular injections of 1 MIU IFN-γ for 1 month and then 3 times per week for up to 6 months as adjuvant to indicated chemotherapy. Results of this study indicated that after IFN-γ treatment, all eight patients exhibited loss of disease signs and symptoms, conversion of sputum test results, and improvement in pulmonary lesions. In both studies IFN-γ treatment was well tolerated.

The studies that have been conducted to date indicate the numerous variables and factors that complicate the ability to definitively establish the efficacy of IFN-γ1b in this clinical indication.

11.4.2
Leishmaniasis

Leishmaniasis is an intracellular parasitic infection. The clinical manifestations of this disease vary, based on the *Leishmania* species causing the infection and the state of the host's immune response. Cutaneous leishmaniasis, the mildest form of the disease, is characterized by single or multiple focal cutaneous lesions and is associated with strong delayed hypersensitivity reactions to leishmanial antigens. This form of the disease usually heals spontaneously. Visceral leishmaniasis (*kala-azar*), the most severe and often fatal form, is characterized by high fever, hepatosplenomegaly, bone marrow suppression, widespread organism dissemination, and profound anemia, leukopenia and thrombocytopenia. The disease is associated with an inability to mount a delayed hypersensitivity response to leishmanial antigen and lymphocytes from these patients cannot produce IFN-γ or IL-2 in response to leishmanial antigen [35]. Clinical trials of IFN-γ in this indication have shown promising results. Apparent cure rates of 60–75% were observed after treatment of refractory visceral leishmaniasis with IFN-γ in combination with the standard chemotherapy for this disease, i.e. pentavalent antimony. The cure rate with antimony alone for 87 historical control patients who had untreated or refractory disease was 43%. Cure was defined as improvement in signs, symptoms and laboratory abnormalities of leishmaniasis and eradication of parasites from splenic aspirates [36–38].

11.4.3
Opportunistic Infections in HIV Disease

Opportunistic infections continue to contribute to morbidity and mortality observed in HIV-infected individuals in spite of the widespread use of highly active antiretroviral therapy (HAART). The scientific basis for consideration of IFN-γ for the treatment of opportunistic infections in HIV-infected individuals is broad ranging from its well-established antimicrobial activities to varied forms of IFN-γ dysregulation in these patients, such as decreased production of IFN-γ and decreased macrophage antimicrobial activity [39]. In a clinical study in which 22 HIV-infected patients were treated with IFN-γ or IL-2, decreased bacterial infections was observed in the IFN-γ-treated group [40].

A subsequent phase 3 study was conducted to rigorously assess the potential benefit of IFN-γ in this clinical indication [39]. In this multicenter, double-blind, placebo-controlled study, 84 HIV-positive patients on stable antiretroviral therapy received either IFN-γ (0.05 mg m^{-2} s.c., 3 times per week for 48 weeks) or placebo. The primary endpoint in this study was incidence of opportunistic infections.

While IFN-γ treated patients did exhibit fewer opportunistic infections in the first 48 weeks (mean of 1.71 versus 3.45 in the placebo group) this decrease was not statistically significant. Similarly, the enhanced survival rate at 3 years observed in the IFN-γ-treated group was also not statistically significant. Despite these disappointing results, the authors were encouraged by the observed trends, the observed inhibitory effects on difficult-to-treat infections such as herpes simplex, cytomegalovirus and mucosal candidiasis, and the safety profile established in this patient population.

Pappas et al. [41] conducted a phase 2 trial to evaluate the antifungal activity of IFN-γ in HIV-infected individuals with cryptococcal meningitis. The 75 patients analyzed in this randomized, double-blind, placebo-controlled multicenter study were HIV-positive or HIV-negative and had newly diagnosed or relapsed cryptococcal meningitis. All patients received standard antifungal therapy for cryptococcal meningitis and either placebo or IFN-γ1b (100 or 200 μg s.c., 3 times per week for 10 weeks). The endpoints included conversion of cerebrospinal fluid fungal cultures from positive to negative at week 2. The results of this study established safety of IFN-γ1b treatment in this patient population. A trend towards more rapid sterilization of cerebrospinal fluid in the IFN-γ-treated group compared to the placebo group was observed; however, this result was not statistically significant. No dose response of IFN-γ1b was observed in this study. The authors concluded that while this study was underpowered to be able to definitively establish the benefit of IFN-γ1b in this patient population, additional studies are warranted.

11.5
Infection Following Serious Trauma

The use of IFN-γ to prevent or ameliorate infection after serious trauma raised considerable interest beginning in the late 1980s [42–50]. This was due to coincident appearance of a series of experiments that demonstrated that IFN-γ could modulate infections in murine models as well as new developments in the understanding of the effects of serious trauma on resistance to infection [42–50].

Late-term infection after serious trauma remains a significant medical problem [48–50]. Despite aggressive use of antibiotics, the use of sterile techniques and development of mechanical barriers have been shown not to be completely ineffective in dealing with this problem [48–50]. Research has demonstrated that some individuals that have undergone severe traumatic event and corrective surgery appear to be immunosuppressed. The main immunologic factors that have been shown to be inhibited are expression of class II histocompatibility antigens on circulating monocytes and opsonic capacity of serum from trauma patients [48–52].

Decreased class II histocompatibility antigen expression on the monocytes of trauma patients could be an indication of inhibited antigen presentation, which could have lead to the decreased ability to handle infections often observed in serious trauma patients [48–52]. At the same time that those results were observed, data were emerging that indicated a fundamental immunoregulatory role for IFN-γ [42–47]. Data were developed that indicated that treatment of animals or monocytes with IFN-γ resulted in enhanced expression of class II histocompatibility antigens. Therefore, it was a logical step to follow-through on these data to postulate that treatment of trauma patients with IFN-γ could enhance their resistance to surgical wound infection. This hypothesis was supported by data that were also emerging that indicated that treatment of experimental animals with IFN-γ could prevent or ameliorate the development of infections with the same Gram-negative pathogens that were the cause of devastating post-trauma surgical wound infections [42–47].

Experimental work was begun and involved studies with a series of animal models of trauma wound infection [42–45]. None of the models could perfectly replicate the conditions of trauma and infection. However, taken together, they suggested that infection after serious trauma could be influenced by treatment with IFN-γ. The models initially used included: (i) IFN-γ treatment of mice that had received an intramuscular injection of *Klebsiella pneumoniae* [44], (ii) IFN-γ treatment of mice that had applied a suture contaminated with *K. pneumoniae* [42], (iii) IFN-γ treatment of mice that had burn wounds infected with *K. pneumoniae* [43], and (iv) IFN-γ treatment of mice that were infected intraperitoneally (i.p.) with *Escherichia coli*, then received a laparotomy and then were infected intramuscularly with *K. pneumoniae* [45]. In all four models, prophylaxis with IFN-γ usually significantly decreased mortality. Treatment after the initiation of infection also decreased mortality, but not to the same extent as prophylaxis. The only exception was mice that had a burn wound infection with *Pseudomonas aeruginosa* [42]. Later it was discovered that *P. aeruginosa* produces a protease that inactivates IFN-γ and if bacteria that were deficient in that enzyme were used, mice could be protected against *P. aeruginosa* infection as well [53].

These studies were later expanded to include a hemorrhagic shock model of surgical wound infection [46, 47]. In this model, rats were subjected to hemorrhagic shock and then resuscitated. Rats maintained in this model do not have infections that respond to antibiotic prophylaxis, which is similar to what is often found in trauma patients [45, 46]. The mice were then infected s.c. with a combination dose of *E. coli* and *Bacteroides fragilis*. The rats were treated with appropriate antibiotics and some were also treated with IFN-γ. Only those rats treated with both antibiotic and IFN-γ were able to resolve the infected abscesses, suggesting that the IFN-γ restored the immune system of the rats to a point where it could eliminate the lowered state of infection produced by antibiotic therapy.

Armed with the supportive information of the animal model studies, clinical studies began. First, it was observed that severely injured trauma patients had decreased IFN-γ production [54]. Second, it was observed that *in vitro* treatment of the monocytes of even the most severely injured patients with IFN-γ resulted in enhanced expression of class II histocompatibility antigens [55]. These data were sufficient to support the development of clinical trials to determine the capacity of IFN-γ treatment to prevent or ameliorate infection in severe trauma patients.

The first human trial was a small (15 patient) "proof-of-concept" study. The patients were treated with three different doses of IFN-γ1b and, consistently, there was upregulation of class II histocompatibility expression on monocytes (unpublished observations). This led to a full-scale clinical trial study involving 213 trauma patients at four different hospitals. The patients began receiving IFN-γ1b for 10 days shortly after admission. Patients with an injury severity score of greater than 20, indicating identifiable bacterial wound contamination and other severe injury signs, were entered into the study and received 100 µg IFN-γ1b s.c. for 10 days after admission to hospital [56]. Standard supportive therapy, including antibiotic administration, was included for all patients. Levels of monocyte class II histocompatibility antigen expression increased significantly in the IFN-treated group compared to placebo controls. Although there was a trend towards decreased interventions for infections and lower death rate due to infection in the IFN-γ-treated group, there were no significant differences noted between the IFN-treated and placebo-treated groups [56]. This led the authors of the study to call for a larger trial with longer treatment periods [56].

A second trial was undertaken involving 416 patients in nine different hospitals. The regimen was similar to that used in the previous trial, except that IFN-γ1b was administered for a longer period of 21 days [57]. No difference in infection rate was observed between IFN- and placebo-treated groups, but it appeared that the death rate due to infection was decreased in the IFN-γ1b-treated group compared to the placebo-treated group [57]. However, after analysis, it became clear that the findings were skewed by the results from one hospital that had the highest number of patients enrolled in the study, as well as the highest infection and death rates of all of the nine hospitals involved in the study. This skewing of the data appeared to be due to multiple confounding variables that could not be fully identified [58]. However, it was clear that the severity and type of injury in that one hospital influenced the outcome of IFN-γ1b treatment. Therefore, the results of the study could not provide adequate support for use of IFN-γ1b to prevent or ameliorate infections in serious trauma patients.

Evaluation of the original clinical study confirmed that IFN-γ1b treatment of severely injured trauma patients did not decrease the rate of infection overall of the patients [59]. Class II histocompatibility antigen expression was enhanced in the patients. Even though the class II histocompatibility antigen expression of the patients was restored, there was no benefit shown in reducing the incidence of infection. Therefore, it appears that effects additional to restoration of class II histocompatibility antigen expression may be required to reduce the rate of infection of trauma patients. Several additional studies demonstrated that IFN-γ treatment of severe trauma patients resulted in enhanced expression of immunological and other trauma-altered physiological parameters, but none were able to show protection from infection or death due to infection in those same patients [60–65].

Therefore, the results of the clinical trials involving IFN-γ and infection after trauma indicate a potential use for IFN-γ that has value for only a subset of patients. It appears that the most severely injured patients benefit most from IFN-γ treatment to ameliorate or prevent serious complications due to infection [56–58]. No studies to date have been carried out to determine which factors involved in serious trauma or just how severely injured the trauma patient must be in order to benefit from IFN-γ treatment. Such studies must be carried out to fully identify the characteristics of the patient population that would benefit from IFN-γ therapy and to allow for design of definitive study for this clinical indication.

11.6
Atopic Dermatitis (AD)

AD is a chronic, relapsing cutaneous disease that occurs most frequently in children, but can also effect adults. This relapsing disease can result in significant physical and psychological morbidity; while numerous treatments are used, many patients do not improve and side-effects of treatments are significant, leading to the need for new effective therapeutics [66, 67].

The clinical trials that were conducted to evaluate the potential of IFN-γ as a therapeutic for this indication were based on the classification of AD as a type 2 helper T lymphocyte (T_h2) cell-mediated disease with patients exhibiting a deficiency in levels of type 1 (T_h1) cytokines. This T_h1/T_h2 imbalance results in increased levels of IL-4 and IL-5, elevated blood eosinophilia, and decreased IFN-γ production [68]. IL-4 induces class switching to IgE production in B cells and IFN-γ inhibits this activity [69]. It was hypothesized that the decreased production of IFN-γ in AD patients could lead to the observed increase in IL-4-induced IgE levels [70], and, therefore, it was hoped that treatment of AD patients with IFN-γ would decrease the T_h2 responses (decrease serum IgE levels and IL-4 levels) and lead to clinical benefit.

The clinical trials included: (i) open-label [71–73], (ii) double-blind placebo controlled [74] and (iii) long-term, open-label follow-up studies [75, 76], and these were summarized in detail by Chang and Stevens [66]. In these studies, IFN-γ was administered s.c. in daily (or every other day) injections over periods of time up to 22 months. The early studies used 100 µg day^{-1} and the double-blind placebo-controlled and long-term follow-up studies used 50 µg m^{-2}. The results of these studies indicate that IFN-γ treatment leads to long-term safety and efficacy in a subset of patients [66]. The observed adverse effects were minimal and could be ameliorated by prophylactic acetaminophen. Observed improvement of clinical parameters included decreases in severity of erythema, pruritis, excoriations, edema/indurations, dryness and reduction in total body surface area involvement. However, these beneficial effects were observed in the absence of effects on serum IgE levels (spontaneous IgE production), while improvement in clinical symptoms did correlate with decreases with absolute white blood cell and eosinophil counts [75, 77, 78]. Since eosiniphilia correlates with clinical severity of AD and eosinophil counts correlate with benefit of IFN-γ treatment [71, 73, 74] Noh and Lee proposed a guideline to recommend IFN-γ therapy for AD patients with blood eosinophil percentages less than 9% and serum IgE levels less than 1500 IU ml^{-1} [78].

The general conclusion from analyses of the completed studies is that s.c. administration of IFN-γ shows clinical benefit in a subset of patients with AD with few adverse events while the immunological mechanism for this benefit still remains unclear.

11.7
Idiopathic Pulmonary Fibrosis (IPF)

IPF is a chronic, progressive and lethal pulmonary fibrotic lung disease. This devastating form of interstitial lung disease is characterized by fibrosis of the lung interstitium, progressive pulmonary insufficiency and a poor survival prognosis [79].

IPF is the most common form of the idiopathic interstitial pneumonias (IIPs) – a group of related, but distinct, interstitial lung diseases. Advances in the characterization of these diseases have led to historical changes in nomenclature and classification complicating the interpretation of results from clinical trials [80]. One of the most significant challenges in the conduct of clinical trials to evaluate and identify effective therapies for this disease continues to involve the ability to identify well-defined patient populations.

The term IPF is reserved (by current consensus statements) for a specific clinical condition associated with diagnostic histological patterns of usual interstitial pneumonia (UIP). The prevalence of this disease is rare (3–20 cases per 100 000 individuals), and more prevalent in males, in older adults and in current or former smokers [81]. IPF has the highest mortality rate of all the diffuse lung diseases. IPF is characterized by insidious onset of dyspnea and abnormal pulmonary function. Most patients die of respiratory insufficiency within 3–8 years with a median survival of 2.5–3.5 years after diagnosis. However, survival varies widely with some patients dying within 1 year of diagnosis and others surviving longer than 6 years [81].

The only proven therapeutic option for patients with IPF is lung transplantation. However, many patients die awaiting transplant [82]. Treatment of IPF is controversial, and current management of IPF involves treatment with agents exhibiting immunosuppressive and/or antifibrotic activities. Common treatments include corticosteroids, and immunosuppressive and cytotoxic agents. No randomized, well-controlled, blinded studies have established efficacy of any of these treatments, and due to the unproven benefit and serious toxicity of these agents, there is a general reluctance to use these treatments until late-stage disease [80, 81, 83]. Hence, there is a critical need for new therapies for this serious disease.

The etiology of IPF has not been established. Selman et al. [81] have proposed that there are at least two pathogenic mechanisms for the development of IPF: (i) an inflammatory pathway in which an early phase of inflammatory pneumonitis is followed by a late phase of fibrosis, and (ii) the alveolar pathway in which alveolar epithelial cell injury and activation leads to fibrotic responses.

Data from nonclinical studies have identified pathways involved in fibrosis and the production and deposition in the lung of extracellular matrix proteins such as procollagens and elastin. Molecules playing roles in these pathways include (i) cytokines and chemokines associated with inflammation, angiogenesis, and cell trafficking, and (ii) growth factors such as transforming growth factor (TGF)-β, platelet-derived growth factor (PDGF) and connective tissue growth factor (CTGF) [84, 85].

It is thought that in response to injury, alveolar epithelial cells release fibrogenic cytokines such as TGF-β, PDGF, TNF-α, IL-1, insulin-like growth factor-1 and basic fibroblast factor. The release of these cytokines may lead to fibroblast proliferation and migration to sites in the lung as well as fibroblast differentiation which may play a role in the chronic nature of IPF [81, 83].

The type 2 T cell response predominates in IPF. Studies of lung tissue and blood from patients with IPF show deficits in IFN-γ as compared to T_h2 cytokines [86]. T_h1 cells produce IFN-γ, IL-2 and TNF-α, which are involved in delayed hypersensitivity responses; T_h2 cells produce IL-4, -5, -6 and -10, which mediate inflammatory and profibrotic responses. T_h1 and T_h2 cytokines have reciprocal modulatory effects, and the balance between their expression is thought to play a central role in the nature and control of the immune response at sites of disease.

Thus, as with AD (see Section 11.6), the reciprocal regulation of type 1 and 2 cytokines formed the basis for evaluation of IFN-γ as a potential therapy for halting and hopefully reversing the IPF disease process. Additional activities attributed to IFN-γ and relevant to this indication include: (i) inhibition of the *in vitro* proliferation of lung fibroblasts and protein synthesis in fibroblasts [87, 88], (ii) suppression of proliferation of fibroblasts and production of collagen synthesis [89–91] and this effect is mediated, partly, by blockade of TGF-β signaling [92], (iii) reduction of tissue myofibroblast numbers [93], (iv) inhibition of collagen synthesis by human fibroblasts *in vitro* [90], (v) enhanced transcription of the c-*met* protooncogene, the receptor for hepatocyte growth factor, a potent mitogen for epithelial cells [94]; (vi) increased expression of MMP-1 message [95], and (vii) downregulation of the gene transcription of TGF-β when used to treat mice with bleomycin-induced lung fibrosis [96].

In 1999, Ziesche et al. [97] reported results from a preliminary study of IFN-γ1b and low-dose prednisolone in patients with IPF. In that study, 18 patients with IPF who had previously failed therapy with corticosteroids were randomized to receive prednisolone (7.5 mg day^{-1}) alone or combined therapy with prednisolone (7.5 mg day^{-1}) and IFN-γ1b (200 µg s.c., 3 times weekly). After 12 months, pulmonary function deteriorated in all nine patients receiving prednisolone alone and improved (increase in total lung capacity and improved arterial oxygenation) in all nine patients receiving IFN-γ1b plus prednisolone. Molecular assessments of lung tissue from all patients prior to treatment showed lack of IFN-γ message and levels of gene transcription for TGF-β1 and CTGF far in excess of levels in normal tissue. After treatment, improvement in lung function in IFN-γ1b-treated patients correlated with significantly reduced levels of transcription of genes for TGF-β1 and CTGF. These results are consistent with the hypothesis that the therapeutic efficacy of IFN-γ was related to its ability to shift the type 1/type 2 balance in favor of the antifibrotic type 1 response.

Importantly, a retrospective analysis of enrolled patients performed by an independent panel demonstrated that 15 of these patients had definite or probable IPF (some of the patients that were enrolled had non-IPF diagnoses, such as nonspecific interstitial pneumonia), and confirmed the improvement in gas exchange and lung volumes in the IFN-γ treated patients [98]. The study by Ziesche et al. [97] formed the basis for a large phase 3 multicenter study of IFN-γ1b in patients with UIP [99]. In this phase 3 study (GIPF-001), a well-defined population of patients (330 patients from 58 centers) with UIP that was unresponsive to corticosteroid therapy was randomized to receive either IFN-γ1b and low-dose prednisolone or placebo and low-dose prednisolone.

The primary efficacy endpoint for this study was progression-free survival, measured from randomization to either disease progression or death. Progression of disease was defined by measures of lung function and gas exchange [changes from baseline of (i) decrease of at least 10% in the predicted forced vital capacity (FVC) or (ii) an increase of at least 5 mmHg in the $P_{(A-a)}O_2$ at rest (alveolar–arterial gradient)]. These parameters were chosen as clinical measures of fibrosis thought to be sensitive to the treatment effects of INF-γ1b based on the earlier clinical study by Ziesche et al. [97]. The study was designed and powered to detect a difference in this composite endpoint rather than in overall survival.

Unfortunately, over a median of 58 weeks of IFN-γ1b treatment, no significant differences were observed in the primary endpoint and no treatment effect was observed on measures of lung function, gas exchange or quality of life. In spite of the lack of improvement in lung disease, there was a trend toward decreased mortality in patients treated with IFN-γ1b. Sixteen of the 162 patients in the IFN-γ1b-treated group died as compared 28 of the 168 in the placebo-treated group. The survival benefit occurred in a subgroup of patients with less severe lung impairment in lung function at baseline who adhered to the protocol.

In attempts to evaluate the effects of IFN-γ1b treatment on blood and lung biomarkers thought to play a role in the pathogenesis of IPF, Streiter et al. [100] conducted a randomized, double-blind placebo-controlled multicenter study in which 32 patients were randomized to receive either s.c. IFN-γ1b ($n = 17$) or placebo ($n = 15$) three times per week for 26 weeks. Bronchoalveolar lavage (BAL) and transbronchial biopsy (TBB) were performed at baseline and at 6 months to measure mRNA transcription in TBB tissue and BAL cells as well as protein levels in BAL fluid and plasma. Safety and efficacy outcome measurements were also conducted. The authors chose to measure genes thought to play a role in IPF (Tab. 11.2); however, the results showed trends, but no significant changes, in expression of genes thought to increase or decrease connective tissue in the IPF lung.

Results of this study demonstrated changes in expression of molecules such as elastin, and types I and II procollagen. IFN-γ1b treatment resulted in the following trends: decreased levels of mRNA for elastin, PDGF-β, type III procollagen and IL-4, and upregulation of mRNA for ITAC/CXCL-11. Measures of protein in BAL fluid and plasma showed decreased levels of ENA-78/CXCL-5, type I procollagen

Tab. 11.2. Proteins associated with fibrosis and infection that are affected by IFN-γ

Protein	Reference
Downregulated	
TGF-β	101
PDGF	102
IL-4	103
IL-8 (IL-8/CXCL-8)	104
IL-13	103
ENA-78/CXCL-5	104
extracellular matrix proteins	90, 91, 105
Upregulated	
IP-10/CXCL-10	106
MIG/CXCL-9	106
ITAC/CXCL-11	106, 107
defensins	108

and PDGF-α, and increased levels of ITAC/CXCL-11. The observed changes in these fibrosis-associated biomarkers are consistent with the nonclinical effects of IFN-γ which formed the rationale for evaluation of IFN-γ in the treatment of this disease. However, in contrast to the observations of Ziesche et al. [97], this biomarker study detected no direct effect of IFN-γ1b treatment on mRNA for TGF-β or CTGF. It is important to note that clinical benefit in survival was shown after 1 year of IFN-γ1b treatment and the biomarker analysis [100] was done after only 6 months of IFN-γ1b treatment. As pointed out by Dauber et al. [109], it is possible that changes in gene expression may be more pronounced after 1 year of therapy and that IFN-γ1b may exert its beneficial effect through alternate pathways with changes in expression of genes that were not measured in the biomarker study.

The increased levels of ITAC/CXCL-11, a molecule that exhibits defensin-like antimicrobial activity (against *E. coli* and *Listeria monocytogenes*) and antiangiogenic activity [110] across lung and plasma compartments, led the authors to hypothesize that the decrease in mortality observed in IFN-γ1b-treated patients could be a result of beneficial effects on the pathogenesis of IPF through multiple mechanistic pathways involving the antimicrobial, antiangiogenic, antifibrotic and immunomodulatory activities of IFN-γ.

Using the database from the GIPF-001 study, a retrospective analysis of the variables that measure clinical outcomes in patients with IPF was performed [110]. The purpose of the analysis was to determine the most appropriate efficacy endpoint for a follow-up phase 3 trial of IFN-γ1b termed the INSPIRE trial. The analyses evaluated characteristics including reliability, validity and sensitivity to treatment effect. The data indicated that in placebo-treated patients, changes in

$P_{(A-a)}O_2$ were not associated with an increased risk of death until an increase of 15 mmHg or more was reached while a 10% or higher decrease in percent of predicted FVC was associated with a 2.4-fold risk of increased mortality. The authors concluded that mortality is the most inclusive endpoint for future trials of IFN-γ1b in this indication and that a 10% or higher decrease of predicted FVC is a reliable and predictive measure of mortality.

The 600 patient INSPIRE trial is currently enrolling patients in 75 sites in North America and Europe. The primary efficacy endpoint is measurement of overall survival time from randomization. The results from this study will be critical in establishing the therapeutic potential of IFN-γ1b in this clinical indication.

11.8
Systemic Sclerosis (SSc)

SSC (or scleroderma) is a rheumatic disease characterized by excessive fibrosis of the skin and other connective tissue-containing organs (lungs, heart, kidney, joints). Analysis of the T_h1/T_h2 cytokine profiles in SSc patients has led to the characterization of this disorder as a T_h2 disease with the presence of high levels of TGF-β. The rationale for the evaluation of IFN-γ as a potential treatment of the fibrosis associated with SSc is the same as that described above for IPF, i.e. the identification of IFN-γ as a T_h1 cytokine that can inhibit the synthesis of collagen and abrogate the pro-fibrotic effects of TGF-β. Hein et al. [111] reported results of a trial in which 9 patients were treated with IFN-γ (50 µg day^{-1}) intramuscularly (i.m.) for 3 days per week over a 12-month period. Significant improvements in total skin score and arterial oxygenation were observed. A subsequent randomized, controlled multicenter study (44 patients) showed mild beneficial effects of IFN-γ treatment (100 µg s.c., 3 times per week for 12 months) on skin sclerosis and disease associated symptoms [112]. Serum levels of type III and IV collagen propeptides and laminin remained unchanged following IFN-γ therapy, whereas type I collagen mRNA levels in the involved skin, and *in vitro* collagen synthesis and mRNA expression in fibroblasts derived from post-IFN treatment biopsies were reduced. Based on these results, it appears that IFN-γ may exert maximal antifibrotic effects in early progressive SSc when collagen synthesis is prominent and when treatment effects on progressive skin involvement would be of benefit.

11.9
Radiation-induced Fibrosis

In 1986, 237 individuals were exposed to β- and γ-irradiation in the Chernobyl Nuclear Power Plant accident and experienced acute radiation disease [113]. Of these,

32 subjects died acutely of cutaneous radiation syndrome; 56 of 115 patients referred for treatment were diagnosed with cutaneous radiation syndrome. Peter et al. [114] reported results of the treatment of six of these patients with low dose IFN-γ (50 µg s.c., 3 times per week for 30 months; once per week for an additional 6 months). The results of this study showed that in all treated patients a significant reduction in radiation fibrosis (skin thickness) compared to a significant increase in fibrosis in the two untreated patients. Although the follow-up of IFN-γ-treated patients 1 year after discontinuation of treatment showed a significant recurrence of fibrosis, the investigators concluded that these results warrant additional evaluation of IFN-γ as a therapeutic for patients receiving radiation therapy or accidental exposure to radiation [114].

Other radiation-induced fibrotic conditions that may benefit from IFN-γ treatment due to inhibition of TGF-β1 include radiation pneumonitis, radiation-induced renal fibrosis and radiation-induced liver fibrosis [115].

11.10
Chronic Hepatitis

Infection with hepatitis viruses causes acute liver disease. In the case of infection with hepatitis B or hepatitis C viruses, in a number of infected individuals, chronic infections can develop and lead to chronic liver damage, potential development of cancer and death [116–118]. According to the Centers for Disease Control, an estimated 3.9 million people in the US have been infected with hepatitis C virus and 2.7 million of these are chronically infected. In the US, this virus infection causes an estimated 10 000–12 000 deaths annually and is the leading indication for liver transplants. Although the use of IFN-α treatment (alone and in combination with ribavirin) to control hepatitis B and C infections in humans is well established, the virus cannot be eliminated and the search for more effective treatments has continued [116–118].

The need for additional treatments led to studies using IFN-γ, with IFN-γ as the sole treatment agent or as adjuvant therapy together with IFN-α or -β.

The first studies were begun by testing the efficacy of IFN-γ therapy in chronic hepatitis B [119]. In this study, patients were treated with 0.1 mg m^{-2} 3 times weekly for 16 weeks and compared to untreated control patients. This study indicated that there were minimal toxic effects of the IFN-γ therapy, but no antiviral effects of IFN-γ therapy were observed. A subsequent study compared the effects of therapy with IFN-γ to therapy with IFN-α for treatment of hepatitis B [120]. IFN-α was found to be a much more effective agent in decreasing viral expression, but it appeared that IFN-γ therapy resulted in additional modulation of immune responses against hepatitis B. Similar results were also found in a clinical trial in children with chronic hepatitis B [121]. Therefore, although IFN-γ treatment did not affect viral marker levels, its use in combination with other therapies might have benefit due to its stimulation of the immune system.

The results of the previous studies led to experiments to determine the effects of combined therapy of IFN-γ and other IFNs against chronic hepatitis B. The concept developed for these experiments was that IFN-γ would enhance the immune response against hepatitis B, while the other IFNs would have an antiviral effect [122, 123]. In addition, it is well established that IFN-γ when used in combination with IFN-α or -β, exerts synergistic antiviral and antiproliferative activities [124].

The first study reported poor tolerance of combined therapy with IFN-α and -γ, and no benefit to resolving the disease process beyond use of IFN-α alone [122]. The second study involved combined short-term treatment of hepatitis B with both IFN-β and -γ. The results of this study suggested that treatment effects of the combined therapy were equivalent to that of treating with IFN-α alone, but that the combination of IFN-β and IFN-γ was much better tolerated by the patients than was IFN-α [123].

A more recent study was focused on the effects of IFN-γ on hepatic fibrosis in a rat model of chemically induced liver fibrosis and in humans with hepatitis B virus-induced fibrosis [125]. The results in both systems indicated that IFN-γ therapy was effective in treatment of fibrosis in the rats and in patients with moderate disease [125]. Therefore, IFN-γ might be very useful for treatment of earlier stages of hepatitis B virus-induced liver fibrosis. This is a very interesting finding in view of potential beneficial effects of IFN-γ therapy in other types of fibrosis described elsewhere in this chapter (Sections 11.7–11.9).

Studies on the use of IFN-γ for therapy of chronic hepatitis C have been more limited. Treatment of chronic hepatitis C with IFN-γ was ineffective [126]. Pretreatment of hepatitis C with IFN-γ before beginning IFN-α treatment resulted in enhanced immunologic activity of the patients [127]. The authors speculated that the enhanced immunologic activity in the patients might contribute to improved viral clearance [127]. Therefore, the results of IFN-γ treatment of chronic hepatitis C are consistent with those obtained after treatment of chronic hepatitis B with IFN-γ. Combination treatments with forms of IFN-α (pegylated and nonpegylated) and ribavirin have been shown to be effective in the treatment of chronic hepatitis C [128, 129]. However, there is a group of patients that fail to achieve a sustained virologic response after first-line treatment with this combination and these are termed "nonresponder" patients. With the possibility of demonstration of enhanced activity, a phase 2 trial of a form of IFN-α termed Infergen (Interferon alfacon-1) and IFN-γ1b with and without ribavirin for the treatment of hepatitis C virus nonresponders is currently ongoing.

The usefulness of IFN-γ therapy for treatment of chronic hepatitis has yet to be established. The results reviewed above have generated additional clinical trials with IFN-γ for treatment of hepatitis, with particular emphasis on combined therapies and effects on liver fibrosis. These trials are ongoing at the time of preparation of this chapter. The determination of the value of IFN-γ in the treatment armamentarium for chronic hepatitis treatment will have to await the completion of the ongoing trials.

11.11
Oncology Indications: Ovarian Cancer

The direct antiproliferative activities and immunomodulatory activities of IFN-γ have led to numerous clinical studies to evaluate its therapeutic potential in a wide variety of malignant conditions. Studies that have focused on the use of IFN-γ as a single-agent treatment for various oncology indications have yielded results that are disappointing as those seen in studies of metastatic renal cell carcinoma [130], melanoma [131, 132]; colorectal cancer [133], leukemias [134], small cell lung cancer [135, 136] and metastatic carcinoid tumor [137].

Reports of the use of IFN-γ in combination with other anticancer therapies have been more encouraging, but are still mixed, showing some failures as in renal cell carcinoma with granulocyte macrophage colony-stimulating factor (GM-CSF) and IL-2 [137] or with IFN-α [139], colorectal cancer with 5-fluorouracil [140], and melanoma with IL-2 [141]; and some promise as in metastatic renal cell carcinoma with vinblastine [142] or IL-2 [143] or IFN-α [144], multiple myeloma with rituximab [145], inoperable hepatocellular carcinoma with GM-CSF [146], and colon cancer with 5-fluorouracil and leukovorin [147]. In the US, to date, no phase 3 oncology trials have provided sufficient evidence of effectiveness to establish the benefit of IFN-γ alone or in combination for the treatment of any form of cancer.

While using IFN-γ in combination with "standard of care" chemotherapy for specific cancers appears to hold most promise, the inherent difficulty with phase 3 trials of this type of design is that the trials take many years to complete in order to obtain endpoint data and during that time the "standard" chemotherapy might change to a more effective therapy, thus leading to the inability of the study data to be used for licensure by health authorities. Such was the case for a phase 3 study of IFN-γ1b in ovarian cancer.

As the third most common cancer in women, in the US, epithelial ovarian cancer represents approximately 20 000 new cases per year and is the leading cause of death from gynecologic cancers [148]. The disease is difficult to detect early and, therefore, most cases are diagnosed at an advanced stage. Epithelial ovarian cancer responds well to chemotherapy. However, while more than 50% of diagnosed patients achieve remission after surgery and primary chemotherapy, the overall 5-year survival rate is less than 40% [148]; hence, the need for improved therapeutic regimens.

Data from *in vitro* and *in vivo* studies suggest benefit of IFN-γ treatment in ovarian cancer [149–154]. Clinical trials showing benefit of IFN-γ in this clinical indication have utilized the i.p. [152], i.v. [153] and s.c. [154] routes of administration. The study of i.p. administered IFN-γ demonstrated achievement of surgically documented benefit as second-line therapy of ovarian cancer. Subsequently, Windbichler et al. [154] reported their results of a randomized phase 3 trial of IFN-γ administered s.c. to women receiving first-line platinum-based chemotherapy (cisplatinum and cyclophosphamide) after surgery for advanced staged ovarian cancer (FIGO stages Ic–III and some stage IV). This study was conducted in multiple sites in Austria, enrolling 148 patients from 1991 to 1997. Patients were randomly assigned to receive either the standard chemotherapy regimen of cisplatin and cyclophosphamide given every 4 weeks or the same chemotherapy plus IFN-γ in a fixed dose and schedule. At 3 years, progression-free survival was improved from 38% in the control group to 51% in the IFN-γ treated group. Complete clinical response was observed in 68% of the patients in the IFN-γ-treated group versus 56% in the control group. Evaluation of safety parameters showed the typical side-effects associated with IFN-γ treatment (flu-like syndrome with fever, headache, fatigue and myalgia).

The protocol was originally designed to enroll 200 patients; however, due to the change in standard treatment from chemotherapy with cisplatin and cyclophosphamide to a platinum plus taxane, the study had to be prematurely terminated [154]. On the basis of these promising results and due to this change in standard treatment, a new phase 3 study termed the "GRACES" (Gamma interferon and Chemotherapy Efficacy Study) study was initiated to evaluate IFN-γ in combination with carboplatin and paclitaxel in patients with advanced ovarian cancer or primary peritoneal cancer. The results of the interim analysis for this trial are expected in mid-2005. It is hoped that the results of this currently ongoing study will establish the evidence of effectiveness needed to establish this clinical use as an FDA approved indication.

11.12
Conclusions

The clinical indications for which IFN-γ has been evaluated for potential efficacy can be grouped into three categories: (i) established clinical benefit (CGD and osteopetrosis), (ii) negative or marginal effect in controlled trials (allergic rhinitis [155], steroid-dependant asthma [156] and rheumatoid arthritis [157, 158]) and (iii) encouraging results, awaiting additional confirmatory controlled clinical trials. This last category includes, but is not limited to, bacterial and fungal infections (including infection following trauma and opportunistic infections in AIDS patients); idiopathic pulmonary fibrosis, chronic hepatitis and ovarian cancer.

IFN-γ has been definitively shown to play a major multifaceted immunoregulatory role in host immune function. The ability to establish uses for IFN-γ in the treatment of clinical disorders may be limited by that very fundamental immunoregulatory role. Even in the areas where there are positive indications for the use of IFN-γ, it has become clear that the best effects are seen for specific subgroups of patients. For example, in infection after trauma and in idiopathic pulmonary fibrosis, the most seriously ill patients had the greatest response to IFN-γ therapy.

If the observed effectiveness of IFN-γ in specific subsets of patients represents a true event, clinical investigations will need to be carefully designed to establish evidence of effectiveness. First, the appropriate subjects for study must be identified, based on a clear understanding of the clinical features and course of the disease. Many of the diseases targeted for treatment with IFN-γ are heterogeneous in nature, resulting in heterogeneity of patients enrolled in the studies. Validation of clinically useful diagnostic criteria will allow for better characterization of these patients and will improve our ability to define the subgroups of patients that might benefit from IFN-γ treatment. Second, given the potent immunomodulatory activities of IFN-γ, a better characterization of a patient's baseline immune responsiveness may help identify the patient who will be more likely to benefit from IFN-γ treatment. Finally, having identified the appropriate study subjects, more accurate assumptions of effect size should lead to an appropriately powered study.

The authors of this chapter have been involved in research with IFN-γ since the time that it was first identified as a unique type of IFN. In the early years, when immunoregulatory effects were suggested for IFN-γ, reviewers often expressed skepticism since IFNs were well known as antiviral substances and thought to be only antiviral substances. It is rewarding to note that IFN-γ is now firmly established as an immunoregulatory agent and has clinical usefulness as a result of its immunoregulatory properties. Clinical trials conducted to date show that IFN-γ has made a significant impact by altering the course of serious human diseases and holds potential yet to be established. As we await the results of ongoing and new trials, the future holds great promise for this important therapeutic.

References

1. A. Isaacs, J. Lindenmann, *Proc Roy Soc Lond B* **1957**, *147*, 258–67.
2. S. Pestka, C. D. Krause, M. R. Walter, *Immunol Rev* **2004**, *202*, 8–32.
3. H. S. Jaffe, S. A. Sherwin. *Interferons as Cell Growth Inhibitors and Antitumor Factors*. Alan R. Liss, New York, **1986**, pp. 509–22.
4. Y. Vilcek, H. C. Kelker, J. Le, Y. K. Yip, in: R. J. Ford, A. L. Maizel (Eds.), *Mediators in Cell Growth and Differentiation*. Raven Press, New York, **1985**, pp. 299–313.
5. J. I. Gallin, J. M. Farber, S. M. Holland, T. B. Nutman, *Ann Inter Med* **1995**, *123*, 216–24.
6. E. Rinderknecht, B. O'Connor, H. Rodriguez, *J Biol Chem* **1984**, *259*, 6790–7.
7. E. Rinderknecht, L. E. Burton, in: H. Kirchner, H. Schellekens (Eds.), *The Biology of the Interferon System*. Elsevier, Amsterdam, **1985**, pp. 397–402.
8. J. I. Gallin, H. L. Malech, *J Am Med Ass* **1990**, *263*, 1533–7.
9. C. F. Nathan, C. R. Horowitz, et al., *Proc Natl Acad Sci USA* **1985**, *82*, 8696–90.
10. C. F. Nathan, G. Kaplan, et al., *N Eng J Med* **1986**, *315*, 6–15.
11. The International Chronic Granulomatous Disease Cooperative Study Group. *N Eng J Med* **1991**, *324*, 509–16.
12. B. E. Marciano, R. Wesley, et al., *Clin Infect Dis* **2004**, *39*, 692–9.
13. R. A. Ezekowitz, M. C. Dinauer, et al., *N Eng J Med* **1988**, *319*, 146–51.

14 J. M. G. Sechler, H. L. Malech, C. J. White, J. I. Gallin, Proc Natl Acad Sci USA **1988**, *85*, 4874–8.
15 F. Shapiro, M. J. Glimche, et al., J Bone Joint Surg Am **1980**, *62*, 384–99.
16 R. M. Rodriguez, L. L. Key, W. L. Reis, Pediatr Res **1993**, *33*, 384–9.
17 L. L. Key, Jr., W. L. Reis, R. M. Rodriguez, H. C. Hatcher, J Pediatr **1992**, *121*, 119–24.
18 L. L. Key, Jr., R. M. Rodriguez, et al., N Eng J Med **1995**, *332*, 1594–9.
19 C. W. Czarniecki, G. Sonnenfeld, APMIS **1993**, *101*, 1–17.
20 H. W. Murray, Am J Med **1994**, *97*, 459–67.
21 J. I. Gallin, J. M. Farber, S. M. Holland, T. B. Nutman. Ann Intern Med **1995**, *123*, 216–24.
22 H. W. Murray, Intensive Care Med **1996**, *22 (Suppl 4)*, S456–61.
23 W. C. Liles, Semin Respir Infect **2001**, *16*, 11–7.
24 C. F. Nathan, G. Kaplan, et al., N Eng J Med **1986**, *315*, 6–15.
25 G. Kaplan, N. K. Mathur, et al., Proc Natl Acad Sci USA **1989**, *86*, 8073–7.
26 G. Kaplan, Immunobiology **1994**, *191*, 564–8.
27 E. P. Sampaio, A. M. Malta, et al., Int J Lepr Other Mycobact Dis **1996**, *64*, 268–73.
28 K. E. Squires, S. T. Brown, et al., J Infect Dis **1992**, *166*, 686–7.
29 S. M. Holland, E. Einstein, et al., N Eng J Med **1994**, *330*, 1348–55.
30 L. E. Bermudez, C. Inderlied, L. S. Young. Antimicrob Agents Chemother **1991**, *35*, 2625–9.
31 R. Condos, W. N. Rom, et al., Lancet **1997**, *349*, 1513–5.
32 W. J. Koh, O. J. Kwon, et al., J Korean Med Sci **2004**, *19*, 167–71.
33 R. Condos, F. P. Hull, et al., Chest **2004**, *125*, 2146–55.
34 R. Suarez-Mendez, I. Garcia-Garcia, et al., BMC Infectious Dis **2004**, *4*, 44.
35 S. G. Reed, P. Scott, Curr Opin Immunol **1993**, *5*, 524–31.
36 R. Badaro, E. Falcoff, et al., N Eng J Med **1990**, *322*, 16–21.
37 S. Sundar, F. Rosenkaimer, H. W. Murray, J Infect Dis **1994**, *170*, 659–62.
38 S. Sundar, F. Rosenkaimer, M. L. Lesser, H. W. Murray, J Infect Dis **1995**, *171*, 992–6.
39 L. A. Riddell, A. J. Pinching, et al., AIDS Res Hum Retroviruses **2001**, *17*, 789–97.
40 P. M. Murphy, C. H. Lane, J. I. Gallin, A. S. Fauci, Ann Intern Med **1988**, *108*, 36–41.
41 P. G. Pappas, B. Bustamante, et al., J Infect Dis **2004**, *189*, 2185–91.
42 M. J. Hershman, G. Sonnenfeld, et al., Microbial Pathogenesis **1988**, *4*, 165–8.
43 M. J. Hershman, G. Sonnenfeld, et al., Interferon Res **1988**, *8*, 367–73.
44 M. J. Hershman, H. C. Polk, Jr., et al., Clin Exp Immunol **1988**, *73*, 406–9.

45 M. J. Hershman, H. C. Polk, Jr., et al., *Immunology* **1988**, *56*, 2412–6.
46 D. H. Livingston, S. H. Appel, G. Sonnenfeld, M. A. Malangoni, *J Surg Res* **1989**, *46*, 322–6.
47 M. A. Malangoni, D. H. Livingston, G. Sonnenfeld, H. C. Polk, Jr., *Arch Surg* **1990**, *125*, 444–6.
48 H. C. Polk, Jr., S. Galandiuk, M. J. Hershman, G. Sonnenfeld, in: A. V. Pollock (Ed.), *Surgical Immunology*. Edward Arnold, Sevenoaks, Kent, **1991**, pp. 254–264.
49 H. C. Polk, Jr., W. G. Cheadle, G. Sonnenfeld, M. J. Hershman, in: H. Jaffe, B. Bucalo, S. Sherwin (Eds.), *Anti-infective Applications of Interferon-Gamma*. Marcel Dekker, New York, **1992**, pp. 29–36.
50 W. G. Cheadle, G. Sonnenfeld, S. R. Wellhausen, H. C. Polk, Jr., in: G. Sonnenfeld, C. Czarniecki, C. Nacy, G. I. Byrne, M. Degré (Eds.), *Cytokines and Resistance to Nonviral Pathogenic Infections*. Biomedical Press, Augusta, GA, **1992**, pp. 85–91.
51 E. Faist, A. Mewes, et al., *Arch Surg* **1988**, *123*, 287–92.
52 D. R. Green, E. Faist, *Immunol Today* **1988**, *9*, 253–5.
53 S. S. Pierangeli, H. C. Polk, Jr., M. J. Parmely, G. Sonnenfeld, *Cytokine* **1993**, *5*, 230–34.
54 D. H. Livingston, S. H. Appel, et al., *Arch Surg* **1988**, *123*, 1309–12.
55 M. J. Hershman, S. H. Appel, et al., *Clin Exp Immunol* **1992**, *77*, 67–70.
56 H. C. Polk, Jr., W. G. Cheadle, et al., *Am J Surg* **1992**, *163*, 191–6.
57 D. J. Dries, G. J. Jurkovich, et al., *Arch Surg* **1994**, *129*, 1031–41.
58 C. N. Mock, D. J. Dries, G. J. Jurkovich, R. V. Maier, *Shock* **1996**, *5*, 235–40.
59 D. H. Livingston, P. A. Loder, et al., *Arch Surg* **1994**, *129*, 172–8.
60 D. J. Dries, J. M. Walenga, D. Hoppensteadt, J. Fareed, *J Interferon Cytokine Res* **1998**, *18*, 327–35.
61 C. Schinkel, K. Licht, et al., *Shock* **2001**, *16*, 329–33.
62 C. Schinkel, K. Licht, et al., *J Trauma* **2001**, *50*, 321–7.
63 R. J. Rentenaar, J. de Metz, et al., *Clin Exp Immunol* **2001**, *125*, 401–8.
64 A. K. Licht, C. Schinkel, et al., *J Interferon Cytokine Res* **2003**, *23*, 149–54.
65 C. Schinkel, *J Interferon Cytokine Res* **2003**, *23*, 341–9.
66 T. T. Chang, S. R. Stevens, *Am J Clin Dermatol* **2002**, *3*, 175–83.
67 J. M. Hanifin, K. D. Cooper, et al., *J Am Acad Dermatol* **2004**, *50*, 391–404.
68 M. Tang, A. Kemp, G. Varigos, *Clin Exp Immunol* **1993**, *92*, 120–4.
69 C. M. Snapper, W. E. Paul, *Science* **1987**, *236*, 944–6.
70 K. Jujo, H. Renz, et al., *J Allergy Clin Immunol* **1992**, *90*, 323–31.
71 U. Reinhold, W. Wehrmann, et al., *Lancet* **1990**, *335*, 1282.

72 M. Boguniewicz, H. S. Jaffe, et al., *Am J Med* **1990**, *88*, 365–70.
73 U. Reinhold, S. Kukel, et al., *J Am Acad Dermatol* **1993**, *29*, 58–63.
74 J. M. Hanifin, L. C. Schneider, et al., *J Am Acad Dermatol* **1993**, *28*, 189–97.
75 S. R. Stevens, J. M. Hanifin, et al., *Arch Dermatol* **1998**, *134*, 799–804.
76 L. C. Schneider, Z. Baz, et al., *Ann Allergy Asthma Immunol* **1998**, *80*, 263–8.
77 I. G. Jang, J. K. Yang, et al., *J Am Acad Dermatol* **2000**, *42*, 1033–40.
78 G. W. Noh, K. Y. Lee, *Allergy* **1998**, *53*, 1202–7.
79 M. Selman, T. King, A. Pardo, *Ann Intern Med* **2001**, *134*, 136–51.
80 J. J. Swigris, W. G. Kuschner, J. L. Kelsey, M. K. Gould, *Chest* **2005**, *127*, 275–83.
81 M. Selman, V. J. Thannickal, et al., *Drugs* **2004**, *64*, 405–30.
82 S. D. Nathan, S. D. Barnett, et al., *Respiration* **2004**, *71*, 77–82.
83 N. Kahlil, R. O'Connor, *Can Med Ass J* **2004**, *171*, 153–60.
84 M. P. Keane, P. M. Henson, R. M. Streiter, in: J. F. Murray, J. A. Nadel, R. Mason, H. Boushey (Eds.), *Textbook of Respiratory Medicine*, 3rd edn. Saunders, Philadelphia, PA, **2000**, pp. 495–539.
85 M. P. Keane, J. A. Belperio, R. M. Streiter, in: M. I. Schwartz, T. E. King (Eds.), *Interstitial Lung Disease*, 4th edn. Decker, Hamilton, Ontario, 2003, pp. 245–75.
86 S. Majumdar, D. Li, T. Ansari, et al., *Eur Respir J* **1999**, *14*, 251–7.
87 J. A. Elias, B. Freundlich, J. A. Kern, J. Rosenbloom, *Chest* **1990**, *97*, 1439–45.
88 R. S. Bienkowski, M. G. Gotkin, *Proc Soc Exp Biol Med* **1995**, *209*, 118–40.
89 A. M. Cornelissen, J. W. Von den Hoff, et al., *Arch Oral Biol* **1999**, *44*, 541–7.
90 A. S. Narayanan, J. Whithey, et al., *Chest* **1992**, *101*, 1326–31.
91 T. Okada, I. Sugie, K. Aisaka, *Lymphokine Cytokine Res* **1993**, *12*, 87–91.
92 O. Eickelberg, A. Pansky, et al., *FASEB J* **2001**, *15*, 797–806.
93 S. D. Oldroyd, G. L. Thomas, et al., *Kidney Int* **1999**, *56*, 2116–27.
94 T. Nagahori, M. Dohi, et al., *Am J Respir Cell Mol Biol* **1999**, *21*, 490–7.
95 K. Tamai, H. Ishikawa, et al., *J Invest Dermatol* **1995**, *104*, 384–90.
96 G. Gurujeyalakshmi, S. N. Giri, *Exp Lung Res* **1995**, *21*, 791–808.
97 R. Ziesche, E. Hofbauer, et al., *N Engl J Med* **1999**, *341*, 1264–9.
98 G. Raghu, K. Brown, et al., *Continuing Education Series*. American Thoracic Society, Atlanta, GA, **2000**, pp. 44–51.

99 G. Raghu, K. K. Brown, et al., *N Eng J Med* **2004**, *350*, 125–33.
100 R. M. Streiter, K. M. Starko, et al., *Am J Respir Crit Care Med* **2004**, *170*, 133–40.
101 J. Varga, A. Olsen, et al., *Eur J Clin Invest* **1990**, *20*, 487–93.
102 J. Ji, L. Si, W. Fang, *Chin Med J* **2001**, *114*, 139–42.
103 F. Q. Wen, X. D. Liu, et al., *Chest* **2003**, *123 (Suppl 1)*, 272–3S.
104 S. Schnyder-Candrian, R. M. Streiter, S. L. Kunkel, A. Waltz, *J Leuk Biol* **1995**, *57*, 929–35.
105 J. Rosenbloom, G. Feldman, B. Freundlich, S. A. Jimenez, *Biochem Biophys Res Commun* **1984**, *123*, 365–72.
106 R. M. Streiter, J. A. Belperio, et al., in: R. M. Streiter, S. L. Kunkel, T. J. Standiford (Eds.), *Chemokines in the Lung*. Marcel Dekker, New York, **2003**, pp. 297–324.
107 K. E. Cole, C. A. Strick, et al., *J Exp Med* **1998**, *187*, 2009–21.
108 A. M. Cole, T. Ganz, et al., *J Immunol* **2001**, *167*, 623–7.
109 J. H. Dauber, K. F. Gibson, N. Kaminski, *Am J Resp Crit Care Med* **2004**, *170*, 107–8.
110 T. E. King, S. Safrin, et al., *Chest* **2005**, *127*, 171–7.
111 R. Hein, J. Behr, et al., *Br J Dermatol* **1992**, *126*, 496–501.
112 A. Grassegger, G. Schuler, et al., *Br J Dermatol* **1998**, *139*, 639–48.
113 A. Barabanova, D. P. Osanov, *Int J Radiat Biol* **1990**, *57*, 775–82.
114 R. U. Peter, P. Gottlober, et al., *Int J Radiat Oncol Biol Phys* **1999**, *45*, 147–52.
115 N. P. Nguyen, J. E. Antoine, et al., *Cancer* **2002**, *95*, 1151–63.
116 H. B. Greenberg, R. B. Pollard, et al., *N Engl J Med* **1978**, *195*, 517–22.
117 O. Weiland, R. Schvarcz, et al., *Scand J Infect Dis* **1989**, *21*, 127–32.
118 G. L. Davis, L. A. Balart, et al., *J Hepatol* **1990**, *11 (Suppl 1)*, 531–5.
119 J. Y. Lau, C. L. Lai, et al., *J Med Virol* **1991**, *34*, 184–7.
120 S. Kakumu, T. Ishikawa, et al., *J Med Virol* **1991**, *35*, 32–7.
121 M. Ruiz-Moreno, M. J. Rua, et al., *Pediatrics* **1992**, *90*, 254–8.
122 V. Carreno, A. Moreno, F. Galiana, F. J. Bartolome, *J Hepatol* **1993**, *17*, 321–5.
123 E. Musch, B. Hogemann, et al., *Hepatogastroenterology* **1998**, *45*, 2282–94.
124 C. W. Czarniecki, C. W. Fennie, D. B. Powers, D. A. Estell, *J Virol* **1984**, *49*, 490–6.
125 H. L. Weng, W. M. Cai, R. H. Liu, *World J Gastroenterol* **2001**, *7*, 42–8.
126 F. Saez-Royuela, J. C. Porres, et al., *Hepatology* **1991**, *13*, 327–31.
127 K. Katayama, A. Kasahara, et al., *J Viral Hepat* **2001**, *8*, 180–5.
128 T. Poynard, P. Marcellin, et al., *Lancet* **1998**, *35*, 1426–32.
129 J. G. McHutchinson, S. C. Gordon, et al., *N Eng J Med* **1998**, *339*, 1485–92.
130 M. E. Gleave, M. Elhilali, et al., *N Engl J Med* **1998**, *338*, 1265–71.

131 U. R. Kleeberg, S. Suciu, et al., *Eur J Cancer* **2004**, *40*, 390–402.
132 J. H. Schiller, M. Pugh, et al., *Clin Cancer Res* **1996**, *2*, 29–36.
133 T. D. Brown, P. J. Goodman, et al., *J Immunother* **1991**, *10*, 379–82.
134 R. M. Stone, D. R. Spriggs, et al., *J Clin Oncol* **1993**, *16*, 159–63.
135 N. van Zandwijk, H. J. Groen, et al., *Eur J Cancer* **1997**, *33*, 1759–66.
136 J. R. Jett, A. W. Maksymiuk, et al., *J Clin Oncol* **1994**, *12*, 2321–6.
137 K. Stuart, D. E. Levy, T. Anderson, *Invest New Drugs* **2004**, *22*, 75–81.
138 M. Schmidinger, G. Steger, et al., *J Immunother* **2001**, *24*, 257–62.
139 P. H. De Mulder, G. Oosterhof, et al., *Br J Cancer* **1995**, *71*, 371–5.
140 N. Pavlidis, C. Nicolaides, et al., *Oncology* **1996**, *53*, 159–62.
141 C. J. Kim, J. K. Taubenberger, et al., *J Immunother. Emphasis Tumor Immunol* **1996**, *19*, 50–8.
142 C. Bacoyiannis, M. A. Dimopoulos, et al., *Oncology* **2002**, *63*, 130–8.
143 M. Schmidinger, G. G. Steger, et al., *Cancer Immunol Immunother* **2000**, *49*, 395–400.
144 A. Fujii, K. Yui-En, et al., *BJU Int* **1999**, *84*, 399–404.
145 S. P. Treon, L. M. Pilarski, et al., *J Immunother* **2002**, *25*, 72–81.
146 W. Reinisch, M. Holub, et al., *J Immunother* **2002**, *25*, 489–99.
147 L. S. Schwartzberg, I. Petak, et al., *Clin Cancer Res* **2002**, *8*, 2488–98.
148 A. Jemal, R. C. Tiwari, et al., *CA Cancer J Clin* **2004**, *54*. 8–29.
149 L. Wall, F. Burke, et al., *Gynecol Oncol* **2003**, *88 (1 Pt 2)*, S149–51.
150 C. Marth, H. Fiegl, et al., *Am J Obstet Gynecol* **2004**, *191*, 1598–605.
151 R. S. Freedman, A. P. Kudelka, et al., *Clin Cancer Res* **2000**, *6*, 2268–78.
152 E. Pujade-Lauraine, J. P. Guastalla, et al., *J Clin Oncol* **1996**, *14*, 343–50.
153 C. E. Welander, H. D. Homesley, S. D. Reich, E. A. Levin, *Am J Clin Oncol* **1988**, *11*, 465–9.
154 G. H. Windbichler, H. Hausmaninger, et al., *Br J Cancer* **2000**, *82*, 1138–44.
155 J. T. Li, J. W. Yunginger, et al., *J Allergy Clin Immunol* **1990**, *85*, 934–40.
156 M. Boguniewicz, L. C. Schneider, et al., *Clin Exp Allergy* **1993**, *23*, 785–90.
157 German Lymphokine Study Group, *Rheumatol Int* **1992**, *12*, 175–85.
158 G. W. Cannon, R. D. Emkey, et al., *J Rheumatol* **1993**, *20*, 1867.

Section D
Measurement of Interferons and Anti-Interferons

12
Measurement of Interferon Activities

Tony Meager

12.1
Introduction

Polypeptide growth factors and cytokines stimulate biological responses by binding to specific cell surface receptors; these trigger cytoplasmic signal transduction pathways and subsequent expression of specific inducible genes. In this respect, the interferon (IFN) family of cytokines, defined by the capacity of its members to induce antiviral activity, is typical. While structural differences exist among members of the IFN family, and despite the existence of different classes of receptor, the IFNs all exert a broadly similar range of biological activities in responsive cells. These activities include antiviral, antiproliferative, antitumor and immunomodulatory activities, which can be observed in appropriate cell culture systems and quantified where dose-dependency to IFN stimulation is manifested. Historically, the activity of IFNs *in vitro* has been measured by their capacity to induce an "antiviral state" in susceptible cell lines. On this basis, quantitative "antiviral assays" have been developed that generate dose–response data from which the relative activity or "potency" of particular IFN preparations may be calculated. The discovery of other activities induced by IFNs has led to a widening of the types of bioassay that can be used for determining potency. For example, antiproliferative assays where IFN activity leads to dose-dependent inhibition of cell proliferation are now regularly carried out. However, the identification of those proteins that are specifically upregulated following IFN stimulation has led to the development of more specific bioassays. The process of inducible protein expression has been further exploited by the development of "reporter gene assays". For these, a promoter region of an IFN-inducible gene is ligated to a foreign gene, such as one that encodes a "measurable" enzyme, e.g. firefly luciferase or alkaline phosphatase, and transfected via plasmid DNA into a suitably IFN-responsive cell line. Expression of the reporter gene in stably transfected cells is dependent on IFN concentration, the assay readout being the measured enzymatic activity, which is directly related to potency. The elucidation of phosphorylated intermediates in IFN signal transduction pathways might also lead to development of additional novel bioassays.

The Interferons: Characterization and Application. Edited by Anthony Meager
Copyright © 2006 WILEY-VCH Verlag GmbH & Co. KGaA, Weinheim
ISBN: 3-527-31180-7

Using appropriate examples, this chapter aims to review the bioassay methods that are currently in use, and those that could potentially be used, for potency testing of IFNs. The advantages and disadvantages of particular bioassay methods will be considered.

12.2
The IFNs: Mechanisms of Action, Protein Induction and Biological Activities

12.2.1
Mechanisms of Action

The basis for understanding how the potency of IFNs is measured is underscored by knowledge of their mechanisms of action, the proteins they induce and their consequent biological activities. These topics have been extensively reviewed by other authors in this book (see Chapter 7) and therefore these areas will be only briefly covered here.

The IFN family includes type I and II IFN molecules: type I IFNs are a mixture of heterogeneous proteins in all animal species, while type II IFN appears to have remained as a single protein throughout evolution. Within the type I IFN family, several subclasses have been characterized; these are normally specified by letters of the Greek alphabet, and include IFN-α, -β, -ε, -κ and -ω in *Homo sapiens*; IFN-δ and -τ, and limitin are found in other mammals [1, 2]. The single type II IFN is commonly designated IFN-γ. Recently, another small family of three related cytokines, interleukin (IL)-28A, -28B and -29, very distantly related to type I IFNs, have been shown to induce IFN-like activities. For this reason, they have been called IFN-λs ($\lambda 1$ = IL-29, $\lambda 2$ = IL-28A and $\lambda 3$ = IL-28B) and referred to collectively as "type III IFNs" [3–5]. It is clear that the potency of these IFN-like cytokines can be determined in bioassays similar to those applied to potency determinations of type I and II IFNs [6]. Thus, they are included for consideration and comparison with the "classical" type I and II IFNs, and will be referred to as type III IFNs (IFN-λs) from here on.

All IFNs exert their activities in cells following initial binding with high affinity to IFN-specific cell surface receptors. The characteristics of these receptors have been extensively reviewed (see Chapter 5). Briefly, all type I IFNs share a common receptor comprised of two different receptor chains, which have been molecularly characterized as belonging to the type II cytokine receptor family (CRF2) [1, 2, 7]. The CRF2 family also contains receptors for IL-10 and its paralogs. Type II IFN or IFN-γ binds to a heterodimeric receptor distinct from that of the type I IFN receptor, but whose subunits are also members of the CRF2 family. Lastly, type III IFNs or IFN-λs bind to another distinct CRF2 heterodimeric receptor; it is comprised of a unique IFN-λ-binding subunit, IL-28Rα or IFNLR, but the other, IL-10R2, is one of the subunits of the functional IL-10 receptor [3, 4, 7]. However, it is not uncommon for different cytokines to share receptor subunits; besides IL-10 and IFN-λs, IL-10R2 is also an integral component of the IL-22 and -26 receptors [7].

Amino acid sequences of IFN receptors of different animal species indicate their coevolution with their cognate IFNs; this has led to some degree of species specificity for most IFNs [1, 2]. For example, while human type I IFN-β and most IFN-α subtypes bind poorly to the mouse type I IFN receptor, human IFN-α1 subtype is active in both human and murine cells, although it has the lowest antiviral activity of all the subtypes in human cells [2, 8]. Curiously, human IFN-α subtypes are active in bovine cells. In contrast, human type II IFN-γ shows high species specificity [9].

IFN binding to receptors occasions structural changes in the receptor molecules leading to stimulation of cognate intracellular signaling pathways (Fig. 12.1). Although the receptor chains do not contain integral protein kinases, "non-receptor" tyrosine kinases of the Janus class are associated with their intracellular domains [1, 2, 10–12]. Phosphorylation of specific tyrosine residues in the intracellular domains by the Janus kinase (JAK) and tyrosine-specific kinase (TYK) kinases leads

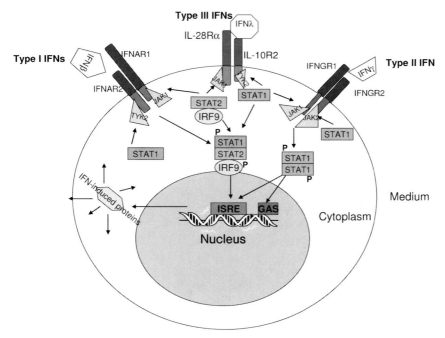

Fig. 12.1. Schematic diagram illustrating the IFN signal transduction pathways. IFNs bind to extracellular domains of their cognate class II cytokine receptors leading to activation of JAK-1 and -2, and TYK-2 tyrosine kinases associated with receptor endodomains. Subsequent phosphorylation of endodomains leads to recruitment of STAT-1 and/or -2–IRF-9, phosphorylation and dimerization of STATs followed by their translocation into the nucleus. These transcription factor complexes bind to the ISRE and/or GAS present in the promoter region of IFN-inducible genes to activate their transcription and the synthesis of IFN-inducible proteins. (This figure also appears with the color plates.)

to recruitment and phoshorylation of members of the "signal tranducers and activators of transcription (STATs)" class of transcription factors, as depicted in Fig. 12.1 [10, 11]. Dimerization of STATs, principally of STAT-1 to -2 in the case of type I IFNs and of two copies of STAT-1 alone for type II IFNs, and pre-bound additional accessory factors, e.g. IFN-regulatory factor (IRF)-9 to STAT-2, leads to their translocation into the nucleus and binding to gene-activating promoter sequences, the IFN-stimulated response element (ISRE) for type I IFNs and the γ-activated sequence (GAS) for type II IFN [10, 11]. However, transcriptional activation of ISRE- and GAS-linked genes by the individual type I and II IFN JAK–STAT signaling pathways is not mutually exclusive, and type II IFN in particular can stimulate some type I IFN-inducible genes. While type III IFNs (IFN-λs) have structurally different receptor subunits to those of type I IFNs, in common with type I IFN receptor chains they are associated with JAK-1 and TYK-2, and appear to trigger the same, or very similar, signal transduction mechanism involving STAT-1:2–IRF9 transcription complexes (Fig. 12.1) [3–5]. Although all three types of IFN appear to work mainly through the JAK–STAT signaling pathways, it is likely that some activities are mediated through alternative pathways [2]. For example, signaling via serine/threonine kinases such as protein kinase C (PKC) has been described as well as several other pathways (see Chapter 7).

12.2.2
Protein Induction

The end result of IFN-induced signaling pathways is the induction and general upregulation of the synthesis of IFN-inducible proteins. These, whose total number probably runs to hundreds [13], include both those proteins induced early after IFN stimulation and those produced at later times, often in response to the actions of "early" IFN-inducible proteins. Many IFN-inducible proteins are induced by both type I and II IFNs, although some may be preferentially induced by either type [14]. Yet other proteins are exclusively induced by either type I or II IFNs [14]. Only a limited range of proteins are as yet known to be induced by type III IFNs (IFN-λs), but these appear to coincide mainly with type I IFN-inducible proteins [3, 4]. In some instances, the strength of induction and/or the rapidity of induction of IFN-inducible proteins may vary according to the type of IFN, or even between members of a type of IFN, e.g. IFN-β can often appear to be a stronger inducer than IFN-α subtypes [13]. In some cases, IFN-inducible proteins are virtually absent from cells before IFN stimulation, while in others the stimulation results in an overall increase in concentration of proteins that are already being expressed. Table 12.1 shows a list of well-characterized IFN-inducible proteins, outlining their functions and their relative inducibility by either type I or II IFN. It should be recognized that cell dependency of induction probably occurs such that only limited subsets of the total IFN-inducible proteins are likely to be present in specific cell types.

Tab. 12.1. IFN-inducible proteins

Protein	Function	Induction by	
		Type I IFN	Type II IFN
2′–5′-OAS	dsRNA-dependent synthesis of ppp(A2′p)nA; activator of RNase L	+	+
PKR	phosphorylation of peptide initiation factor eIF-2α	+	+
MHC class I (HLA-A, -B, -C)	antigen presentation to cytotoxic T lymphocytes	+	+
MHC class II (HLA-DR)	antigen presentation to T helper lymphocytes	+	+
IDO	tryptophan catabolism	–	+
Guanylate-binding proteins (GBP; γ67)	GTP, GDP binding	+	+
Mx	specific inhibition of influenza virus	+	–
IRF-1/ISGF-2	transcription factor	+	+
IRF-2	transcription factor	+	–
IP-10	related to chemotactic IL-8-like cytokines	+	+
Metallothionein	metal detoxification	+	+
TNF receptors	mediate TNF action	+	+
IL-2 receptors	mediate IL-2 action	–	+
ICAM-1	endothelial cell adhesion protein	–	+
Ig FcR	Ig binding by macrophages/neutrophils	+	+
Thymosin β_4	induction of terminal transferase in B lymphocytes	+	?
β_2-microglobulin	immune function	+	+
Nitric oxide synthetase	production of nitric oxide from arginine; increased microbicidal activity in macrophages	–	+
200 family	from cluster of six genes; p204 is located in nucleolus	+	–
6-16	unknown; extracellular	+	+
1-8/9-27	cell surface proteins	+	+

12.2.3
Biological Activities

12.2.3.1 Antiviral Activity

In natural infections, viruses provoke the synthesis of IFNs, which are secreted from infected cells and act in a paracrine manner to induce antiviral activity in surrounding cells. Depending on the nature of their genomes, i.e. RNA or DNA, viruses have evolved a number of replication strategies. Despite this, there appear to be a number of common replicative stages in the diverse replication cycles of viruses that are vulnerable to inhibition by certain IFN-inducible "antiviral" proteins. One of the most studied of these "antiviral" proteins is an enzyme known as 2′–5′-oligoadenylate synthetase (2′–5′-OAS) which, in the presence of double-stranded (ds)RNA (often an intermediate of viral RNA synthesis), catalyzes the formation of an unusual oligonucleotide, ppp(A2′p)nA or 2′–5′-oligoadenylate synthetase (2′–5′-OAS), which is a necessary cofactor for activation of a latent endoribonuclease, RNase L [15–18]. Once activated RNase L degrades viral mRNAs and thus inhibits viral protein synthesis. Small RNA viruses, such as Mengo virus or murine encephalomyocarditis virus (EMCV), whose replication is "cytoplasmic", are among the most susceptible viruses to the 2′–5′-OAS/RNase L antiviral mechanism [19, 20].

Another IFN-inducible enzyme with antiviral potential is a protein kinase activated by dsRNA, now designated PKR, which in the active state phosphorylates the peptide initiation factor, eIF-2, involved in polyribosomal translation of viral mRNA – phospho-eIF-2 is inactive and therefore viral protein synthesis is inhibited [15–17]. Rhabdoviruses, dsRNA viruses such a vesicular stomatitis virus (VSV), appear to be among those most affected by the PKR/phospho-eIF-2 antiviral mechanism [16, 21]. Both the "2′–5′-OAS" and "PKR" systems are rapidly induced by type I IFNs, but in comparison their induction by type II IFN is slower, and by type III IFNs, probably weaker. A third "antiviral" protein with GTPase activity, known as Mx (in mice) or MxA (in humans), is strongly induced by type I IFNs, to a lower extent by type III IFNs, but not at all by type II IFN [3, 4, 6, 17, 22]. In murine cells, Mx has been shown to specifically inhibit the nuclear stage of influenza viral nucleic acid synthesis. In human cells, where MxA is cytoplasmically located, MxA has been shown to interfere with the replication of ortho- and paramyxoviruses and viruses of the Bunyaviridae family, probably mainly by binding and "trapping" viral nucleoproteins so that they become unavailable for the generation of new virus particles [23, 24].

12.2.3.2 Antiproliferative Activity

Although the IFN-inducible "2′–5′-OAS/RNase L" and "PKR/eIF-2" systems are involved in inhibiting viral protein synthesis, there has been growing evidence they also inhibit cellular protein synthesis, thus leading to inhibition of cell division [11, 25–27]. For example, levels of 2′–5′-OAS and RNase L enzymes have been found to be high in growth-arrested cells, suggesting they are involved in regulation of proliferation [28]. PKR has also been implicated in IFN-induced antiproliferative activity [27]. Cyclin-dependent kinase (CDK)-2, which mediates the phosphorylation of retinoblastoma (Rb) protein in the cell cycle of proliferation, is inhibited by type I IFNs and this could also contribute to antiproliferative activity [29–31]. Additionally, several other IFN-induced manifestations, including increased cell rigidity [32], depletion of essential metabolites [33, 34] and suppression of oncogenes [35], could be involved the antiproliferative activity of IFN. The IFN-inducible transcription factor IRF-1 (Tab. 12.1), which is involved in the expression of a number of IFN-inducible proteins, including 2'–5'-OAS and PKR, is also implicated as a tumor suppressor and regulator of proliferation rates [36, 37]. IFNs probably also induce apoptosis in some cells (see Chapter 8).

12.2.3.3 Immunomodulatory Activity

While direct antiproliferative activity of IFNs may form part of their antitumor activities, the regulation of various immune responses by IFNs *in vivo* can also be significantly involved in killing of tumor cells. IFNs upregulate immune responses mainly by increasing the expression of cell surface molecules, e.g. β_2-microglobulin, class I and II MHC antigens, and immunoglobulin G (IgG) Fc receptor (FcR), crucially required for "immune" recognition [8, 38–40]. All types of IFN variably stimulate increased expression of class I MHC antigens, but only type II IFN stimulates the *de novo* synthesis of class II MHC antigen, HLA-DR, required for triggering both humoral and cell-mediated immunity [9, 41, 42]. Additionally, the expression of cellular adhesion molecules, such as intercellular adhesion molecule (ICAM)-1 [43, 44] and other cell surface marker molecules, e.g. 9-27/Leu13 [45, 46], is stimulated by IFN, particularly type II IFN. Type I IFNs also act on natural killer (NK) cells to increase their cytotoxicity against virally infected and tumor cell targets [47, 48].

12.3
Measurement of Biological Activities of IFNs

12.3.1
General Considerations

It is evident from the preceding section that IFNs exert a range of biological activities, many of which are demonstrable in *in vitro* cell culture systems. Measurement of an activity relies on the capacity of the system (assay) to respond to the IFN in a concentration-dependent manner and the ability to make quantitative measurements of the "responding" parameter, thus creating analyzable dose–response data. Put another way, an IFN may be quantified "if an assay generates a measurable parameter that reproducibly increases with increasing concentration of IFN". In general, measurements of activity made from such assays are of relative activity or potency – the "strength" in biological activity units – of a particular test IFN preparation in relation to that of a reference preparation (standard) of the same IFN, the potency of which has been preassigned. It is important for assay validity that the test IFN preparation behaves identically in the assay to the IFN standard. This is necessary to ensure that when the test and standard are compared in the assay, any difference in measured response reflects only the difference in their respective concentrations. However, with complex IFN protein molecules, even identity between identically prepared and highly purified preparations can be difficult to attain/prove; they could contain different impurity and/or contaminant levels. "Less pure" IFN preparations will always contain impurities and contaminants, e.g. other cytokines and endotoxin, which could, in theory, have effects on IFN activity and, in practice, often do. Therefore, it is best if the measured parameter is exclusively related to IFN concentration. Where this is not possible, steps to evaluate and then minimize the "interference" due to contaminants should be taken.

12.3.2
Biological Standards for IFNs

It is evident that well-characterized, representative IFN preparations (standards), which are available in the long term, are required to calibrate the biological assays used to quantify the potency of test preparations of IFNs. In addition, since biological assays are inherently variable – they are sensitive to cell culture and general assay conditions – such standards serve to qualify/monitor the performance of individual assays [49–51]. The WHO has for many years supported an international program of research and development of biological standards for cytokines, including human IFNs. This program has become of increasing importance with the growing use of therapeutic IFN products in the clinic. The preparation of lyophilized IFN standards in sealed glass ampoules took place at the Medical Research Council (MRC) laboratories, UK, up until 1975; from then on, at the National Institute for Biological Standards and Control (NIBSC), UK, and at the National Institutes of Health (NIH) of the US and Japan [50]. With technological advances in the production and purification of IFNs, the quality and purity of the IFN materials going into IFN standards has improved markedly, with most standards now containing highly purified recombinant IFNs [51]. Each candidate standard undergoes extensive evaluation in WHO international collaborative studies involving the participation of laboratories expert in the performance of bioassays for IFNs. Each participant titrates the candidate standards in their own bioassays and sends assay raw (dose–response) data back to the organizing institution for statistical analysis and calculation of geometric mean potency. Subsequently, following receipt of the study report and recommendations on the suitability of candidate standards to serve as International Standards (ISs), the WHO, through its Expert Committee on Biological Standardization (ECBS), has established ISs for many types and subtypes of human IFN, together with some other animal IFN (Tab. 12.2) [51, 52].

Tab. 12.2. WHO ISs for IFNs

Preparation	Product code	Status	Potency/ampoule (IU)
Human IFN			
IFN-α leukocyte	94/784[a]	first	11000
IFN-α1 rDNA (*E. coli*)	83/514[a]	first	8000
IFN-α1/8 rDNA (*E. coli*)	95/572[a]	first	27000
IFN-α2a rDNA (*E. coli*)	95/650[a]	second	63000
IFN-α2b rDNA (*E. coli*)	95/566[a]	second	70000
IFN-α2c rDNA (*E. coli*)	95/580[a]	first	40000
IFN-αn1 lymphoblastoid	95/568[a]	second	38000
IFN-αn3 leukocyte	95/574[a]	first	60000
IFN-αcon1 rDNA (*E. coli*)	94/786[a]	first	100000
IFN-ω rDNA (CHO cell)	94/754[a]	first	20000
IFN-β, glycosylated, rDNA (CHO cell)	00/572[a]	third	40000
IFN-β, Ser17, rDNA (*E. coli*)	Gxb02-901-535[b]	first	6000
IFN-γ rDNA (*E. coli*)	Gxg01-902-535[b]	second	80000
Murine (mouse) IFN			
IFN-α/β	Gu02-901-511[b]	second	10000
IFN-α	Ga02-901-511[b]	first	12000
IFN-β	Gb02-902-511[b]	first	15000
IFN-γ	Gg02-901-533[b]	first	1000
Rabbit IFN			
IFN-α/β	G-019-902-528[b]	first	10000
Chick IFN			
IFN-β	67/18[a]	first	80

[a] These international standards are available from NIBSC, Blanche Lane, South Mimms, Hertfordshire EN6 3QG, UK.
[b] These international standards are available from The Research Resources Branch, National Institute of Allergy and Infectious Diseases, NIH, Bethesda, MD 20205, USA.

These WHO ISs are the result of considerable effort to prepare and establish appropriate biological reference materials for particular IFN types and subtypes and, therefore, represent a valuable resource. ISs, as primary standards with potency assigned in International Units (IU), should be used to calibrate working standards (in-house or national standards), which are well-characterized and as close as possible in purity and form to the IS (see Box 1). The latter can then be used to calibrate further assays, so preserving the stock of the primary IS [49–51].

Although a reasonably extensive range WHO ISs for human IFNs now exists (Tab. 12.2), it has not been practicable to prepare an IS for every molecular species of human IFN, e.g. for each of the 12 IFN-α subtypes. Where an IS for a particular IFN is not available, e.g. for human IFN-α6, calibration of working standards (for

> **Box 1**
>
> That the IS and the working standard contain identical or very similar IFN is necessary for compliance with the central tenet of biometric validity of assays, which states: *"When two or more IFN preparations are being compared, they must behave identically in the assay for the assay to be truly valid"*. In this case, where the IS serves as primary reference preparation, the working standard must behave as if it were a more concentrated or more dilute solution of the IS preparation. Stated differently, if like is compared with like in the same assay system, under the same conditions, then any difference in the measured response from the assay system should reflect only the difference in concentration. Dose–response lines for the IS and the working standard must be parallel for biometric validity [53]. Usually, the IS and working standard are compared in assays performed on at least five separate occasions and the geometric mean potency of the working standard, expressed in IU, assigned to it [54]. The working standard should then be used to routinely calibrate all further assays to estimate the potency in IU of IFN preparations containing the same IFN as the working standard. *[It has to be emphasized that the potency of IFN cannot be expressed in molar concentrations or mass units since molarity/mass is a measure of the physical quantity of material and, thus, does not reflect its biological activity. For example, the potency of an IFN preparation held under different conditions may vary significantly as a function of time without any change in molarity/mass. It is imperative therefore to express potency in IU of biological activity.]*

IFN-α6) should be made with the IS for the human IFN preparation to which IFN-α6 shows most similarity. In this case, an IS for a single human IFN-α subtype, e.g. IFN-α2b, should be chosen and not one for a mixture of IFN-α subtypes, e.g. leukocyte IFN [51, 54, 55]. The converse will be true when establishing a working standard for a leukocyte IFN preparation. The availability of ISs for animal IFNs is currently limited to a few species (Tab. 12.2).

12.3.3
Practical Considerations for IFN Preparations

There are many ways in which IFN preparations are presented for potency testing. Variation can occur not only in the type of IFN, but also in the constituents of the solution in which IFN is present. Clinical-grade or commercially supplied, highly purified IFN products are usually formulated in simple buffered salt solutions, which may contain excipient protein, e.g. human serum albumin (HSA), to stabilize IFN activity, or as a "bulking agent" for lyophilization [51]. In contrast, IFN preparations/samples obtained from stimulated cell cultures or clinical samples, including sera, plasmas, cerebral spinal fluid, bronchial lavage fluids, etc., are generally more complex. Often the IFN present will be in low concentration, may be comprised of a mixture of IFN types and/or subtypes, and contain other cytokines and substances which have modulating effects on the sensitivity and performance of assays. Serum samples for example may contain substances that act additively or synergistically with IFN, or interfere with assay responsiveness in general.

Lyophilized IFN preparations, including WHO ISs for IFN and clinical grade therapeutic IFN products, have demonstrated long-term thermal stability of their activity [49, 51]. However, once reconstituted, the resultant solutions of these IFN preparations and those other preparations maintained in liquid solutions are not guaranteed to maintain their full activity, especially at temperatures above 20 °C (room temperature). Generally, any potential loss of activity is prevented by storing IFN solutions as frozen liquid, preferably in small aliquots, at −70 °C. For example, for a working standard of IFN, a suitably large volume should be subdivided in appropriately sized aliquots and frozen at −70 °C, thus maintaining a plentiful stock for future continuous use. Since most IFN are hydrophobic proteins that can adsorb to unsiliconized borosilicate glass surfaces and therefore activity can be lost in this way, plastic containers or siliconized glass containers are recommended for storage. Alternatively, adsorption may be prevented by adding a strongly hydrophobic protein, such as bovine casein [56].

A major problem with clinical grade IFN products from an assay point of view is that they are very highly potent. Their potencies are usually 1 MIU or greater and therefore they require considerable dilution before titration in assays. The accuracy of the IFN potency determination is therefore subject to the accuracy with which the off-plate dilution can be made. This may involve several 10-fold dilution steps. Appropriate volumes of dilution medium, normally cell culture medium containing bovine serum, should be chosen and dilution at each stage carried out with micropipettes calibrated for precise delivery of the dilution volume [54, 55]. At the other end of the scale, samples of body fluids such as sera often contain very low concentrations of IFN. Here, dilution of sample may reduce the already low IFN concentration to below the detection limit of the assay. However, if no dilution is made, the constituents in the sera probably will interfere with the performance of the assay. Thus, potency estimations of IFN activity in such samples remain "challenging" to the IFN assayist.

12.3.4
Antiviral Assays

A wide variety of mammalian cell lines has been established in culture. The use of empirical approaches to determine sensitivity to IFNs and susceptibility to infection by viruses has enabled researchers to choose those combinations of cell line and virus that are suitable for the development of antiviral assays [52, 54, 55, 57]. Experience has shown that several adherent human and other mammalian cell lines yield, when coupled with a suitable "challenge" virus, highly sensitive antiviral assays (Tab. 12.3). However, it is worth knowing that many tumor-derived cell lines have defective IFN response systems, probably resulting from deficiencies in known effector molecules in the IFN-stimulated signal transduction pathway, e.g. lack of TYK-2 in a mutant adenocarcinoma cell line [58] or lack of STAT-1 and -2,

Tab. 12.3. Antiviral assay cell/virus combinations

Cell line	Virus	IFN
Madin-Darby bovine kidney (MDBK)	VSV	human, bovine, etc: type I IFN-α
Human lung carcinoma (A549)	VSV or EMCV	human: types I, II, and III
Human amniotic cell line (WISH)	VSV, EMCV, SFV or SINV	human: types I and II (not III)
Human hepatocyte carcinoma (HepG2)	EMCV or VSV	human: types I, II, and III
Human larynx carcinoma (Hep2)	EMCV, VSV or MV	human: types I, II, and III
Human foreskin diploid fibroblast (FS4, MRC5)	VSV, EMCV or MV	human: types I and II (not III)
Human amnion-derived (FL)	VSV, EMCV, MV or SINV	human: types I and II (III not tested)
Human glioblastoma cell line (2D9)	EMCV	human: types I and II (not III)
Human glioblastoma cell line (LN319)	EMCV	human: types I, II and III
Murine fibroblastic cell line (L929)	VSV or EMCV	murine: types I and II

and p48 (IRF-9) in certain melanoma cell lines [59]. In addition, cell lines such as HEK 293 [60] which express adenoviral E1A (normally required for early adenoviral transcription) are resistant to IFN since E1A reduces p48 (IRF-9) and sequesters p330/CREB-binding protein (CBP) required for transcriptional activation through STAT-2 [11]. In other cases, the underlying deficiencies responsible for defective IFN responses remain unknown [61]. Such cell lines are weakly protected from lysis by viruses such as VSV and EMCV, even in the presence of relatively high concentrations of IFN ([61, 62] and Meager, unpublished), and are thus unsuitable for antiviral assays. In a recent study [6], it was shown that, excluding those with the known defects in IFN responsiveness, about 20 human cell lines were responsive to type I IFNs, but only about a third of those to type III IFNs when the same challenge virus, EMCV, was used. Since type I and III IFNs appear to stimulate the same JAK–STAT signaling pathway [3–5] (see Chapter 6), lack of type III IFN receptors seems the most likely explanation for unresponsiveness, although this has yet to be confirmed. Alternatively, the establishment of the antiviral state by type I IFNs may require additional signaling pathways not found, or not stimulated, in those cell lines unresponsive to type III IFNs [6]. Overall, type III IFNs appeared to act more weakly in inducing antiviral activity to type I IFNs, even in cell lines that were most responsive to type III IFNs [6].

For all antiviral assays, there is a requirement to quantify the inhibitory activity of IFN on viral propagation or replicative processes. Several "read-out" parameters, including reduction of virus yields, reduction of viral cytopathic effect (CPE), reduction of viral plaque formation, reduction of viral protein or RNA synthesis, etc., are amenable to quantification [54, 55, 57, 63]. The virus yield reduction assay [54, 55, 57, 63, 64], which involves infectivity (plaque) assays of virus progeny obtained from cells treated with each serial dilution of IFN, can be informative and quantitative, but is tedious, cumbersome and expensive to perform. Reduction of expression of viral antigens in IFN-treated cells may also be measured by immunoassays [65, 66]. However, for practical reasons, most antiviral assays now involve the more straightforward procedure of measuring the capacity of IFN to protect cells against the CPE of a lytic virus over a range of IFN concentrations [54, 55, 57, 63, 64, 67]. This type of antiviral assay is commonly known as the CPE reduction (CPER) assay. The basic procedure for carrying out a CPER assay is outlined in Box 2. The staining methods used to process CPER assays are by no means perfect; they require washing steps that often remove cells and which might have an adverse impact on dose–responses. Direct addition of chemical reagents that are converted by cellular activities to colored products can overcome this potential problem. The classical example of this is the metabolic reduction of tetrazolium salts such as MTT (3-[4,5-dimethylthiazol-2-yl]-2,5-diphenyl tetrazolium bromide) in viable cells to a colored (purple) formazan product [68]. More recently, MTS (3-[4,5-dimethylthiazol-2-yl]-5-[3-carboxymethoxyphenyl]-2-[4-sulphophenyl]-2H-tetrazolium, inner salt) [69], a tetrazolium salt that is taken up by live cells and which, in the presence of electron acceptors such as phenazine methosulphate or menadione sodium bisulfite (soluble vitamin K_3), is metabolically reduced to a soluble red formazan product, has been preferred to MTT as the soluble formazan diffuses out of cells into the medium in which its absorbance can be determined at 492 nm after 1–2 h incubation at 37 °C without further interventions. In practice, the MTS method only works satisfactorily for cells that are efficiently killed by the challenge virus, e.g. 2D9 by EMCV, since if any live cells remain they will produce red formazan into the medium, i.e. a high background will be generated and the dose–response line will be correspondingly shallow. However, it has been shown that, with suitable virus and cell combinations, the MTS method of processing antiviral assays gives as good as dose–response data as staining methods and even better sensitivity ([71] and Meager, unpublished).

12.3.5
Antiproliferative Assays

Among the enormous number established cell lines that are grown in culture, many are now known to be susceptible to the antiproliferative activity of IFNs

> **Box 2**
>
> **Basic stages of the CPER assay**
>
> - Wells of 96-well microtiter plates are seeded with sufficient numbers of cells of an adherent, *IFN-sensitive*, cell line to form confluent monolayers.
> - Serial dilutions of IFN preparations are incubated with the cells to induce the antiviral state (usually requires overnight incubation).
> - Cells are challenged with a cytolytic virus until maximum CPE (usually about 24 h required) is manifested in "virus" control (no IFN added) cells; cells in the "cell" controls (no virus added) should appear totally alive and healthy, whereas cells in the virus controls should appear dead.
> - At this stage, viable, IFN-protected cells are stained with a vital dye[a] to determine CPE levels and, following spectrophotometric measurements of dye absorbance, readings are plotted graphically to generate dose–response curves from which potency of IFN preparations in terms of units of antiviral activity may be calculated [54, 55].
>
> [a] The majority of CPER assays are still terminated by staining and fixing cell monolayers. Several vital stains, i.e. those that stain only, or preferentially, live cells, are available. The violet dyes, such as crystal or gentian violet prepared as 1–2% solutions in 20% ethanol/PBS, are often used for this purpose, but tend to be very messy [54, 63]. For this reason, naphthol blue black (alternatively called amido black or amido blue black) dissolved in sodium acetate/acetic acid buffer for staining, followed by cell fixation with acidified formalin solution, is recommended [55]. Some cell lines, e.g. Hep2 and A549, tend to give high backgrounds in the "virus" controls because of the stickiness of dead cells. These high backgrounds may affect the slope of the dose–response lines and therefore adversely impact on the outcome of antiviral assays. To avoid high backgrounds, cell lines such as human glioblastoma-derived 2D9 [70], the dead cells of which are removed by washing/staining, should be used.

[72]. There are, however, some notable exceptions such as normal diploid fibroblast lines, which are virtually nonresponsive. Furthermore, the antiproliferative effects are variable according to the type of IFN. Type I IFNs have been the most studied of the IFNs regarding antiproliferative activity, which has been shown in a range of primary, e.g. hematopoietic (myeloid) precursor cells [73], and transformed/tumor-derived cell lines [72, 74–80]. Nevertheless, in most cases where antiproliferative activity has been demonstrated, in comparison with the induction of antiviral activity, relatively much larger concentrations of type I IFNs are required. IFN-β has usually been found to be more effective than IFN-α [6, 72, 74, 77, 78, 80]. However, there are a few exceptional cases, such as the Burkitt's lymphoma-derived B cell line Daudi [81], which is highly sensitive to type I IFN; IFN-α has greater inhibitory activity than IFN-β in this case [30, 62, 72, 75, 82–84]. In contrast, Daudi cell proliferation is insensitive to inhibition by type II [83] and III IFNs [3, 6]. A typical

> **Box 3**
>
> **Basic stages of an antiproliferative assay using the Daudi cell line**
>
> - Type I IFNs are appropriately diluted in 96-well microtiter plates and Daudi cells[a], obtained from exponentially proliferating cultures, added to all wells at 80 000 cells well^{-1} (8×10^5 cells ml^{-1}).
> - After incubation at 37 °C for 3–4 days, the cells are pulsed with [^{3}H] thymidine at 0.5–1.0 µCi well^{-1} for 4–6 h at 37 °C.
> - Cells are harvested onto filter mats and uptake of [^{3}H] thymidine determined by liquid scintillation counting [83, 85].
> - The data are represented as counts per minute (c.p.m.) for each well against the (log) reciprocal dilution of IFN standard or test sample.
> - Potencies of IFN are estimated from the antiproliferative dose–response lines of test samples relative to that of the IFN standard.
>
> [a] Although the Daudi cell line originated from a single source [81], there are now numerous sublines, which may vary in sensitivity to IFN, used in different testing laboratories. The sublines have probably arisen due to nonuniform culturing methods and conditions. It should be noted that probably most of these sublines are continuously secreting a range of growth factors and cytokines, which is a feature of cultured B cells [86].

antiproliferative assay protocol using Daudi cells is outlined in Box 3. While type II IFN can have antiproliferative activity equivalent to or stronger than type I IFN in certain cell lines ([41] and Meager, unpublished), only weak antiproliferative activity for type III IFNs has as yet been shown and in only one cell line [6].

While the Daudi cell line has proved very useful for antiproliferative assays, it is evident that its proliferation may also be influenced by cytokines other than type I IFN. Some of the older WHO ISs, e.g. 69/19, First International Reference Preparation for Human Leukocyte IFN (no longer distributed; see Tab. 12.2), were prepared from rather crudely purified material and thus contain a range of cytokines, e.g. IL-1β, -6, -8, etc. Following titration of 69/19 in a Daudi antiproliferative assays, it was observed that slope of the antiproliferative dose response line for 69/19 differed significantly from those of other IFN-α preparations, e.g. recombinant IFN-α2a, that did not contain "cytokine contaminants" [51].

As alternatives to Daudi for antiproliferative assays, other suspension cell lines that have proved useful include the human promyelocytic leukemia HL60 cell line, the M1 myeloid cell line [31], and human megakaryocytic cell lines such as Dami [87] and MEG-01 [88, 89]. These cell lines are not growth factor dependent, although their proliferation may be subject to autocrine regulation, e.g. by IL-6 [88]. In addition to these, there are, however, certain growth factor-dependent cell lines, e.g. the human erythroleukemic TF-1 cell line [90] and the human megakaryoblastic UT-7/EPO cell line [91], whose proliferation is inhibited by type I IFN. For instance, the granulocyte macrophage colony-stimulating factor (GM-CSF)-dependent proliferation of TF-1 is inhibited by type I IFN, thus providing the basis for an "antiproliferative-type" assay [92]. Alternatively, erythropoietin (EPO)-dependent proliferation of both TF-1 and UT-7/EPO is inhibited by type I IFN [91, 92]. Such assays have been termed "anticytokine bioactivity assays" by Mire-Sluis et al. [92], but are essentially similar in format, performance and processing to other antiproliferative assays. They would appear to be suitable for purified type I IFN preparations, but their regulation by a variety of cytokines suggests that they may be inappropriate for titration of "cytokine-containing" IFN preparations, e.g. samples derived from stimulated cell cultures.

With regard to adherent cell lines, several glioblastoma-derived cell lines have been shown to reasonably sensitive to the antiproliferative activity of human IFN-β and to a lower degree to IFN-α [6, 74, 79], and thus may be suitable for developing antiproliferative assays.

12.3.6
IFN-inducible Protein Assays

12.3.6.1 Bioimmunoassays

In addition to antiviral assays and antiproliferative assays, several alternative bioassays based on the principle of detecting the expression of individual IFN-inducible proteins by immunoassay have been developed. These assays are commonly referred to as bioimmunoassays – either cells in which the induced protein is expressed are chemically fixed and the protein quantified by binding of protein-specific antibody followed by anti-antibody–enzyme conjugate and substrate conversion to a colored metabolite or the induced protein is extracted from cells and quantified in a protein-specific immunoassay. The choice of IFN-inducible proteins for bioimmunoassays includes cell surface proteins such as class I and II MHC antigens [9, 39–41], and ICAM-1 and other cellular adhesion molecules [43, 44]. With the availability of specific monoclonal antibodies (MAbs) to these surface antigens, it has been possible to develop bioimmunoassays that measure increased binding of antigen-specific MAbs in response to IFN stimulation. For example, dose-dependent HLA-DR induction by type II IFN in the human colon adenocarcinoma cell line, COLO 205, has been demonstrated and has led to the development of a novel enzyme-linked bioimmunoassay that utilizes anti-HLA-DR MAb as the primary detection antibody [42]. Similarly, dose-dependent increased expression of β_2-microglobulin, the common component of class I MHC antigens, by type I or type II IFN may be quantified by bioimmunoassay in the human lung carcinoma cell line, A549 [66], or in the vascular endothelial cell $\times$A549 hybrid cell line, EA-hy926 [93] in which β_2-microglobulin expression was found to increase over a wide range of IFN concentrations, with highest sensitivity to type II IFN (Fig. 12.2) (Meager, unpublished). Additionally, IFN-mediated upregulation of expression of cellular adhesion molecules such as ICAM-1 and vascular cell adhesion molecule

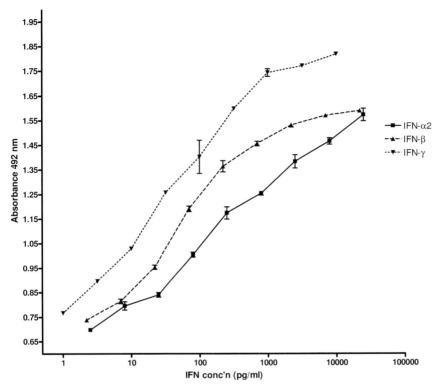

Fig. 12.2. The β_2-microglobulin inducing activity of IFN-α2, -β and -γ measured in a bioimmunoassay using the EAhy296 cell line [93]. Dilutions of stock IFN preparations to give the concentrations shown were added to duplicate wells and β_2-microglobulin levels at the cell surface determined by immunoassay 48 h later. Note the relatively high "background" levels of β_2-microglobulin expressed constitutively.

(VCAM)-1 [44] or the 17-kDa 9-27/leu13 expression [45] could be considered for bioimmunoassay development. Although bioimmunoassays have the advantage of not requiring viruses (as for antiviral assays), the basal level of expression of cell surface antigens can be relatively high and variable leading to problems with reproducibility. Another significant drawback is that, like most other biological assays for IFN, they often lack specificity, i.e. they may respond to all types of IFN and, in some instances, to other cytokines, such as IL-1 and tumor necrosis factor (TNF)-α [44].

As alternatives to cell surface antigens, use has been made of IFN-induced intracellular proteins for developing bioimmunoassays. Protein Mx (mouse) or MxA (human) is just one of several "antiviral" proteins induced by type I and III IFNs; it mediates antiviral activity, especially against ortho- and paramyxoviruses and bunyaviruses [6, 17, 22–24]. In susceptible cell lines, protein Mx/MxA is induced relatively quickly after IFN stimulation and persists for many hours thereafter. Al-

> **Box 4**
>
> **MxA assay procedure (after [97])**
>
> - A549 cells are seeded into wells of 96-well microtiter plates to form semi-confluent monolayers after overnight incubation at 37 °C.
> - Serial dilutions of IFN are prepared in culture medium in separate microtiter plates and then transferred to A549 cell assay plates for a further 24-h incubation at 37 °C.
> - Following which culture medium is removed, cells are lysed with a "lysis" buffer and plates with lysates frozen at −70 °C.
> - Thawed lysates are added to wells of ELISA plates coated with anti-MxA antibody, incubated for 2 hr with "detector" biotinylated anti-MxA antibody, washed, incubated for 1 h with streptavidin–horseradish peroxidase conjugate, washed and TMB (3,3′,5,5′-tetramethylbenzidine dihydrochloride) substrate solution added.
> - Following color development in the ELISA, the enzyme reaction is terminated with 2 M sulfuric acid and absorbances in wells read spectrophotometrically at 450 nm.

though it has GTPase activity [94], it is more readily measured by ELISA or chemiluminescent immunoassays using specific MAb reagents [95, 96]. A sensitive bioimmunoassay for type I IFN based on induced expression of MxA in human A549 cells has been developed by [97]. An outline of the procedures used for this bioimmunoassay is given in Box 4. In contrast to many other bioimmunoassays for IFN, this MxA-based bioimmunoassay is reported to have low background, a wide dynamic range and to be unaffected by other cytokines [97]. Although the A549 cell line from the American Type Cell Collection (ATCC) is highly responsive to type I and, to a lower degree, to type III IFNs, not all "A549" cell lines in laboratories around the world, especially those at high serial passage levels, produce sufficient MxA for the ELISA. In a recent survey [6] of human tumor-derived cell lines, a glioblastoma cell line, LN319, was also found to be suitable for the type I/III IFN-induced MxA bioimmunoassay, as illustrated in Fig. 12.3. Such bioimmunoassays have been readily adapted to the measurement of anti-type I IFN neutralizing antibodies ([98]; see also Chapter 13).

12.3.6.2 Enzyme Expression and Reporter Gene Assays

In a few instances, the enzymatic activity of IFN-inducible enzymes, e.g. 2–5A synthetase, PKR, RNase L and MxA GTPase, can be directly measured and form the basis of bioassays. For example, a bioassay for IFN-α has been developed based on induction and quantification of 2–5A synthetase; this bioassay is reportedly much less sensitive to IFN-β and type II IFN or IFN-γ [99]. In contrast, type II IFN specifically induces the enzyme indoleamine 2,3-dioxgenase (IDO), which converts L-tryptophan to N-formyl kynurenine (NFK). Thus, a type II IFN-specific bioassay

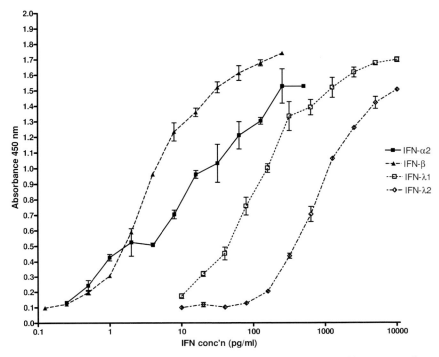

Fig. 12.3. The dose–response relationship of protein MxA to IFN concentration in the human LN319 glioblastoma cell line [6, 111]. Two-fold serial dilutions of IFN-α2b, -β, -λ1 and -λ2 were titrated in duplicate, and MxA concentrations determined by MxA-specific ELISA after 24 h incubation. Note the weaker effectiveness of the IFN-λs compared with IFN-α2b or -β in stimulating MxA synthesis.

has been developed based on measurement of kynurenine, the hydrolyzed product of NFK, in the IFN-γ-stimulated human glioblastoma cell line, 2D9 [70]. However, a significant drawback of such bioassays is that they require several manipulations for quantification of the enzyme product to be performed.

The emergence of recombinant DNA technologies has permitted the now readily achievable preparation of hybrid DNA sequences containing IFN-inducible promoters coupled to encoded well-characterized "immunoassay" enzymes, such as luciferase, alkaline phosphatase or horseradish peroxidase; their enzyme products are more easily measurable than the IFN-inducible enzymes cited above. The construction of reporter gene assays is dependent on the stable transfection of IFN-responsive cell lines with DNA plasmids containing IFN-inducible promoters coupled to encoded enzymes and selectable markers, e.g. for neomycin or puromycin resistance. Once a stably antibiotic resistant cell line has been created, it forms the cell substrate for IFN-mediated activation of plasmid-encoded IFN-inducible promoter-driven enzyme synthesis (Fig. 12.4). In theory, the enzyme concentration

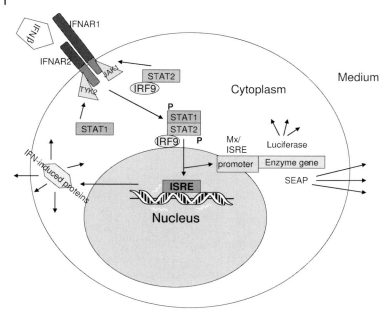

Fig. 12.4. Schematic diagram illustrating the type I IFN signal transduction pathway and basis for reporter gene assays. IFN-β binds to extracellular domains of its cognate class II cytokine receptor leading to activation of JAK-1 and TYK-2 tyrosine kinases associated with receptor endodomains. Subsequent phosphorylation of endodomains leads to recruitment of STAT-1 and -2, phosphorylation and dimerization of STATs, followed by their translocation into the nucleus. There, STATs combine with another DNA-binding protein, IRF-9 or p48, and this transcription factor complex binds to the ISRE present in the promoter region of IFN-β-inducible genes to activate their transcription. For reporter gene assays, the ISRE, or other IFN-inducible promoter, e.g. Mx, are present in DNA plasmids and are fused to enzyme genes, such as luciferase or SEAP. Binding of the transcription factor complex to these plasmid-encoded IFN-inducible promoters leads to synthesis of enzyme mRNA and enzyme protein, the latter being quantified by addition of appropriate substrates. (This figure also appears with the color plates.)

in the cells or, following secretion from the cells, in the culture medium is directly proportional to the IFN concentration used to stimulate cells. Therefore, quantitative assays based on this concept should be achievable, and could offer improved assay simplicity, selectivity and reliability [100–102].

In the first described reporter gene assay, the mouse Mx promoter was linked to the human growth hormone (hGH) gene and this hybrid DNA used to transfect the African Green monkey cell line, Vero [100]. The hGH that accumulated in type I IFN-stimulated, transfected Vero cells was measured by a hGH-specific radioimmunoassay and therefore this particular reporter gene assay has something in common with bioimmunoassays (Section 12.3.6.1). However, substitution of the hGH gene with the firefly luciferase gene provided a more direct readout system in Vero cells [102, 103]. The luciferase activity accumulated in type I IFN-stimulated transfected Vero was determined following cell lysis in the presence of protease inhibitor aprotinin, addition of a luciferase assay reagent (luciferin) and measurement of luminescence in a luminometer [102]. A linear dose–response relationship was found between 1 and 16 IU ml^{-1} for IFN-β. The transfected cells did not respond to type II IFN (IFN-γ) or a number of other cytokines. A transfected A549 cell line expressing luciferase driven by type I IFN-inducible protein MxA (generously provided by Dr. G. Adolf, Bender and Co., Vienna, Austria) has also provided a sensitive bioassay for type I and III IFNs, which can be completed in only 6 h from the point of addition of IFN ([6] and Meager, unpublished). However, special white microtiter plates with shielded wells are needed for low backgrounds and prevention of well-to-well crosstalk and, due to the use of relatively expensive reagents, can be less economical to perform than many other bioassays for IFN. Similar reporter gene assays have also been developed for mouse IFN [101], e.g. using the mouse L-M cell line transfected with Mx promoter–luciferase gene hybrid DNA (generously provided by Dr G. Adolf) (Meager, unpublished). A protocol for this type of reporter gene assays is illustrated in Box 5.

Cell lines stably transfected with IFN-inducible promoter sequences linked to the secreted alkaline phosphatase (SEAP) gene [104] provide useful alternatives to luciferase (Fig. 12.4); however, but, since in these assays SEAP slowly accumulates in the cell culture medium (without phenol red to avoid interference in color development), they take longer (24–48 h) to complete. Advantageously, however, p-nitrophenylphosphate (NPP), the SEAP substrate, may be directly added to wells, leading to the production of yellow product within 30–60 min. Color development may be terminated with 3 M sodium hydroxide and absorbances measured at 405 nm in an ELISA reader. Our experience of such SEAP-based reporter gene assays suggests they are much less sensitive than those based on luciferase (Meager, unpublished), although they have proved highly reproducible for both type I and III IFNs [6]. This lower sensitivity may be turned into an advantage for highly concentrated IFN preparations in that far fewer "off-plate dilutions" will be required to generate dose–response lines than for ultra-sensitive reporter gene assays or antiviral assays (see Section 12.3).

In another example of a reporter gene assay, cells of a human glioblastoma cell line were stably transfected with a glial fibrillary protein (GFAP) promoter linked to a lacZ (Escherichia coli β-galactosidase) reporter gene [105]. In contrast to increased expression of luciferase or SEAP, GFAP is normally constitutively produced and the transfected cells respond to IFN acting in an inhibitory manner, leading to dose-dependent reduction of β-galactosidase activity. Levels of β-galactosidase were determined by conversion of an added fluorophore to a product readily estimated by fluorimetry. This reporter gene assay was most sensitive to IFN-α, then to IFN-β and least sensitive to type II IFN (IFN-γ); of other cytokines tested, only TNF-α induced an appreciable change in β-galactosidase expression [105].

Box 5

Protocol for reporter gene assay with MxA–luciferase transfected A549 cells

- For this assay, A549/93D7 cells are seeded at 5×10^5 ml^{-1} in 100 µl well^{-1} *colorless* Dulbecco's modified minimal essential medium (DMEM) supplemented with 10% heat-inactivated fetal calf serum (FCS) in *white* 96-well culture plates (special white microtiter plates with shielded wells are required for low backgrounds and prevention of well-to-well crosstalk) and incubated overnight at 37 °C.
- The culture medium is removed and 100 µl well^{-1} of DMEM culture medium containing 2% heat-inactivated FCS added to the wells.
- Serial titrations of human type I IFNs are performed in the wells and plates returned to the incubator.
- After 6 h, 100 µl of LucLite reagent[a] (PerkinElmer Sciences) is added per well according to the manufacturer's instructions.
- The plates are sealed (Topseal adhesive) and dark adapted before measuring luminescence using a luminometer.

[a] Luciferin is converted to adenyl-luciferin by luciferase in the presence of ATP. Adenyl-luciferin reacts with molecular oxygen to emit light and produce adenyl-oxyluciferin. The emission of light is usually extremely short and this originally imposed a significant limitation since only one assay plate could be handled at a time. However, several substances have been identified that modify the enzyme reaction to produce a long-lasting light emission. These substances have been combined with luciferin and developed as commercially available reagents, e.g. LucLite (Perkir Elmer).

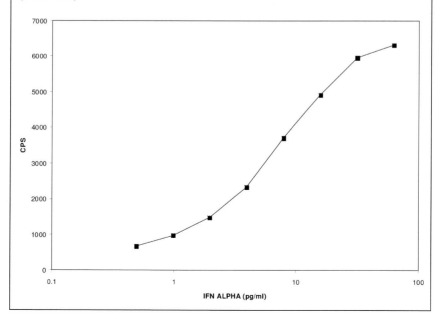

12.3.7
Assays Based on Intracellular Signaling Intermediates

All type I IFNs, IFN-α, -β, -ε, -κ and -ω, bind to a common receptor of the class II cytokine receptor category comprised of two chains, IFN-αR1 and -αR2, whose endodomains are associated with JAK-1 and TYK-2 protein tyrosine kinases, respectively (Fig. 12.1) [10, 11]. Type I IFN binding induces aggregation of receptor chains, bringing JAK-1 and TYK-2 together to cause their activation. Activated JAK-1 and TYK-2 then phosphorylate receptor chain endodomains on specific tyrosine residues, which leads to the recruitment and phosphorylation of transcription factors designated STATs, in this case STAT-1 and -2, causing their dimerization. In combination with an accessory p48 DNA-binding protein (IRF-9), STAT-1:2 heterodimers translocate to the nucleus and bind to a common nucleotide sequence designated the "ISRE" of type I IFN-inducible genes [10, 11, 106]. This signaling "cascade" may be exploited for the design and development of quantitative assays of IFN activity. Various individual stages in such cascades are amenable to quantitative measures of activation, e.g. receptor endodomain phosphorylation by determination of phosphotyrosine residues, phosphorylation of intracellular proteins and STAT cytoplasm–nuclear translocation.

The first type of assay that may be applied to this signaling cascade is the so-called kinase receptor activation (KIRA) assay [107, 108], in which specific phosphotyrosines present in the activated receptor endodomains are quantified. As depicted in Fig. 12.5, it utilizes two separate microtiter plates – one for growth factor/cytokine stimulation of receptor-bearing intact cells, and the other for receptor capture and phosphotyrosine ELISA using antibodies that recognize phosphotyrosine (Fig. 12.5). In principle, the KIRA assay format has potential as a rapid and specific bioassay for type I IFN potency determinations. However, type I IFN receptor numbers per cell are frequently low and there might therefore be insufficient P-Y residues for a strong signal-to-noise ratio, necessary for manifestation of quantifiable dose–responses, to be achieved. Possibly, the paucity of receptors could be overcome by transfecting cells with plasmids that constitutively express, or overexpress, specific receptor chains, although this would be more difficult to accomplish since IFN receptors are comprised of two or more chains [1, 2, 7, 12].

A second type of assay that would measure the phosphorylation of intracellular proteins by specific antibody reagents could be set up in a similar manner to KIRA assays. Such assays have been designated phosphospecific antibody cell-based ELISA (PACE). PACE has the advantage that, following direct permeabilization of cells, the ELISA steps may be performed in the same microtiter plate. At this point in time, however, it is not clear how applicable PACE for IFN determinations would be.

Following IFN stimulation STATs are phosphorylated, dimerize and then translocate to the nucleus (Fig. 12.1). The process of translocation is detectable and quantifiable, and thus can provide the basis of an assay. For example, plated cells are incubated overnight, stimulated with IFN, and fixed and permeabilized

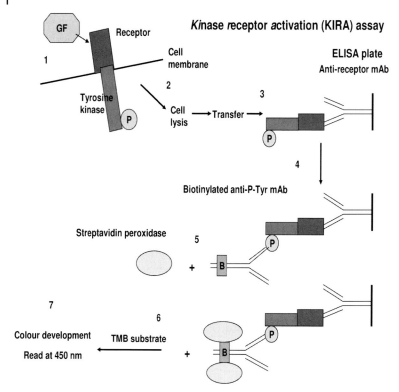

Fig. 12.5. Schematic flow diagram of the KIRA assay. Receptor activation is triggered by growth factor (GF) binding resulting in autophosphorylation of tyrosine residue(s) in the receptor endodomain (step 1). Following a brief incubation, cells are lysed (step 2) and lysates transferred to an ELISA plate coated with anti-receptor MAb (step 3). Phosphorylated receptor is quantified by addition of biotinylated anti-P-Tyr MAb (step 4), followed by addition of streptavidin peroxidase conjugate (step 5), addition of peroxidase substrate (TMB; step 6) with appropriate washings between steps. Following color development and termination of enzyme activity with 2 M sulfuric acid, absorbances are read at 450 nm (step 7). (This figure also appears with the color plates.)

between 0.5–1.0 h later. STAT complexes are stained by a primary anti-STAT antibody followed by a secondary fluorescent antibody and Hoechst 33342, which defines the nuclear–cytoplasmic boundary. Using a cytoplasmic–nuclear translocation algorithm and Array Scan II Technology (Cellomics) to compare IFN-stimulated with untreated cells, STAT translocation into the nucleus can be quantified (Dr Stephen Indelicato, Schering-Plough, personal communication). Although the hardware involved is sophisticated and expensive, assays based on STAT translocation have shown great promise for high throughput and precision.

12.3.8
Assay Design and Data Analysis

The majority of assays reviewed in the preceding sections are based on cell cultures evenly distributed in 96-well microtiter plates. However, the introduction of untoward biases resulting from, for example, uneven cell distribution, "nonuniformity" of wells of 96-well microtiter plates and defects in the precision of delivery of multichannel micropipettes can contribute to variable responses across plates. In some cases, the outer wells of 96-well microtiter plates give responses significantly different from inner wells, either due to uneven spraying of wells in manufacture or due to exposure of cells in these outer wells to more rapid fluctuations in temperature and pH. Experience in my laboratory [109] has shown that, even if wells in columns 1 and 12 are excluded, titrations performed in the wells of columns 2 and 11 are still more likely to produce results that are "outliers" than columns 3–10. For example, if wells in columns 2 and 3 are used as duplicates, the responsiveness of cells to IFN can often be higher in column 3 compared with column 2. The untoward variations in response at the plate edges has led some investigators to use only the inner 60 wells for titrations, even though this means usually fewer samples can be titrated on each plate and that the number of serial dilutions of each sample may be more restricted.

The likely manifestation of "plate effects" can have an impact on decisions regarding positions of samples and standards. Ideally, when several preparations are tested they should be titrated on several plates such that the positions of individual samples relative to the standard are varied. The position of the standard can also influence results and therefore this should be varied too. For instance, an international study to assign potencies to TNF-α preparations found that designs in which each preparation is tested on several plates in any assay with independent serial dilutions of the preparation on each plate generally gave more accurate and reproducible estimates of potency [109]. Complete randomization of samples in terms of plate positions is usually not feasible, but efforts should be made to carefully design assays to permit more valid use of classical methods of analysis. One way of checking intra-assay variability and thus the validity of results in each assay is, using independent dilutions, to titrate one preparation twice, e.g. as coded duplicates, say B and G where G is identical to B except for code. The ratio of the potency of B to that of G should approach 1.0. Large deviations from 1.0 indicate that biases have been introduced or occurred in the assay and may compromise its overall validity.

As far as is possible, samples of different IFN types should not be run in parallel in the same assay and, especially, not together on the same microtiter tray. Assay results will only be valid when "like" is compared with "like" and the dose–response curves are parallel [50, 53]. In general, the measured response parameter for each dilution of a preparation titrated in an assay is plotted graphically (normally on the y-axis) to generate dose–response curves. They are frequently sigmoidal in shape with only the "middle" portion forming the linear part of the curve.

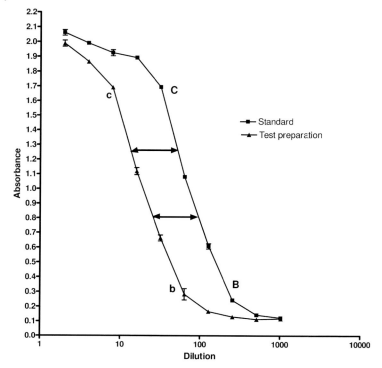

Fig. 12.6. Graphical plot of sigmoidal log dose–response curve in a comparison of two IFN preparations that behave similarly in an antiviral assay. The potency of one preparation to the other is represented by the amount of one preparation which gives the same measured response as a measured quantity of the other, i.e. the horizontal distance between the linear part of the curves B–C (standard) and b–c (test preparation).

The slopes of these linear portions for two or more preparations of the same IFN (identical in molecular structure) should be parallel such that the displacement, i.e. horizontal distance, between two linear portions represents the relative potency of one preparation to another (Fig. 12.6). If, in the example given in Fig. 12.6, B–C is the linear portion of the dose–response curve of the IFN standard and c–b that of a test preparation, then the horizontal distance (displacement) between the two lines is a measure of their relative potencies. This distance may be computed at several points on the lines and interpolated as a value in International Units per milliliter (IU ml^{-1}) for the test preparation from the assigned potency of the standard. For many assays, raw dose–response data are transformed to generate log dose (transformed)–response lines to yield as extended a linear range as feasible. Estimates of relative potency are then obtained as the displacement of parallel log dose–response lines. Potency estimates of the same preparation titrated on several different microtiter plates can then be combined as geometric means and comparisons among them made using analysis of variance of the logs of the estimates [110].

For most graphical plots of dose–response data, it is also possible to define an endpoint. This is frequently taken to be the 50% point of measurements of the response parameter that lies in the range or scale of measurements spanning empirically determined positive and negative controls. For example, in the CPER assay (Section 12.3.4) the positive "cell control" (no IFN, no virus) will generate the highest absorbance reading and the negative "virus control" (no IFN, plus virus) the lowest reading. The dilution of IFN that generates the 50% endpoint between these two readings is directly related to the potency of the preparation, which may be expressed in Laboratory Units per milliliter (LU ml^{-1}) [54]. Using the 50% endpoint on the dose–response curve of the standard allows its potency to be given in LU ml^{-1} and also gives a measure of how sensitive the assay is – if the potency in LU ml^{-1} is less than its assigned potency in IU ml^{-1}, then the assay is of lower sensitivity than if the converse, i.e. more than its assigned potency, were found. Once the potency of the standard has been estimated in LU ml^{-1}, it is possible to calculate the potency of other IFN preparations by multiplying the ratio of the potency of the test preparation in LU ml^{-1} to that of the standard in LU ml^{-1} by the potency of the standard in IU ml^{-1} [54]. For example, if in a particular assay the potency of a standard of assigned potency of 5000 IU ml^{-1} is 10 000 LU ml^{-1} and that of the test preparation 20 000 LU ml^{-1}, then the potency of the test preparation is $20\,000/10\,000 \times 5000 = 10\,000$ IU ml^{-1}. While this type of endpoint analysis is readily applicable to most graphically plotted assay data and provides a simple means for potency calculation, there can be some shortcomings that make it less reliable than the parallel line displacement method of analysis. For instance, the maximum (cell control) and minimum (virus control) readings may vary significantly from plate to plate; the readings on individual dose–response curves may variably approach, but not exceed, the maximum and minimum readings, or may variably exceed these at one end of the scale or both ends. Translated this means that the 50% endpoint determined from individual dose–response curves may not correspond to the 50% endpoint between the maximum and minimum readings, leading to untoward errors and biases in potency estimates. Furthermore, parallelism of dose–response curves is not determined, leaving open the question of biometric validity of results.

12.4
Regulatory Landscape

Both historically and by definition, IFNs have been characterized by their antiviral activity. Biological assays that measure antiviral activity were thus chronologically the first to be developed and have been widely used and accepted by regulatory agencies for potency testing of therapeutic IFN products. The assigned potencies in IU of all WHO ISs of IFNs were derived from data derived from antiviral assays.

Although the assays described based on the induction of other characteristic IFN activities, e.g. antiproliferative activity and inducible protein expression, have been shown to be valid for potency determinations, generate results that can be highly correlated with those generated by more "classical" antiviral assays, and may have certain advantages with respect to precision, reliability and speed of performance, there may be some resistance from regulatory authorities in accepting them for clinical product characterization and assignment of potency values. This usually stems from unfamiliarity with these novel assays, rather than from prejudices held against them, and sometimes results in Regulatory Authorities asking for data from both "classical" antiviral assays bioassays and any new assay used. In some cases this may be initially justified, e.g. the results from a reporter gene assay do not indicate the intended biological activity of an IFN for therapeutic purposes. Nevertheless, if a strong correlation between potency in the reporter gene assays and potency in a "classical" antiviral assays bioassay is demonstrated when the product is in preclinical evaluation studies, then acceptance of the reporter gene assays data should follow for licensing and marketing authorization of product for clinical investigations and treatments. Additional data from classical bioassays should no longer be necessary.

Acknowledgments

I thank Paula Dilger, Stella Williams and Rose Gaines Das for useful contributions and discussions in the preparation of this chapter.

References

1 Pestka S, Krause DK, Walter MR. 2004. Interferons, interferon-like cytokines, and their receptors. *Immunol Rev* 202, 8–32.
2 Meager A. 2005. Viral inhibitors and immune response mediators: the interferons. In: Meyers R (Ed.), *Encyclopedia of Molecular Cell Biology and Molecular medicine: Fundamentals and Applications.* Wiley-VCH, Weinheim, 387–421.
3 Kotenko SV, Gallagher G, Baurin VV, et al. 2003. IFN-λs mediate antiviral protection through a distinct class II cytokine receptor complex. *Nat Immunol* 4, 69–77.
4 Sheppard P, Kindsvogel W, Xu W, et al. 2003. IL-28, IL-29 and their class II cytokine receptor IL-28R. *Nat Immunol* 4, 63–68.
5 Dumoutier L, Tounsi A, Michaels T, et al. 2004. Role of the interleukin-28 receptor tyrosine residues for the antiproliferative activity of IL-29/IFN-λ1: similarities with type I interferon signaling. *J Biol Chem* 279, 32269–32274.
6 Meager A, Visvalingam K, Dilger P, et al. 2005. Biological

activity of interleukins-28 and -29: comparison with type I interferons. *Cytokine 31*, 109–118.

7. LANGER JA, CUTRONE EC, KOTENKO S. **2004**. The class II cytokine receptor (CRF2) family: overview and patterns of receptor–ligand interactions. *Cytokine Growth Factor Rev 15*, 33–48.

8. DEMAEYER E, DEMAEYER-GUIGNARD J. **1988**. *Interferons and Other Regulatory Cytokines*. Wiley Interscience, New York.

9. DEMAEYER E, DEMAEYER-GUIGNARD J. **1998**. Interferon gamma. In: MIRE-SLUIS AR, THORPE R (Eds.), *Cytokines*. Academic Press, San Diego, CA, p. 391.

10. IHLE J, KERR I. **1995**. JAKs and STATs in signaling by the cytokine receptor superfamily. *Trends Genet 11*, 69–74.

11. STARK GR, KERR IM, WILLIAMS BR, et al. **1998**. How cells respond to interferons. *Annu Rev Biochem 67*, 227–264.

12. MOGENSEN KE, LEWERNZ M, REBOUL J, et al. **1999**. The type I interferon receptor: structure, function, and evaluation of a family business. *J Interferon Cytokine Res 19*, 1069–1098.

13. DA SILVA AJ, BRICKELMAIER M, MAJEAU GR, et al. **2002**. Comparison of the gene expression patterns induced by treatment of human umbilical vein endothelial cells with IFN-α2b vs. IFN-β1a: understanding the functional relationship between distinct type I interferons that act through a common receptor. *J Interferon Cytokine Res 22*, 173–188.

14. SEN GC, LENGYEL P. **1992**. The interferon system: a bird's eye view of its biochemistry. *J Biol Chem 267*, 5017–5020.

15. LENGYEL P. **1982**. Biochemistry of interferons and their actions. *Annu Rev Biochem 51*, 251–282.

16. SAMUEL CE. **1987**. Interferon induction of the antiviral state: proteins induced by interferons and their possible roles in the antiviral mechanisms of action. In: PFEFFER LM (Ed.), *Interferon Actions*. CRC Press, Boca Raton, FL, p. 110.

17. STAEHELI P. **1990**. Interferon-induced genes and the antiviral state. *Adv Virus Res 38*, 147–200.

18. ZHOU A, HASSEL BA, SILVERMAN RH. **1993**. Expression cloning of 2–5A-dependent RNAase: a uniquely regulated mediator of interferon action. *Cell 72*, 753–765.

19. RICE AP, DUNCAN R, HERSHEY JWB, et al. **1985**. Double-stranded RNA-dependent protein kinase and 2–5A system are both activated in interferon-treated, encephalomyocarditis virus-infected HeLa cells. *J Virol 54*, 894–898.

20. KUMAR R, CHOUBEY D, LENGYEL P, SEN GC. **1988**. Studies on the role of the $2'$–$5'$-oligonucleotide synthetase–RNase L pathway in β-interferon-meditated inhibition of encephalomyocarditis virus replication. *J Virol 62*, 3175–3181.

21. GUPTA SL, HOLMES SL, MEHRA LL. **1982**. Interferon action against reovirus: activation of interferon-induced protein kinase in mouse L929 cells upon reovirus infection. *Virology 120*, 495–499.

22. RONNI T, MELEN K, MALYGIN A, JULKUNEN J. **1993**. Control of IFN-inducible MxA gene expression in human cells. *J Immunol 141*, 1715–1726.

23. HALLER O, KOCHS G. **2002**. Interferon-induced mx proteins: dynamin-like GTPases with antiviral activity. *Traffic 3*, 710–717.

24 LEROY M, BAISE E, PIRE G, et al. **2005**. Resistance of paramyxoviridae to type I interferon-induced Bos tarus Mx1 dynamin. *J Interferon Cytokine Res 25*, 192–201.

25 MEURS EF, GALABRU J, BARBER GN, et al. **1993**. Tumor suppressor function of the interferon-induced double-stranded RNA-activated protein kinase. *Proc Natl Acad Sci USA 90*, 232–236.

26 BARBER GN, JAGUS, MEURS EF, et al. **1995**. Molecular mechanisms responsible for malignant transformation by regulatory and catalytic domain variants of the interferon-induced enzyme RNA-dependent protein kinase. *J Biol Chem 270*, 17423–17428.

27 TAN S-L, KATZE MG. **1999**. The emerging role of the interferon-induced PKR protein kinase as an apoptotic effector: a new face of death? *J Interferon Cytokine Res 19*, 543–554.

28 LENGYEL P. **1993**. Tumor-suppressor genes: news about the interferon connection. *Proc Natl Acad Sci USA 90*, 6593–6594.

29 RESNITZKY D, TIEFENBRUN N, BERISSI H, KIMCHI A. **1992**. Interferons and interleukin-6 suppress phosphorylation of the retinoblastoma protein in growth sensitive hemopoietic cells. *Proc Natl Acad Sci USA 89*, 402–406.

30 ZHANG K, KUMAR R. **1994**. Interferon-α inhibits cyclin E and cyclin-D1-dependent CDK-2 kinase activity associated with RB and E2F in Daudi cells. *Biochem Biophys Res Commun 200*, 522–528.

31 QIN X-Q, RUNKEL L, DECK C, et al. **1997**. Interferon-β induces S phase accumulation selectively in human transformed cells. *J Interferon Cytokine Res 17*, 355–367.

32 WANG E, PFEFFER LM, TAMM I. **1981**. Interferon-α induces a protein kinase C-ε (PKC-ε) gene expression and a 4.7 kb PKC-ε related transcript. *Proc Natl Acad Sci USA 78*, 6281–6285.

33 SEKAR V, ATMAR VJ, JOSHI AR, et al. **1983**. Inhibition of ornithine decarboxylase in human fibroblast cells by type I and II interferons. *Biochem Biophys Res Commun 114*, 950–954.

34 DE LA MAZA L, PETERSON E. **1988**. Dependence of the *in vitro* antiproliferative activity of recombinant gamma IFN on the concentration of tryptophan in the medium. *Cancer Res 48*, 346–350.

35 CONTENTE S, KENYON K, RIMOLDI D, FRIEDMAN R. **1990**. Expression of gene *rrg* is associated with reversion of NIH3T3 transformed by LTR-c-H-*ras Science 249*, 796–798.

36 KRÖGER A, KÖSTER M, SCHROEDER K, et al. **2002**. Activities of IRF-1. *J Interferon Cytokine Res 22*, 5–14.

37 BOUKER KB, SKAAR TC, RIGGINS RB, et al. **2005**. Interferon regulatory factor-1 (IRF-1) exhibits tumor suppressor activities in breast cancer associated with caspase activation and induction of apoptosis. *Carcinogenesis 26*, 1527–1535.

38 HERON I, HOKLAND M, BERG K. **1978**. Enhanced expression of β2-microglobulin and HLA antigens on lymphoid cells by type I interferons. *Proc Natl Acad Sci USA 75*, 6215–6219.

39 FELLOUS M, KAMOUN M, GRESSER I, BONO R. **1979**. Enhanced expression of HLA antigens and β2-microglobulin on interferon-treated human lymphoid cells. *Eur J Immunol 9*, 446–449.

40 HOKLAND M, BERG K. 1981. Interferon enhances the antibody-dependent cellular cytotoxicity (ADCC) of human polymorphonuclear leukocytes. *J Immunol* 127, 1585–1588.

41 BILLIAU A. 1996. Interferon-γ: biology and role in pathogenesis. *Adv Immunol* 62, 61–130.

42 GIBSON UEM, KRAMER SM. 1989. Enzyme-linked bioimmunoassay for IFN-γ by HLA-DR induction. *J Immunol Methods* 125, 105–113.

43 BOUILLON M, AUDETTE M. 1993. Transduction of retinoic acid and γ-interferon signal for intercellular adhesion molecule-1 expression on human tumor cell lines: evidence for the late-acting involvement of protein kinase C inactivation. *Cancer Res* 53, 826–832.

44 MEAGER A. 1996. Bioimmunoassays for proinflammatory cytokines involving cytokine-induced cellular adhesion molecule expression in human glioblastoma cell lines. *J Immunol Methods* 190, 235–244.

45 DEBLANDRE GA, MARINX OP, EVANS SS, et al. 1995. Expression cloning of an interferon-inducible 17-kDA membrane protein implicated in the control of cell growth. *J Biol Chem* 270, 23860–23866.

46 LEVY S, TODD SC, MAECKER HT. 1998. CD81 (TAPA-1): a molecule involved in signal transduction and cell adhesion in the immune system. *Annu Rev Immunol* 16, 89–109.

47 BIRON CA, NGUYEN KB, PIEN GC, et al. 1999. Natural killer cells in antiviral defense: function and regulation by innate cytokines. *Annu Rev Immunol* 17, 189–220.

48 BIRON CA. 2001. Interferons alpha and beta as immune regulators – a new look. *Immunity* 14, 661–664.

49 PESTKA S, MEAGER A. 1997. Interferon standardization and designations. *J Interferon Cytokine Res 17 (Suppl 1)*, S9–S14.

50 MEAGER A. 1998. Biological standardization of interferons and other cytokines. In: SUBRAMANIAN G (Ed.), *Biosparation and Bioprocessing*. Wiley-VCH, Weinheim, vol. II, p. 255.

51 MEAGER A, GAINES DAS R, ZOON K, MIRE-SLUIS A. 2001. Establishment of new and replacement biological standards for human interferon alpha and omega. *J Immunol Methods* 257, 17–33.

52 MEAGER A. 2002. Biological assays for interferons. *J Immunol Methods* 261, 21–26.

53 BANGHAM DR. 1983. Assays and standards. In: GRAY CC, JAMES VHT (Eds.), *Hormones and Blood*, 3rd edn. Academic Press, London, vol. 5, p. 256.

54 MEAGER A. 1987. Quantification of interferons by anti-viral assays and their standardization. In: CLEMENS MJ, MORRIS AG, GEARING AJH (Eds.), *Lymphokines and Interferons: A Practical Approach*, IRL Press, Oxford, p. 129.

55 MEAGER A. 2003. Assays for antiviral activity. In: DE LEY M (Ed.), *Cytokine Protocols (Methods in Molecular Biology 249)*. Humana Press, Totowa, NJ, p. 121.

56 MEAGER A. 2005. Biological standardization of human interferon beta: establishment of a replacement World Health Organization international standard for human glycosylated

interferon beta. *J Immunol Methods* in press/published online Sept. '05

57 GROSSBERG SE, SEDMAK JJ. **1984**. Assays of interferons. In: BILLIAU A (Ed.), *Interferon: 1. General and Applied Aspects.* Elsevier, Amsterdam, p. 189.

58 PELLEGRINI S, JOHN J, SHEARER M, et al. **1989**. Use of a selectable marker regulated by interferon to obtain mutations in the signaling pathway. *Mol Cell Biol* 9, 4605–4612.

59 WONG LH, KRAMER KG, HATZINISIRIOU I, et al. **1997**. Interferon-resistant human melanoma cells are deficient in ISGF3 components, STAT1, STAT2, and p48-ISGF3 gamma. *J Biol Chem* 272, 28779–28784.

60 GRAHAM F, SMILEY J, RUSSELL WC, NAIRN R. **1977**. Characteristic of a human cell line transformed by DNA from human adenovirus type 5. *J Gen Virol* 36, 59–74.

61 STOJDL DF, LICHTY B, KNOWLES S, et al. **2000**. Exploiting tumor-specific defects in the interferon pathway with a previously known oncolytic virus. *Nature Med* 6, 821–825.

62 FOSTER GR, RODRIGUES O, GHOUZE F, et al. **1996**. Different relative activities of human cell-derived interferon-α subtypes: IFN-α8 has very high antiviral potency. *J Interferon Cytokine Res* 16, 1027–1033.

63 LEWIS JA. **1987**. Biological assay of interferons. In: CLEMENS MJ, MORRIS AG, GEARING AJH (Eds.), *Lymphokines and Interferons: A Practical Approach.* IRL Press, Oxford, p. 73.

64 STEWART II WE. **1979**. *The Interferon System.* Springer, Vienna.

65 JULKUNEN I, LINNAVUORI K, HOVI T. **1982**. Sensitive interferon assay based on immunoenzymatic quantification of viral antigen synthesis. *J Virol Methods* 5, 85–91.

66 HERMODSSON S, STRANNEGÅRD Ö, JEANSSON S. **1984**. Bioimmunoassays (BIAs) of human interferon. *Proc Soc Exp Med* 175, 44–51.

67 ARMSTRONG JA. **1981**. Cytopathic effect inhibition assay for interferon: microculture plate assay. *Methods Enzymol* 78(A), 381–387.

68 MOSMANN T. **1983**. Rapid colorimetric assay for the cellular growth and survival: application to proliferation and cytotoxicity assays. *J Immunol Methods* 65, 55–63.

69 BUTTKE TM, MCCUBREY JA, OWEN TC. **1993**. Use of an aqueous soluble tetrazolium/formazan assay to measure viability and proliferation of lymphokinedependent cell lines. *J Immunol Methods* 157, 233–240.

70 DÄUBENER W, WAGANAT N, PILZ K, et al. **1994**. A new, simple, bioassay for human IFN-γ. *J Immunol Methods* 168, 39–47.

71 KHABAR KSA, AL-ZOGHAIB F, DZIMIRI M, et al. **1996**. MTS interferon assay: a simplified cellular dehydrogenase assay for interferon activity using a water soluble tetrazolium salt. *J Interferon Cytokine Res* 16, 31–33.

72 BORDEN EC, HOGAN TF, VOELKEL JG. **1982**. Comparative antiproliferative activity *in vitro* of natural interferons α and β for diploid and transformed human cells. *Cancer Res* 42, 4948–4953.

73 RIGBY WF, BALL ED, GUYRE PM, FANGER MW. **1985**. The

effect of recombinant-DNA-derived interferons on the growth of myeloid progenitor cells. *Blood 65*, 858–861.

74 DICK RS, HUBBELL HR. **1987**. Sensitivities of human glioma cell lines to interferons and double-stranded RNAs individually and in synergistic combinations. *J Neurooncol 5*, 331–338.

75 HILFENHAUS J, DAMM H, KARGES HE, MANTHEY KF. **1976**. Growth inhibition of human lymphoblastoid Daudi cells *in vitro* by interferon preparations. *Arch Virol 51*, 87–97.

76 SHEARER M, TAYLOR-PAPADIMITRIOU J. **1987**. Regulation of cell growth by interferon. *Cancer Metast Rev 6*, 199–221.

77 JOHNS TG, MACKAY IR, CALLISTER KA, et al. **1992**. Antiproliferative potencies of interferons on melanoma cell lines and xenografts: higher efficacy of interferon beta. *J Natl Cancer Inst 84*, 1185–1190.

78 CORADINI D, BIFFI A, PIRRONELLO E, DI FRONZO G. **1994**. The effect of alpha-, beta- and gamma-interferon on the growth of breast cancer cell lines. *Anticancer Res 14*, 1779–1784.

79 GARRISON JI, BERENS ME, SHAPIRO JR, et al. **1996**. Interferon-beta inhibits proliferation and progression through S phase of the cell cycle in five glioma cell lines. *J Neurooncol 30*, 213–223.

80 DAMDINSUREN B, NAGANO H, SAKON M, et al. **2003**. Interferon-beta is more potent than interferon-alpha in inhibition of human hepatocellular carcinoma cell growth when used alone and in combination with anticancer drugs. *Ann Surg Oncol 10*, 1184–1190.

81 KLEIN E, KLEIN G, NADKARNI JS, et al. **1967**. Surface IgM kappa specificity on a Burkitt lymphoma line *in vitro* and in derived culture lines. *Cancer Res 28*, 1300–1310.

82 ADAMS A, STRANDER H, CANTELL K. **1975**. Sensitivity of the Epstein–Barr virus transformed human lymphoid cell lines to interferon. *J Gen Virol 28*, 207–217.

83 NEDERMAN T, KARLSTRÖM E, SJÖDIN L. **1990**. An *in vitro* bioassay for quantitation of human interferons by measurements of antiproliferative activity on a continuous human lymphoma cell line. *Biologicals 18*, 29–34.

84 MCNURLAN M, CLEMENS M. **1986**. Inhibition of cell proliferation by interferons: relative contributions of changes in protein synthesis and breakdown to growth control of human lymphoblastoid cells. *Biochem J 237*, 871–876.

85 WADHWA M, BIRD C, DILGER P, et al. **2000**. Quantitative biological assays for individual cytokines. In: BALKWILL F (Ed.), *Cytokine Cell Biology: A Practical Approach*. Oxford University Press, Oxford, p. 207.

86 PISTOIA V. **1997**. Production of cytokines by human B cells in health and disease. *Immunol Today 18*, 343–350.

87 MARTYRÉ M-C, WIETZERBIN J. **1994**. Characterisation of specific functional receptors for HuIFN-α on a human megakaryocytic cell line (Dami): expression related to differentiation. *Br J Haematol 86*, 244–252.

88 KELLAR KL, HOOPER WC, BENSON JM. **1995**. MEG-01s cells have receptors for and respond to IL-3, IL-6, and SCF. *Exp Hematol 23*, 557–564.

89 OGURA M, MORISHIMA Y, ONO R, et al. **1985**. Establishment of

a novel human megakaryoblastic leukemia cell line, MEG-01, with positive Philadelphia chromosome. *Blood 66*, 1384–1392.

90 Kitamura T, Tojo A, Kuwaki T, et al. **1989**. Identification and analysis of human erythropoietin receptors on a factor-dependent cell line, TF-1. *Blood 73*, 375–380.

91 Komatsu N, Yamamoto M, Fujita H, et al. **1993**. Establishment and characterisation of an erythropoietin-dependent subline, UT-7/EPO, derived from human leukemia cell line, UT-7. *Blood 82*, 456–464.

92 Mire-Sluis AR, Page L, Meager A, et al. **1996**. An anticytokine bioactivity assay for interferons-alpha, -beta, and -omega. *J Immunol Methods 195*, 55–61.

93 Edgell, C-JS, McDonald CC, Graham JB. **1983**. Permanent cell line expressing human factor VIII-related antigen established by hybridisation. *Proc Natl Acad Sci USA 80*, 3734–3737.

94 Horisberger MA. **1992**. Interferon-induced human protein MxA is a GTPase which binds transiently to cellular proteins. *J Virol 66*, 4705–4709.

95 Towbin H, Schmitz A, Jakshies D, et al. **1992**. A whole blood immunoassay for the interferon-inducible human Mx protein. *J Interferon Res 12*, 67–74.

96 Oh S-K, Luhowskyj S, Witt P, et al. **1994**. Quantitiation of interferon-induced Mx protein in whole blood lysates by an immunochemiluniescent assay: elimination of protease activity of cell lysates *in toto*. *J Immunol Methods 176*, 79–91.

97 Files JG, Gray JL, Do LT, et al. **1998**. A novel sensitive and selective bioassay for human type I interferons. *J Interferon Cytokine Res 18*, 1019–1024.

98 Pungor Jr E, Files JG, Gabe JD, et al. **1998**. A novel bioassay for the determination of neutralizing antibodies to IFN-β1. *J Interferon Cytokine Res 18*, 1025–1030.

99 Uno K, Sato T, Takada Y, et al. **1998**. A bioassay for serum interferon based on induction of 2′–5′-oligoadenylate synthetase activity. *J Interferon Cytokine Res 18*, 1011–1018.

100 Lleonart R, Näf D, Browning H, et al. **1990**. A novel, quantitative bioassay for type I interferon using a recombinant indicator cell. *Biotechnology 8*, 1263–1267.

101 Lewis J. **1995**. A sensitive biological assay for interferons. *J Immunol Methods 185*, 9–15.

102 Canosi U, Mascia M, Gazza L, et al. **1996**. A highly precise reporter gene bioassay for type I interferon. *J Immunol Methods 199*, 69–76.

103 Brasier AR, Tate JE, Habener JF. **1989**. Optimised use of the firefly luciferase assay as a reporter gene in mammalian cell lines. *Biotechniques 7*, 1116–1122.

104 LaFleur DW, Nardelli B, Tsareva T, et al. **2001**. Interferon-kappa, a novel type I interferon expressed in human keratinocytes. *J Biol Chem 276*, 39765–39771.

105 Hammerling U, Bongcam-Rudloff E, Setterblad M, et al. **1998**. The β-Gal interferon assay: a new, precise, and sensitive method. *J Interferon Cytokine Res 18*, 451–460.

106 Taniguchi T, Takaoka A. **2002**. The interferon α/β system in antiviral responses: a multimodal machinery of gene

regulation by the IRF family of transcription factors. *Curr Opin Immunol 14*, 111–116.
107 SADICK MD, INTINOLI A, QUAMBY V, et al. **1999**. Kinase receptor activation (KIRA): a rapid and accurate alternative to end-point bioassays. *J Pharm Biomed Anal 19*, 883–891.
108 SADICK MD. **1999**. Kinase receptor activation (KIRA): a rapid accurate alternative to endpoint bioassays. *Dev Bio Stand 97*, 121–133.
109 GAINES DAS RE, MEAGER A. **1995**. Evaluation of assay designs for assays using microtiter plates: results of a study of *in vitro* bioassays and immunoassays for tumor necrosis factor (TNF). *Biologicals 23*, 285–297.
110 GAINES DAS RE, TYDEMAN MS. **1982**. Iterative weighted regression analysis of bioassays and immunoassays. *Comput Programs Biomed 15*, 13–20.
111 VAN MEIR EG, KIKUCHI T, TADA M, et al. **1994**. Analysis of the p53 gene and its expression in human glioblastoma cells. *Cancer Res 54*, 649–652.

13
The Development and Measurement of Antibodies to Interferon

Sidney E. Grossberg and Yoshimi Kawade

13.1
Introductory Perspective

Soon after the discovery of interferon (IFN), the use of polyclonal antibodies (PAbs) to neutralize the ability of IFN to induce antiviral resistance in cultured cells or tissue preparations made it possible to define the different major types of IFN as α, β and γ, and to distinguish their animal species of origin [1–4]. The development of monoclonal Abs (MAbs) led to the identification of epitopes or molecular domains that appear to play a vital role in achieving the effects of biologically active molecules through their binding to specific receptors, e.g. of human IFN-α2 [5], -β [6–8] and -γ [9].

The growing recognition of Ab formation in patients administered biologically active proteins derived from human genes by recombinant DNA technology for the treatment of very different kinds of diseases has made these issues of immediate clinical concern. It is now appreciated that many patients can develop neutralizing Abs (NAbs) to such proteins, become resistant to therapy, and undergo relapse. Inasmuch as an imposing amount of scientific literature has developed about the development of Abs in patients receiving various forms of IFN-α and -β, this chapter will concentrate on the observations and problems revealed by experimental, theoretical and clinical studies on the Abs to these type I IFNs. Although the administration of anti-IFN-γ Abs can prevent the occurrence of disease or alleviate its manifestations in animal models of inflammation, immunity, cancer, transplant rejection and delayed-type hypersensitivity, a definitive use of IFN-γ as a clinically useful therapeutic agent remains elusive [10]. Some of the lessons learned about Abs to IFN-α and -β may be applicable to other cytokines and therapeutic human proteins.

13.1.1
Immunological Perspective

The IFNs are self-antigens, i.e. they are protein or glycoprotein gene products normally found in the body of an individual. Although the adaptive immune response is a critical component of host defense against infection and, therefore, essential for normal health under most circumstances, such responses can be elicited by antigens in normal tissues or their products. Ordinarily, lymphocytes during their maturation stages are screened *in vivo* for reactivity with self-antigens, and those that do respond usually undergo apoptosis and are eliminated. The normal circumstance is the development of tolerance to self-antigens, such as the IFNs, resulting in the failure to mount an immune response to such antigens. Abs, which are the secreted form of B lymphocyte receptors, can functionally be divided into two categories – those that inhibit or neutralize biological activity and those that manifest some degree of binding in an *in vitro* immunoassay, but fail to neutralize; the latter are sometimes referred to as binding Abs, but are more properly designated as non-NAbs inasmuch as all Abs bind to antigen [11].

Each B lymphocyte carries about 50 000 Ab molecules on its surface, specific for a single epitope. The human immune repertoire contains hundreds of millions of B cells, each potentially of different specificity. Ab-binding sites that can interact with specific epitopes are formed by variable regions of the heavy and light immunoglobulin (Ig) chains, designated the hypervariable or the complementarity-determining regions, that can recognize quite small areas on a sugar or protein, e.g. a peptide containing between 5 and 8 amino acids, even though the mass of an IgG molecule is 160 000. The typical immune response involves numerous B lymphocyte clones, producing Abs having different specificities and affinities, and resulting in a polyclonal response [11].

13.1.2
Antibodies to Self-antigens

Abs that react with a variety of self-antigens can occasionally be found in the serum of normal, nonimmunized individuals and in patients suffering from infectious and inflammatory diseases – so-called auto-Abs [12–18]. Prolonged, high-dose therapy with self-antigens, in particular cytokines, hormones, growth factors or clotting factors, can induce Ab formation that can inhibit or neutralize their desired therapeutic biological activities. The following examples are illustrative. In patients with hemophilia, 20–40% treated with the clotting factor VIII can develop inhibitory Abs [19]. The effect and pharmacokinetics of the hormone human insulin can be altered by the formation of anti-insulin Abs [20, 21]. The treatment of anemia with recombinant human erythropoietin in patients with chronic renal disease led to the production of Abs that interfered with the therapeutic stimulation of erythropoesis, with potentially lethal consequences [22, 23]. Granulocyte macrophage colony-stimulating factor (GM-CSF) therapy induced in a large proportion of patients Abs that inhibited its therapeutic effects [24]. Other instances as well as the possible factors and mechanisms involved have been discussed [25]. The development of NAbs to IFN was first reported in 1981 in a patient being treated with natural IFN-β for a nasopharyngeal carcinoma [26]. The large literature on the development of Abs to IFN-α and -β in patients is summarized in Section 13.5.

13.2
NAbs

The term "neutralization" was introduced to describe the eradication by Abs of the lethal or otherwise severe effects in animals of microbial toxins [27], of viruses [28] and subsequently of toxins from a variety of other biological sources. The power of a serum to neutralize quantitatively has usually been arbitrarily defined by the nature of the test, i.e. the effect achieved in a particular biological system, either in the intact animal, in an organ or tissue preparation, or in cells in culture. Such highly specific PAbs nullified, inhibited or reduced, i.e. neutralized, the activity of biologically active IFN molecules by interfering with the binding to the two-component receptor for the type I IFNs, IFNAR-1 and -2, and for the type II (IFN-γ) receptor, IFNGR-1 and -2 [29]. More recently, the use of neutralizing MAbs has made possible the determination of epitopes or domains of IFN molecules that must interact with these receptors to achieve IFN's diverse biologic effects (see Section 13.4).

Neutralization by Ab depends upon (i) its ability to bind, (ii) the strength of its binding (affinity) and (iii) the location of binding on the IFN molecule. The relationship is not stoichiometric (Section 13.2.2). The measure of neutralizing potency of an Ab preparation is carried out in a biological assay most conveniently performed in cells in culture, the varieties of which are described by Meager in Chapter 12. Any cell culture system can be used for IFN Ab bioassay, provided that the assay is suitably sensitive to IFN action – a very important factor (Section 13.2.4). The most commonly used quantitative IFN bioassays measure as endpoint some degree of induced cellular antiviral resistance, the classical hallmark of IFN action, or the amount of an induced host protein, e.g., MxA, a dynamin family GTPase that has a role in antiviral resistance to influenza virus, a myxovirus [30, 31]. The most frequently used antiviral assays employ such cell–virus combinations as A549 human bronchioloalveolar carcinoma cells and encephalomyocarditis virus [32], or vesicular stomatitis virus or Sindbis virus in human WISH or amnion-derived FL cells [33]. The measurement of the degree of antiviral effect (usually to a 50% endpoint) by the uptake of a vital dye, and by the use of computerized automated spectrophotometry and appropriate software to calculate the endpoints, construct the curves, and evaluate statistically the data, can make the results obtained entirely objective. Similarly, the amount of IFN-induced MxA protein in a series of wells of A549 cells can be objectively measured by quantitative immunoassay with potent anti-MxA Ab to achieve a 90% reduction endpoint [30, 31]; this bioassay has the advantages [34] of avoiding the use of viruses and being somewhat more rapid. Other bioassays, e.g. by measuring the antiproliferative effects of IFN [35], have not been widely used. As described in Chapter 12, an appropriate endpoint is arbitrarily chosen for the IFN effect, in comparison with controls, in the rectilinear portion of the typically sigmoidal dose–response curve (Fig. 13.1), an endpoint which operationally defines one Laboratory Unit (LU) per unit volume, since the dose–response curve is concentration dependent [36]. The relative sensitivity of an IFN bioassay is measured by its relative ability (in LU)

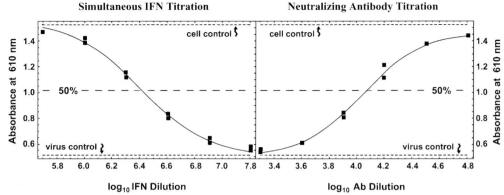

Fig. 13.1. Characteristic dose–response curves in a neutralization bioassay. Shown on the left is the titration results in duplicate of HuIFN-β1a (Rebif), performed simultaneously with the titration of NAb, shown on the right, in a patient's serum measured by the constant IFN method. The bioassay is the objective, cytopathic effect, naphthol blue black dye-uptake procedure, utilizing A549 human lung carcinoma cells and encephalomyocarditis virus [32]. The 50% endpoint (median value between the cell control and virus control values), indicated by the broken line is taken to be 1 LU mL^{-1}. The IFN target dose is 10 LU mL^{-1} and the observed antigen dose measured in the simultaneous IFN titration is 5.25 LU mL^{-1}. The absorbance values are determined in a spectrophotometric plate reader, and the data then processed by a computer program created by Leslie D. Grossberg that provides the dose–response curves, statistical analyses of different parameters, calculation and final adjustment of the titers (t values) according to the formula $t = f(n-1)/9$ (see text), reportable as TRU mL^{-1} [38]. The t value of NAb in this serum sample is $\log_{10}$ 3.74 or 5500 TRU mL^{-1}. This example also illustrates the need for adjustment by the above formula in the circumstances not uncommonly encountered where the target IFN dose and the antigen dose actually measured are different, thereby allowing the expression of titer by a well-defined unit, i.e. TRU mL^{-1}.

to detect the unitage assigned in International Units (IU) to the homologous WHO IFN International Standard (IS) Preparation [36, 37]. The analyses below (Section 13.2.3) always refer to the intrinsic sensitivity of the bioassay for the particular IFN and not to measurement of Ab; IFN sensitivity importantly affects the results of Ab testing. An assay can detect more activity than that assigned to an IFN IS by a factor of 2–3 (high sensitivity), or less activity by a factor of 5–10 (low sensitivity) [38]. Data on the sensitivity of the IFN bioassay should be provided in every report.

13.2.1
Neutralization Bioassay Design

To determine the neutralizing potency of an Ab preparation, be it MAbs or PAbs in serum or other body fluids, two approaches can be used: (i) the constant IFN method, in which a chosen fixed concentration of IFN is mixed with varying dilutions of Ab, and (ii) the constant Ab method, by which a fixed concentration of Ab is mixed with varying concentrations of IFN. In both designs, the endpoint is taken where the IFN activity is reduced to give 1 LU mL^{-1}. The two designs do not differ in principle and can be combined to create a grid or checkerboard design. In an initial effort to standardize the neutralization assay two decades ago, the WHO [39] recommended an operational approach based on the studies by Kawade [40] to utilize as antigen the amount of biological activity represented as 10 LU mL^{-1} of IFN which would be neutralized to 1 LU mL^{-1} – the endpoint of most IFN bioassays [39]. This recommendation has been repeatedly endorsed by WHO and other international committees [41–45] as the general approach to be taken for the measurement of IFN neutralization by PAbs and MAbs [41–45]. Although the WHO recommendation was generally accepted as an approach, the methodology required further explanation and refinement based on more extensive theoretical analyses and experimental data [36–38, 46–50] (see the following sections).

13.2.1.1 The Constant IFN Method

The constant IFN method is the bioassay design that has been most frequently employed. The following description outlines the procedure. A dilution of IFN is selected on the basis of previous titrations to provide 20 LU mL^{-1} (initial concentration) to be mixed with an equal volume of a series of dilutions, usually 2-fold, of the Ab preparation. Serum should previously have been heated at 56 °C for 30 min to inactivate nonspecific factors such as complement components. Following incubation, usually 1 h at 37 °C, the IFN–Ab mixtures are placed on IFN-sensitive cells for an additional prolonged incubation, usually overnight, and the remainder of the IFN bioassay procedure completed. The relationship thereby obtained between the dose of Ab as the serum dilution and the IFN response, i.e. the serum dilution that gives 1 LU mL^{-1} of remaining IFN activity, the usual bioassay endpoint, permits the determination of the neutralization endpoint from the Ab–IFN dose–response curve. Since the apparent IFN antigen input unitage may vary from titration to titration, it is essential to perform simultaneous measurement of the IFN units actually used in that day's test. A typical set of titration results is illustrated in Fig. 13.1. Appropriate control cultures include (i) cell controls not treated with IFN, (ii) virus controls, if appropriate, and (ii) a serum control in which the lowest dilution of serum under test is placed on cells to monitor possible serum cytotoxicity. Figure 13.2 illustrates results obtained with a different bioassay from that shown in Fig. 13.1, using different IFN antigen doses. For various reasons (see below), IFN doses in IU mL^{-1} should not be used in the neutralization test unless there is the unusual circumstance of demonstrated absolute identity between IU and LU in each titration.

The preferred way to state the result as an index of neutralization is a titer (t) calculated by the formula:

$t = f(n-1)/9$

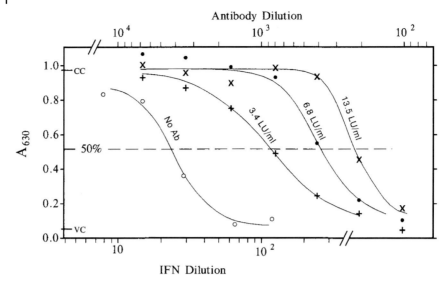

Fig. 13.2. Neutralization curves obtained with the constant IFN method for HuIFN-β antiserum employing the bioassay with Sindbis virus in FL cells with different doses of antigen. The scale on the upper abscissa applies to the three neutralization curves to the right, changing their direction (from that in Fig. 13.1) to make them comparable to the simultaneous IFN titration curve (No Ab) to which the lower abscissa scale applies. CC = absorbance of cell controls; VC = absorbance of virus controls. (Reproduced from [37] with permission.)

where f is the reciprocal of the Ab dilution achieving the endpoint and n is the IFN concentration (in LU mL^{-1}) measured in the same titration. (The divisor is 9, not 10, because the endpoint is 1 LU mL^{-1}, not 0 [36].) The resulting titer is expressed as a whole number in Tenfold Reduction Units (TRU) per milliliter [36–38]. In the case of monoclonal Ab purified so that its concentration of Ig is known, its neutralizing potency, as the reduction of 10 to 1 LU mL^{-1}, can be expressed on a weight basis as microgram of Ig protein per unit volume, or as a micromolar concentration.

13.2.1.2 The Constant Antibody Method

For Abs of low potency, the constant IFN method described above using 10 LU mL^{-1} of IFN as antigen will require low dilutions of Ab to effect significant neutralization. In some circumstances the constant Ab method may be better than the constant IFN method: (i) to determine small degrees of neutralization, (ii) to characterize monoclonal Abs and (iii) to analyze mixtures containing different antigenic types of IFN [47–49]. Serum samples at low dilutions often cause cytotoxicity and other cellular effects, and therefore must be used of dilutions at 1:10 or more – a condition that makes it difficult to detect neutralization by samples having low Ab titers. Theoretical calculations based on Kawade's simplest model predict that the constant IFN method described in the preceding section will fail to determine the Ab titer as less than about 1:10 or 1:20 unless serum dilutions lower than 1:10 are used. The use of less IFN as antigen might remedy this situation, but the assay result would be less reliable.

The constant Ab method is useful for assaying Abs of low titer. The procedure of the constant Ab method essentially is to carry out an IFN titration in the presence of a certain fixed concentration of Ab, by adding a series of IFN concentrations to a constant dilution of Ab to determine the IFN concentration (LU mL^{-1}) that gives the endpoint (1 LU mL^{-1} of IFN). Examples of results from different laboratories calculated with this method are illustrated elsewhere [37]. The value of the titer t is calculated by the formula as before [i.e. $t = f(n-1)/9$]. Theoretical neutralization curves in the Constant Proportion mode (see Section 13.2.2) are shown in Fig. 13.3 for six Abs having different titers, all assayed at a 1:10 dilution. It should be possible to screen the sera of patients by setting the lowest serum dilution at 1:10 to determine unequivocally low Ab titers such as $t = 2$ or 5, provided the IFN bioassay is of sufficient precision. The theory based on Kawade's simplest model predicts that the constant Ab method is about 10 times more sensitive than the constant

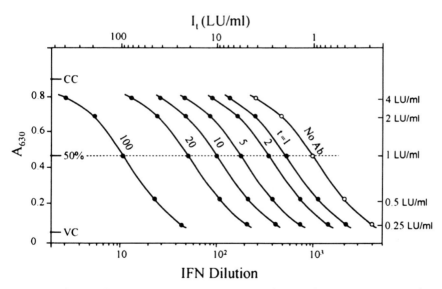

Fig. 13.3. Theoretical neutralization curves calculated for data obtainable with the constant Ab method for six Ab preparations (each diluted 1:10), having titers $t = 1, 2, 5, 10, 20$ and 100 (Constant Proportion mode), assayed by a dye-uptake IFN bioassay measuring viral cytopathic effect [37]. The direction of the neutralization curves is the same as the control IFN titration curve and the curves are exactly parallel to each other. It = total IFN in the mixture with Ab; CC = absorbance of cell controls; VC = absorbance of virus controls. This method makes it possible to detect very low NAb titers.

IFN method, and that the neutralization curves for the Constant Proportion mode are always parallel to the control IFN curve without Ab (Fig. 13.3).

13.2.2
Theoretical Analyses

Kawade [40, 46–49] proposed a model of the IFN neutralization reaction termed the "simplest model", which assumes that the Ab-binding site and IFN molecule interact in a 1:1 relationship such that the bound IFN is totally inactive. This model permits the use of the Law of Mass Action to correlate the extent of neutralization with Ab concentration. The theoretical construct predicts that if the Ab affinity is high, the circumstance of equimolar neutralization applies, an Ab action designated the Fixed Amount mode [37]. In this case, an amount of Ab would neutralize 10 to 1 LU mL^{-1}, 20 to 11 LU mL^{-1} and 100 to 91 LU mL^{-1}. If, however, the Ab has low affinity, the Ab is predicted to neutralize IFN with a set ratio of reduction of added IFN to the active IFN remaining, a circumstance described as the Constant Proportion mode. In that case, an amount of Ab would neutralize 10 to 1 LU mL^{-1}, 20 to 2 LU mL^{-1} and 100 to 10 LU mL^{-1}. Theoretical neutralization curves for high- and low-affinity Abs differ markedly in their slopes, and thus help to distinguish between the two modes [37]. The theoretical neutralization curves for Kawade's simplest model using the constant IFN method obtained for a high-affinity Ab and a low-affinity Ab against a wide range of IFN antigen doses showed that the curves of the high-affinity Ab are much steeper in slope than those for the low-affinity Ab. The theoretical neutralization curves for the simplest model using the constant Ab method showed the low-affinity curves to be parallel to the control IFN curve and the high-affinity curves always to be much steeper [37].

The affinity of an Ab, expressed as an association constant K, is, of course, unvarying. It should be appreciated that the apparent Ab affinity, defined in the context here, is high or low in relation to the molar concentration of free IFN ($[I_f]$) left unbound to Ab, such that the apparent affinity is low when $K < 1/[I_f]$ and high when $K > 1/[I_f]$. The $[I_f]$ can be experimentally varied, so that a given Ab may act either in the Constant Proportion mode (low affinity) or in the Fixed Amount mode (high affinity), depending on the ratio of Ab and IFN in the mixture [50]. Since, in most neutralization assays, $[I_f]$ at the endpoint (1 LU mL^{-1}) is usually very small, most Abs will therefore be low in their apparent affinity by this definition, and their neutralization will be in the Constant Proportion mode. Even when an Ab has a very high association constant (K value), or the bioassay is very low in sensitivity and the equilibrium is in the Fixed Amount mode, the mode will shift and will eventually reach the Constant Proportion mode if more Ab is added and the level of the remaining IFN is reduced. Then complete elimination of biological activity by the Ab will, in general, be difficult to obtain because at low levels of effector molecules, Ab action is likely to be in the Constant Proportion mode, leaving some remaining activity, however small it may be, that may exert some biological activity. This circumstance may apply to *in vivo* conditions.

If Ab were to act in the Fixed Amount mode, as many investigators seem to imagine, different quantities of Ab will be needed to neutralize the same LU of IFN in different assays resulting in different values of the titer. This problem is eliminated with the Kawade method, assuming the Constant Proportion mode, which seems to apply to most Abs.

It should be cautioned that the term "affinity" denotes the strength of binding of one molecule to another, such as an Ab molecule to a monovalent antigen or single epitope, and is distinct from "avidity" sometimes mentioned in the literature. The latter is the sum total of the strength of binding of two molecules to one another at multiple sites, which is applicable to viruses, bacteria or other entities having multiple repetitive epitopes [11] and not applicable to the IFNs that have single, non-repetitive epitopes.

13.2.3
Experimental Analyses

Following the original studies of Kawade [40, 46–49] describing the theoretical, experimental, and functional aspects of the neutralization of IFNs, experimental data were provided by many laboratories involved in an international collaborative study on human sera – one containing Abs to human IFN-α and the other containing Abs to human IFN-β [37, 38]. The two serum preparations were then established as WHO Reference Reagents [51] to be used to determine the functionality of the neutralization assay and not to calibrate it. Figure 13.2 illustrates results obtained with the constant IFN method using different concentrations of IFN-β antigen. Analyses of the data clearly supported the Constant Proportion hypothesis applicable to Ab with low affinity in which the serum Abs reduced IFN activity in a set ratio of added-to-residual biologically active IFN, provided that the bioassay was suitably sensitive to IFN. A greater appreciation of the theoretical constructs can be had by referring to the detailed explanations given in the appendices of these two papers [37, 38].

13.2.4
Standardization and the Reporting of Neutralization Results

The need for a commonly accepted way to report the results of neutralization tests has been repeatedly cited [34, 39, 41–44, 52–54]. The results of IFN neutralization have been defined and reported in the scientific literature in various ways, in addition to that previously recommended by WHO. Table 13.1 lists a dozen examples of such remarkably varied and sometimes quite imaginative approaches. The relationship between IU and LU, i.e. the relative sensitivity of the bioassay used, has sometimes been stated in publication, but often not. Neutralizing potency has most often been reported as titers but sometimes as units, often not defined.

There has been much controversy both with regard to the relative incidence of

Tab. 13.1. The variety of ways IFN Ab neutralization results have been reported in the scientific literature

1.	Amount of Ab required to reduce 10 to 1 LU mL^{-1} [36–38]
2.	Amount of AB required to reduce 10 to 1 IU mL^{-1} [55]
3.	IFN neutralizing units mL^{-1}, defined as the mean of the product of the serum dilution inhibiting the protective effect of IFN times the IFN concentration present in the corresponding well [56, 57]
4.	Serum dilution giving 50% reduction of 50 U mL^{-1} [58]
5.	Final serum dilution that reduces 3 to 1 LU mL^{-1}, with an additional calculation, not described, to express the titer in terms of IU [59]
6.	Dilution causing significant reduction (more than 3 times the standard deviation of 3 IU) [60]
7.	Highest 2-fold serum dilution (as 1 neutralizing unit) completely inhibiting 5 IU mL^{-1} [61, 62]
8.	Serum dilution that neutralizes 2–20 U mL^{-1}, as a neutralizing unit defined as the amount of Ab neutralizing 1 U of IFN [63]
9.	Serum dilution after incubation with 5–20 IU mL^{-1} yielding a 50% endpoint multiplied by a correction factor f to obtain the neutralization titer NT$_{50}$, adjusted to 10 LU mL^{-1}, where f = LU mL^{-1} divided by 10 LU mL^{-1} [64]
10.	ND$_{50}$, that concentration of Ab required to yield one-half maximal inhibition of activity at a cytokine concentration high enough to elicit maximum response, e.g. 50 U [65]
11.	Percent neutralization of 3, 10 or 100 U of IFN [66]
12.	Number of IFN units neutralizable by the Ab [67]

Abs after treatment with particular IFN preparations and whether loss of clinical response results when the Ab concentration in serum exceeds a critical value. These controversies reflect the fact that there has been no universally accepted way either to measure anti-IFN NAbs or to express their concentration. It is now well appreciated that patients injected repeatedly with human IFN preparations may develop NAbs that can abolish or reduce the beneficial effects of the treatment (Section 13.5). It is, therefore, important to establish a common method of reporting NAb levels, by extending the recommendations by the WHO concerning the general design of neutralization tests and how to report the data therefrom. A WHO international collaborative study on human sera with Abs against human IFN-α and -β made it possible to test the theoretical and experimental constructs of Kawade et al. [46–50] concerning the neutralization reaction with data obtained in different bioassay systems in different laboratories and to obtain many data points for statistical evaluation with bioassays having a great range of sensitivity to IFN. Those analyses substantiated and extended the original conclusions of Kawade that the reaction mode of serum Abs was that of the Constant Proportion mode (explained above) by which Ab reduces IFN in a set ratio of added-to-residual biologically active IFN, a consequence of the low molar concentration of free IFN at the neutralization endpoint (1 LU mL^{-1} or about 10^{-12} M). The statistical evidence indicated that the titer based on the use of IFN antigen doses in LU was more constant in numerous assays having widely different IFN sensitivity carried out on the same Ab preparations than did the titer based on IFN doses in IU. The titer based on the use of IFN antigen doses in LU when the assays varied in sensitivity by a factor of less than about 3.5, i.e. if the bioassay measured one-third or less the number of units assigned to the appropriate WHO IS [38]. (Assay sensitivity is determined as the relative proportion of LU actually measured in relation to the assigned value in IU of a WHO IS reference preparation, i.e. if a WHO IS has an assigned value of 10 000 IU and the assay measures 5000 units as LU, the relative IFN sensitivity of that assay is 0.5 or one-half that of an assay which measures 10 000 units in LU.) As the premise of low Ab affinity may not be applicable when the bioassay sensitivity is very low, i.e. when the molar concentration of IFN at the titration endpoint is high, it is desirable to utilize a bioassay with relatively high sensitivity.

These results, obtained in multiple laboratories, support the recommendation that the preferred way to state the index of neutralization of Abs is a titer (t), calculated by the formula given above, i.e. $t = f(n-1)/9$, where f is the reciprocal of the Ab dilution achieving the endpoint and n is the IFN concentration measured in that day's titration. The Tenfold Reduction Unit (TRU) of neutralization was proposed [38] (Section 13.2.5) to express the quantity or unitage of IFN NAb. The determination of the index of neutralization as described herein, such that the proposed derivative term TRU can be properly used, should help make results in different laboratories employing different bioassay systems more readily comparable and interpretable, provided that the bioassays are sufficiently sensitive to IFN. (The proposal to utilize the Kawade approach to reporting NAbs to IFN-β in patients with multiple sclerosis was formally endorsed by an international group of neurologists and other investigators meeting in London in 2003 [45].)

13.2.5
A Solution to the Problem of NAb Unitage

Many different methods have been used to measure anti-IFN-NAbs, and their concentrations have been calculated and expressed in a variety of ways (Tab. 13.1). The unfortunate consequence is that results provided by one laboratory can be difficult to compare with those from another, and thus the interpretation of such results is made uncertain and the IFN neutralization titers from different laboratories reported in the scientific literature can be difficult to correlate. Since patients may receive subcutaneous or intramuscular injections of an IFN preparation several times a week for even years (Section 13.5), and a significant proportion of these patients will develop NAbs against the IFN product administered, some in consequence will lose any benefits from the treatment. Apparent titers may fluctuate and clinical decisions may be influenced by such information. The analyses by Kawade et al. over the past two decades have considered a large range of variables, among them the amount of IFN antigen used in the test, the sensitivity to IFN of the bioassay system and how to express the results – points that have been confirmed by experimental data from international collaborative studies [37, 38]. A very important point is that these analyses and thus the recommendation below do not assume a particular bioassay methodology, and thus would be applicable to the MxA bioassay [30, 31, 34] or suitable antiviral bioassays. General recommendations about IFN neutralization had been published by WHO, but those were not explicit about bioassay design or how to compute and report the results.

The procedure outlined below was unanimously recommended by the Standards Committee of the International Society for Interferon and Cytokine Research for adoption by the field to be used for the design and reporting the results of neutralization tests, especially of serum samples from patients with NAbs to IFNs.

(1) Utilize a quantitative IFN bioassay that measures an IFN-inducible product or effect in human cells and has the usual level of sensitivity to IFN (see Item 2 below). The IFN bioassay may involve the induction of a cellular gene product, such as the MxA protein (a GTPase), or antiviral effects assessed in terms of cytopathogenicity or reduction in virus yield. In all these assays, an arbitrary endpoint is chosen, for example, a 50% reduction in viral cytopathic effect; the IFN concentration at the endpoint is defined as 1 LU mL^{-1}.

(2) Determine the sensitivity of the chosen bioassay. Titrate a preparation of the IFN concerned that has been calibrated in IU mL^{-1} against the homologous WHO IFN IS. It is desirable that the endpoint in the assay, i.e. 1 LU mL^{-1}, should correspond to no more than about 4 IU mL^{-1}. If the assay proves to be very insensitive, such that the amount of IFN used is relatively much larger than in most other assay systems, the results are less likely to be comparable with those in other laboratories.

(3) In the NAb test, use a concentration of IFN as antigen that is expected to be 10 LU mL^{-1}, based on previous titrations with the IFN preparation concerned. Carry out a simultaneous titration of the IFN in parallel with the Ab test in order to measure the amount of IFN antigen actually used. If this antigen concentration proves to be somewhat more or less than 10 LU mL^{-1}, as often happens, a correction should be made in calculating the titer of the Abs (see Item 5 below).

(4) Test the selected concentration of IFN against a range of dilutions of the patient's serum and determine the Ab dilution at which the indicated amount of IFN is reduced to 1 LU mL^{-1}.

(5) Enter the results into the equation $t = f(n-1)/9$, where t is the titer of Ab expressed in TRU mL^{-1}, f is the reciprocal of the Ab dilution at the bioassay endpoint and n is the actual IFN concentration in LU mL^{-1} used in the test

as measured in the simultaneous control IFN titration. Results should be reported as TRU mL^{-1}.

For the next few years it will be very helpful if investigators calculate and report anti-IFN NAb data not only in TRU mL^{-1}, but also as they may have done before for the sake of continuity with their previously published data for comparison and possible reassessment.

13.3
Immunoassays for Non-NAbs

Various *in vitro* immunoassays for IFNs have been derived for the measurement of Abs or receptors that bind IFNs but fail to reflect neutralization of biological activity. Some of these are qualitative, such as Western immunoblot, or relatively quantitative, such as ELISAs of various designs, radioimmunoprecipitation assay [68, 69], a column-based assay [70] and chemofluorescence [71], among others [66, 69, 72, 73]. ELISA immunoassays can be structured as direct or indirect, with the former based on binding the IFN antigen directly to the plastic tube or plate, or the latter a sandwich or capture configuration, in which anti-IFN Ab is bound to the plastic to which the IFN antigen is then bound and the unknown (human) Ab is then added, to be detected by heterologous species (anti-human) labeled Ab for appropriate subsequent quantitation; various permutations of the latter detection process have been described [68]. Immunoassays require the use of Abs that recognize only the IFN to be quantified, utilizing, as appropriate, radiolabeled pure IFN radiolabeled Ab, or enzyme-linked or biotinylated Ab. The general methodology has been reviewed by Meager [68, 69], who has helpfully described the details in constructing immunoassays and identified the pitfalls, as well as the required controls and how to interpret the assay. There is no standardized ELISA for IFNs, and there are dozens of combinations of Abs, reagents and conditions that may be used. Since immunoassays measure immunoreactive mass, it is difficult to prove the assumption that native IFN molecules are recognized by anti-IFN Abs, and therefore every immunoassay should be calibrated only with the IFN type, subtype or mutant recombinant and calibrated against a positive Ab with a specific IFN type or recombinant mutant. The results of immunoassays should be interpreted with caution and considered as alternative assays; they must not be considered as substitutes for biological assays unless they can be properly validated [68, 69].

Non-NAbs detected in immunoassays have been often referred to as binding Abs (BAb), a term which does not distinguish them from the NAbs that must obviously also bind to IFN molecules in order to effect neutralization – a preferable term would be non-NAbs; some MAbs neutralize and some bind, but do not neutralize (e.g. [68]). Such *in vitro* immunoassays measure antigenic mass reported as weight, which may be useful for approximating the amount of IFN present either in a preparation or in the circulation, but do not necessarily measure the amount of biologically active IFN molecules or NAbs. IFNs vary in their stability, and there is a correspondence between weight and biological activity (as in most other pharmaceutical products), but Ab tests in which IFNs may be diluted in nonstabilizing solutions or performed under destabilizing conditions may make the measurement of NAbs by *in vitro* immunoassays highly problematic. Some reports and descriptions of methods by some commercial suppliers of ELISA kits may claim incorrectly the measurement of, or correlation with, biologically active IFN molecules and NAbs on the basis that they have been calibrated against either reference or even non-reference NAbs or against international IFN standard preparations that are standards only for biological activity, not content of IFN mass. Such claims must be substantiated by experimental evidence, for example, by correlating detectable binding with the progressive inactivation of biological activity of an IFN by heat or other means [74] and may very much depend on the particular monoclonal Ab. Non-NAbs in patients' sera have been reported to appear earlier than NAbs (Section 13.5). Such correlations may depend very much upon the nature of the immunoassay as well as the particular MAb used as capture Ab in the test [75], and require comparisons of large numbers of Ab-positive and -negative sera tested both by IFN immunoassay and neutralization tests. There are no WHO standard reference reagents, such as MAb, or standard conditions established for any *in vitro* immunoassay for IFNs, so that currently there is no basis for the reporting of international units of IFN by immunoassay or of the unitage of non-NAbs.

13.4
Epitope Analysis

The development of neutralizing and non-neutralizing MAbs made it possible to pose the question as to which part of the IFN molecule interacted with the then putative IFN receptor, which for IFN-β was initially identified as a linear epitope in the region of the IFN-β molecule between amino acid residues 32 and 56 [8]. Subsequently the type I IFN-α/β receptor moieties IFNAR-1 and -2 were identified and cloned (reviewed in [29]). When the crystal structure of the IFN-β1a molecule was solved [76], it was possible to undertake systematic mutational mapping of sites that are important for receptor binding and functional activity [77]. The binding properties of nine MAbs were defined on these IFN molecular mutants having alanine substituted at targeted, surface-exposed residues and were correlated with MAb neutralizing potency [78]. Locations of receptor and anti-IFN-β MAb binding sites on the IFN-β molecule are illustrated in Fig. 13.4. Whereas some neutralizing MAbs directed to the AB3 and AB2 regions of the molecule and inhibited IFN-β/IFNAR-2 complex formation, suggesting that interference with receptor binding constituted their mechanism of neutralization, two MAbs that bound to sites involving the BC, C1 and C2 regions remote from the IFNAR-2-binding site also inhibited IFN-β/IFNAR-2 complex formation and demonstrated potent neutralizing

(A) Areas for IFNAR-1 and IFNAR-2 binding

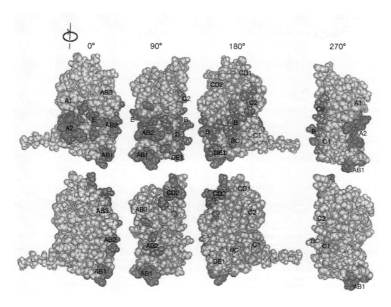

(B) Mutated residues reducing mAb binding

Fig. 13.4. Location of the binding sites of receptor and of anti-IFN-β MAbs on the IFN-β molecule. The three-dimensional crystal structure of IFN-β is shown in space-filling models. Four different views of the molecule are shown, as indicated at the top. (A) The positions of amino acid residues important for IFNAR-1 and -2 receptor chain binding are shown in blue and red, respectively. The regions occupied by the various alanine substitution mutants (A1–E) are labeled on the molecule. (B) The positions of residues, which when mutated to alanine abrogate or reduce the binding of anti-IFN-β MAbs, are highlighted in different colors on the IFN-β molecule. Shown in pale and dark blue are those that respectively abrogate or reduce the binding of MAbs Bio 1, Bio 2, Bio 5, A1 and A5. Those MAbs abrogating binding (Bio 4, A7 or Bio 6) are shown, respectively, in red, dark green, and purple. The sites abrogating (BC and C1) or reducing (C2) by MAb B-02 binding are shown in pale and dark green. (Reproduced with modifications from [78] with permission.) For identity of the MAbs and mutants the reader is referred to the original publication [78]. (This figure also appears with the color plates.)

activity. Thus, it is possible for some Abs that bind at a distance from the receptor binding site could also neutralize, presumably by causing IFN conformational distortion, resulting in interference with its binding to receptor [78]. Comparable information on submolecular localization of neutralizing MAbs is not available for IFN-α or -γ.

13.5
Development of Antibodies during IFN Therapy

IFN Abs, so-called natural Abs, have been detected in serum of normal individuals [60, 79, 82], patients with autoimmune diseases [17, 83], viral infections [84], tumors [85], and transplants [86]. Such Abs have also been referred to as auto-Abs [11, 14, 68] since they are produced against self-antigens by unknown mechanisms, but considered to be distinctive, at least in common parlance, from the Abs engendered by parenteral injections of large doses of IFNs for therapeutic reasons. In early experiments to show the importance of endogenous, induced IFN during viral infection [87], injection of Abs to mouse IFN made the viral infection worse and enhanced lethality.

Some important technical factors should be considered. Serum samples to be analyzed for Ab to IFN should be taken at least 24 h, preferably 48 h or more, after the last injection of IFN, since circulating IFN may affect the apparent NAb or non-NAb Ab titer, as shown for anti-IFN-α Abs [56, 62]. The matter of nonspecific inhibitors of IFN has been raised [88, 89]. Since IgG is heat-stable, it is important to heat serum prior to bioassay for NAbs at 56 °C for 30 min (or 60 °C for 20 min), which should denature such inhibitory substances [31, 36].

The levels of Abs to IFN-α and -β may not only appear and disappear, especially in patients with low titers, but can also fluctuate during therapy [90–92], perhaps by clearance of immune complexes by the reticuloendothelal system. In 20–30% of patients given Betaseron, NAb titers were significantly reduced or became undetectable (titer of less than 20) over time. Whether this phenomenon might be related to switching in Ig isotype or subclass, epitope spreading, alterations in antigen-presenting cells, induction of tolerance or other mechanisms remains unknown.

The types and subtypes of Igs may vary. Anti-IFN-α NAbs are usually Ig of the IgG type; non-NAbs, which develop before NAbs, can be of the IgM type [56]. Analysis of Ig subclasses in patients receiving IFN-β1b (Betaseron) [93] showed that in patients with NAb, IgG2 and IgG4 occurred more frequently than in patients with non-NAb, with the NAb titer correlating strongly with the IgG4 level; further, the median levels of IgG1 and 4 as well as of total IgG appeared to be significantly higher in patients with NAb than in those with non-NAb.

The time of appearance of Abs to IFN is variable, ranging from 1 month to many months after therapy is initiated, more usually 3–4 months. In hepatitis C patients treated with IFN-α, NAbs appeared in 6.6% during the first month, 73% during the first 3 months, and in the remaining 20% during the first 8 months [94]; of those that developed non-NAbs, 67% did so in the first 4 months and almost all did within the first 6 months. Patients given a single subtype of IFN-α may develop Abs to that subtype that do not neutralize other IFN-α subtypes [95–99]. In most patients with relapsing-remitting multiple sclerosis, IFN-β non-NAb (binding) Abs appeared after 3 months of therapy, but Ab formation may be delayed for a couple of years [90–92].

The results of Ab studies in patients over the past two decades have at times been difficult to compare and interpret because of the very different ways Ab testing has been carried out in the many laboratories involved (Section 13.2.4). The lack of standardization of any *in vitro* immunoassay for non-NAbs Abs (Section 13.3) as well as the appreciation of factors like sensitivity, specificity and reliability of the bioassays for NAbs have contributed to the variability of results [34, 36–38].

There are a variety of factors that may contribute to IFN immunogenicity. They include the frequency of IFN injection, duration of IFN treatment, as well as the formulation (e.g. pH, stabilizers, other additives) or presence of aggregates in the IFN product being administered. The route of administration may also be a factor, at least based on the experience gained from the immunization of animals, where subcutaneous injection of antigen can be more effective than intramuscular (or intravenous) routes, depending on the antigen. Particular circumstances relating to the development in patients of Abs to IFN-α and -β are discussed below.

13.5.1
Antibodies to IFN-α

Long-term treatment with recombinant IFN-α2 in patients with hairy cell leukemia or chronic myelogenous leukemia resulted in a higher incidence of Abs than in patients treated for shorter periods [56, 61, 100]. The incidence of Ab formation among patients treated with different subtypes of IFN-α may differ [101]. It has been thought that patients with different forms of cancer produce Abs less frequently than those suffering from infectious diseases [102]. The immunogenicity of IFN-α2a was about 10 times more immunogenic than the IFN-α2b preparation attributable to oxidation and aggregation of the IFN-α2a product during storage following its purification and formulation [103]. Abs to IFN-α can affect its therapeutic efficacy at least in some patients suffering with chronic myelogenous leukemia, hairy cell leukemia, renal carcinoma, non-Hodgkin's lymphoma, melanoma, B cell leukemia, cryoglobulinemia, and hepatitis [57, 58, 94, 97, 100, 104–115]. Although there have been attempts to relate the magnitude of Ab titers as well as the time of their development to interference by NAbs interfered with therapeutic effect, the considerable differences in the way Ab titers are reported and in the sensitivity of the assays used in different studies (as discussed above) has made interpretation of such correlations difficult and controversial [38, 97, 109–111, 113–118].

Patients who had developed Abs and had become resistant to recombinant IFN-α preparations could subsequently be treated effectively with native IFN-α lymphoblastoid or leukocyte products [57, 108, 115, 119, 120], probably related to the multiple IFN-α subtypes present in the latter preparations that were not neutralized by NAbs that had formed as a result of the prior, single recombinant IFN-α subtype therapy. It is of interest that Abs to IFN-α have not been reported to be associated with adverse clinical sequelae [61, 116], such as immune complex formation. What the clinical significance is of non-NAbs to different IFN-α subtypes has not been established.

Tab. 13.2. Comparison of characteristics of recombinant IFN-β products and NAb responses during treatment of relapsing-remitting multiple sclerosis

	Rebif (IFN-β1a)	Avonex (IFN-β1a)	Betaseron[a] (IFN-β1b)
Antibody induction (%)	12.5–28	2–6	28–47
Injection site reaction (%)	66	4	85
IFN producer cells	CHO	CHO	E. coli
Molecular weight (kDa)	22–24	22–24	18.5
Specific activity (MIU mg^{-1})	300	300	32
Formulation			
pH	3.8	7.2	7.2
buffer	acetate	phosphate	phosphate
stabilizer	HSA/mannitol	HSA	HSA
Administration	s.c.	i.m.	s.c.
Dose	22 and 44 µg (3/week)	30 µg (1/week)	250 µg (3/week)

[a] Betaseron in North America is the same as Betaferon in Europe
CHO = Chinese hamster ovary cell culture; HSA = human serum albumin; i.m. = intramuscular; s.c. = subcutaneous.

13.5.2
Antibodies to IFN-β

Three commercial preparations of IFN-β have been introduced into clinical practice in the past decade, primarily for the treatment of relapsing-remitting multiple sclerosis and sometimes for secondary progressive multiple sclerosis. These products have been the subject of a large number of clinical trials. Shown in Tab. 13.2 are some of the salient characteristics of these preparations, their modes of administration and the range of induced Ab formation reported in clinical trials that included assessment of NAbs, the results of which have been tabulated by Giovannoni et al. [121]. The factors that may contribute to immunogenicity include the presence of aggregates, nature (or absence) of glycosylation, nature of product formulation, dose, route of administration, frequency of administration, manifestation of the autoimmune process in multiple sclerosis and differences in bioassay type, assay design, calculation of results as well as in the ways titers are reported. Although it would appear that the subcutaneous administration of the molecularly altered IFN-β1b (Betaseron, Betaferon with serine in place of cysteine at position 17) is the most immunogenic, followed by the IFN-β1a (Rebif) given subcutaneously and then the intramuscularly injected IFN-β1a (Avonex) as the least immu-

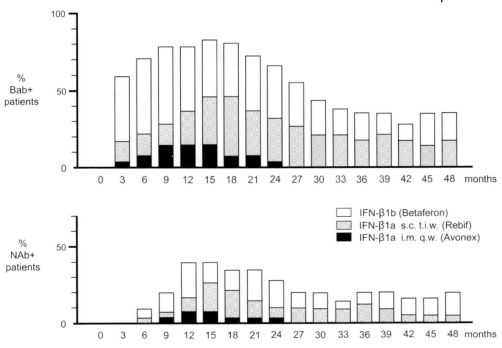

Fig. 13.5. Time course of development of non-NAbs, designated as BAbs, and of NAbs to subcutaneous (s.c.) IFN-β1b (Betaferon), intramuscular (i.m.) IFN-β1a (Avonex) and s.c. IFN-β1a (Rebif) by study month [white bars = s.c. IFN-β1b (Betaferon), black bars = i.m. IFN-β1a (Avonex); gray bars = s.c. IFN-β1a (Rebif)]. (Reproduced with modifications from [92] with permission.)

nogenic, other factors relating to dose and formulation may also contribute to the differences in immunogenicity [92, 122–124] (see Tab. 13.2).

As illustrated in Fig. 13.5, NAb formation becomes detectable between 3 and 18 months after initiation of therapy, appearing earliest with IFN-β1b (Betaseron, Betaferon) with a plateau reached in about 6 months, and somewhat later with IFN-β1a (Rebif), compared with IFN-β1a (Avonex), for which it took 9–15 months to achieve a plateau [61, 92, 125–127]. Generally, the clinically negative effect of NAbs became evident at about 2 years of therapy as indicated by relapse rate and brain lesions detected by magnetic resonance imaging; the relapse rate was 62% higher and the number of active lesions almost 5 times greater in NAb-positive patients than in those without detectable NAb. Thus, on the basis of several studies [125–130], a consistent correlation between the presence of NAbs and decreased clinical efficacy has been demonstrated, although not an all-or-none effect.

NAbs to the different IFN-β products were cross-reactive [131, 132] and the level of NAbs appeared to be somewhat lower tested against IFN-β1b than against IFN-β1a

[132]. The development of NAbs to IFN-β1a (Avonex) resulted in Nab titer-dependent reduction in neopterin, and $β_2$-microglobulin induction [125]. In patients treated with IFN-β1b (Betaseron), the appearance of NAbs correlated with a reversal of the IFN effect on MxA [129] and on natural killer cell number and activity [133]. Similarly, during treatment with IFN-β1a (Rebif), NAbs led to a reversion to pretreatment levels of MxA and TRAIL (tumor necrosis factor-related apoptosis-inducing ligand), which also has potent T cell-mediated, anti-inflammatory properties [134]. NAb-positive patients can revert to negative Ab status over a 2- to 5-year follow-up [126, 127, 135–138], more usually in patients with relatively low titers of NAbs, but occasionally also in those with higher titers. Serum samples from patients treated with IFN-β1a and IFN-β1b, which contained polyclonal NAbs as well as non-NAbs, were shown to bind to 12mer peptides at the N-terminus (residues 1–12) as well as to a peptide (residues 151–162) close to the C-terminus [67]. Persistent NAb levels have been shown to reduce the induction of MxA protein and mRNA [129, 139, 140] in the peripheral blood mononuclear cells from multiple sclerosis patients injected about 12 h before with IFN-β, indicating a clear correlation between the presence of NAbs and reduced bioavailability of the drug; the higher than 1:20 titer cut-off, as reportedly calculated by the Kawade method, correlated to a high degree with reduced bioavailability. While the MxA protein may not be involved in the pathogenesis of multiple sclerosis, matrix metalloproteinases (MMPs) have been postulated to act as effector molecules in the disease, especially the gelatinases MMP-2 and -9, which long-term IFN-β therapy reduces, and then NAbs reverse, with MMP-9 showing a quantitative correlation with NAb titers [141].

Strategies are being pursued to prevent or reduce the formation of anti-IFN Abs, e.g. by the use of monthly methylprednisolone [142, 143]. Immune complex formation or other adverse affects of circulating Abs to IFN-β preparations has not been reported. Although there has been considerable controversy about the significance of the development of NAbs to IFN-β in the treatment of multiple sclerosis [45, 144], the growing appreciation is that the occurrence of NAbs, especially at significant and persistent levels, should be taken into account along with clinical measures of the loss of therapeutic benefit [45, 92, 145–149].

13.6
Summary

The development of Abs to IFNs in patients with hepatitis, multiple sclerosis or cancers of different types treated with various IFN-α and -β products has been a matter of increasing clinical concern. The areas of concern include the relative incidence of NAbs induced by different products and how the levels of NAbs in serum relate to the loss of therapeutic clinical response. The need for standardization of the measurement of NAbs has been repeatedly cited over the past two decades, and should be readily achievable with the knowledge now available to make the results from different laboratories more readily comparable and interpretable.

References

1 PAUCKER, K. **1965**. *J Immunol 94*, 371.
2 FINTER, N. B. (Ed.). **1966**. *Frontiers of Biology: 2. Interferons*. Saunders, Philadelphia, PA.
3 GROSSBERG, S. E. **1972**. *New Engl J Med 287*, 13, 79, 122.
4 GROSSBERG, S. E. **1979**. *J Biol Stand 7*, 383.
5 CEBRIAN, M., YAGUE, E., DE LANDAZURI, M. O., et al. **1987**. *J Immunol 138*, 484.
6 REDLICH, P. N., GROSSBERG, S. E. **1989**. *J Immunol 143*, 1887.
7 REDLICH, P. N., GROSSBERG, S. E. **1990**. *Eur J Immunol 20*, 1933.
8 REDLICH, P. N., HOEPRICH, JR., P. E., COLBY, C. B., GROSSBERG, S. E. **1991**. *Proc Natl Acad Sci USA 88*, 4040.
9 ZIAI, M. R., IMBERTI, L., KOBAYASHI, M., PERUSSIA, B., TRINCHIERI, G., FERRONE, S. **1986**. *Cancer Res 46*, 187.
10 SCHREIBER, G. H., SCHREIBER, R. D. **2003**. In: THOMSON, A. W., LOTZE, M. T. (Eds.), *The Cytokine Handbook*, 4th edn. Academic Press, London, vol. I, p. 567.
11 JANEWAY, JR., C. A., TRAVERS, P., WALPORT, M., SHLOMCHIK, **2001**. *Immunobiology 5: The Immune System in Health and Disease*. Garland, New York.
12 BENDTZEN, K., SVENSON, M., JONSSON, V., HIPPE, E. **1990**. *Immunol Today 11*, 167.
13 GRIBBEN, J. G., DEVEREUX, S., THOMAS, N. S., et al. **1990**. *Lancet 335*, 434.
14 AVRAMEAS, S. **1991**. *Immunol Today 12*, 154.
15 BENDTZEN, K., HANSEN, M. B., DIAMANT, M., ROSS, C., SVENSON, M. **1994**. *J Interferon Res 14*, 157.
16 KEARNEY, J. F. **1994**. *J Interferon Res 14*, 151.
17 MEAGER, A. **1997**. *J Interferon Cytokine Res 17 (Suppl 1)*, S51.
18 MEAGER, A., VINCENT, A., NEWSOM-DAVIS, J., WILLCOX, N. **1997**. *Lancet 350*, 1596.
19 LUSHER, J. M. **2000**. *Haematologica 85*, 2.
20 FINEBERG, S., GALLOWAY, I. J., FINEBERG, N., et al. **1983**. *Diabetologia 25*, 465.
21 MEAGER, A. **1994**. *J Interferon Res 14*, 181.
22 PECES, R., DE LA TORRE, M., ALCAZAR, R., URRA, J. M. **1996**. *N Engl J Med 335*, 523.
23 CASADEVALL, N., NATAF, J., VIRON, B., et al. **2002**. *N Eng J Med 346*, 469.
24 REVOLTELLA, R. P. **1998**. *Biotherapy 10*, 321.
25 SCHELLEKENS, H. **2003**. *Neurology 61*, S11.
26 VALLBRACHT, A., TREUNER, T., FLEHMING, B., JOESTER, K. E., NIETHAMMER, C. **1981**. *Nature 287*, 496.
27 VON BEHRING, E., KITASATO, S. **1890**. *Dt Med Wochenscr 16*, 1113.
28 STERNBERG, G. M. **1892**. *Trans Ass Am Physic 7*, 68.
29 MOGENSEN, K. E., LEWERENZ, M., REBOUL, J., LUTFALLA, G., UZE, G. **1999**. *J Interferon Cytokine Res 19*, 1069.
30 FILES, J. G., GRAY, J. L., DO, L. T., et al. **1998**. *J Interferon Cytokine Res 18*, 1019.

31 PUNGOR, E., JR., FILES, J. G., GABE, J. D., et al. **1998**. *J Interferon Cytokine Res 18*, 1025.
32 GROSSBERG, S. E., TAYLOR, J. L., SIEBENLIST, R. E., JAMESON, P. **1986**. In: ROSE, N. R., FRIEDMAN, H., FAHEY, J. L. (Eds.), *Manual of Clinical Immunology*, 3rd edn. American Society for Microbiology, Washington, DC, pp. 295.
33 MEAGER, A. M. **2002**. *J Immunol Methods 261*, 21.
34 DEISENHAMMER, F., SCHELLEKENS, H., BERTOLOTTO, A. **2004**. *J Neurol 251 (Suppl. 2)*, 11/31.
35 PRÜMMER, O., PROZSOLT, F. **1994**. *J Interferon Res 14*, 193.
36 GROSSBERG, S. E., KAWADE, Y. **1997**. *Biotherapy 10*, 93.
37 GROSSBERG, S. E., KAWADE, Y., KOHASE, M., YOKOYAMA, H., FINTER, N. **2001**. *J Interferon Cytokine Res 21*, 729.
38 GROSSBERG, S. E., KAWADE, Y., KOHASE, M., KLEIN, J. P. **2001**. *J Interferon Cytokine Res 21*, 743.
39 BERG, L., BILLIAU, A., DEMAEYER, E., et al. **1983**. *WHO Tech Rep Ser 687*, 35.
40 KAWADE, Y. **1980**. *J Interferon Res 1*, 61.
41 ANDZHAPARIDZE, O. G., AYRES, J. J., DE MAEYER, E., et al. **1988**. *WHO Tech Rep Ser 771*, 37.
42 GROSSBERG, S. E. **1988**. *J Interferon Res 8(6)*, v–vii.
43 GROSSBERG, S. E. **1990**. *Prog Oncol 13*, 17.
44 GROSSBERG, S. E. **2003**. *Neurology 61 (Suppl 5)*, S21.
45 PACHNER, A. R. **2003**. *Neurology 61 (Suppl 5)*, S1.
46 KAWADE, Y., WATANABE, Y. **1984**. *J Interferon Res 4*, 571.
47 KAWADE, Y., WATANABE, Y. **1985**. *Immunology 56*, 489.
48 KAWADE, Y. **1985**. *Immunology 56*, 497.
49 KAWADE, Y. **1986**. *Methods Enzymol 119*, 558.
50 KAWADE, Y., FINTER, N., GROSSBERG, S. E. **2003**. *J Immunol Methods 28*, 127.
51 WHO Report of the Expert Committee on Biological Standarization **1996**. *WHO Tech Rep Ser 858*, 11.
52 SCHELLEKENS, H., RYFF, J. C., VAN DER MEIDE, P. H. **1997**. *J Interferon Cytokine Res 17 (Suppl 1)*, S5–S8.
53 ANTONELLI, G. **1997**. *J Interferon Cytokine Res 17 (Suppl 1)*, S39.
54 ANTONELLI, G., DIANZANI, F. **1999**. *Eur Cytokine Netw 10*, 413.
55 SPIEGEL, R. J., JACOBS, S. L., TRUEHAFT, M. W. **1989**. *J Interferon Res 9*, S17.
56 VON WUSSOW, P., JAKSCHIES, D., FREUND, M., DEICHER, H. **1989**. *J Interferon Res 9*, S25–S31.
57 VON WUSSOW, P., PRALLE, H., HOCHKEPPEL, D., et al. **1991**. *Blood 78*, 38.
58 OBERG, K., ALM, G. V. **1989**. *J Interferon Res 9*, S45.
59 LAROCCA, A. P., STEVEN, C. L., STEPHEN, G. M., COLBY, C. B., BORDEN, E. C. **1989**. *J Interferon Res 9*, S51, S60.
60 ROSS, C., SVENSON, M., HANSEN, M. B., VEJISGAARD, G. L., BENDTZEN, K. **1995**. *J Clin Invest 95*, 1974.
61 ANTONELLI, G., CURRENTI, M., TURRIZIANI, O., DIANZANI, F. **1991**. *J Infect Dis 163*, 882.
62 ANTONELLI, G., SIMEONI, E., ARTINI, M., TURRIZIANI, O., DIANZANI, F. **1997**. *Hepatol Res 8*, 149.

63 PROTZMAN, W. P., JACOBS, S. L., MINNICOZZI, M., ODEN, E. M., KELSEY, D. K. **1984**. *J Immunol Methods* 75, 317.
64 PRÜMMER, O. **1993**. For the Delta-P Study Group. *Cancer* 71, 1828.
65 R & D Systems. Technical Information.
66 BENDTZEN, K. **2003**. *Neurology* 61, S6.
67 GNEISS, C., REINDL, M., BERGER, T., LUTTEROTTI, A., EHLING, R., EGG, R., DEISENHAMMER, F. **2004**. *J Interferon Cytokine Res* 24, 283.
68 MEAGER, A. **1987**. In: CLEMENTS, M. J., MORRIS, A. G., GEARING, A. J. H. (Eds.), *Lymphokines and Interferons: A Practical Approach*. IRL Press, Oxford, P. 105.
69 MEAGER, A. **2000**. In: BALKWILL, F. (Ed.), *Cytokine Cell Biology: A Practical Approach*. Oxford University Press, Oxford, p. 193.
70 ROSS, C., CLEMMESEN, K. M., SVENSON, M., et al. **2000**. *Ann Neurol* 48, 706.
71 GOLGHER, R. R., REDLICH, P. N., TOTTI, D. O. S. B., GROSSBERG, S. E. **1999**. *J Interferon Cytokine Res* 19, 995.
72 BRICKELMAIER, M., HOCHMAN, P. S., BACIU, R., CHAO, B., CUERVO, J. H., WHITTY, A. **1999**. *J Immunol Methods* 227, 121.
73 PACHNER, A. R. **2003**. *Neurology* 61, 1444.
74 SEDMAK, J. J., SIEBENLIST, R., GROSSBERG, S. E. **1985**. *J Interferon Res* 5, 397.
75 PACHNER, A. R., OGER, J., PALACE, J. **2003**. *Neurology* 61, S18.
76 KARPUSAS, M., NOLTE, M., BENTON, C. B., MEIER, W., GOELZ, S. **1997**. *Proc Natl Acad Sci USA* 94, 11813.
77 RUNKEL, L., DEDIOS, C., KARPUSAS, M., et al. **2000**. *Biochemistry* 39, 2538.
78 RUNKEL, L., DEDIOS, C., KARPUSAS, M., et al. **2001**. *J Interferon Cytokine Res* 21, 931.
79 DE MAEYER-GUIGNARD, J., DE MAEYER, E. **1986**. *J Immunol* 136, 1708.
80 ROSS, C., HANSEN, M. B., SCHYBERT, T., BERG, K. **1990**. *Clin Exp Immunol* 82, 57.
81 CARUSO, A., BONFANTI, C., COLOMBRIOITA, D., DE FRACESCO, H., DE RANGO, C., FORESTI, I., et al. **1990**. *J Immunol* 144, 685.
82 SEDMAK, J. J., GROSSBERG, S. E. **1989**. *J Interferon Res* 9 (Suppl 1), S61.
83 PANEM, S., CHECK, I. J., HENRIKSEN, D., VILCEK, J. **1982**. *J Immunol* 129, 1.
84 IKEDA, Y., TODA, G., HASHIMOTO, N., et al. **1991**. *Clin Exp Immunol* 85, 80.
85 TROWN, P. W., KRAMER, M. J., DENNIN, R. A., JR., et al. **1983**. *Lancet* 1, 81.
86 PRÜMMER, O., BUNJES, D., WIESNETH, M., ARNOLD, R., PORZSOLT, F., HEIMPEL, H. **1994**. *Bone Marrow Transplant* 14, 483.
87 GRESSER, I., TOVEY, M. G., BANDU, T. M., et al. **1976**. *J Exp Med* 144, 135.
88 MEDENICA, R. D., MUKERJEE, S., HUSCHART, T., CORBITT, W. **1994**. *J Clin Apheresis* 9, 216.
89 HAVREDAKI, M., BARONA, F. **1985**. *Jpn J Med Sci Biol* 38, 107.

90 Kivisakk, P., Alm, G. V., Tian, W. Z., Matusevicius, D., Fredrikson, S., Link, H. **1997**. *Mult Scler* 3, 184.
91 Antonelli, G., Bagnato, F., Pozzilli, C., et al. **1998**. *J Interferon Cytokine Res* 18, 345.
92 Perini, P., Calabrese, M., Biasi, G., Gallo, P. **2004**. *J Neurol* 251, 305.
93 Deisenhammer, F., Reindl, M., Berger, T. **2001**. *J Interferon Cytokine Res* 21, 167–171.
94 Giannelli, G., Antonelli, G., Fera, G., et al. **1994**. *Clin Exp Immunol* 97, 4.
95 Viscomi, G. C., Grimaldi, M., Palazzini, E., Silvestri, S. **1995**. *Med Res Rev* 15, 445.
96 Ronnblom, L. E., Janson, E. T., Perers, A., Oberg, K. E., Alm, G. V. **1992**. *Clin Exp Immunol* 89, 330.
97 Brand, C. M., Leadbeater, L., Budiman, R., Lechner, K., Gisslinger, H. **1994**. *Br J Haematol* 86, 216.
98 Antonelli, G., Gianelli, G., Currenti, M., et al. **1995**. *Int Hepatol Commun* 4, 232.
99 Nolte, K. U., Frank, R., Guenther, G., et al. **1994**. *J Interferon Res* 14 (Suppl 1), S118.
100 Figlin, R. A., Iltri, M. **1988**. *Semin Hematol* 25, 9.
101 Steis, R. G., Smith II, J. W., Urba, W. J., et al. **1988**. *N Engl J Med* 318, 1409.
102 Jacobs, S., Friedman, R. M., Nagabhushan, T. L., et al. **1989**. *J Interferon Res* 9 (Suppl 2), S292.
103 Hochuli, E. **1997**. *J Interferon Cytokine Res* 17 (Suppl 1), S15.
104 Lok, A. S., Lai, C. L., Leung, E. K. **1990**. *Hepatology* 12, 1266.
105 Itri, L. M., Sherman, H. I., Palleroni, A. U., et al. **1987**. *J Interferon Res* 9 (Suppl), 59.
106 Quesada, J. R., Rios, A., Swanson, D., Trown, P., Gutterman, J. U. **1985**. *J Clin Oncol* 3, 1522.
107 von Wussow, P., Freund, M., Block, B., Diedrich, H., Poliwoda, H., Deicher, H. **1987**. *Lancet* 2, 635.
108 Casato, M., Lagan, B., Antonelli, G., Dianzani, F., Bonomo, L. **1991**. *Blood* 78, 3142.
109 Milella, M., Antonelli, G., Santantonio, T., Currenti, M., Monno, L., Mariano, N., et al. **1993**. *Liver* 13, 146.
110 Antonelli, G., Gianelli, G., Pistello, M., Maggi, F., Vatteroni, L., Currenti, M., et al. **1994**. *J Interferon Res* 14, 211.
111 McSweeney, E. N., Giles, F. J., Goldstone, A. H. **1994**. *J Interferon Res* 14, 191.
112 Prümmer, O., Bunjies, R., Arnold, B., Hertenstein, R., Wiesneth, R., Heimpel, H. **1994**. *J Interferon Res* 14, S118.
113 Bonetti, P., Diodati, G., Drago, C., et al. **1994**. *J Hepatol* 20, 416.
114 Antonelli, G., Gianelli, G., Currenti, M., et al. **1996**. *Clin Exp Immunol* 104, 384.
115 Roffi, L., Mels, G. C., Antonelli, G., et al. **1995**. *Hepatol* 21, 645.
116 Craxi, A., Di Marco, V., Volpes, R., Palazzo, U. **1988**. *Hepatogastroenterology* 35, 304.

117 BROOK, M. G., McDONALD, J. A., KARAYANNIS, P., et al. **1989**. *Gut 30*, 1116.
118 PRÜMMER, O., PROZSOLT, F. **1994**. *J Interferon Res 14*, 193.
119 CATANI, L., GUGLIOTTA, L., ZAULI, G., et al. **1992**. *Haematol 77*, 318.
120 VON WUSSOW, P., HARTMAN, F., FREUND, P., POLIWODA, H., DEICHER, H. **1988**. *Lancet 1*, 882.
121 GIOVANNONI, G., MUNSCHAUER III, F. E., DEISENHAMMER, F. **2002**. *J Neurol Neurosurg Psychiatry 73*, 465.
122 BERTOLOTTO, A., MALUCCHI, S., SALA, A., et al. **2002**. *J Neurol Neurosurg Psychiatry 73*, 148.
123 MALUCCHI, S., SALA, A., GILLI, F., et al. **2004**. *Neurology 62*, 2031.
124 BERTOLOTTO, A. **2004**. *Curr Opin Neurol 17*, 241.
125 RUDICK, R. A., SIMONIAN, N. A., ALAM, J. A., et al. **1998**. *Neurology 50*, 1266.
126 IFNβ MS Study Group. **1996**. *Neurology 47*, 889.
127 RICE, G. **2001**. *Arch Neurol 58*, 1297.
128 MYHR, K. M., ROSS, C., NYLAND, H. I., et al. **2000**. *Neurology 55*, 1569.
129 DEISENHAMMER, F., REINDL, M., HARVEY, J., et al. **1999**. *Neurology 52*, 1239.
130 PRISMS-4. **2001**. *Neurology 56*, 1628.
131 KHAN, O. A., DHIB-JALBUT, S. S. **1998**. *Neurology 51*, 1698.
132 BERTOLOTTO, A., MALUCCHI, S., MILANO, E. **2000**. *Immunopharmacology 48*, 95.
133 KASTRUKOFF, L. F., MORGAN, N. G., ZECCHINI, D., et al. **1999**. *Neurology 52*, 351.
134 WANDINGER, K.-P., LUNEMANN, J. D., WENGERT, O., et al. **2003**. *Lancet 361*, 2036.
135 PETKAU, J., WHITE, R. **1997**. *Mult Scler 3*, 402.
136 ARNASON, B. G., TOSCAS, A., DAYAL, A., et al. **1997**. *J Neurol Transm Suppl 49*, 117.
137 PRICE, C. **1997**. *Br Med J 314*, 600.
138 RICE, G. P., PASZNER, B., OGER, J., et al. **1999**. *Neurology 52*, 1277.
139 VALLITTU, A. M., HALMINEN, M., PELTONIEMI, J., et al. **2002**. *Neurology 58*, 1786.
140 BERTOLOTTO, A., GILLI, F., SALA, A., et al. **2003**. *Neurology 60*, 634.
141 GILLI, F., BERTOLOTTO, A., SALA, A., et al. **2004**. *Brain 127*, 259.
142 POZZILLI, C., ANTONINI, G., BAGNATO, F., et al. **2002**. *J Neurol 249*, 50.
143 GIOVANNONI, G. **2003**. *Neurology 61*, S13.
144 WOLINSKY, J. S., TOYKA, K., KAPPOS, L., GROSSBERG, S. E. **2003**. *Lancet 2*, 4.
145 ROSS, C., CLEMMESEN, K. M., SVENSON, M., et al. **2000**. *Ann Neurol 48*, 706.
146 SORENSEN, S. **2003**. *Neurology 61*, S27.
147 BERTOLOTTO, A. **2004**. *Curr Opin Neurol 17*, 241.
148 GIOVANNONI, G. **2004**. *J Neurol Neurosurg Psychiatry 75*, 1234.
149 BERTOLOTTO, A., SALA, A., MALUCCHI, S., et al. **2004**. *J Neurol Neurosurg Psychiatry 75*, 1294.

Index

1
17-kDa 9-27/leu13   355

2
2′-5′-oligoadenylate synthetase   210 ff.
– (2′-5′-OAS)   150
– – ribonuclease (RNase) L   231 ff.
2-5A synthetase   356

4
4E-BP1   218

9
96-well microtiter plates   363, 363 ff.

a
A46R protein   250
A52R protein   250
A549 cell line   355 f., 359, 377
activating protein   1 (AP-1)   35
AD   see atopic dermatitis
Adenovirus   248
– E3 protein   248
– EIA protein   248
– VAI RNA   248
African green monkey cell line, Vero   358
AK-1   358
Akt   218
alkaline phosphatase   357
antiapoptotic effect   219 ff.
antibodies to IFN-α   391
antibodies to IFN-β   392
antibodies to self-antigen   376
antiproliferative activity   351, 366
antiproliferative assay   351, 352, 353 ff., 354
antitumor agent   277
antiviral   227 ff.
– 2′-5′-OAS183
– 2′-5′-OAS/RNase L   186
– ADAR-1   186, 188
– IFITs   188
– IRS-P13K   183
– ISG-20   186, 188
– JAK-STAT   183
– Mx   187
– MxA   183
– p38MAPK   183
– P56   186
– PKR   183, 186
– PLSCR-1   186, 188
– RNase L   183
– TRAIL   186, 188
– Viperin   186
antiviral activity   365
IFN-λ   142
antiviral agent   277
antiviral assay   349 ff., 351 f., 354, 359, 364 ff., 366, 377
– challenge virus   349
antiviral protection   152
antiviral response   58, 141 ff., 150 f., 153
apoptosis   207 ff., 211 ff.
areas for IFNAR-1 and IFNAR-2 binding   389
arenavirus   242 ff.
ARF   209 ff.
assay   345, 349
assay design and data analysis   363
assays based on intracellular signaling intermediates   361
ATF   11
ATF4   214
atopic dermatitis (AD), IgE   321
– Th1 response   321
– Th2 response   321
AU-rich element   15
avian Newcastle disease virus   232 ff.

b

B18R protein 249 ff.
bacille Calmette-Guerin 125
basal cell carcinoma (BCC) 292
BFP 127
BICP0 protein 251
bioimmunoassays 354 ff., 358
biological activities, antiproliferative activity 344 ff.
– antitumor activity 345
– antiviral activity 344
– immunomodulatory activity 345
BP see lipid-binding protein
Bunyamwera virus 244
Bunyavirus 242 ff.
Burkitt's lymphoma-derived B cell line Daudi 352

c

C protein 232 ff.
cAMP response element modulation protein (CREM) activating transcription factor (ATF) 39
cancer 157, 220
caspase activation and recruitment domains (CARD) 59
cell differentiation 278
cell proliferation 278
cell viabilities 208
CHB see chronic hepatitis B
CHC 288
CHD 290
chimeric receptor 122 ff.
chromosome see YAC
chronic granulomatous disease (CGD) 311
chronic hepatitis (CHB) 286
– Hepatitis B virus (HBV) 327
– Hepatitis C virus (HCV) 327
chronic myeloic leukemia (CML) 294
cis-trans peptidyl-prolyl isomerase (PPIase) activity 46
class I MHC (HLA) molecules 153
class II 113 ff.
class II cytokine receptor family (CRF2), CRF2 cytokines 142 ff.
– ligands 142
clinical grade, IFN products 348 ff.
clinical samples, bronchial lavage fluids 348
clinical trials, IFN-γ 309 ff.
clinical trials of IFN-γ 309
clotting factor VIII 376
CML 294
coded duplicates 363
condyloma acuminata 296

constant antibody method 380
constant IFN method 379
coronavirus 246
CPE 351, 352
CPE reduction (CPER) assay 351 f., 365
CpG oligodeoxynucleotides (ODNs) 57 f.
CREB-binding protein (CBP/po300) 40 f.
CRF2 see class II cytokine receptor family
CRID (coding region instability determinant) elements 15
CrkL pathway, C3G-Rap1 179
– STAT-5 179
cyclophilin B (CypB) 46
cytokines 113
Cytomegalovirus, gB protein 251
– IE86 protein 251
– pp65 protein 251
cytotoxic cell activity 278

d

Dami 353
Daudi cell line 353
DCs 152
Dengue virus 153, 244
development of antibodies during IFN therapy 390
dose-response curve 363 ff.
double-stranded (ds) RNA 210, 212
double-stranded (ds) RNA-activated protein kinase (PKR) 150

e

E1A protein 248
E2 glycoprotein 246
E3 protein 248
E6 protein 252
E7 protein 252
EAHy296 355
EBER-1 RNA 252
EBNA-2 protein 252
Ebola virus 232 ff.
– VP35 protein 242
EBV see Epstein-Barr virus (EBV)
eIF-2α 212 ff., 214
eIF-4E 218
ELISA 356, 359, 361
– luciferase 357
– PACE 361
EMCV see encephalomyocarditis virus (EMCV)
encephalomyelitis disseminata 297
encephalomyocarditis virus (EMCV) 123, 153, 377
endpoint analysis 365

enhanceosome 40
enzyme expression and reporter gene assay 356
epigenetic regulation, chromatin 88, 89
– DNA methylation 88, 89
– DNase I hypersensitivity 89
– γ promoter 89
– histone acetylation 88, 89
epitope 375
epitope analysis 388
Epstein-Barr virus (EBV) 47, 251 ff., 295, 297
– EBER-1RNA 252
– EBNA-2 protein 252
– LMP-1 protein 252
Escherichia coli β-galactosidase 359
experimental analyses of neutralization 383

f
FDA-approved indications for IFNs 310 ff.
fibroblast IFN 3 f.
filovirus 232 ff.
flavivirus 244
FLICE-inhibitory protein (vFLIP) 252
fluorescence emission spectra 125 ff.
fluorescence resonance energy transfer (FRET) 126
foot and mouth disease virus 244

g
β-galactosidase 359
GAS116 ff., 209
– γ-activated sequence 150
GATA-3 39
GFP 127
glial fibrillary protein (GFAP) 359
glioblastoma cell line 354, 356, 357
glycoprotein gB 251
glycosylation, post-translational modifications IFNs 23
granulocyte macrophage colony-stimulating factor (GM-CSF) 376
growth-inhibitory 188
– AIM-2 190
– Bak 190
– Bax 190
– Bcl-1 190
– C3G-Rap1 189
– caspases 190
– CDK 2 189
– CML 189
– Crk 189
– E2F 189
– FADD 190
– Fas 190
– HIN-200 190
– IFI-16 190
– IFIX-α1 190
– IP6K2 190
– IRF-1 190
– IRF-5 190
– MNDA 190
– c-*myc* 189
– p15 189
– P202-E2F 190
– p21 189
– p27 189
– p38MAPK 189
– PML 189
– Rap1 189
– STAT-5 189
– TRAIL 189
– Vav 189
– XAF-1 190

h
hairy cell leukemia (HCL) 290
HBV see Hepatitis B virus
HCL see hairy cell leukemia
HCV see Hepatitis C virus
HDV see Hepatitis δ virus
Hela cells 216
Hendra virus 232 ff.
– V protein 241
Hepadnavirus 248
Hepatitis B virus (HBV) 248, 286 f.
– chronic (CHB) 286
Hepatitis C virus (HCV) 59, 244 ff.
– chronic (CHC) 157, 288
– Core protein 245
– NS3/4A protein 245
– NS3A protein 245
– NS5A protein 245
Hepatitis δ virus (HDV) 290
Herpes simplex virus 250 ff.
– B-ICP0 protein 251
– ICP0 protein 251
– ICP34.5 protein 251
– US11 protein 251
high-mobility-group protein [HMGI(Y)] 4, 40
HIV see human immunodeficiency virus
HIV-1 247
horseradish peroxidase 357
HPIV-2, V protein 236 ff., 237
HPIV-3, C protein 236 ff., 237
HPV see human papillomavirus
human colon adenocarcinoma cell line, COLO 205 354
human cytomegalovirus 250 ff.

human erythroleukemic TF-1 cell line 353
human erythropoietin 376
human glioblastoma cell line 357, 359
human herpesvirus 8 251 ff.
– Kaposi's Sarcoma-associated herpesvirus (KSHV) 251 ff.
human immunodeficiency virus (HIV) 247, 295
– TAR RNA 248
– TAT protein 247
human insulin 376
human lung carcinoma cell line, A549 354
human megakaryoblastic UT-7/EPO cell line 353, 354
human megakaryocytic cell lines, Dami 353
– MEG01 353
human papillomavirus (HPV) 295 f.
human parainfluenza virus type 2 232 ff.
human parainfluenza virus type 3 232 ff.
human promyelocytic leukemia HL60 353
human simplex virus (HSV) type I 47
human T cell leukemia virus 295
human tumor-derived cell line 356

i
ICAM-1 vascular cell adhesion molecule (VCAM)-1 355
ICP0 protein 251
ICP34.5 protein 251
ideopathic pulmonary Fibrosis (IPF) 322
IE86 protein 251
IFN activation pathway 46
IFN bioassays, antiviral assay 377
– MxA 377
IFN (Interferon) 51, 73 ff.
– activation 35
– amplification loop 59
– autocrine and paracrine manner 152
– divergence 20
– evolution of family 156
– expansion 20
– expression 20
– fibroblast IFN 3 f.
– gene subfamily 16
– genes 85
– IFN-α 3 ff.
– immune response 35
– induced mutation 21
– leukocyte IFN 3
– leukocyte-derived 75
– long-acting 78
– lymphoblastoid 76
– natural mutation 21
– production 75 ff.

– promoter 20
– protein 73 ff.
– purification 75 ff.
– resistance to pathogens 35
– STAT signaling 26
– structure 73 f.
– toxicity 298
– virus-mediated activation 41
IFN production, NIPC 167
– pDCs 167
IFN products, clinical-grade 348
– excipient protein 348
IFN receptor 39
IFN receptor recognition peptide 168
IFN regulatory factor see IRF
IFN signal transduction pathway 50, 358
– JAK-1 and -2, and TYK-2 tyrosine kinases 341 ff.
IFN therapy 220
IFN-α 16, 21 ff., 24, 35, 46 f., 57, 73, 75, 146, 153, 154, 157, 289, 290
– induction 55
– leucocyte IFN 7 ff.
– pharmacokinetics 284
– promoter 11 ff.
– therapy 288 ff.
– viral infection 40, 144
– virus-mediated induction 15
IFN-α/β 39
IFN-β 7, 16, 21 f., 24, 35, 46 f., 51, 53, 58 f., 73, 146, 151, 153, 154
– immunoregulation 21
– induction 55, 57
– pharmacokinetics 285
– production 21
– promoter 11 ff., 40, 59
– response 21
– viral infection 144
– virus-inducible 39
IFN-β1, therapy 297
IFN-δ 10, 16, 24
IFN-ε 9, 16, 24
IFN-γ 217
– adverse reactions 311
– aerosolized 311
– antitumor defenses 144
– Cell-DC crosstalk 97
– DC presentation 97 ff.
– gene structure and regulation 86, 88
– IFN-γ/β 311
– immunity 144
– in T_h cell development 95 f.
– mRNA 85
– PBL-γ 311

– role on tumor development and growth, angiogenesis   100 ff.
– signal transduction   90 ff.
IFN-γ-activated factor (GAF)   39
IFN-induced signaling pathways   342
IFN-inducible proteins   341, 342 ff., 354 ff.
– 2′-5′-OAS343
– 2′-5′-oligoadenylate synthetase (2′-5′-OAS)   344
– antigens   354
– antiproliferative activity   345
– assays   354
– Cyclin-dependent kinase (CDK)-2   345
– encephalomyocarditis virus (EMCV)   344
– ICAM-1   343, 354
– intercellular adhesion molecule (ICAM)-1   345
– IRF-1   345
– MHC class I (HLA-A, -B, -C)   343
– MHC class II (HLA-DR)   343
– β_2-microglobulin   343, 354
– β_2-microglobulin, class I and II MHC antigens   345
– Mx   343 f.
– MxA   344
– peptide initiation factor, eIF-2   344
– PKR   343 f.
– RNase L   344
– vesicular stomatitis virus (VSV)   344
IFN-κ   16
– intron   8
– transcriptional regulation   8
IFN-λ   141, 151, 154
– antiviral activity   142, 153
– genomic structure   145
– viral infection   144
IFN-λ antiviral system   156
– murine   154
IFN-λ receptor, IL-28R   146 ff.
IFN-λ receptor complex, signaling   148
IFN-λ1   142 f.
IFN-λ2   142 f.
IFN-ν   10, 16
IFN-ω   8, 16, 22
IFN-producing cells (IPC)   4
IFN-θ   9, 16, 24
IFN-regulatory factor   see IRF
IFN-regulatory factor (IRF) binding sites   11
IFN-responsive gene   150
IFN-specific cell surface receptors, type II cytokine receptor family (CRF2)   340
IFN-stimulated gene factor 3 (ISGF-3)   39, 142, 150

IFN-stimulated genes (ISGs)   19, 39, 59
IFN-stimulated responsive element (ISRE)   39, 116 ff., 209, 358
IFN-τ   9, 22
– pregnancy   16
– trophoblast IFN   20
IFNAR receptor   4, 20
IFNAR-1   74
– Gb$_2$ 174
– Gb$_3$ 174
IFNAR-2, IFN-α   172
– IFN-α2   170
– IFN-β   173
– IRRP-1   168
– IRRP-3   168
Iκbα kinase complex (KK)   55
IKKε   41 ff., 58, 59
immune response   35, 50
immune stimuli   144
immunity   144
immunoassays for non-NAbs   387
immunomodulation
– B cell   191
– B lymphocyte   192
– DCs   192
– granulocytes   191
– IFN-β knockout   190
– IFNAR-1-deficient   190
– macrophages   191
– MHC class I   192
– myeloid   190
– NK cells   192
– plenic architecture   191
– T lymphocyte   192
– TNF-α   190
index of neutralization   379
indoleamine 2,3-dioxygenase (IDO)   356
inducible protein expression   366
infection following serious trauma   318
– *Bacteroides fragilis*   319
– *Escherichia coli*   319
– *Klebsiella pneumoniae*   319
– *Pseudomonas aeruginosa*   319
influenza virus   57, 153, 242 ff.
– NS1 protein   243
innate immunity   227
interferon   see IFN
– interferon regulatory factor   see IRF
– interleukin   142, 228 ff.
– IL-1 receptor family   50
– IL-15   39
– IL-28A   113 ff., 127
– IL-29   113 ff.
– IL-1β   228 ff.

interleukin-Iβ (IL-Iβ) 228 ff.
International Units (IU) 364, 378
intron 6
IPC see IFN producing cell
IPF see ideopathic pulmonary fibrosis
IRAK1 53, 57
IRAK4 57
IRF (IFN regulatory factor) 35, 41, 51, 150, 342
– apoptosis 41
– cell cycle 41
– tumor suppression 41
IRF-1 217
IRF-3 15, 40 ff., 51, 57, 59, 228 ff., 230 ff.
– autoinhibitory interaction 41
– autoinhibitory structure 46
IRF-5 46 f., 57
IRF-7 47 ff., 57, 58
– transactivation activity 49
IRF-9 350, 358
IRF-association domain (IAD) 41
IRF family 40, 48
IRSpathway, 4E-BP1 181
– eIF-4E 181
– Grb-2 181
– IRS-1 179
– IRS-2 179
– mTOR 181
– p70S6 181
– PI3K 179
ISGF-3 see IFN-stimulated gene factor 3
ISGs see IFN-stimulated genes
ISRE see IFN-stimulated responsive element
IU 365

j

JAK see Janus kinase
JAK-1 361
JAK-1 tyrosine kinase 148 ff.
JAK-STAT signal transduction 142 ff., 175, 177 f., 350
Janus Kinase (JAK) 114
– JAK-1 114 ff.
– JAK-2 114 ff.
– signal transducers and activators of transcription (STAT) 230 ff.
– STAT signaling 39
– TYK-2 114 ff., 116
Jun 11
c-Jun N-terminal kinase (JNK) 55

k

K1L protein 249 ff.
Kaposi sarcoma (KS) 292

Kaposi's Sarcoma-associated herpesvirus (KSHV) 251 ff.
Kawade 382
kinase receptor activation (KIRA) assay 361
KIRA assay 361, 361 ff., 362
KS-associated herpesvirus (KSHV) 293
KSHV, vFLIP 252
– vIRFs 252

l

Laboratory Unit (LU) 365, 377
laryngeal papillomatosis 295
latent membrane protein 252
LBP see lipid-binding protein
leishmaniasis 317
leprosy 314
leukemic reticuloendotheliosis 290
leukocyte IFN 3 f.
limitin see IFN-θ
lipid-binding protein (LBP) 53
lipopolysaccharide (LPS) 53, 144
LPS see lipopolysaccharide
luciferase 358, 359, 360
luciferin 359, 360

m

M1 myeloid cell line 353
MAbs 377
malignant melanoma 292
MAPK 219
MAPK kinase 55
MAPK pathway 181
MCF-7 208
Measles virus 232 ff.
– V protein 239
measurement of antibodies to interferon 375
measurement of biological activities
– antiproliferative activity 351
– antiproliferative assays 351
– antiviral assay 351
– antiviral assay cell/virus combinations 350
– CPE 351
– CPE reduction (CPER) assay 351
– EMCV 350, 351
– JAK-STAT signaling pathway 350
– MTS(3-[4,5-dimeth ylthiazol-2-yl]-5-[3-carboxymethoxyphenyl]-2-[4-sulphophenyl]-2H-tetrazolium inner salt 351
– MTT (3-[4,5-dimethylthiazol-2-yl]-2,5-diphenyl tetrazolium bromide) 351
– SFV 350
– SINV 350
– viral cytopathic effect (CPE) 351
– virus yield reduction assay 351

– VSV   350
measurement of interferon activities
– antiviral assay   349
– biological standards for IFNs   345
– general considerations   345
– introduction   339
– practical considerations for IFN preparations   348
MEG01   353
melanoma differentiation-associated Gene-5 (mda-5)   59
MHC class I   116 ff.
β_2-microglobulin   355
mitogen-activated protein kinase (MAPK)   53
monoclonal Abs (MAbs)   375
MS   297
– secondary-progressive MS (SPMS)   298
mTOR   218 ff.
MTS(3-[4,5-dimethylthiazol-2-yl]-5-[3-carboxymethoxypehnyl]-2-[4-sulphophenyl]-2H-tetrazolium inner salt   351
MTT (3-[4,5-dimethylthiazol-2-yl]-2,5-diphenyl tetrazolium bromide)   351
multiple myeloma (plasma cell myeloma)   293
multiple sclerosis (MS), relapsing-remitting MS(RRMS)   297
Mumps virus   232 ff.
– V protein   238
mutated residues reducing mAb binding   389
MV virus, V protein   239
Mx protein   150 f., 231 ff.
MxA   355, 356, 357, 359, 377
MxA assay   356
MxA-luciferase   360
Mxo   355, 358
Mycobacterium avium infection   315
Mycobacterium infections   314
mycosis fungoides (MF)   296
MyD88 adaptor-like protein (MAL)   53
MyD88 induction   57

n

N1L protein   250
NAbs   *see* neutralizing antibodies
natural IFN-producing cells (NIPC)   144, 152
natural killer (NK) cells   278
NDV *see* Newcastle disease virus
negative regulatory domains (NRDI), promoter   40
neutralization   377
neutralization bioassay design   378
neutralization by Ab   377
neutralizing antibodies (NAbs)   285, 375
neutralizing antibody titration   378

Newcastle disease virus (NDV)   47 ff., 75, 238
NF-κB   11, 15, 35, 40, 53, 213, 218, 227 ff.
NF-κB signaling pathway   51 ff.
NF-κB-regulating factor (NRF)   40
NHL   295
Nipah virus, P protein   241
– V protein   241
– W protein   241
NIPCs   *see* natural IFN-producing cells
p-nitrophenylphosphate (NPP)   359
non-Hodgkin's lymphoma (NHL)   295
nonsegmented negative-strand RNA viruses   232
nonstructural protein   240
NS1 protein   232 ff., 240
NS2 protein   232 ff., 240
NS3/4A protein   245
NS3A protein   245
NS5A   245 f.
NSs, NSα protein   242 ff.
nuclear factor (NF)-κB   *see* NF-κB
nuclear factor of activated T cells (NF-AT)   35

o

oncology indications   329
oncolytic virus   214
opportunistic infections in HIV disease   317
orthomyxovirus   242 ff.
Osteopetrosis   313

p

P protein   232 ff.
p53   209 ff., 211
PAbs   *see* polyclonal antibodies
PACE   361
PACT/RAX   212
papillomavirus, E6 protein   252
– E7 protein   252
parallelism   365
paramyxovirus   232 ff.
PARP   208
pathogen-associated molecular patterns (PAMPs)   50
pharmacokinetics   284 f.
phosphatidylinositol-3-kinase   218, 240
phosphatidylinositol-3-kinase pathway (PI3K)   51
phosphospecific antibody cell-based   361
picornaviruses   244
PKC, PKC-δ 182
PKR   212 ff., 228 ff., 356
plasma cell myeloma   *see* multiple myeloma
plasmacytoid DCs (pDCs)   144
plate effect   363

poliovirus 244
polyclonal antibodies (PAbs) 375, 377
positive regulatory domains (PRDs) 11, 40
post-transcriptional regulation 90
potency 346, 348 f., 363 ff.
potency determination 361
poxvirus 248
pp65 251
PRD *see* positive regulatory domains
PRDI-like elements 15
pro-inflammatory cytokines 57 f.
pro-inflammatory genes 51 ff.
proliferation 208
prostate cancer 211
protein induction 342
protein kinase B 218
protein kinase C 240
protein synthesis 216
pseudogenes 3 ff.
psoralen and UV A (PUVA) therapy 296

r

Rabies virus 232 ff.
– P protein 242
radiation-induced fibrosis 326 f.
RANTES 39, 51
rapamycin 218 ff.
receptor complex 114 ff.
receptor-interacting protein (RIP-1) 51
receptors, "non-receptor" tyrosine kinases of the Janus, JAK 341
– TYK 341
recommended NAb procedure 386
reference preparation 346
regulatory landscape 365
renal cell carcinoma (RCC) 291
reovirus 247
reporter gene assay 358, 359, 360, 366
reporting of neutralization results 383
respiratory syncytial virus 232 ff.
retinoic-inducible gene (rig)-I, signaling 58
retrovirus 247
rhabdovirus 232 ff.
ribonuclease (RNase) 210 ff.
rift valley fever virus, NSα protein 242 ff.
RNA helicase activity, DExD/H box 59
RNase 356
RSV, nonstructural protein 240

s

SARS (severe acute respiratory syndrome) 18
– coronavirus (SARS-CoV) 246 ff.
scaffold/matrix associated regions (S/MARs) 21

secreted alkaline phosphatase (SEAP) 358 f.
segmented negative-strand RNA viruses 242
self-antigen 376
Sendai Virus 41 ff., 75, 153, 232 ff.
– C protein 233 f.
– V protein 233 f.
severe acute respiratory syndrome *see* SARS
SFV 350
signal transducer and activator of transcription *see* STAT
signal transduction 218 ff.
simian virus 5 232 ff.
simplest model 382
simultaneous IFN titration 378
Sindbis virus (SINV) 153, 350, 377
Sjögren's syndrome 299
solution to the problem of NAb unitage 385
standard 363, 364
standardization of neutralization results 383
STAT knockout mice 178
STAT (signal transducer and activator of transcription) 26, 39, 114, 148 ff., 342, 349, 361
– cytoplasmic-nuclear translocation 362
– STAT translocation 362
STAT-1 349, 358, 361
STAT-2 349, 358, 361
SV5, V protein 234 ff., 236
systemic sclerosis (SSc) 326

t

T-bet 39
TAK1 (transforming growth factor-β activated kinase) 55
TAK1-binding protein (TAB2) 55
TAR RNA 248
TAT protein 247
TBK1 41, 58, 59
Tenfold reduction Units (TRU) 380
theoretical analyses of neutralization 382
thermal stability 349
TIRAP 53
TLR family 50
TLR-3, signaling 51
TLR-4, signaling 53
TLR-7, signaling 57, 58
TLR-dependent pathway 35 ff., 58
TLR-independent pathway 35 ff.
TLRs (Toll-like receptors) 192, 228 ff.
TNF *see* tumor necrosis factor
Toll receptor 50
Toll-interacting region (TIR) domain 50
Toll-like receptors *see* TLRs
TRAF6 51 ff., 57

TRAIL   116 ff., 207 ff., 215 ff., 220
transformed/tumor-derived cell line   352
– African green monkey cell line, Vero   358
– glioblastoma-derived cell lines   354
– human colon adenocarcinoma cell line COLO 205   354
– human lung carcinoma cell line, A549   354
– mouse L-M cell line   359
transforming growth factor   15
transforming growth factor-β activated kinase *see* TAK1
translation-initiation factor eIF-2α 151
TRIF-related adaptor molecule (TRAM)   55
TRIF/-TICAM-1   51
TRL-9, signaling   58
Tuberculosis (TB)   315
tumor necrosis factor (TNF)-α 53, 207 ff., 228 ff., 355, 363
TYK-2   117, 358, 361
type I IFN   4 ff., 20, 22 f., 46, 57 f., 142, 277 ff., 342, 350 ff., 354 ff., 359 ff.
– biological effects   277
– clinical applications   277 ff., 285
– cloned   76
– conserverd amino acid residues   22
– constitutive expression   11
– disulfide bond formation   24
– evolution   15
– gene-regulatory regions   11
– genetics   3 ff.
– IFN-α 340
– IFN-β 340
– IFN-δ 4 ff.
– IFN-ε 4 ff., 340
– IFN-κ 4 ff., 6, 340
– IFN-ν 4 ff.
– IFN-ω 4 ff., 340
– IFN-θ 4 ff.
– IFN-τ 4 ff.
– inducible expression   11
– inflammation   144
– intron   6
– murine type I Ifn genes   6
– mutation   21
– promoter   11 ff.
– secondary structure   22
– secretion   58
– stimulation   41
– structure   3 ff., 24, 73
– therapeutic potential   157
– type I IFN production   4 ff.
– vertebrate   15
– viral   35
type II IFN   142, 342, 352 f., 359

– IFN-γ 4, 340
– stimulation   41
type III IFN   142, 350 f., 353, 355, 359
– biological activities   152
– IFN-λ 4 ff., 141, 340

u
ubiquitin conjugating enzyme complex   55
upstream stimulatory activity (USA)   40
US11 protein   251

v
V protein   232 ff.
Vaccinia virus   248 ff.
VAI RNA   248
variable response   363
Vav
– DNA-PK   182
– Ku80   182
– p38MAPK   182
– PKC-υ 182
– Rac1   182
– STAT-1   182
– TYK-2   182
Vero cells   359
vesicular stomatitis virus (VSV)   47 ff., 153, 350, 377
viral cytopathic effect (CPE)   351
viral defense   227
viral infection   11, 40, 51, 58, 141, 144, 286, 295
viral replication   152, 278
vIRFs   252
virus infection   41 f., 154
virus response element (VRE)   11
virus yield reduction assay   351
VP35 protein   232 ff.
VSV   *see* vesicular stomatitis virus

w
W protein   232 ff.
West Nile virus   244 ff.
– nonstructural protein   246
WHO, biological standards   346
– chick IFN   347
– human IFN   347
– International Standard (IS)   346
– murine (mouse) IFN   347
– rabbit IFN   347
– WHO international collaborative studies   346
– WHO ISs for IFNs   347, 365
WHO ISs   347 ff., 353
– International Units (IU)   347

– IS working standard 348
– leukocyte IFN 348
– working standards 347 f.
WISH or amnion-derived FL cells 377

y

YAC, yeast artiticial 118
yeast artificial chromosome 118 ff.
yin-yang-1 39